An Introduction to Polynomial and Semi-Algebraic Optimization

This is the first comprehensive introduction to the powerful moment approach for solving global optimization problems (and some related problems) described by polynomials (and even semi-algebraic functions). In particular, the author explains how to use relatively recent results from real algebraic geometry to provide a systematic numerical scheme for computing the optimal value and global minimizers. Indeed, among other things, powerful positivity certificates from real algebraic geometry allow one to define an appropriate hierarchy of semidefinite (sum of squares) relaxations or linear programming relaxations whose optimal values converge to the global minimum. Several specializations and extensions to related optimization problems are also described.

Graduate students, engineers and researchers entering the field can use this book to understand, experiment and master this new approach through the simple worked examples provided

JEAN BERNARD LASSERRE is Directeur de Recherche at the LAAS-CNRS laboratory in Toulouse and a member of the Institute of Mathematics of Toulouse (IMT). He is a SIAM Fellow and in 2009 he received the Lagrange Prize, awarded jointly by the Mathematical Optimization Society (MOS) and the Society for Industrial and Applied Mathematics (SIAM).

Cambridge Texts in Applied Mathematics

All titles listed below can be obtained from good booksellers or from Cambridge University Press. For a complete series listing, visit www.cambridge.org/mathematics.

An Introduction to Polynomial and Semi-Algebraic Optimization

JEAN BERNARD LASSERRE
LAAS-CNRS and Institut de Mathématiques, Toulouse, France

CAMBRIDGE
UNIVERSITY PRESS

CAMBRIDGE
UNIVERSITY PRESS

University Printing House, Cambridge CB2 8BS, United Kingdom

Cambridge University Press is part of the University of Cambridge.

It furthers the University's mission by disseminating knowledge in the pursuit of education, learning and research at the highest international levels of excellence.

www.cambridge.org
Information on this title: www.cambridge.org/9781107630697

First published 2015

A catalogue record for this publication is available from the British Library

Library of Congress Cataloguing in Publication data
Lasserre, Jean-Bernard, 1953- author.
An introduction to polynomial and semi-algebraic optimization / Jean Bernard Lasserre,
LAAS-CNRS and Institut de mathématiques, Toulouse, France.
pages cm. – (Cambridge texts in applied mathematics)
Includes bibliographical references and index.
ISBN 978-1-107-06057-9 (Hardback) – ISBN 978-1-107-63069-7 (Paperback)
1. Polynomials.   2. Mathematical optimization.   I. Title.
QA161.P59L37 2015
512.9´422–dc23   2014031786

ISBN 978-1-107-63069-7 Paperback

To my daughter Julia, and to Carole ...

Contents

Preface

This book has benefited from several stimulating discussions, especially with
M. Anjos, G. Blekherman, R. E. Curto, E. de Klerk, L. A. Fialkow, J. W.
Helton, M. Kojima, S. Kuhlmann, M. Laurent, M. Marshall, T. Netzer,
J. Nie, P. Parrilo, D. Pasechnich, D. Plaumann, V. Powers, M. Putinar,
F. Rendl, B. Reznick, C. Riener, M.-F. Roy, C. Scheiderer, K. Schmüdgen,
M. Schweighofer, B. Sturmfels, T. Theobald, K. Toh, and V. Vinnikov, during
several visits and/or workshops in particular at the *Mathematical Sciences
Research Institute* (MRSI, Berkeley), the *Institute of Pure and Applied
Mathematics* (IPAM, Los Angeles), the *American Institute of Mathematics*
(AIM, Palo Alto), the *Fields Institute for Research in Mathematical Sciences*
(Toronto), the *Institute for Mathematics and its Applications* (IMA, Minneapo-
lis), the *Oberwolfach Institute*, the *Isaac Newton Institute* (Cambridge, UK),
the *Institute for Computational and Experimental Research in Mathematics*
(ICERM, Providence) and the *Institute for Mathematical Sciences* (IMS,
Singapore). I gratefully acknowledge financial support from the above insti-
tutes which made these visits possible.

I also want to thank G. Blekherman, F. Bugarin, M. Ghasemi, N. Gravin,
R. Laraki, M. Laurent, M. Mevissen, T. Netzer, M. Putinar, S. Robins,
P. Rostalski, C. Savorgnan, and T. Phan Thanh, as some of the results presented
in this book have been obtained with their collaboration. I would like to
thank Cynthia Vinzant who kindly provided the picture on the book cover.
Finally, special thanks are also due to my colleague Didier Henrion at LAAS
(especially for our collaboration on the software GloptiPoly which he made
user friendly with free access), as well as to J.-B. Hiriart-Urruty, J. Renegar,
S. Sorin, and M. Todd for their constant and friendly support.

For part of this research, financial support from the (French) ANR (GEOLMI
project), the Simone and Cino Del Duca foundation (OPTIGACOM project),
and the Gaspar Monge Program for Optimization and Operations Research

(from the Fondation Mathématique Jacques Hadamard (FMJH)) is gratefully acknowledged. And last but not least, I want to thank the CNRS institution for providing a very nice and pleasant working environment at the LAAS-CNRS laboratory in Toulouse.

Toulouse, June 2014   *Jean B. Lasserre*

Symbols

$\mathbb{N}$, the set of natural numbers
$\mathbb{Z}$, the set of integers
$\mathbb{Q}$, the set of rational numbers
$\mathbb{R}$, the set of real numbers
$\mathbb{R}_+$, the set of nonnegative real numbers
$\mathbb{C}$, the set of complex numbers

$\leq$, less than or equal to
$\leqq$, inequality "$\leq$" or equality "$=$"
$\mathbf{A}$, matrix in $\mathbb{R}^{m \times n}$
$\mathbf{A}_j$, column j of matrix $\mathbf{A}$
$\mathbf{A} \succeq 0$ ($\succ 0$), $\mathbf{A}$ is positive semidefinite (definite)
x, scalar $x \in \mathbb{R}$
$\mathbf{x}$, vector $\mathbf{x} = (x_1, \ldots, x_n) \in \mathbb{R}^n$
$\boldsymbol{\alpha}$, vector $\boldsymbol{\alpha} = (\alpha_1, \ldots, \alpha_n) \in \mathbb{N}^n$
$|\boldsymbol{\alpha}| = \sum_{i=1}^{n} \alpha_i$ for $\boldsymbol{\alpha} \in \mathbb{N}^n$
$\mathbb{N}_d^n, \subset \mathbb{N}^n$, the set $\{\, \boldsymbol{\alpha} \in \mathbb{N}^n : |\boldsymbol{\alpha}| \leq d \,\}$
$\mathbf{x}^{\boldsymbol{\alpha}}$, monomial $\mathbf{x}^{\boldsymbol{\alpha}} = (x_1^{\alpha_1} \cdots x_n^{\alpha_n})$ $\mathbf{x} \in \mathbb{C}^n$ or $\mathbf{x} \in \mathbb{R}^n$, $\boldsymbol{\alpha} \in \mathbb{N}^n$
$\mathbb{R}[x]$, ring of real univariate polynomials
$\mathbb{R}[\mathbf{x}] = \mathbb{R}[x_1, \ldots, x_n]$, ring of real multivariate polynomials
$(\mathbf{x}^{\boldsymbol{\alpha}})$, $\boldsymbol{\alpha} \in \mathbb{N}^n$, canonical monomial basis of $\mathbb{R}[\mathbf{x}]$
$V_{\mathbb{C}}(I) \subset \mathbb{C}^n$, the algebraic variety associated with an ideal $I \subset \mathbb{R}[\mathbf{x}]$
$\sqrt{I}$, the radical of an ideal $I \subset \mathbb{R}[\mathbf{x}]$
$\sqrt[\mathbb{R}]{I}$, the real radical of an ideal $I \subset \mathbb{R}[\mathbf{x}]$
$I(V_{\mathbb{C}}(I)) \subset \mathbb{C}^n$, the vanishing ideal $\{\, f \in \mathbb{R}[\mathbf{x}] : f(\mathbf{z}) = 0, \ \forall \mathbf{z} \in V_{\mathbb{C}}(I) \,\}$
$V_{\mathbb{R}}(I) \subset \mathbb{R}^n$ (equal to $V_{\mathbb{C}}(I) \cap \mathbb{R}^n$), the real variety associated with an ideal $I \subset \mathbb{R}[\mathbf{x}]$
$I(V_{\mathbb{R}}(I)) \subset \mathbb{R}[\mathbf{x}]$, the real vanishing ideal $\{\, f \in \mathbb{R}[\mathbf{x}] : f(\mathbf{x}) = 0, \forall \mathbf{x} \in V_{\mathbb{R}}(I) \,\}$

$\mathbb{R}[\mathbf{x}]_t \subset \mathbb{R}[\mathbf{x}]$, vector space of real multivariate polynomials of degree at most t

$\sum[\mathbf{x}]_t \subset \mathbb{R}[\mathbf{x}]_{2t}$, the convex cone of SOS polynomials of degree at most $2t$

$\mathbb{R}[\mathbf{x}]^*$, vector space of linear forms on $\mathbb{R}[\mathbf{x}]$

$\mathbb{R}[\mathbf{x}]_t^*$, vector space of linear forms on $\mathbb{R}[\mathbf{x}]_t$

$\mathbf{y} = (y_\alpha)$, $\alpha \in \mathbb{N}^n$, real moment sequence indexed in the canonical basis of $\mathbb{R}[\mathbf{x}]$

$\mathbf{M}_d(\mathbf{y})$, moment matrix of order d associated with the sequence $\mathbf{y}$

$\mathbf{M}_d(g\,\mathbf{y})$, localizing matrix of order d associated with the sequence $\mathbf{y}$ and $g \in \mathbb{R}[\mathbf{x}]$

$P(g) \subset \mathbb{R}[\mathbf{x}]$, preordering generated by the polynomials $(g_j) \subset \mathbb{R}[\mathbf{x}]$

$Q(g) \subset \mathbb{R}[\mathbf{x}]$, quadratic module generated by the polynomials $(g_j) \subset \mathbb{R}[\mathbf{x}]$

co $\mathbf{X}$, convex hull of $\mathbf{X} \subset \mathbb{R}^n$

$B(\mathbf{X})$, space of bounded measurable functions on $\mathbf{X}$

$C(\mathbf{X})$, space of bounded continuous functions on $\mathbf{X}$

$\mathcal{M}(\mathbf{X})$, vector space of finite signed Borel measures on $\mathbf{X} \subset \mathbb{R}^n$

$\mathcal{M}(\mathbf{X})_+ \subset \mathcal{M}(\mathbf{X})$, space of finite (nonnegative) Borel measures on $\mathbf{X} \subset \mathbb{R}^n$

$\mathcal{P}(\mathbf{X}) \subset \mathcal{M}(\mathbf{X})_+$, space of Borel probability measures on $\mathbf{X} \subset \mathbb{R}^n$

$L_1(\mathbf{X}, \mu)$, Banach space of functions on $\mathbf{X} \subset \mathbb{R}^n$ such that $\int_{\mathbf{X}} |f| d\mu < \infty$

$L_\infty(\mathbf{X}, \mu)$, Banach space of measurable functions on $\mathbf{X} \subset \mathbb{R}^n$ such that $\|f\|_\infty := \mathrm{ess}\,\sup |f| < \infty$

$\sigma(\mathcal{X}, \mathcal{Y})$, weak topology on $\mathcal{X}$ for a dual pair $(\mathcal{X}, \mathcal{Y})$ of vector spaces

$\mu_n \Rightarrow \mu$, weak convergence for a sequence $(\mu_n)_n \subset \mathcal{M}(\mathbf{X})_+$

$\nu \ll \mu$, ν is absolutely continuous with respect to to μ (for measures)

$\uparrow$, monotone convergence for nondecreasing sequences

$\downarrow$, monotone convergence for nonincreasing sequences

SOS, sum of squares

LP, linear programming (or linear program)

SDP, semidefinite programming (or semidefinite program)

GMP, generalized moment problem (or GPM, generalized problem of moments)

SDr, semidefinite representation (or semidefinite representable)

KKT, Karush–Kuhn–Tucker

CQ, constraint qualification

LMI, linear matrix inequality

b.s.a., basic semi-algebraic

b.s.a.l., basic semi-algebraic lifting

l.s.c., lower semi-continuous

u.s.c., upper semi-continuous

1
Introduction and message of the book

1.1 Why polynomial optimization?

Consider the global optimization problem:

$$\mathbf{P}: \qquad f^* := \inf_{\mathbf{x}} \{ f(\mathbf{x}) \ : \ \mathbf{x} \in \mathbf{K} \} \qquad (1.1)$$

for some feasible set

$$\mathbf{K} := \{ \mathbf{x} \in \mathbb{R}^n \ : \ g_j(\mathbf{x}) \geq 0, \quad j = 1, \ldots, m \}, \qquad (1.2)$$

where $f, g_j : \mathbb{R}^n \to \mathbb{R}$ are some continuous functions.

If one is only interested in finding a *local* (as opposed to *global*) minimum then **P** is a Nonlinear Programming (NLP) problem for which several methods and associated algorithms are already available.

But in this book we insist on the fact that **P** is a *global* optimization problem, that is, f^* is the *global* minimum of f on **K**. In full generality problem (1.1) is very difficult and there is no general purpose method, even to approximate f^*.

However, and this is one of the messages of this book, if one now restricts oneself to *Polynomial Optimization*, that is, optimization problems **P** in (1.1) with the restriction that:

$$f \text{ and } g_j : \mathbb{R}^n \to \mathbb{R} \text{ are all polynomials, } j = 1, \ldots, m,$$

then one may approximate f^* as closely as desired, and sometimes solve **P** exactly. (In fact one may even consider *Semi-Algebraic Optimization*, that is,

1

when f and g_j are semi-algebraic functions.) That this is possible is due to the conjunction of two factors.

- On the one hand, *Linear Programming* (LP) and *Semidefinite Programming* (SDP) have become major tools of convex optimization and today's powerful LP and SDP software packages can solve highly nontrivial problems of relatively large size (and even linear programs of extremely large size).

- On the other hand, remarkable and powerful representation theorems (or positivity certificates) for polynomials that are positive on sets like **K** in (1.2) were produced in the 1990s by real algebraic geometers and, importantly, the resulting conditions can be checked by solving appropriate semidefinite programs (and linear programs for some representations)!

And indeed, in addition to the usual tools from *Analysis, Convex Analysis* and *Linear Algebra* already used in optimization, in Polynomial Optimization *Algebra* may also enter the game. In fact one may find it rather surprising that algebraic aspects of optimization problems defined by polynomials have not been taken into account in a systematic manner earlier. After all, the class of linear/quadratic optimization problems is an important subclass of Polynomial Optimization! But it looks as if we were so familiar with linear and quadratic functions that we forgot that they are polynomials! (It is worth noticing that in the 1960s, Gomory had already introduced some algebraic techniques for attacking (pure) linear integer programs. However, the algebraic techniques described in the present book are different as they come from *Real Algebraic Geometry* rather than pure algebra.)

Even though Polynomial Optimization is a restricted class of optimization problems, it still encompasses a lot of important optimization problems. In particular, it includes the following.

- Continuous convex and nonconvex optimization problems with linear and/or quadratic costs and constraints, for example

$$\inf_{\mathbf{x}} \{ \mathbf{x}^T \mathbf{A}_0 \mathbf{x} + \mathbf{b}_0^T \mathbf{x} : \mathbf{x}^T \mathbf{A}_j \mathbf{x} + \mathbf{b}_j^T \mathbf{x} - c_j \geq 0, \quad j = 1, \dots, m \},$$

for some scalars c_j, $j = 1, \dots, m$, and some real symmetric matrices $\mathbf{A}_j \in \mathbb{R}^{n \times n}$ and vectors $\mathbf{b}_j \in \mathbb{R}^n$, $j = 0, \dots, m$.
- 0/1 optimization problems, modeling a Boolean variable $x_i \in \{0, 1\}$ via the quadratic polynomial constraint $x_i^2 - x_i = 0$. For instance, the celebrated MAXCUT problem is the polynomial optimization problem

$$\sup_{\mathbf{x}} \{ \mathbf{x}^T \mathbf{A} \mathbf{x} : x_i^2 - x_i = 0, \quad i = 1, \ldots, n \},$$

where the real symmetric matrix $\mathbf{A} \in \mathbb{R}^{n \times n}$ is associated with some given graph with n vertices.

- Mixed-Integer Linear and NonLinear Programming (MILP and MINLP), for instance:

$$\inf_{\mathbf{x}} \left\{ \mathbf{x}^T A_0 \mathbf{x} + \mathbf{b}_0^T \mathbf{x} \quad : \quad \mathbf{x}^T A_j \mathbf{x} + \mathbf{b}_j^T \mathbf{x} - c_j \geq 0, \ j = 1, \ldots, m; \right.$$
$$\mathbf{x} \in [-M, M]^n;$$
$$\left. x_k \in \mathbb{Z}, \ k \in J \right\},$$

for some real symmetric matrices $A_j \in \mathbb{R}^{n \times n}$, vectors $\mathbf{b}_j \in \mathbb{R}^n$, $j = 0, \ldots, m$, and some subset $J \subseteq \{1, \ldots, n\}$. Indeed, it suffices to model the constraint $x_i \in [-M, M]$, $i \notin J$, with the quadratic inequality constraints $M - x_i^2 \geq 0$, $j \notin J$, and the integrality constraints $x_k \in \mathbb{Z} \cap [-M, M]$, $k \in J$, with the polynomial equality constraints:

$$(x_k + M) \cdot (x_k + M - 1) \cdots x_k \cdot (x_k - 1) \cdots (x_k - M) = 0, \qquad k \in J.$$

1.2 Message of the book

We have already mentioned one message of the book.

- Polynomial Optimization indeed deserves a special treatment because its algebraic aspects can be taken into account in a systematic manner by invoking powerful results from real algebraic geometry.

But there are other important messages.

1.2.1 Easyness

A second message of the book which will become clear in the next chapters, is that the methodology for handling polynomial optimization problems **P** as defined in (1.1) is rather simple and easy to follow.

- Firstly, solving a polynomial optimization problem (1.1) is trivially equivalent to solving

$$f^* = \sup_{\lambda} \{ \lambda : f(\mathbf{x}) - \lambda \geq 0, \quad \forall \mathbf{x} \in \mathbf{K} \}, \tag{1.3}$$

which, if f is a polynomial of degree at most d, is in turn equivalent to solving

$$f^* = \sup_{\lambda} \{ \lambda : f - \lambda \in C_d(\mathbf{K}) \}, \tag{1.4}$$

where $C_d(\mathbf{K})$ is the convex cone of polynomials of degree at most d which are nonnegative on $\mathbf{K}$. But (1.4) is a finite-dimensional *convex* optimization problem. Indeed, a polynomial $f : \mathbb{R}^n \to \mathbb{R}$ of degree d is encoded by its vector $\mathbf{f} \in \mathbb{R}^{s(d)}$ of coefficients (e.g. in the usual canonical basis of monomials), where $s(d) := \binom{n+d}{n}$ is the dimension of the vector space of polynomials of degree at most d (that is the number of monomials $\mathbf{x}^\alpha = x_1^{\alpha_1} \cdots x_n^{\alpha_n}$, $\alpha \in \mathbb{N}^n$, such that $\sum_{i=1}^n \alpha_i \leq d$). And so $C_d(\mathbf{K})$ is a finite-dimensional cone which can be viewed as (or identified with) a convex cone of $\mathbb{R}^{s(d)}$. Therefore (with some abuse of notation) (1.4) also reads

$$f^* = \sup_{\lambda} \left\{ \lambda : \mathbf{f} - \lambda \begin{pmatrix} 1 \\ 0 \\ \vdots \\ 0 \end{pmatrix} \in C_d(\mathbf{K}) \right\}, \tag{1.5}$$

a convex conic optimization problem in $\mathbb{R}^{s(d)}$. Note in passing that the convex formulations (1.3) and (1.5) are proper to the global optimum f^* and are *not* valid for a local minimum $\hat{f} > f^*$! However, (1.5) remains hard to solve because in general there is no simple and *tractable* characterization of the convex cone $C_d(\mathbf{K})$ (even though it is finite dimensional).

Then the general methodology that we use follows a simple idea. We first define a (nested) increasing family of convex cones $(C_d^\ell(\mathbf{K})) \subset C_d(\mathbf{K})$ such that $C_d^\ell(\mathbf{K}) \subset C_d^{\ell+1}(\mathbf{K})$ for every ℓ, and each $C_d^\ell(\mathbf{K})$ is the projection of either a polyhedral cone or the intersection of a subspace with the convex cone of positive semidefinite matrices (whose size depends on ℓ). Then we solve the hierarchy of conic optimization problems

$$\rho_\ell = \sup_{\lambda} \left\{ \lambda : \mathbf{f} - \lambda \begin{pmatrix} 1 \\ 0 \\ \vdots \\ 0 \end{pmatrix} \in C_d^\ell(\mathbf{K}) \right\}, \quad \ell = 0, 1, \ldots \tag{1.6}$$

For each fixed ℓ, the associated conic optimization problem is convex and can be solved efficiently by appropriate methods of convex optimization. For instance, by using some appropriate *interior points* methods, (1.6) can be solved to arbitrary precision fixed in advance, in time polynomial in its input size. As the $C_d^\ell(\mathbf{K})$ provide a nested sequence of inner approximations of $C_d(\mathbf{K})$, $\rho_\ell \leq \rho_{\ell+1} \leq f^*$ for every ℓ. And the $C_d^\ell(\mathbf{K})$ are chosen so as to ensure the convergence $\rho_\ell \to f^*$ as $\ell \to \infty$. So depending on which type of convex approximation is used, (1.6) provides a *hierarchy* of linear or semidefinite programs (of increasing size) whose respective associated sequences of optimal values both converge to the desired global optimum f^*.

- Secondly, the powerful results from Real Algebraic Geometry that we use to justify the convergence $\rho_\ell \to f^*$ in the above methodology, are extremely simple to understand and could be presented (without proof) in undergraduate courses of Applied Mathematics, Optimization and/or Operations Research. Of course their proof requires some knowledge of sophisticated material in several branches of mathematics but we will *not* prove such results, we will only *use* them! After all, the statement in Fermat's theorem is easy to understand and this theorem may be used with no need to understand its proof.

  For illustration and to give the flavor of one such important and powerful result, we will repeatedly use the following result which states that a polynomial f which is (strictly) positive on $\mathbf{K}$ (as defined in (1.2) and compact) can be written in the form

$$f(\mathbf{x}) = \sigma_0(\mathbf{x}) + \sum_{j=1}^{m} \sigma_j(\mathbf{x})\, g_j(\mathbf{x}), \qquad \forall \mathbf{x} \in \mathbb{R}^n, \tag{1.7}$$

  for some polynomials σ_j that are *Sums of Squares* (SOS). By SOS we mean that each σ_j, $j = 0, \ldots, m$, can be written in the form

$$\sigma_j(\mathbf{x}) = \sum_{k=1}^{s_j} h_{jk}(\mathbf{x})^2, \qquad \forall \mathbf{x} \in \mathbb{R}^n,$$

  for finitely many polynomials h_{jk}, $k = 1, \ldots, s_j$.

  As one may see, (1.7) provides f with a *certificate* of its positivity on $\mathbf{K}$. This is because if $\mathbf{x} \in \mathbf{K}$ then $f(\mathbf{x}) \geq 0$ follows immediately from (1.7) as $\sigma_j(\mathbf{x}) \geq 0$ (because σ_j is SOS) and $g_j(\mathbf{x}) \geq 0$ (because $\mathbf{x} \in \mathbf{K}$), for all j. In other words, there is no need to check the positivity of f on $\mathbf{K}$ as one may read it directly from (1.7)!

- Finally, the convex conic optimization problem (1.4) has a *dual* which is another finite-dimensional convex conic optimization problem. And in fact this classical duality of convex (conic) optimization captures and illustrates the beautiful duality between *positive polynomials* and *moment problems*. We will see that the dual of (1.4) is particularly useful for extracting global minimizers of $\mathbf{P}$ when the convergence is finite (which, in addition, happens generically!). Depending on which type of positivity certificate is used we call this methodology the *moment-LP* or *moment-SOS* approach.

1.2.2 A general methodology

The class of polynomial optimization problems contains "easy" convex problems (e.g. Linear Programming and convex Quadratic Programming) as well as NP-hard optimization problems (e.g. the MAXCUT problem already mentioned).

> Still, the general methodology presented in this book *does not distinguish* between easy convex problems and nonconvex, discrete and mixed-integer optimization problems!

This immediately raises the following issues.

How effective can a general purpose approach be for addressing problems which can be so different in nature (e.g. convex, or continuous but possibly with a nonconvex and nonconnected feasible set, or discrete, or mixed-integer, etc.)?

- *Should we not specialize the approach according to the problem on hand, with ad hoc methods for certain categories of problems?*
- *Is the general approach reasonably efficient when applied to problems considered easy? Indeed would one trust a general purpose method designed for hard problems, and which would not behave efficiently on easy problems?*

Indeed, a large class of convex optimization problems are considered "easy" and can be solved efficiently by several ad hoc methods of convex optimization. Therefore a highly desirable feature of a general purpose approach is the ability somehow to *recognize* easier convex problems and behave accordingly (even if this may not be as efficient as specific methods tailored to the convex case).

A third message of this book is that this is indeed the case for the moment-SOS approach based on semidefinite relaxations, which uses representation results of the form (1.7) based on SOS. This is not the case for the moment-LP approach based on LP-relaxations, which uses other representation results.

In our mind this is an important and remarkable feature of the moment-SOS approach. For instance, and as already mentioned, a Boolean variable x_i is not treated with any particular attention and is modeled via the quadratic equality constraint $x_i^2 - x_i = 0$, just one among the many other polynomial equality or inequality constraints in the definition of the feasible set $\mathbf{K}$ in (1.2). Running a local minimization algorithm of continuous optimization with such a modeling of a Boolean constraint would not be considered wise (to say the least)! Hence this might justify some doubts concerning the efficiency of the moment-SOS approach by lack of specialization. Yet, and remarkably, the resulting semidefinite relaxations

- still provide the strongest relaxation algorithms for hard combinatorial optimization problems, and
- *recognize* easy convex problems as in this latter case convergence is even finite (and sometimes at the first semidefinite relaxation of the hierarchy)!

Of course, and especially in view of the present status of semidefinite solvers, the moment-SOS approach is still limited to problems of modest size; however, if symmetries or some structured sparsity in the problem data are detected and taken into account, then problems of much larger size can be solved.

1.2.3 Global optimality conditions

Another message of this book is that in the moment-SOS approach, *generically* the convergence $\rho_\ell \to f^*$ as $\ell \to \infty$ is finite (and genericity will be given a precise meaning)!

And so in particular, generically, solving a polynomial optimization problem (1.1) on a compact set $\mathbf{K}$ as in (1.2) reduces to solving a *single* semidefinite program.

But of course and as expected, the size of the resulting semidefinite program is not known in advance and can be potentially large.

Moreover, the powerful Putinar representation (1.7) of polynomials that are (strictly) positive on $\mathbf{K}$ also holds generically for polynomials that are only nonnegative on $\mathbf{K}$. And this translates into *global optimality* conditions that must be satisfied by global minimizers, provided that each global minimizer satisfies standard well-known constraint qualification, strict complementarity and second-order sufficiency conditions in nonlinear programming. Again, such conditions hold generically.

Remarkably, these optimality conditions are the perfect analogues in (non-convex) polynomial optimization of the Karush–Kuhn–Tucker (KKT) optimality conditions in convex programming. In particular, and in contrast to the KKT conditions in nonconvex programming, the constraints that are important but not active at a global minimizer still play a role in the optimality conditions.

1.2.4 Extensions

The final message is that the above methodology can also be applied in the following situations.

- To handle *semi-algebraic* functions, a class of functions much larger than the class of polynomials. For instance one may handle functions like

$$f(\mathbf{x}) := \sqrt{\min[q_1(\mathbf{x}), q_2(\mathbf{x})] - \max[q_3(\mathbf{x}), q_4(\mathbf{x})] + (q_5(\mathbf{x}) + q_6(\mathbf{x}))^{1/3}},$$

  where the q_i are given polynomials.
- To handle extensions like *parametric* and *inverse* optimization problems.
- To build up polynomial *convex underestimators* of a given nonconvex polynomial on a box $\mathbf{B} \subset \mathbb{R}^n$.
- To approximate as closely as desired, sets defined with quantifiers, for example the set

$$\{ \mathbf{x} \in \mathbf{B} : f(\mathbf{x}, \mathbf{y}) \leq 0 \text{ for all } \mathbf{y} \text{ such that } (\mathbf{x}, \mathbf{y}) \in \mathbf{K} \},$$

where $\mathbf{K} \subset \mathbb{R}^{n+p}$ is a set of the form (1.2), and $\mathbf{B} \subset \mathbb{R}^n$ is a simple set.

1.3 Plan of the book

The book is divided into three main parts.

Part I is introductory and Chapter 2 is dedicated to presenting basic and important results on the representation of polynomials that are positive on a subset $\mathbf{K}$ of $\mathbb{R}^n$. This problem of real algebraic geometry has a nice *dual facet*, the so-called $\mathbf{K}$-moment problem in Functional Analysis and Probability. And so results on the algebraic side are complemented with their dual counterparts on the moment side. Interestingly, convex duality in optimization (applied to appropriate convex cones) nicely captures this duality. Chapter 3 describes another characterization of polynomials nonnegative on a closed set $\mathbf{K}$ which is of independent interest and is *dual* to the characterizations of Chapter 2. Now knowledge on $\mathbf{K}$ is from moments of a measure supported on $\mathbf{K}$ rather than from polynomials that describe the boundary of $\mathbf{K}$. These two dual points of view are exploited in Chapter 4 to provide explicit outer and inner approximations of the cone of polynomials nonnegative on $\mathbf{K}$.

Part II is dedicated to polynomial and semi-algebraic optimization. It describes how to use results of Part I to define hierarchies of convex relaxations whose optimal values provide monotone sequences of lower bounds which converge to the global optimum. Depending on the type of representation (or positivity certificate) used, one obtains a hierarchy of linear programs or semidefinite programs. Their respective merits and drawbacks are analyzed, especially in the light of global optimality conditions. In particular, we describe a global optimality condition which is the exact analogue in nonconvex polynomial optimization of the celebrated KKT optimality conditions in convex optimization.

Using the representation results described in Chapter 3, one also obtains a hierarchy of eigenvalue problems which provide a monotone sequence of upper bounds which converges to the global optimum. Notice that most (primal-type) minimization algorithms provide sequences of upper bounds on the global minimum (as they move from a feasible point to another feasible point) but in general their convergence (if it eventually takes place) is guaranteed to a local minimum only. We also (briefly) describe how to use sparsity or symmetry to reduce the computational burden associated with the hierarchy of relaxations. It is worth noticing that the extension from polynomials to semi-algebraic functions (in both the objective function and the description of the feasible set) enlarges significantly the range of potential applications that can be treated.

Part III describes some specializations and extensions.

- *Convex polynomial optimization* to show that the moment-SOS approach somehow "recognizes" some classes of easy convex problems; in particular the hierarchy of semidefinite relaxations has finite convergence. Some properties of convex polynomials and convex basic semi-algebraic sets are also described and analyzed.

- *Parametric polynomial optimization*, that is, optimization problems where the criterion to minimize as well as the constraints that describe the feasible set, may depend on some parameters that belong to a given set. The ultimate and difficult goal is to compute or at least provide some information and/or approximations on the global optimum and the global minimizers, viewed as *functions* of the parameters. Hence there is a qualitative jump in difficulty as one now searches for *functions* on some domain (an infinite-dimensional object) rather than a vector $\mathbf{x} \in \mathbb{R}^n$ (a finite-dimensional object). With this in mind we describe what we call the "joint+marginal" approach to parametric optimization and show that in this context the moment-SOS approach is well suited for providing good approximations (theoretically as closely as desired) when the parametric optimization problem is described via polynomials and basic semi-algebraic sets.

- *Inverse polynomial optimization* where given a point $\mathbf{y} \in \mathbf{K}$ (think of an iterate of some local minimization algorithm) and a polynomial criterion f, one tries to find a polynomial $\tilde{f}$ as close as possible to f and for which $\mathbf{y}$ is a global minimizer on $\mathbf{K}$. This problem has interesting potential theoretical and practical applications. For instance, suppose that $\mathbf{y} \in \mathbf{K}$ is a current iterate of a local optimization algorithm to minimize f on $\mathbf{K}$, and suppose that solving the inverse problem provides a new criterion $\tilde{f}$ relatively close to f. Should we spend (expensive) additional effort to obtain a better iterate or should we stop (as $\mathbf{y}$ solves an optimization problem close to the original one)? In addition, the inverse problem is also a way to measure how ill-conditioned is the (direct) optimization problem.

- *Convex underestimators.* For difficult large scale nonlinear problems and particularly for mixed-integer nonlinear programs (MINLP), the only practical way to approximate the global minimum is to explore an appropriate Branch and Bound search tree in which exploration is guided by lower bounds computed at each node of the tree. The quality of lower bounds is crucial for the overall efficiency of the approach. In general and for obvious reasons, efficient computation of lower bounds is possible only on some appropriate convex relaxation of the problem described at the current node. A standard way to obtain a convex relaxation is to replace the nonconvex objective function with a convex underestimator; similarly, an inequality constraint $g(\mathbf{x}) \leq 0$ is replaced with $\tilde{g}(\mathbf{x}) \leq 0$ for some convex underestimator $\tilde{g}$ of g. Therefore deriving *tight* convex underestimators is of crucial importance for the quality of the resulting lower bounds. We show that the moment-SOS approach is particularly well suited to obtaining tight convex underestimators.

- *Polynomial optimization on sets defined with quantifiers.* In this context one is given a set $\mathbf{K} := \{ (\mathbf{x}, \mathbf{y}) : g_j(\mathbf{x}, \mathbf{y}) \geq 0, \; j = 1, \ldots, m \}$ for some polynomials $g_j, \; j = 1, \ldots, m$. Given a set $\mathbf{B} \subset \mathbb{R}^n$ (typically a box or an ellipsoid) the goal is to approximate as closely as desired the set

$$\mathbf{K_x} := \{ \mathbf{x} \in \mathbf{B} : f(\mathbf{x}, \mathbf{y}) \leq 0 \text{ for all } \mathbf{y} \text{ such that } (\mathbf{x}, \mathbf{y}) \in \mathbf{K} \},$$

by a set $\mathbf{K_x^\ell} \subset \mathbf{K_x}$ simply defined by $\mathbf{K_x^\ell} = \{ \mathbf{x} : h_\ell(\mathbf{x}) \leq 0 \}$ for some polynomial h_ℓ. We show how to obtain a sequence of polynomials (h_ℓ) of increasing degree $\ell \in \mathbb{N}$, such that the Lebesgue volume $\mathrm{vol}(\mathbf{K_x} \setminus \mathbf{K_x^\ell})$ tends to zero as $\ell \to \infty$. And so any optimization problem involving the set $\mathbf{K_x}$ (difficult to handle) can be approximated by substituting $\mathbf{K_x}$ with $\mathbf{K_x^\ell}$ for some sufficiently large ℓ.

- *A nonconvex generalization of the Löwner–John problem.* We investigate properties of the sublevel set $\mathbf{G} := \{\,\mathbf{x} : g(\mathbf{x}) \leq 1\,\}$ associated with a quasi-homogeneous polynomial g. We then solve a nonconvex generalization of the celebrated Löwner–John problem. Namely, with d a fixed even integer, we want to determine an homogeneous polynomial g of degree d such that its sublevel set $\mathbf{G}$ contains a given (not necessarily convex) set $\mathbf{K}$ and has minimum volume among all such sets $\mathbf{G} \supset \mathbf{K}$; when $d = 2$ this problem has a unique optimal solution, the celebrated Löwner–John ellipsoid. We show that our generalization is a convex optimization problem which also has a unique optimal solution characterized in terms of contact points.

PART I

Positive polynomials and moment
problems

2

Positive polynomials and moment problems

In one dimension, the ring $\mathbb{R}[x]$ of real polynomials in a single variable has the fundamental property (Theorem 2.3) that every nonnegative polynomial $p \in \mathbb{R}[x]$ is a sum of squares of polynomials, that is,

$$p(x) \geq 0, \quad \forall x \in \mathbb{R} \quad \Leftrightarrow \quad p(x) = \sum_{i=1}^{k} p_i(x)^2, \quad \forall x \in \mathbb{R},$$

for finitely many polynomials $(p_i) \subset \mathbb{R}[x]$ (in short p is SOS). In multiple dimensions, however, it is possible for a polynomial to be nonnegative without being a sum of squares. In fact, the only cases (n, d) where all nonnegative polynomials are also SOS are $(1, d)$ (nonnegative univariate polynomials), $(n, 2)$ (nonnegative quadratic polynomials) and $(2, 4)$ (nonnegative quartic bivariate polynomials). In his famous address in a Paris meeting of mathematicians in 1900, Hilbert posed important problems for mathematics to be addressed in the twentieth century, and in his 17th problem he conjectured that every nonnegative polynomial can always be written as a sum of squares of rational functions. This conjecture was later proved to be correct by Emil Artin in 1926, using the Artin–Schreier theory of real closed fields.

An immediate extension is to consider characterizations of polynomials p that are nonnegative on a basic semi-algebraic set $\mathbf{K}$, i.e., a set defined by polynomial inequalities $g_j(\mathbf{x}) \geq 0$, $j = 1, \ldots, m$. In particular, one searches for (tractable) characterizations of the convex cone $C_d(\mathbf{K}) \subset \mathbb{R}[\mathbf{x}]_d$ of polynomials of degree at most d, nonnegative on $\mathbf{K}$. By characterization, we mean a representation of p in terms of the g_j such that the nonnegativity of p on $\mathbf{K}$ follows immediately from the latter representation in the same way that

p is obviously nonnegative when it is a sum of squares. In other words, this representation of p can be seen as a *certificate* of nonnegativity of p on **K**.

In this chapter, we review key representation theorems for nonnegative (or positive) polynomials and their dual counterparts for the **K**-moment problem. We will see later how this beautiful duality between positive polynomials and moments is nicely captured by standard duality in convex optimization applied to appropriate convex cones. These results form the theoretical basis on which are based the (semidefinite or linear) convex relaxations that we later define for solving the polynomial optimization problem (1.1).

2.1 Sum of squares representations and semidefinite optimization

In this section we show that if a nonnegative polynomial has a sum of squares (or in short, SOS) representation then one can compute this representation using semidefinite optimization methods. Given that semidefinite optimization problems are efficiently solved from both theoretical and practical points of view, it follows that we can compute an SOS decomposition of a nonnegative polynomial, if it exists, efficiently.

Let $\mathbb{R}[\mathbf{x}]$ denote the ring of real polynomials in the variables $\mathbf{x}=(x_1, \ldots, x_n)$. A polynomial $p \in \mathbb{R}[\mathbf{x}]$ is an SOS if it can be written as

$$\mathbf{x} \mapsto p(\mathbf{x}) = \sum_{j \in J} p_j(\mathbf{x})^2, \qquad \mathbf{x} \in \mathbb{R}^n,$$

for some finite family of polynomials $(p_j : j \in J) \subset \mathbb{R}[\mathbf{x}]$. Notice that necessarily the degree of p must be even, and also the degree of each p_j is bounded by half that of p.

Denote by $\Sigma[\mathbf{x}] \subset \mathbb{R}[\mathbf{x}]$ the space of SOS polynomials. For any two real symmetric matrices $\mathbf{A}, \mathbf{B}$, recall that $\langle \mathbf{A}, \mathbf{B} \rangle$ stands for trace$(\mathbf{AB})$, and the notation $\mathbf{A} \succeq 0$ stands for $\mathbf{A}$ is positive semidefinite. Finally, for a multi-index $\boldsymbol{\alpha} \in \mathbb{N}^n$, let $|\boldsymbol{\alpha}| := \sum_{i=1}^{n} \alpha_i$, and let $\mathbb{N}_d^n := \{ \boldsymbol{\alpha} \in \mathbb{N}^n : |\boldsymbol{\alpha}| \leq d \}$. With $\mathbf{x}^T$ (or sometimes $\mathbf{x}'$) denoting the transpose of a vector $\mathbf{x}$, consider the vector

$$
\begin{aligned}
\mathbf{v}_d(\mathbf{x}) &= (\mathbf{x}^{\boldsymbol{\alpha}}), \quad \boldsymbol{\alpha} \in \mathbb{N}_d^n \\
&= (1, x_1, \ldots, x_n, x_1^2, x_1 x_2, \ldots, x_{n-1} x_n, x_n^2, \ldots, x_1^d, \ldots, x_n^d)^T,
\end{aligned}
$$

$$(2.1)$$

of all monomials $\mathbf{x}^\alpha$ of degree less than or equal to d, which has dimension $s(d) := \binom{n+d}{d}$. These monomials form a canonical basis of the vector space $\mathbb{R}[\mathbf{x}]_d$ of polynomials of degree at most d.

Proposition 2.1 *A polynomial $g \in \mathbb{R}[\mathbf{x}]_{2d}$ has a sum of squares decomposition (i.e., is SOS) if and only if there exists a real symmetric and positive semidefinite matrix $\mathbf{Q} \in \mathbb{R}^{s(d) \times s(d)}$ such that $g(\mathbf{x}) = \mathbf{v}_d(\mathbf{x})^T \mathbf{Q} \mathbf{v}_d(\mathbf{x})$, for all $\mathbf{x} \in \mathbb{R}^n$.*

Proof Suppose there exists a real symmetric $s(d) \times s(d)$ matrix $\mathbf{Q} \succeq 0$ for which $g(\mathbf{x}) = \mathbf{v}_d(\mathbf{x})^T \mathbf{Q} \mathbf{v}_d(\mathbf{x})$, for all $\mathbf{x} \in \mathbb{R}^n$. Then by the spectral decomposition, $\mathbf{Q} = \mathbf{H}\mathbf{H}^T$ for some $s(d) \times k$ matrix $\mathbf{H}$, and thus,

$$g(\mathbf{x}) = \mathbf{v}_d(\mathbf{x})^T \mathbf{H}\mathbf{H}^T \mathbf{v}_d(\mathbf{x}) = \sum_{i=1}^{k} (\mathbf{H}^T \mathbf{v}_d(\mathbf{x}))_i^2, \qquad \forall \mathbf{x} \in \mathbb{R}^n.$$

Since $\mathbf{x} \mapsto (\mathbf{H}^T \mathbf{v}_d(\mathbf{x}))_i$ is a polynomial, then g is expressed as a sum of squares of the polynomials $\mathbf{x} \mapsto (\mathbf{H}^T \mathbf{v}_d(\mathbf{x}))_i$, $i = 1, \ldots, k$.

Conversely, suppose that g of degree $2d$ has an SOS decomposition $g(\mathbf{x}) = \sum_{i=1}^{k} h_i(\mathbf{x})^2$ for some family $\{ h_i : i = 1, \ldots, k \} \subset \mathbb{R}[\mathbf{x}]$. Then necessarily, the degree of each h_i is bounded by d. Let $\mathbf{h}_i$ be the vector of coefficients of the polynomial h_i, i.e., $h_i(\mathbf{x}) = \mathbf{h}_i^T \mathbf{v}_d(\mathbf{x})$, $i = 1, \ldots, k$. Thus,

$$g(\mathbf{x}) = \sum_{i=1}^{k} \mathbf{v}_d(\mathbf{x})^T \underbrace{\mathbf{h}_i \mathbf{h}_i^T}_{\mathbf{Q}} \mathbf{v}_d(\mathbf{x}) = \mathbf{v}_d(\mathbf{x})^T \mathbf{Q} \mathbf{v}_d(\mathbf{x}), \qquad \forall \mathbf{x} \in \mathbb{R}^n,$$

with $\mathbf{Q} \in \mathbb{R}^{s(d) \times s(d)}$, $\mathbf{Q} := \sum_{i=1}^{k} \mathbf{h}_i \mathbf{h}_i^T \succeq 0$, and the proposition follows. □

Given an SOS polynomial $g \in \mathbb{R}[\mathbf{x}]_{2d}$, the identity $g(\mathbf{x}) = \mathbf{v}_d(\mathbf{x})^T \mathbf{Q} \mathbf{v}_d(\mathbf{x})$ for all $\mathbf{x}$, provides linear equations that the coefficients of the matrix $\mathbf{Q}$ must satisfy. So if one writes

$$\mathbf{v}_d(\mathbf{x})\mathbf{v}_d(\mathbf{x})^T = \sum_{\alpha \in \mathbb{N}_{2d}^n} \mathbf{B}_\alpha \, \mathbf{x}^\alpha,$$

for appropriate $s(d) \times s(d)$ real symmetric matrices $\mathbf{B}_\alpha$, $\alpha \in \mathbb{N}_{2d}^n$, then

checking whether the polynomial $\mathbf{x} \mapsto g(\mathbf{x}) = \sum_\alpha g_\alpha \mathbf{x}^\alpha$ is SOS reduces to solving the *semidefinite optimization (feasibility) problem*.

Find $\mathbf{Q} \in \mathbb{R}^{s(d) \times s(d)}$ such that:

$$\mathbf{Q} = \mathbf{Q}^T, \quad \mathbf{Q} \succeq 0, \quad \langle \mathbf{Q}, \mathbf{B}_\alpha \rangle = g_\alpha, \qquad \forall \alpha \in \mathbb{N}_{2d}^n, \tag{2.2}$$

a tractable convex optimization problem (in fact, a semidefinite optimization problem) for which efficient software packages are available.

Indeed, up to arbitrary precision $\epsilon > 0$ fixed, a semidefinite program can
be solved in a computational time that is polynomial in the input size of the
problem. Observe that the size $s(d) = \binom{n+d}{n}$ of the semidefinite program (2.2)
is bounded by n^d.

On the other hand, *nonnegativity* of a polynomial $g \in \mathbb{R}[\mathbf{x}]$ can be checked
also by solving a single semidefinite program. Indeed, if g is nonnegative on $\mathbb{R}^n$
then it can be written as a sum of squares of rational functions, or equivalently,
clearing denominators,

$$h\,g = f, \qquad (2.3)$$

for some nontrivial SOS polynomials $h, f \in \Sigma[\mathbf{x}]$, $k = 1, \ldots, r$, and there
exist bounds on the degree of h and f. Conversely, if there exist nontrivial
SOS polynomials $h, f \in \Sigma[\mathbf{x}]$ solving (2.3), then obviously g is nonnegative.
Indeed suppose that $g(\mathbf{x}_0) < 0$ for some $\mathbf{x}_0$ (and so $g(\mathbf{x}_0) < 0$ in some open
neighborhood $\mathbf{B}(\mathbf{x}_0)$ of $\mathbf{x}_0$). Since $hg = f$ and f is SOS then necessarily
$f = 0$ on $\mathbf{B}(\mathbf{x}_0)$, and so $f(\mathbf{x}) = 0$ for all $\mathbf{x} \in \mathbb{R}^n$, which in turn implies that
$h = 0$, in contradiction with h, f being nontrivial.

Therefore, using (2.2) for h and f in (2.3), finding a certificate $h, f \in$
$\Sigma[\mathbf{x}]$ for nonnegativity of g reduces to solving a single semidefinite program.
Unfortunately, the available bounds for the size of this semidefinite program
are far too large for practical implementation. This is what makes the sum of
squares property very attractive computationally, in contrast with the weaker
nonnegativity property which is much harder to check (if not impossible).

Example 2.2   Consider the bivariate polynomial in $\mathbb{R}[\mathbf{x}] = \mathbb{R}[x_1, x_2]$

$$\mathbf{x} \mapsto \quad f(\mathbf{x}) = 2x_1^4 + 2x_1^3 x_2 - x_1^2 x_2^2 + 5x_2^4, \qquad \mathbf{x} \in \mathbb{R}^2.$$

Suppose that we want to check whether f is SOS. As f is homogeneous, we
attempt to write f in the form

$$
\begin{aligned}
f(x_1, x_2) &= 2x_1^4 + 2x_1^3 x_2 - x_1^2 x_2^2 + 5x_2^4 \\
&= \begin{pmatrix} x_1^2 & x_2^2 & x_1 x_2 \end{pmatrix} \begin{bmatrix} q_{11} & q_{12} & q_{13} \\ q_{12} & q_{22} & q_{23} \\ q_{13} & q_{23} & q_{33} \end{bmatrix} \begin{pmatrix} x_1^2 \\ x_2^2 \\ x_1 x_2 \end{pmatrix} \\
&= q_{11}\,x_1^4 + q_{22}\,x_2^4 + (q_{33} + 2q_{12})\,x_1^2 x_2^2 + 2q_{13}\,x_1^3 x_2 + 2q_{23}\,x_1 x_2^3,
\end{aligned}
$$

for some $\mathbf{Q} \succeq 0$. In order to have an identity, we obtain

$$\mathbf{Q} \succeq 0, \quad q_{11} = 2, \quad q_{22} = 5, \quad q_{33} + 2q_{12} = -1, \quad 2q_{13} = 2, \quad q_{23} = 0.$$

In this case we easily find the particular solution

$$0 \preceq \mathbf{Q} = \begin{bmatrix} 2 & -3 & 1 \\ -3 & 5 & 0 \\ 1 & 0 & 5 \end{bmatrix} = \mathbf{H}\mathbf{H}^{T}, \quad \text{for} \quad \mathbf{H} = \frac{1}{\sqrt{2}} \begin{bmatrix} 2 & 0 \\ -3 & 1 \\ 1 & 3 \end{bmatrix},$$

and so

$$f(x_1, x_2) = \frac{1}{2}(2x_1^2 - 3x_2^2 + x_1 x_2)^2 + \frac{1}{2}(x_2^2 + 3x_1 x_2)^2,$$

which is indeed an SOS.

2.2 Representation theorems: univariate case

In this section we review the major representation theorems for nonnegative univariate polynomials for which the results are quite complete. Let $\mathbb{R}[x]$ be the ring of real polynomials in the single variable x, and let $\Sigma[x] \subset \mathbb{R}[x]$ be its subset of SOS polynomials, i.e., polynomials that are sums of squares of elements of $\mathbb{R}[x]$. We first prove that if $p \in \mathbb{R}[x]$ is nonnegative, then $p \in \Sigma[x]$.

Theorem 2.3   *A polynomial $p \in \mathbb{R}[x]$ of even degree is nonnegative if and only if it can be written as a sum of squares of other polynomials, i.e., $p(x) = \sum_{i=1}^{k} h_i(x)^2$, with $h_i \in \mathbb{R}[x]$, $i = 1, \ldots, k$.*

Proof   Clearly, if $p(x) = \sum_{j=1}^{k} [h_j(x)]^2$, then $p(x) \geq 0$ for all $x \in \mathbb{R}$.

Conversely, suppose that a polynomial $p \in \mathbb{R}[x]$ of degree $2d$ (and with highest degree term $p_{2d} x^{2d}$) is nonnegative on the real line $\mathbb{R}$. Then, the real roots of p should have even multiplicity, otherwise p would alter its sign in a neighborhood of a real root. Let λ_j, $j = 1, \ldots, r$, be its real roots with corresponding multiplicity $2m_j$. Its complex roots can be arranged in conjugate pairs, $a_\ell + ib_\ell$, $a_\ell - ib_\ell$, $\ell = 1, \ldots, t$ (with $i^2 = -1$). Then p can be written in the form:

$$x \mapsto \quad p(x) = p_{2d} \prod_{j=1}^{r} (x - \lambda_j)^{2m_j} \prod_{\ell=1}^{t} \left((x - a_\ell)^2 + b_\ell^2 \right).$$

Note that the leading coefficient p_{2d} needs to be positive. By expanding the terms in the products, we see that p can be written as a sum of squares of $k = 2^t$ polynomials. (In fact, one may also show that p is a sum of only two squares.) □

We next concentrate on polynomials $p \in \mathbb{R}[x]$ that are nonnegative on an interval $I \subset \mathbb{R}$. Moreover, the three cases $I = (-\infty, b]$, $I = [a, \infty)$ and $I = [a, b]$ (with $a, b \in \mathbb{R}$) all reduce to the basic cases $I = [0, \infty)$ and $I = [-1, +1]$ using the changes of variable

$$f(x) := p(b - x), \quad f(x) := p(x - a), \text{ and } f(x) := p\left(\frac{2x - (a + b)}{b - a}\right),$$

respectively.

The representation results that one obtains depend on the particular choice of the polynomials used in the description of the interval. The main result in the one-dimensional case can be summarized in the next theorem.

Theorem 2.4 *Let $p \in \mathbb{R}[x]$ be of degree n.*

(a) *Let $g \in \mathbb{R}[x]$ be the polynomial $x \mapsto g(x) := 1 - x^2$. Then $p \geq 0$ on $[-1, 1]$ if and only if*

$$p = f + g h, \quad f, h \in \Sigma[x],$$

and with both summands of degree less than $2n$.

(b) *Let $x \mapsto g_1(x) := 1 - x$, $x \mapsto g_2(x) := 1 + x$, $x \mapsto g_3(x) := g_1(x)g_2(x)$. Then, $p \geq 0$ on $I = [-1, 1]$ if and only if*

$$p = f_0 + g_1 f_1 + g_2 f_2 + g_3 f_3, \quad f_0, f_1, f_2, f_3 \in \Sigma[x].$$

In addition, all the summands have degree at most n, and $f_1, f_2 = 0$, if n is even, whereas $f_0, f_3 = 0$, if n is odd.

The Goursat transform $f \to \hat{f}$ on $\mathbb{R}[x]$, of a polynomial $f \in \mathbb{R}[x]$, is defined by:

$$x \mapsto \hat{f}(x) := (1 + x)^m f\left(\frac{1 - x}{1 + x}\right), \quad x \in \mathbb{R}.$$

Using the Goursat transform, the case $[0, \infty)$ reduces to the case $[-1, 1]$ and we have the following theorem.

Theorem 2.5 *Let $p \in \mathbb{R}[x]$ be nonnegative on $[0, +\infty)$. Then*

$$p = f_0 + x f_1,$$

for two SOS polynomials $f_0, f_1 \in \Sigma[x]$ and the degree of both summands is bounded by the degree of p.

Example 2.6 Let $p \in \mathbb{R}[x]$ be the univariate polynomial $x \mapsto p(x) = x^5 - 2x^3 + x + 1$. To check whether p is nonnegative on $[0, \infty)$ one searches for two SOS polynomials $f_0, f_1 \in \Sigma[x]$ of degree 4 such that $p = f_0 + xf_1$. Equivalently, one searches for two real symmetric and positive semidefinite 3×3 matrices

$$\mathbf{F}_0 = \begin{bmatrix} a & b & c \\ b & d & e \\ c & e & f \end{bmatrix} \succeq 0, \quad \mathbf{F}_1 = \begin{bmatrix} A & B & C \\ B & D & E \\ C & E & F \end{bmatrix} \succeq 0,$$

such that (with $\mathbf{v}_2(x)^T = (1, x, x^2)$)

$$p(x) = \mathbf{v}_2(x)^T \mathbf{F}_0 \, \mathbf{v}_2(x) + x \, \mathbf{v}_2(x)^T \mathbf{F}_1 \, \mathbf{v}_2(x), \qquad \forall x \in \mathbb{R}.$$

That is, $\mathbf{F}_0$ and $\mathbf{F}_1$ satisfy $\mathbf{F}_0, \mathbf{F}_1 \succeq 0$ and the linear equations:

$$a = 1, \ 2b + A = 1, \ d + 2c + 2B = 0, \ 2e + D + 2C = -2, \ f + 2E = 0, \ F = 1.$$

A particular solution is $a = 1$, $b = c = d = e = f = 0$ and $A = F = 1 = -C$, $B = D = E = 0$, which yields

$$p(x) = 1 + x \, (x^2 - 1)^2, \qquad \forall x \in \mathbb{R},$$

and certifies that p is nonnegative on $[0, +\infty)$.

It is important to emphasize that Theorem 2.4 (respectively Theorem 2.5) explicitly uses the specific representation of the interval $[-1, 1]$ (respectively $[0, +\infty)$) as the basic semi-algebraic set: $\{x : 1 - x \geq 0; \ 1 + x \geq 0\}$ or $\{x : 1 - x^2 \geq 0\}$ (respectively $\{x : x \geq 0\}$).

In the case where $[-1, 1] = \{x \in \mathbb{R} : h(x) \geq 0\}$, with h not equal to the polynomial $x \mapsto (1+x)(1-x)$, a weaker result is obtained in the next section.

The next result also considers the interval $[-1, 1]$ but provides another representation that does not use SOS.

Theorem 2.7 *Let $p \in \mathbb{R}[x]$. Then $p > 0$ on $[-1, 1]$ if and only if*

$$p = \sum_{i+j \leq d} c_{ij} \, (1 - x)^i (1 + x)^j, \quad c_{ij} \geq 0, \qquad (2.4)$$

for some nonnegative scalars (c_{ij}) and some sufficiently large d.

Theorem 2.7 leads to a linear optimization feasibility problem to determine the coefficients c_{ij} in the representation (2.4). On the other hand, notice that the representation (2.4) is guaranteed only if p is (strictly) positive on $[-1, 1]$.

2.3 Representation theorems: multivariate case

In this section we consider the multivariate case. As already mentioned, a nonnegative polynomial $p \in \mathbb{R}[\mathbf{x}]$ does not necessarily have an SOS representation. For instance, the two polynomials

$$
\begin{aligned}
\mathbf{x} \mapsto p(\mathbf{x}) &= x_1^2 x_2^2 (x_1^2 + x_2^2 - 1) + 1, \\
\mathbf{x} \mapsto p(\mathbf{x}) &= x_1^2 x_2^2 (x_1^2 + x_2^2 - 3x_3^2) + 6x_3^6,
\end{aligned}
$$

are nonnegative but they do not have an SOS representation. On the other hand, nonnegative quadratic polynomials, and nonnegative fourth degree homogeneous polynomials of three variables have an SOS representation.

The following celebrated theorem due to Polyá provides a certificate of positivity for homogeneous polynomials that are positive on the simplex.

Theorem 2.8 (Polyá) *If $p \in \mathbb{R}[\mathbf{x}]$ is homogeneous and $p > 0$ on $\mathbb{R}_+^n \setminus \{\mathbf{0}\}$, then for sufficienly large $k \in \mathbb{N}$, all nonzero coefficients of the polynomial $\mathbf{x} \mapsto (x_1 + \cdots + x_n)^k p(\mathbf{x})$ are positive.*

The next result is a specialized form of Hilbert's 17th problem and provides a certificate of positivity for homogeneous polynomials that are positive everywhere except at $\mathbf{x} = 0$.

Theorem 2.9 *If $p \in \mathbb{R}[\mathbf{x}]$ is homogeneous and $p > 0$ on $\mathbb{R}^n \setminus \{0\}$ then there exists $d \in \mathbb{N}$ such that $\|\mathbf{x}\|^{2d} p$ is SOS.*

We next characterize when a semi-algebraic set described by polynomial inequalities, equalities and nonequalities is empty. In order to achieve this, we need the following definition.

Definition 2.10 For $F := \{f_1, \ldots, f_m\} \subset \mathbb{R}[\mathbf{x}]$, and a set $J \subseteq \{1, \ldots, m\}$ we denote by $f_J \in \mathbb{R}[\mathbf{x}]$ the polynomial $\mathbf{x} \mapsto f_J(\mathbf{x}) := \prod_{j \in J} f_j(\mathbf{x})$, with the convention that $f_\emptyset = 1$. The set

$$
P(f_1, \ldots, f_m) := \left\{ \sum_{J \subseteq \{1, \ldots, m\}} q_J f_J \; : \; q_J \in \Sigma[\mathbf{x}] \right\} \tag{2.5}
$$

is called (by algebraic geometers) the *preordering* generated by $f_1, \ldots, f_m$.

We first state an important and powerful result.

Theorem 2.11 (Stengle)   *Let k be a real closed field, and let*

$$F := (f_i)_{i \in I_1}, \quad G := (g_i)_{i \in I_2}, \quad H := (h_i)_{i \in I_3} \subset k[\mathbf{x}]$$

be finite families of polynomials. Let

(a) $P(F)$ *be the preordering generated by the family F,*
(b) $M(G)$ *be the set of all finite products of the g_i, $i \in I_2$ (the empty product being the constant polynomial 1), and*
(c) $I(H)$ *be the ideal generated by H, i.e., $I(H) = \{ \sum_{i \in I_3} p_i \, h_i :$ $p_i \in \mathbb{R}[\mathbf{x}], \ i \in I_3 \}$.*

Consider the semi-algebraic set

$$\mathbf{K} = \{ \mathbf{x} \in k^n \ : \quad f_i(\mathbf{x}) \geq 0, \quad \forall i \in I_1; \quad g_i(\mathbf{x}) \neq 0, \quad \forall i \in I_2;$$
$$h_i(\mathbf{x}) = 0, \quad \forall i \in I_3 \}.$$

The set $\mathbf{K}$ is empty if and only if there exist $f \in P(F)$, $g \in M(G)$ and $h \in I(H)$ such that

$$f + g^2 + h = 0. \tag{2.6}$$

The polynomials $f, g, h \in k[\mathbf{x}]$ in (2.6) provide a Stengle *certificate* of $\mathbf{K} = \emptyset$. In (2.6), there is also a (very large) bound on the degree of $g \in M(G)$, the degree of the SOS weights $q_J \in \Sigma[\mathbf{x}]$ in the representation (2.5) of $f \in P(F)$, as well as the degree of the weights $p_j \in \mathbb{R}[\mathbf{x}]$ in the representation of $h = \sum_j p_j h_j$.

Therefore in view of Proposition 2.1, in principle checking the existence of a certificate (f, g, h) requires solving a *single* semidefinite program. But unfortunately, the available degree bound being huge, the size of such a semidefinite program is far too large for practical implementation. This is no surprise as the class of sets $\mathbf{K}$ described in Theorem 2.11 is very large and contains extremely complicated sets (e.g. highly nonconvex and possibly disconnected).

Stengle and Farkas certificates

The Stengle certificate $f, g, h \in k[\mathbf{x}]$ in (2.6) for $\mathbf{K} = \emptyset$, is a nonlinear generalization of the celebrated Farkas Lemma (or theorem of the alternative) in linear algebra which provides a certificate of emptyness for a polyhedral set $\mathbf{K} = \{ \mathbf{x} : \mathbf{A}\mathbf{x} \leq \mathbf{b} \}$ (for some matrix $\mathbf{A} \in \mathbb{R}^{m \times n}$ and some vector $\mathbf{b} \in \mathbb{R}^m$).

In fact, a Farkas certificate is a particularly simple Stengle certificate for convex polyhedra. Indeed, a Farkas certificate of $\emptyset = \{ \mathbf{x} : \mathbf{A}\mathbf{x} \leq \mathbf{b} \}$ is a nonnegative vector $\mathbf{u} \in \mathbb{R}^m_+$ such that $\mathbf{u}^T \mathbf{A} = 0$ and $\mathbf{u}^T \mathbf{b} < 0$. So let $\mathbf{x} \mapsto f_j(\mathbf{x}) = (\mathbf{b} - \mathbf{Ax})_j$, $j = 1, \dots, m$, $\mathbf{x} \mapsto g_1(\mathbf{x}) = 1$ (so that $M(G)$ is identical to the constant polynomial 1) and $\mathbf{x} \mapsto h_1(\mathbf{x}) = 0$ (so that $I(H)$ is the 0 polynomial). The polynomial

$$\mathbf{x} \mapsto f(\mathbf{x}) := \frac{-1}{\mathbf{u}^T \mathbf{b}} \sum_{j=1}^{m} u_j \, (\mathbf{b} - \mathbf{Ax})_j, \qquad \mathbf{x} \in \mathbb{R}^n,$$

is an element of $P(F)$ because $u_j \geq 0$ for every $j = 1, \dots, m$ and $\mathbf{u}^T \mathbf{b} < 0$. But then

$$1 + f(\mathbf{x}) = 1 - \frac{1}{\mathbf{u}^T \mathbf{b}} \sum_{j=1}^{m} u_j (\mathbf{b} - \mathbf{Ax})_j = 1 - \frac{\mathbf{u}^T \mathbf{b} - \mathbf{u}^T \mathbf{Ax}}{\mathbf{u}^T \mathbf{b}} = 0,$$

that is, (2.6) holds.

We next consider basic semi-algebraic sets, that is, semi-algebraic sets defined by finitely many inequalities only. As a direct consequence of Theorem 2.11, one obtains the following result.

Theorem 2.12 (Stengle's Positivstellensatz and Nullstellensatz)   *Let k be a real closed field, $f \in k[\mathbf{x}]$, and let*

$$\mathbf{K} = \{\, \mathbf{x} \in k^n : \ f_j(\mathbf{x}) \geq 0, \ j = 1, \dots, m \,\}.$$

(a) **Nichtnegativstellensatz** $f \geq 0$ *on* $\mathbf{K}$ *if and only if there exist $\ell \in \mathbb{N}$ and $g, h \in P(f_1, \dots, f_m)$ such that $fg = f^{2\ell} + h$.*
(b) **Positivstellensatz** $f > 0$ *on* $\mathbf{K}$ *if and only if there exist $g, h \in P(f_1, \dots, f_m)$ such that $fg = 1 + h$.*
(c) **Nullstellensatz** $f = 0$ *on* $\mathbf{K}$ *if and only if there exist $\ell \in \mathbb{N}$ and $g \in P(f_1, \dots, f_m)$ such that $f^{2\ell} + g = 0$.*

Again, as for Theorem 2.11, there is also a bound on ℓ and the degree of the SOS weights $q_J \in \Sigma[\mathbf{x}]$ in the representation (2.5) of $g, h \in P(f_1, \dots, f_m)$. This bound depends only on the dimension n and on the degree of the polynomials $(f, f_1, \dots, f_m)$. Therefore in principle, checking the existence of a certificate (l, g, h) in Theorem 2.12(a)–(c) requires solving a single semidefinite program (but of huge size). In practice, one fixes an a priori (much smaller) degree bound and solves the corresponding semidefinite program. If the latter has a feasible solution then one obtains a certificate $g, h \in P(f_1, \dots, f_m)$. However, such a certificate is *numerical* and so can be obtained only up to some machine precision, because of numerical inaccuracies inherent to semidefinite programming solvers.

2.4 Polynomials positive on a compact basic semi-algebraic set

In this section, we restrict our attention to compact basic semi-algebraic sets $\mathbf{K} \subset \mathbb{R}^n$ and obtain certificates of positivity on $\mathbf{K}$ that have certain algorithmic advantages. In fact, Putinar's Positivstellensatz below is the key result that we will later use extensively to solve polynomial optimization problems.

2.4.1 Representations via sums of squares

The first representation result, known as Schmüdgen's Positivstellensatz, was an important breakthrough as it was the first to provide a simple characterization of polynomials positive on a compact basic semi-algebraic set $\mathbf{K}$, and with *no* additional assumption on $\mathbf{K}$ (or on its description).

Theorem 2.13 (Schmüdgen's Positivstellensatz) *Let $(g_j)_{j=1}^m \subset \mathbb{R}[\mathbf{x}]$ be such that the basic semi-algebraic set*

$$\mathbf{K} := \{\, \mathbf{x} \in \mathbb{R}^n : g_j(\mathbf{x}) \geq 0, \ j = 1, \ldots, m \,\} \tag{2.7}$$

is compact. If $f \in \mathbb{R}[\mathbf{x}]$ is strictly positive on $\mathbf{K}$ then $f \in P(g_1, \ldots, g_m)$, that is,

$$f = \sum_{J \subseteq \{1,\ldots,m\}} \sigma_J \, g_J, \quad \text{for some SOS } \sigma_J \in \Sigma[\mathbf{x}], \tag{2.8}$$

where $g_J = \prod_{j \in J} g_j$, for all $J \subseteq \{1, \ldots, m\}$.

Theorem 2.13 is a very powerful result and (2.8) provides a (Schmüdgen) certificate of positivity on $\mathbf{K}$ for f. But note that the number of terms in Schmüdgen's representation (2.8) is exponential in the number of polynomials that define the set $\mathbf{K}$. However, a major improvement is possible under a relatively weak assumption on the polynomials that define the compact set $\mathbf{K}$. Associated with the finite family $(g_j) \subset \mathbb{R}[\mathbf{x}]$, the set

$$Q(g) := \left\{ \sigma_0 + \sum_{j=1}^m \sigma_j \, g_j : \sigma_j \in \Sigma[\mathbf{x}], \ j = 0, \ldots, m \right\} \tag{2.9}$$

is called the *quadratic module* generated by the family (g_j).

Assumption 2.14 *Let $(g_j)_{j=1}^m \subset \mathbb{R}[\mathbf{x}]$. There exists $u \in Q(g)$ such that the level set $\{\, \mathbf{x} \in \mathbb{R}^n : u(\mathbf{x}) \geq 0 \,\}$ is compact.*

Theorem 2.15 (Putinar's Positivstellensatz)   *Let* $\mathbf{K} \subset \mathbb{R}^n$ *be as in (2.7) and let Assumption 2.14 hold. If* $f \in \mathbb{R}[\mathbf{x}]$ *is strictly positive on* $\mathbf{K}$ *then* $f \in Q(g)$*, that is,*

$$f = \sigma_0 + \sum_{j=1}^{m} \sigma_j \, g_j, \qquad (2.10)$$

for some SOS polynomials $\sigma_j \in \Sigma[\mathbf{x}]$, $j = 0, 1, \ldots, m$.

The SOS polynomials $(\sigma_j) \subset \Sigma[\mathbf{x}]$ provide a *Putinar certificate* of nonnegativity on $\mathbf{K}$ for the polynomial f. In contrast to Theorem 2.13, the number of terms in Putinar's representation (2.10) is linear in the number of polynomials that define $\mathbf{K}$, a crucial improvement from a computational point of view. The condition on $\mathbf{K}$ (Assumption 2.14) is not very restrictive. For instance, it is satisfied in the following cases:

1. all the g_i are affine and $\mathbf{K}$ is compact (hence a polytope);
2. the set $\{\mathbf{x} \in \mathbb{R}^n : g_j(\mathbf{x}) \geq 0\}$ is compact for some $j \in \{1, \ldots, m\}$.

Also, suppose that we know some $N > 0$ such that $\mathbf{K} \subset \{\mathbf{x} \in \mathbb{R}^n : \|\mathbf{x}\|^2 \leq N\}$. Let $\mathbf{x} \mapsto g_{m+1}(\mathbf{x}) := N - \|\mathbf{x}\|^2$. Adding the quadratic constraint $g_{m+1}(\mathbf{x}) \geq 0$ in the definition (2.7) of $\mathbf{K}$ does not change $\mathbf{K}$ as this last constraint is redundant. But with this new representation, $\mathbf{K}$ satisfies the required condition in Theorem 2.15.

The following theorem provides further understanding of the condition in Theorem 2.15.

Theorem 2.16   *Let* $(g_j)_{j=1}^{m} \subset \mathbb{R}[\mathbf{x}]$, *assume that* $\mathbf{K} \subset \mathbb{R}^n$ *defined in (2.7) is compact, and let* $Q(g)$ *be as in (2.9). The following conditions are equivalent.*

(a) *There exist finitely many* $p_1, \ldots, p_s \in Q(g)$ *such that the level set* $\{\mathbf{x} \in \mathbb{R}^n : p_j(\mathbf{x}) \geq 0, \; j = 1, \ldots, s\}$ *which contains* $\mathbf{K}$ *is compact and* $\prod_{j \in J} p_j \in Q(g)$ *for all* $J \subseteq \{1, \ldots, s\}$.

(b) *Assumption 2.14 holds.*

(c) *There exists* $N \in \mathbb{N}$ *such that the polynomial* $\mathbf{x} \mapsto N - \|\mathbf{x}\|^2 \in Q(g)$.

(d) *For all* $p \in \mathbb{R}[\mathbf{x}]$, *there is some* $N \in \mathbb{N}$ *such that both polynomials* $\mathbf{x} \mapsto N + p(\mathbf{x})$ *and* $\mathbf{x} \mapsto N - p(\mathbf{x})$ *are in* $Q(g)$.

Degree bound The following result provides a bound on the degree of the weights $(\sigma_j) \subset \Sigma[\mathbf{x}]$ in the representation (2.10). For $f \in \mathbb{R}[\mathbf{x}]$ written $\mathbf{x} \mapsto f(\mathbf{x}) = \sum_\alpha f_\alpha \mathbf{x}^\alpha$ let

$$\|f\|_0 := \max_\alpha |f_\alpha| \frac{\alpha_1! \cdots \alpha_n!}{|\alpha|!}. \qquad (2.11)$$

Theorem 2.17 (Degree bound) *Let* $\mathbf{K} \subset \mathbb{R}^n$ *in (2.7) satisfy Assumption 2.14 and assume that* $\emptyset \neq \mathbf{K} \subset (-1, 1)^n$. *Then there is some* $c > 0$ *such that for all* $f \in \mathbb{R}[\mathbf{x}]$ *of degree* d, *and strictly positive on* $\mathbf{K}$ *(i.e., such that* $f^* := \inf\{ f(\mathbf{x}) : \mathbf{x} \in \mathbf{K} \} > 0$*), the representation (2.10) holds with*

$$\deg \sigma_j \, g_j \; \leq \; c \, \exp\left(\left(d^2 n^d \, \frac{\|f\|_0}{f^*} \right)^c \right), \qquad \forall j = 1, \dots, m.$$

A matrix version of Putinar's Positivstellensatz For $f \in \mathbb{R}[\mathbf{x}]$ written $\mathbf{x} \mapsto f(\mathbf{x}) = \sum_\alpha f_\alpha \, \mathbf{x}^\alpha$, let $\|f\|_0$ be as in (2.11).

Definition 2.18   (a) The norm of a real symmetric matrix-polynomial $\mathbf{F} \in \mathbb{R}[\mathbf{x}]^{p \times p}$ is defined by $\|\mathbf{F}\| := \sup_{\|\xi\|=1} \|\xi^T \mathbf{F}(\mathbf{x}) \xi\|_0$.
   (b) A real symmetric matrix-polynomial $\mathbf{F} \in \mathbb{R}[\mathbf{x}]^{p \times p}$ is said to be a sum of squares (in short SOS) if $\mathbf{F} = \mathbf{L} \mathbf{L}^T$ for some $q \in \mathbb{N}$ and some real matrix-polynomial $\mathbf{L} \in \mathbb{R}[\mathbf{x}]^{p \times q}$.

Let $\mathbf{I}$ denote the $p \times p$ identity matrix.

Theorem 2.19   *Let* $\mathbf{K} \subset \mathbb{R}^n$ *be the basic semi-algebraic set in (2.7) and let Assumption 2.14 hold. Let* $\mathbf{F} \in \mathbb{R}[\mathbf{x}]^{p \times p}$ *be a real symmetric matrix-polynomial of degree* d. *If for some* $\delta > 0$, $\mathbf{F}(\mathbf{x}) \succeq \delta \mathbf{I}$ *for all* $\mathbf{x} \in \mathbf{K}$, *then*

$$\mathbf{F}(\mathbf{x}) \; = \; \mathbf{F}_0(\mathbf{x}) + \sum_{j=1}^m \mathbf{F}_j(\mathbf{x}) \, g_j(\mathbf{x}) \qquad (2.12)$$

for some SOS matrix-polynomials $\mathbf{F}_j$, $j = 0, \dots, m$, *and*

$$\deg \mathbf{F}_0, \; \deg \mathbf{F}_1 g_1, \dots, \deg \mathbf{F}_m g_m \; \leq \; c \left(d^2 n^d \frac{\|\mathbf{F}\|}{\delta} \right)^c,$$

for some constant c *that depends only on the polynomials* (g_j).

Obviously, Theorem 2.19 is a matrix-polynomial analogue of Theorem 2.15. In fact, one might have characterized the property $\mathbf{F}(\mathbf{x}) \succeq \delta \mathbf{I}$ on $\mathbf{K}$ by using Theorem 2.15 as follows. Let $\mathbf{S} := \{ \xi \in \mathbb{R}^p : \xi^T \xi = 1 \}$ and notice that the compact basic semi-algebraic set $\mathbf{K} \times \mathbf{S} \subset \mathbb{R}^n \times \mathbb{R}^p$ satisfies Assumption 2.14 whenever $\mathbf{K}$ does. If $f \in \mathbb{R}[\mathbf{x}, \xi]$ denotes the polynomial $(\mathbf{x}, \xi) \mapsto \xi^T \mathbf{F}(\mathbf{x})\xi$, then

$$\mathbf{F} \succeq \delta\, \mathbf{I} \quad \text{on } \mathbf{K} \quad \Leftrightarrow \quad f \geq \delta \quad \text{on } \mathbf{K} \times \mathbf{S},$$

and so by Theorem 2.15, $\mathbf{F} \succeq \delta\, \mathbf{I}$ on $\mathbf{K}$ implies

$$\xi^T \mathbf{F}(\mathbf{x})\, \xi \;=\; \sigma_0(\mathbf{x}, \xi) + \sum_{j=1}^{m} \sigma_j(\mathbf{x}, \xi)\, g_j(\mathbf{x}) + \sigma_{m+1}(\mathbf{x}, \xi)\, (\xi^T \xi - 1), \quad (2.13)$$

for some SOS polynomials $\sigma_j \in \Sigma[\mathbf{x}, \xi]$, $j = 0, \ldots, m$, and some polynomial $\sigma_{m+1} \in \mathbb{R}[\mathbf{x}, \xi]$.

However, in general, the degree of weights in the representations (2.12) and (2.13) is not the same. It is not clear which one should be preferred.

2.4.2 Testing membership in $Q(g)$ via semidefinite programs

Both Theorems 2.13 and 2.15 have significant computational advantages. Indeed, from Proposition 2.1, given a polynomial $f > 0$ on $\mathbf{K}$, checking whether f has the representation (2.10) with an a priori bound on the degree of the unknown SOS polynomials (σ_j), reduces to solving a semidefinite optimization problem. To see this, let $g_0 \in \mathbb{R}[\mathbf{x}]$ be the constant polynomial equal to 1, and $v_j := \lceil (\deg g_j)/2 \rceil$ for all $j = 0, \ldots, m$. Write

$$\mathbf{x} \mapsto g_j(\mathbf{x})\, \mathbf{v}_{d-v_j}(\mathbf{x}) \mathbf{v}_{d-v_j}(\mathbf{x})^T = \sum_{\alpha \in \mathbb{N}^n_{2d}} \mathbf{C}^j_\alpha\, \mathbf{x}^\alpha, \qquad j = 0, \ldots, m,$$

for some real symmetric matrices $(\mathbf{C}^j_\alpha)$, $j = 0, \ldots, m$, of appropriate size. Then testing whether $f \in \mathbb{R}[\mathbf{x}]_{d_0}$ can be written as in (2.10) with a degree bound $2(d - v_j)$ on the SOS weights σ_j, $j = 0, \ldots, m$, reduces to checking whether there exist real symmetric matrices $\mathbf{X}_j$, $j = 0, \ldots, m$, of appropriate size, such that

$$\left. \begin{aligned} f_\alpha &= \sum_{j=0}^{m} \langle \mathbf{C}^j_\alpha, \mathbf{X}_j \rangle, & \forall \alpha \in \mathbb{N}^n_{d_0} \\ 0 &= \sum_{j=0}^{m} \langle \mathbf{C}^j_\alpha, \mathbf{X}_j \rangle, & \forall \alpha \in \mathbb{N}^n_{2d}, \; |\alpha| > d_0 \\ \mathbf{X}_j &\succeq 0, \quad j = 0, \ldots, m. \end{aligned} \right\} \qquad (2.14)$$

Observe that checking whether the system of linear matrix inequalities (2.14) has a feasible solution $(\mathbf{X}_0, \ldots, \mathbf{X}_m)$ reduces to solving a semidefinite program.

Indeed suppose that (2.14) has a feasible solution $\mathbf{X}_0 \succeq 0, \ldots, \mathbf{X}_m \succeq 0$. Then multiplying both sides with $\mathbf{x}^\alpha$ and summing up yields

$$f(\mathbf{x}) = \sum_\alpha f_\alpha \mathbf{x}^\alpha = \sum_{j=0}^m \left\langle \left(\sum_\alpha \mathbf{C}_\alpha^j \mathbf{x}^\alpha \right), \mathbf{X}_j \right\rangle$$

$$= \sum_{j=0}^m g_j(\mathbf{x}) \left\langle \mathbf{v}_{d-v_j}(\mathbf{x}) \mathbf{v}_{d-v_j}(\mathbf{x})^T, \mathbf{X}_j \right\rangle$$

$$= \sum_{j=0}^m g_j(\mathbf{x}) \left(\sum_{t=1}^{s_j} (\mathbf{q}_{jt}^T \mathbf{v}_{d-v_j}(\mathbf{x}))^2 \right),$$

where, since $\mathbf{X}_j \succeq 0$ for all $j = 0, \ldots, m$, we have used the spectral decomposition $\mathbf{X}_j = \sum_{t=1}^{s_j} \mathbf{q}_{jt} \mathbf{q}_{jt}^T$ for some vectors $(\mathbf{q}_{jt})$ and some $t_j \in \mathbb{N}$. Next, introducing the SOS polynomials

$$\mathbf{x} \mapsto \sigma_j(\mathbf{x}) := \sum_{t=1}^{s_j} (\mathbf{q}_{jt}^T \mathbf{v}_{d-v_j}(\mathbf{x}))^2 \in \Sigma[\mathbf{x}], \qquad j = 0, \ldots, m,$$

one obtains (2.10).

Example 2.20 Let $n = 2$, $\mathbf{x} \mapsto f(\mathbf{x}) = x_1^3 - x_1^2 + 2x_1x_2 - x_2^2 + x_2^3$, and
$$\mathbf{K} = \{\mathbf{x} \in \mathbb{R}^2 : g_1(\mathbf{x}) = x_1 \geq 0, \ g_2(\mathbf{x}) = x_2 \geq 0, \ g_3(\mathbf{x}) = x_1 + x_2 - 1 \geq 0\}.$$

To check whether $f \geq 0$ on $\mathbf{K}$ we attempt to write $f = \sum_{j=0}^3 \sigma_j g_j$, where $g_0 = 1$ and each SOS $\sigma_j \in \Sigma[\mathbf{x}]$ has degree 2, that is,

$$\mathbf{x} \mapsto \sigma_j(\mathbf{x}) := (1, x_1, x_2) \, \mathbf{X}_j \, (1, x_1, x_2)^T,$$

for some real symmetric positive semidefinite matrices $\mathbf{X}_j \succeq \mathbf{0}$, $j = 0, 1, 2, 3$. For instance,

$$\mathbf{X}_0 = \mathbf{0}, \mathbf{X}_1 = \begin{bmatrix} 0 & 0 & 0 \\ 0 & 0 & 0 \\ 0 & 0 & 1 \end{bmatrix}, \mathbf{X}_2 = \begin{bmatrix} 0 & 0 & 0 \\ 0 & 1 & 0 \\ 0 & 0 & 0 \end{bmatrix}, \mathbf{X}_3 = \begin{bmatrix} 0 & 0 & 0 \\ 0 & 1 & -1 \\ 0 & -1 & 1 \end{bmatrix},$$

is a feasible solution of (2.14) and so,

$$f(\mathbf{x}) = x_2^2 x_1 + x_1^2 x_2 + (x_1 - x_2)^2 (x_1 + x_2 - 1)$$

which certifies that f is nonnegative on $\mathbf{K}$.

2.4.3 An alternative LP-based representation

We next present an alternative representation which is not based on SOS polynomials. We make the following assumption.

Assumption 2.21 *The set* **K** *in (2.7) is compact and the polynomials* $(g_j)_{j=0}^m$ *(with $g_0 = 1$) generate the algebra* $\mathbb{R}[\mathbf{x}]$.

Note that if $(1, g_j)_{j=1}^m$ do not generate $\mathbb{R}[\mathbf{x}]$, we add redundant inequalities as follows. Let $\underline{x}_k \leq \inf\{x_k : \mathbf{x} \in \mathbf{K}\}$ for all $k = 1, \ldots, n$. Then, with $g_{m+k}(\mathbf{x}) := x_k - \underline{x}_k$, we introduce the additional (redundant) constraints $g_{m+k}(\mathbf{x}) \geq 0, k = 1, \ldots, n$, in the definition (2.7) of $\mathbf{K}$, and reset $m := m + n$. With this new equivalent definition, Assumption 2.21 holds.

For every $j = 1, \ldots, m$, let $\overline{g}_j := \sup_{\mathbf{x} \in \mathbf{K}} g_j(\mathbf{x})$ (this is well defined because **K** is compact), and let $(\widehat{g}_j)_{j=1}^m$ be the polynomials g_j, normalized with respect to **K**, that is, for $j = 1, \ldots, m$

$$\widehat{g}_j := \begin{cases} g_j/\overline{g}_j & \text{if } \overline{g}_j > 0, \\ g_j & \text{if } \overline{g}_j = 0. \end{cases} \tag{2.15}$$

We next let $G := (0, 1, \widehat{g}_1, \ldots, \widehat{g}_m, 1 - \widehat{g}_1, \ldots, 1 - \widehat{g}_m) \subset \mathbb{R}[\mathbf{x}]$, and let $\Delta_G \subset \mathbb{R}[\mathbf{x}]$ be the set of all products of the form $q_1 \cdots q_k$, for polynomials $(q_j)_{j=1}^k \subset G$, and integer $k \geq 1$. Denote by C_G the *cone* generated by Δ_G, i.e., $f \in C_G$ if

$$f = \sum_{\alpha, \beta \in \mathbb{N}^m} c_{\alpha\beta} \, \widehat{g}_1^{\alpha_1} \cdots \widehat{g}_m^{\alpha_m} (1 - \widehat{g}_1)^{\beta_1} \cdots (1 - \widehat{g}_m)^{\beta_m},$$

for finitely many nonnegative coefficients $(c_{\alpha\beta}) \subset \mathbb{R}_+$ (and with $\widehat{g}_k^0 = 1$), or using the vector notation

$$\widehat{\mathbf{g}} = \begin{pmatrix} \widehat{g}_1 \\ \vdots \\ \widehat{g}_m \end{pmatrix} \in \mathbb{R}[\mathbf{x}]^m, \quad \mathbf{1} - \widehat{\mathbf{g}} = \begin{pmatrix} 1 - \widehat{g}_1 \\ \vdots \\ 1 - \widehat{g}_m \end{pmatrix} \in \mathbb{R}[\mathbf{x}]^m,$$

$f \in C_G$ if f has the compact form

$$f = \sum_{\alpha, \beta \in \mathbb{N}^m} c_{\alpha\beta} \, \widehat{\mathbf{g}}^\alpha \, (\mathbf{1} - \widehat{\mathbf{g}})^\beta. \tag{2.16}$$

Equivalently, $f \in C_G$ if f is a polynomial of $\mathbb{R}[\widehat{g}_1, \ldots, \widehat{g}_m, 1 - \widehat{g}_1, \ldots, 1 - \widehat{g}_m]$, with nonnegative coefficients.

Theorem 2.22 *Let* $(g_i)_{i=1}^m \subset \mathbb{R}[\mathbf{x}]$, $\mathbf{K} \subset \mathbb{R}^n$ *be as in (2.7), and let Assumption 2.21 hold.*

If $f \in \mathbb{R}[\mathbf{x}]$ *is strictly positive on* **K**, *then* $f \in C_G$. *Equivalently, (2.16) holds for finitely many nonnegative coefficients* $(c_{\alpha\beta}) \subset \mathbb{R}_+$, $\alpha, \beta \in \mathbb{N}^m$.

The case of polytopes If the g_j are all affine (so that **K** is a convex polytope) then Theorem 2.22 simplifies and we obtain a generalization of Theorem 2.7 for $[-1, 1] \subset \mathbb{R}$ to a polytope in $\mathbb{R}^n$.

Theorem 2.23 *Let $g_j \in \mathbb{R}[\mathbf{x}]$ be affine for every $j = 1, \ldots, m$, and assume that **K** in (2.7) is compact with a nonempty interior. If $f \in \mathbb{R}[\mathbf{x}]$ is strictly positive on **K** then*

$$f = \sum_{\alpha \in \mathbb{N}^m} c_\alpha \, g_1^{\alpha_1} \cdots g_m^{\alpha_m}, \qquad (2.17)$$

for finitely many nonnegative scalars (c_α).

Notice that Theorem 2.23 is of the same flavor as Theorem 2.22, except it does not require introduction of the polynomials $1 - \widehat{g}_j$, $j = 1, \ldots, m$.

Testing membership in C_G In contrast to Theorems 2.13 and 2.15, the representation (2.16) (or (2.17)) involves some nonnegative scalar coefficients $(c_{\alpha\beta})$ (or (c_α)) rather than SOS polynomials in $\Sigma[\mathbf{x}]$. Determining membership in C_G using Theorem 2.22 and with an a priori bound t on $|\alpha| + |\beta|$, leads to a linear optimization feasibility problem for which extremely efficient software packages are available. Indeed, expanding both sides of (2.16) in the monomial basis $(\mathbf{x}^\alpha)$, $\alpha \in \mathbb{N}_{2t}^n$, and equating each coefficient, yields linear constraints on the $c_{\alpha\beta}$ (plus the additional nonnegativity constraints). On the other hand, it involves products of arbitrary powers of the g_i and $(1 - g_j)$, a highly undesirable feature. In particular, the presence of large binomial coefficients is a source of ill-conditioning and numerical instability.

Example 2.24 Let f be the univariate polynomial $x \mapsto f(x) := 9 + 3x + x^2 - 8x^3$. Suppose that one wishes to test whether f is positive on the interval $[0, 1]$ (i.e., the trivial polytope $\{x : x \geq 0; 1 - x \geq 0\}$ on the real line $\mathbb{R}$). Let us try to obtain a certificate of positivity (2.17) with $|\alpha| \leq 3$. This yields the linear system

$$0 \leq \mathbf{c} = (c_\alpha) \in \mathbb{R}^{10} : 9 = c_0 + c_{01}, \quad 3 = c_{10} - c_{10} + c_{11} - 2c_{02} + c_{12} - 3\,c_{03},$$

$$1 = c_{20} - c_{11} + c_{02} + c_{21} - 2\,c_{12} + 3\,c_{03}, \quad -8 = c_{30} - c_{21} + c_{12} - c_{03}.$$

The vector $\mathbf{c} = (3, 2, 6, 0, 7, 0, 0, 8, 0, 0)$ is a feasible solution and so we obtain

$$f(x) = 3 + 2x + 6(1 - x) + 7x(1 - x) + 8x^2(1 - x),$$

which is a certificate of positivity of f on $[0, 1]$.

Remarks There are three features that distinguish the case $n > 1$ from the case $n = 1$ treated in the previous section.

1. Theorems 2.13, 2.15, and 2.22, all deal with *compact* sets **K**, whereas Theorem 2.4 can handle the (noncompact) interval $[0, \infty)$.
2. In Theorems 2.13, 2.15, and 2.22, p is restricted to be *strictly positive*, instead of *nonnegative* in Theorem 2.4.
3. In Theorems 2.13, 2.15, and 2.22, nothing is said on the **degree** of the polynomials involved in (2.8), (2.10), or in (2.16), whereas in Theorem 2.4, the degree is bounded and known in advance. In fact, bounds exist for the representations (2.8) and (2.10), see for example Theorem 2.17. However, they are not practical from a computational viewpoint. This is the reason why Theorems 2.13, 2.15, and 2.22 do not lead to a polynomial time algorithm to check whether a polynomial f is positive on **K**.

2.5 Polynomials nonnegative on real varieties

In this section, we introduce representations of polynomials on a real variety. The first result considers an arbitrary real variety, whereas the second considers a finite variety associated with a zero-dimensional ideal I of $\mathbb{R}[\mathbf{x}]$ which is radical.

Recall that $I \subset \mathbb{R}[\mathbf{x}]$ is an ideal if whenever $f \in I$ then $fg \in I$ for every $g \in \mathbb{R}[\mathbf{x}]$. For a background on basic definitions and results of algebraic geometry the interested reader is referred to for example Adams and Loustaunau (1994) and Basu et al. (2003). Let $V \subset \mathbb{R}^n$ be the real variety defined by:

$$V := \{ \mathbf{x} \in \mathbb{R}^n : g_j(\mathbf{x}) = 0, \quad j = 1, \ldots, m \}, \qquad (2.18)$$

for some family of real polynomials $(g_j) \subset \mathbb{R}[\mathbf{x}]$. Given $f \in \mathbb{R}[\mathbf{x}]$ and $\epsilon > 0$, let $f_{\epsilon r} \in \mathbb{R}[\mathbf{x}]$ be the polynomial:

$$f_{\epsilon r} := f + \epsilon \sum_{k=0}^{r} \sum_{i=1}^{n} \frac{x_i^{2k}}{k!}, \qquad \epsilon \geq 0, \quad r \in \mathbb{N}. \qquad (2.19)$$

Theorem 2.25 *Let $V \subset \mathbb{R}^n$ be as in (2.18), and let $f \in \mathbb{R}[\mathbf{x}]$ be nonnegative on V. Then, for every $\epsilon > 0$, there exists $r_\epsilon \in \mathbb{N}$ and nonnegative scalars $(\lambda_j)_{j=1}^{m}$, such that, for all $r \geq r_\epsilon$,*

$$f_{\epsilon r} + \sum_{j=1}^{m} \lambda_j \, g_j^2 \quad \text{is SOS.} \qquad (2.20)$$

In addition, $\|f - f_{\epsilon r}\|_1 \to 0$, as $\epsilon \downarrow 0$ (and $r \geq r_\epsilon$).

Theorem 2.25 is a denseness result which provides a certificate of nonnegativity of f on V. Notice that in contrast with Theorems 2.13 and 2.15 (letting an equality constraint be two reverse inequality constraints), Theorem 2.25 makes *no* assumption on the variety V; in addition one has *scalar* multipliers (λ_j) in (2.20) instead of SOS multipliers in (2.8) or (2.10). On the other hand, the former theorems state that if V is compact and f is nonnegative on V, then $f + \epsilon = f_{\epsilon/n,0}$ has the sum of squares representation (2.8) (or 2.10)), and $\|f - f_{\epsilon/n0}\|_\infty \to 0$ as $\epsilon \downarrow 0$, instead of the weaker $\|f - f_{\epsilon r}\|_1 \downarrow 0$ in Theorem 2.25.

Polynomials nonnegative on a finite variety   A zero-dimensional ideal $I \subset \mathbb{R}[\mathbf{x}]$ is an ideal such that the associated variety

$$V_{\mathbb{C}}(I) := \{ \mathbf{x} \in \mathbb{C}^n \ : \ g(\mathbf{x}) = 0 \quad \forall g \in I \}$$

is *finite*. In such a case, the quotient ring $\mathbb{R}[\mathbf{x}]/I$ is a finite-dimensional $\mathbb{R}$ vector space whose dimension is larger than $|V_{\mathbb{C}}(I)|$ and equal to $|V_{\mathbb{C}}(I)|$ if and only if I is radical.[1] This is an important special case when one deals with discrete sets as in discrete optimization, for instance when the set $\mathbf{K} \subset \mathbb{R}^n$ consists of the grid points (x_{ij}) that are solutions of the polynomial equations

$$\mathbf{K} = \left\{ \mathbf{x} \in \mathbb{R}^n : \ \prod_{j=1}^{2r_i} (x_i - x_{ij}) = 0, \quad i = 1, \dots, n \right\}. \tag{2.21}$$

Binary (or 0/1) optimization deals with the case when $r_i = 1$ for all $i = 1, \dots, n$ and $x_{ij} \in \{0, 1\}$.

Theorem 2.26   *Let* $I = \langle h_1, \dots, h_{m_2} \rangle$ *be a zero-dimensional ideal of* $\mathbb{R}[\mathbf{x}]$ *with associated (finite) variety* $V_{\mathbb{C}}(I) \subset \mathbb{C}^n$. *Let* $V_{\mathbb{R}}(I) := V_{\mathbb{C}}(I) \cap \mathbb{R}^n$, $(g_j)_{j=1}^{m_1} \subset \mathbb{R}[\mathbf{x}]$ *and*

$$\mathbf{K} := \{ \mathbf{x} \in V_{\mathbb{R}}(I) : \ g_j(\mathbf{x}) \geq 0, \quad j = 1, \dots, m_1 \}. \tag{2.22}$$

Then $f \in \mathbb{R}[\mathbf{x}]$ *is nonnegative on* $\mathbf{K}$ *if and only if*

$$f = \sigma_0 + \sum_{j=1}^{m_1} \sigma_j \, g_j + \sum_{\ell=1}^{m_2} \psi_\ell \, h_\ell \tag{2.23}$$

for some polynomials $(\psi_\ell)_{\ell=1}^{m_2} \subset \mathbb{R}[\mathbf{x}]$ *and some SOS polynomials* $(\sigma_j)_{j=0}^{m_1} \subset \Sigma[\mathbf{x}]$.

---

[1] Given an ideal $I \subset \mathbb{R}[\mathbf{x}]$, the set $\sqrt{I} := \{ f \in \mathbb{R}[\mathbf{x}] : \ \exists s \in \mathbb{N} \text{ s.t. } f^s \in I \}$ is called the radical of I, and I is radical if $I = \sqrt{I}$.

In the case where $\mathbf{K}$ is the set (2.21), the degree of the polynomials (ψ_ℓ) is bounded by $(\sum_{\ell=1}^{n} r_\ell) - n$.

For finite varieties, observe that Theorem 2.26 provides a representation result stronger than that of Theorem 2.15. In particular, f is not required to be strictly positive and sometimes a degree bound is also available.

2.6 Representations with sparsity properties

In this section, we introduce a *sparse representation* for polynomials f nonnegative on a basic semi-algebraic set $\mathbf{K}$, when there is a *weak coupling* between some subsets of variables in the polynomials g_j that define the set $\mathbf{K}$, and f. By weak coupling (to be detailed later) we mean that

(a) each polynomial in the definition of $\mathbf{K}$ contains a few variables only, and
(b) the polynomial f is a sum of polynomials, each also containing a few variables only.

This sparse representation is very important from a computational viewpoint because it translates into smaller semidefinite programs for computing the SOS polynomials that define the representation. In fact, given the current state of semidefinite optimization, it is absolutely critical to exploit sparsity in order to solve problems involving a large number of variables.

With $\mathbb{R}[\mathbf{x}] = \mathbb{R}[x_1, \ldots, x_n]$, let the index set $I_0 := \{1, \ldots, n\}$ be the union $\cup_{k=1}^{p} I_k$ of p subsets of indices I_k, $k = 1, \ldots, p$ (with possible overlaps). Let $n_k = |I_k|$. Let $\mathbb{R}[\mathbf{x}(I_k)]$ denote the ring of polynomials in the n_k variables $\mathbf{x}(I_k) = \{x_i : i \in I_k\}$, and so $\mathbb{R}[\mathbf{x}(I_0)] = \mathbb{R}[\mathbf{x}]$.

Assumption 2.27 *Let $\mathbf{K} \subset \mathbb{R}^n$ be as in (2.7). A scalar $M > 0$ is known such that $\|\mathbf{x}\|_\infty < M$ for all $\mathbf{x} \in \mathbf{K}$.*

Note that under Assumption 2.27, we have $\sum_{i \in I_k} x_i^2 \leq n_k M^2$, $k = 1, \ldots, p$, and therefore, in the definition (2.7) of $\mathbf{K}$, we add the p redundant quadratic constraints

$$0 \leq g_{m+k}(\mathbf{x}) := n_k M^2 - \sum_{i \in I_k} x_i^2, \quad k = 1, \ldots, p, \qquad (2.24)$$

and set $m' = m + p$, so that $\mathbf{K}$ is now defined by:

$$\mathbf{K} := \{\mathbf{x} \in \mathbb{R}^n : g_j(\mathbf{x}) \geq 0, \quad j = 1, \ldots, m'\}. \qquad (2.25)$$

Note that $g_{m+k} \in \mathbb{R}[\mathbf{x}(I_k)]$, for all $k = 1, \ldots, p$.

We now define what is called a *structured sparsity pattern*.

Assumption 2.28 (Structured sparsity pattern)   *Let* $\mathbf{K} \subset \mathbb{R}^n$ *be as in (2.25).*
The index set $J = \{1, \ldots, m'\}$ *is partitioned into* p *disjoint sets* J_k, $k = 1, \ldots, p$, *and the collections* $\{I_k\}$ *and* $\{J_k\}$ *satisfy the following.*

   (a) *For every* $j \in J_k$, $g_j \in \mathbb{R}[\mathbf{x}(I_k)]$, *that is, for every* $j \in J_k$, *the constraint* $g_j(\mathbf{x}) \geq 0$ *only involves the variables* $\mathbf{x}(I_k) = \{x_i : i \in I_k\}$.
   (b) *The objective function* $f \in \mathbb{R}[\mathbf{x}]$ *can be written as*

$$f = \sum_{k=1}^{p} f_k, \quad \text{with } f_k \in \mathbb{R}[\mathbf{x}(I_k)], \quad k = 1, \ldots, p. \qquad (2.26)$$

Under such a sparsity pattern one would like positivity of polynomials f on $\mathbf{K}$ (with f and $\mathbf{K}$ as in Assumption 2.28) to be certified in a manner that preserves this sparsity. And indeed this is the case under some conditions. The main result about sparsity is as follows.

Theorem 2.29   *Let* $\mathbf{K} \subset \mathbb{R}^n$ *be as in (2.25) (i.e.,* $\mathbf{K}$ *is as in (2.7) with the additional redundant quadratic constraints (2.24)). Let Assumptions 2.27 and 2.28 hold and in addition assume that for every* $k = 1, \ldots, p - 1$,

$$\left(I_{k+1} \cap \left(\cup_{j=1}^{k} I_j \right) \right) \subseteq I_s \quad \text{for some } s \leq k. \qquad (2.27)$$

If $f \in \mathbb{R}[\mathbf{x}]$ *is strictly positive on* $\mathbf{K}$, *then*

$$f = \sum_{k=1}^{p} \left(q_k + \sum_{j \in J_k} q_{jk} g_j \right), \qquad (2.28)$$

for some SOS polynomials $(q_k, q_{jk}) \subset \mathbb{R}[\mathbf{x}(I_k)]$, $k = 1, \ldots, p$.

The key property (2.27) that allows the sparse representation (2.28) is called the *running intersection property*. When this property holds true, the absence of coupling of variables in the original data is preserved in the representation (2.28). So Theorem 2.29 provides a representation that is more specific than that of Theorem 2.15. Interestingly, the running intersection property (2.27) is well known in graph theory, graphical models, and Bayesian networks in which nodes are random variables (with a conditional probability distribution) and arcs provide statistical dependence between nodes.
   Let us illustrate Theorem 2.29 with an example.

Example 2.30 Let $\mathbf{x} = (x_1, \ldots, x_5)$ and

$$
\begin{aligned}
\mathbf{x} \mapsto g_1(\mathbf{x}) &= 1 - x_1^2 - x_2^2 \\
\mathbf{x} \mapsto g_2(\mathbf{x}) &= 1 - x_2^4 x_3^4 - x_3^2 \\
\mathbf{x} \mapsto g_3(\mathbf{x}) &= 1 - x_3^2 - x_4^2 - x_5^2 \\
\mathbf{x} \mapsto f(\mathbf{x}) &= 1 + x_1^2 x_2^2 + x_2^2 x_3^2 + x_3^2 - x_3^2 x_4^2 - x_3^2 x_5^2 \\
\mathbf{K} &= \{ \mathbf{x} : g_1(x_1, x_2), g_2(x_2, x_3), g_3(x_3, x_4, x_5) \geq 0 \}.
\end{aligned}
$$

Then $I_1 = \{1, 2\}$ meaning that the polynomial g_1 only involves variables x_1, x_2, $I_2 = \{2, 3\}$ and $I_3 = \{3, 4, 5\}$. Moreover, $f = f_1 + f_2 + f_3$, with $f_1 = 1 + x_1^2 x_2^2$, $f_2 = x_2^2 x_3^2$ and $f_3 = x_3^2 - x_3^2 x_4^2 - x_3^2 x_5^2$.

Let us check property (2.27). For $k = 1$, $I_2 \cap I_1 = \{2\} \subset I_1$. For $k = 2$, $I_3 \cap (I_1 \cup I_2) = \{3\} \subset I_2$ and thus property (2.27) holds. Thus, Theorem 2.29 allows the sparse representation

$$
\begin{aligned}
f(\mathbf{x}) &= 1 + x_1^4 x_2^2 + x_1^2 x_2^4 + x_1^2 x_2^2 g_1(\mathbf{x}) + x_2^6 x_3^6 + x_2^2 x_3^4 \\
&\quad + x_2^2 x_3^2 g_2(\mathbf{x}) + x_3^4 + x_3^2 g_3(\mathbf{x}).
\end{aligned}
$$

2.7 Moment problems

Most of the results of this section are the *dual* facets of those described in previous sections. Indeed, the problem of representing polynomials that are positive on a set $\mathbf{K} \subset \mathbb{R}^n$ has a dual facet in functional analysis, the **K**-moment problem of characterizing sequences of reals that are moment sequences of some finite Borel measure supported on **K**. Moreover, as we shall see, this beautiful duality is nicely captured by standard duality in convex optimization, applied to some appropriate convex cones of $\mathbb{R}[\mathbf{x}]$. We review basic results for the (real) moment problem and also particularize to some specific important cases as in previous sections.

Recall that for every $\mathbf{x} \in \mathbb{R}^n$, $\alpha \in \mathbb{N}^n$, the notation $\mathbf{x}^\alpha$ stands for the monomial $x_1^{\alpha_1} \cdots x_n^{\alpha_n}$ of $\mathbb{R}[\mathbf{x}]$. A Borel measure on $\mathbb{R}^n$ is a nonnegative and countably additive set function on the Borel σ-field $\mathcal{B}$ of $\mathbb{R}^n$ (i.e., the σ-algebra generated by the open sets of $\mathbb{R}^n$). The *support* of a Borel measure μ on $\mathbb{R}^n$, denoted supp μ, is a closed set, the complement in $\mathbb{R}^n$ of the largest open set $O \subset \mathbb{R}^n$ such that $\mu(O) = 0$.

Atomic measure An *atom* $A \in \mathcal{B}$ (relative to a Borel measure μ) is a Borel set such that $\mu(A) > 0$ and $\mu(B) = 0$ or $\mu(A \setminus B) = 0$ whenever $B \in \mathcal{B}$ and $B \subset A$. In fact, as $\mathbb{R}^n$ is a separable metric space, if A is an atom for μ then there exists $\mathbf{x} \in A$ such that $\mu(A \setminus \{\mathbf{x}\}) = 0$. Therefore an *s*-atomic measure is a finite Borel measure on $\mathbb{R}^n$ supported on s points (atoms), that is, a positive linear combination $\sum_{i=1}^{s} \lambda_i \, \delta_{\mathbf{x}(i)}$, of Dirac measures $(\delta_{\mathbf{x}(i)})$ at some points $(\mathbf{x}(i)) \subset \mathbb{R}^n$, with positive weights λ_i, $i = 1, \ldots, s$.

Definition 2.31   (a) The full moment problem. Let $(g_i)_{i=1}^{m} \subset \mathbb{R}[\mathbf{x}]$ and let $\mathbf{K} \subset \mathbb{R}^n$ be the set defined by:

$$\mathbf{K} := \{ \mathbf{x} \in \mathbb{R}^n : g_j(\mathbf{x}) \geq 0, \; j = 1, \ldots, m \}.$$

Given an infinite sequence of real numbers $\mathbf{y} = (y_\alpha) \subset \mathbb{R}$, $\alpha \in \mathbb{N}^n$, is $\mathbf{y}$ a $\mathbf{K}$-moment sequence, i.e., does there exist a finite Borel measure μ supported on $\mathbf{K}$, such that

$$y_\alpha = \int_{\mathbf{K}} \mathbf{x}^\alpha \, d\mu, \qquad \forall \, \alpha \in \mathbb{N}^n ? \tag{2.29}$$

(b) The truncated moment problem. Given (g_i), $\mathbf{K} \subset \mathbb{R}^n$ as above, a finite subset $\Delta \subset \mathbb{N}^n$, and a finite sequence of real numbers $\mathbf{y} = (y_\alpha)$, $\alpha \in \Delta$, is $\mathbf{y}$ a $\mathbf{K}$-moment sequence, i.e., does there exist a measure μ supported on $\mathbf{K}$, such that

$$y_\alpha = \int_{\mathbf{K}} \mathbf{x}^\alpha \, d\mu, \qquad \forall \, \alpha \in \Delta ? \tag{2.30}$$

Definition 2.32   In both full and truncated cases, a measure μ as in (2.29) or (2.30), is said to be a *representing* measure of the sequence $\mathbf{y}$.

If the representing measure μ is unique then μ is said to be *determinate* (i.e., determined by its moments), and *indeterminate* otherwise.

Example 2.33   For instance, the probability measure μ on the real line $\mathbb{R}$, with density with respect to the Lebesgue measure given by

$$x \mapsto f(x) := \begin{cases} (x \sqrt{2\pi})^{-1} \exp\left(-\ln(x)^2/2\right) & \text{if } x > 0, \\ 0 & \text{otherwise,} \end{cases}$$

called the log-normal distribution, is *not* determinate. Indeed, for each a with $-1 \leq a \leq 1$, the probability measure with density

$$x \mapsto \quad f_a(x) := f(x)\,[\,1 + a \sin(2\pi \ln x)\,]$$

has exactly the same moments as μ.

On the other hand, and as a consequence of the Stone–Weierstrass theorem, finite Borel measures on compact sets are moment determinate.

The above moment problem encompasses all the classical one-dimensional (real) moment problems of the twentieth century, namely the Hamburger problem which refers to $\mathbf{K} = \mathbb{R}$ and $(y_\alpha)_{\alpha \in \mathbb{N}} \subset \mathbb{R}$, the Stieltjes problem (with

$\mathbf{K} = \mathbb{R}_+$, and $(y_\alpha)_{\alpha \in \mathbb{N}} \subset \mathbb{R}_+)$ and the Hausdorff problem (with $\mathbf{K} = [a, b]$, and $(y_\alpha)_{\alpha \in \mathbb{N}} \subset \mathbb{R}$).

The multi-dimensional moment problem is significantly more difficult than the one-dimensional case for which the results are fairly complete. This is because, in view of Theorem 2.34 below, obtaining conditions for a sequence to be moments of a representing measure with support on a given subset $\Omega \subseteq \mathbb{R}^n$, is related to characterizing polynomials that are nonnegative on Ω. When the latter characterization is available, it will translate into conditions on the sequence. But as already mentioned, and in contrast to the univariate case, polynomials that are nonnegative on a given set $\Omega \subseteq \mathbb{R}^n$ have no simple characterization. A notable exception is the case of compact basic semi-algebraic sets as detailed in Section 2.4. Thus, for instance, the full multi-dimensional $\mathbf{K}$-moment problem is still unsolved for general sets $\mathbf{K} \subset \mathbb{R}^n$, including $\mathbf{K} = \mathbb{R}^n$.

Before we proceed further, we first state the important Riesz–Haviland theorem.

The Riesz linear functional Given a real sequence $\mathbf{y} = (y_\alpha) \subset \mathbb{R}$, introduce the so-called *Riesz linear functional* $L_\mathbf{y} : \mathbb{R}[\mathbf{x}] \to \mathbb{R}$, defined by:

$$f \left(= \sum_{\alpha \in \mathbb{N}^n} f_\alpha \mathbf{x}^\alpha \right) \quad \mapsto \quad L_\mathbf{y}(f) = \sum_{\alpha \in \mathbb{N}^n} f_\alpha \, y_\alpha, \qquad \forall f \in \mathbb{R}[\mathbf{x}]. \quad (2.31)$$

Theorem 2.34 (Riesz–Haviland) *Let* $\mathbf{y} = (y_\alpha)$, $\alpha \in \mathbb{N}^n$, *and let* $\mathbf{K} \subseteq \mathbb{R}^n$ *be closed. There exists a finite Borel measure* μ *on* $\mathbf{K}$ *such that*

$$\int_\mathbf{K} \mathbf{x}^\alpha \, d\mu = y_\alpha, \qquad \forall \alpha \in \mathbb{N}^n, \quad (2.32)$$

if and only if $L_\mathbf{y}(f) \geq 0$ *for all polynomials* $f \in \mathbb{R}[\mathbf{x}]$ *nonnegative on* $\mathbf{K}$.

Note that Theorem 2.34 is not very practical as we do not have an explicit characterization of polynomials that are nonnegative on a general closed set $\mathbf{K} \subset \mathbb{R}^n$. However, in Section 2.4 we have seen some nice representations for the subclass of compact basic semi-algebraic sets $\mathbf{K} \subset \mathbb{R}^n$. Theorem 2.34 will serve as our primary proof tool in the next sections.

As we are only interested in the *real* moment problem, we next introduce the basic concepts of moment and localizing matrices in the real case $\mathbb{R}^n$. (However, these concepts also have their natural counterparts in $\mathbb{C}^n$, with the usual scalar product $\langle \mathbf{u}, \mathbf{v} \rangle = \sum_j \overline{u}_j v_j$.)

2.7.1 Moment and localizing matrix

Moment matrix Given an $s(2d)$ sequence $\mathbf{y} = (y_\alpha)$, let $\mathbf{M}_d(\mathbf{y})$ be the real symmetric *moment* matrix of dimension $s(d)$, with rows and columns labeled by $\alpha \in \mathbb{N}_d^n$, and constructed as follows:

$$\mathbf{M}_d(\mathbf{y})(\alpha, \beta) = L_\mathbf{y}(\mathbf{x}^\alpha \mathbf{x}^\beta) = y_{\alpha+\beta}, \qquad \forall \alpha, \beta \in \mathbb{N}_d^n, \qquad (2.33)$$

with $L_\mathbf{y}$ defined in (2.31). Equivalently, $\mathbf{M}_d(\mathbf{y}) = L_\mathbf{y}(\mathbf{v}_d(\mathbf{x})\,\mathbf{v}_d(\mathbf{x})^T)$ where the latter slight abuse of notation means that we apply $L_\mathbf{y}$ to each entry of the matrix $\mathbf{v}_d(\mathbf{x})\mathbf{v}_d(\mathbf{x})^T$. Equivalently, $\mathbf{M}_d(\mathbf{y}) = \sum_{\alpha \in \mathbb{N}_{2d}^n} \mathbf{C}_\alpha^0 \, y_\alpha$, where the real symmetric matrices $\mathbf{C}_\alpha^0, \alpha \in \mathbb{N}_{2d}^n$, have been defined in Section 2.4.2.

Let us consider an example with $n = d = 2$. In this case, $\mathbf{M}_2(\mathbf{y})$ becomes

$$\mathbf{M}_2(\mathbf{y}) = \begin{bmatrix} y_{00} & | & y_{10} & y_{01} & | & y_{20} & y_{11} & y_{02} \\ - & - & - & - & - & - & - & - \\ y_{10} & | & y_{20} & y_{11} & | & y_{30} & y_{21} & y_{12} \\ y_{01} & | & y_{11} & y_{02} & | & y_{21} & y_{12} & y_{03} \\ - & - & - & - & - & - & - & - \\ y_{20} & | & y_{30} & y_{21} & | & y_{40} & y_{31} & y_{22} \\ y_{11} & | & y_{21} & y_{12} & | & y_{31} & y_{22} & y_{13} \\ y_{02} & | & y_{12} & y_{03} & | & y_{22} & y_{13} & y_{04} \end{bmatrix}.$$

$\mathbf{M}_d(\mathbf{y})$ defines a bilinear form $\langle., .\rangle_\mathbf{y}$ on $\mathbb{R}[\mathbf{x}]_d$ as follows:

$$\langle p, q\rangle_\mathbf{y} := L_\mathbf{y}(pq) = \langle \mathbf{p}, \mathbf{M}_d(\mathbf{y})\,\mathbf{q}\rangle = \mathbf{p}^T \mathbf{M}_d(\mathbf{y})\,\mathbf{q}, \qquad \forall \mathbf{p}, \mathbf{q} \in \mathbb{R}^{s(d)},$$

where again, $p, q \in \mathbb{R}[\mathbf{x}]_d$, and $\mathbf{p}, \mathbf{q} \in \mathbb{R}^{s(d)}$ denote their respective vectors of coefficients.

If the sequence $\mathbf{y}$ has a representing measure μ (see Definition 2.32) then for every $q \in \mathbb{R}[\mathbf{x}]$,

$$\langle \mathbf{q}, \mathbf{M}_d(\mathbf{y})\,\mathbf{q}\rangle = L_\mathbf{y}(q^2) = \int_{\mathbb{R}^n} q^2 \, d\mu \geq 0, \qquad (2.34)$$

so that $\mathbf{M}_d(\mathbf{y}) \succeq \mathbf{0}$. It is also immediate to check that if the polynomial q^2 is expanded as $q(\mathbf{x})^2 = \sum_{\alpha \in \mathbb{N}^n} h_\alpha \mathbf{x}^\alpha$, then

$$\langle \mathbf{q}, \mathbf{M}_d(\mathbf{y})\,\mathbf{q}\rangle = L_\mathbf{y}(q^2) = \sum_{\alpha \in \mathbb{N}^n} h_\alpha y_\alpha.$$

In the one-dimensional case $n = 1$, the moment matrix $\mathbf{M}_d(\mathbf{y})$ is a Hankel matrix.

Not every sequence $\mathbf{y}$ that satisfies $\mathbf{M}_d(\mathbf{y}) \succeq \mathbf{0}$ has a representing measure μ on $\mathbb{R}^n$. This is in contrast with the one-dimensional case where a full sequence $\mathbf{y}$ such that $\mathbf{M}_d(\mathbf{y}) \succeq \mathbf{0}$ for all $d = 0, 1, \ldots$, has a representing measure on $\mathbb{R}$

(and so **y** solves the Hamburger moment problem). But this is not necessarily true for a finite sequence as the next example shows.

Example 2.35   In the univariate case $n = 1$, let $\mathbf{y} \in \mathbb{R}^5$ be the truncated sequence $\mathbf{y} = (1, 1, 1, 1, 2)$, hence with associated (Hankel) moment matrix

$$
\mathbf{M}_2(\mathbf{y}) = \begin{bmatrix} 1 & 1 & 1 \\ 1 & 1 & 1 \\ 1 & 1 & 2 \end{bmatrix}.
$$

One may easily check that $\mathbf{M}_2(\mathbf{y}) \succeq 0$, but it turns out that **y** has no representing Borel measure μ on the real line $\mathbb{R}$.

Sufficient conditions of existence of a representing measure   There exist some sufficient conditions on **y** or on the associated moment matrix $\mathbf{M}_d(\mathbf{y})$ which ensure existence of a representing measure.

Theorem 2.36   *Let* $\mathbf{y} = (y_\alpha)$, $\alpha \in \mathbb{N}^n_{2d}$, *be such that* $\mathbf{M}_d(\mathbf{y}) \succeq 0$. *If* rank $\mathbf{M}_d(\mathbf{y}) \leq 6$ *if* $d = 2$, *or* rank $\mathbf{M}_d(\mathbf{y}) \leq 3d - 3$ *if* $d \geq 3$, *then* **y** *has an atomic representing measure supported on* rank $\mathbf{M}_d(\mathbf{y})$ *points.*

Observe that the condition in Theorem 2.36 is stated in terms of finitely many moments only.

Proposition 2.37   *Let* **y** *be a sequence indexed in the basis* $\mathbf{v}_\infty(\mathbf{x})$, *which satisfies* $\mathbf{M}_d(\mathbf{y}) \succeq 0$, *for all* $d = 0, 1, \ldots$.

  (a) *If the sequence* **y** *satisfies*

$$
\sum_{k=1}^{\infty} \left[L_\mathbf{y}(x_i^{2k}) \right]^{-1/2k} = +\infty, \qquad i = 1, \ldots, n, \qquad (2.35)
$$

then **y** *has a determinate representing measure on* $\mathbb{R}^n$.
  (b) *If there exist* $c, a > 0$ *such that*

$$
| y_\alpha | \leq c \, a^{|\alpha|}, \qquad \forall \, \alpha \in \mathbb{N}^n, \qquad (2.36)
$$

then **y** *has a determinate representing measure with support contained in the box* $[-a, a]^n$.

In the one-dimensional case (2.35) is called Carleman's condition. In contrast to the rank condition of Theorem 2.36, the conditions (2.35) and (2.36) are concerned with infinitely many moments y_α.

The moment matrix also has the following interesting properties.

Proposition 2.38 *Let $d \geq 1$, and let $\mathbf{y} = (y_\alpha) \subset \mathbb{R}$ be such that $\mathbf{M}_d(\mathbf{y}) \succeq 0$. Then*

$$| y_\alpha | \leq \max \left[y_0, \max_{i=1,\ldots,n} L_\mathbf{y}(x_i^{2d}) \right], \qquad \forall 0 \neq \alpha \in \mathbb{N}_{2d}^n.$$

In addition, rescaling $\mathbf{y}$ so that $y_0 = 1$, and letting $\tau_d := \max_{i=1,\ldots,n} L_\mathbf{y}(x_i^{2d})$,

$$| y_\alpha |^{\frac{1}{|\alpha|}} \leq \tau_d^{\frac{1}{2d}}, \qquad \forall 0 \neq \alpha \in \mathbb{N}_{2d}^n.$$

Localizing matrix Given a polynomial $u \in \mathbb{R}[\mathbf{x}]$ with coefficient vector $\mathbf{u} = (u_\gamma)$, we define the *localizing* matrix with respect to $\mathbf{y}$ and u, to be the matrix $\mathbf{M}_d(u\,\mathbf{y})$ with rows and columns indexed by $\alpha \in \mathbb{N}_d^n$, and defined by:

$$\mathbf{M}_d(u\,\mathbf{y})(\alpha, \beta) = L_\mathbf{y}(u(\mathbf{x})\mathbf{x}^\alpha \mathbf{x}^\beta) = \sum_{\gamma \in \mathbb{N}^n} u_\gamma\, y_{\gamma + \alpha + \beta}, \qquad \forall \alpha, \beta \in \mathbb{N}_d^n. \tag{2.37}$$

Equivalently, $\mathbf{M}_d(u\,\mathbf{y}) = L_\mathbf{y}(u\,\mathbf{v}_d(\mathbf{x})\mathbf{v}_d(\mathbf{x})^T)$, where again this slight abuse of notation means that $L_\mathbf{y}$ is applied entrywise to the matrix $\mathbf{u}(\mathbf{x})\mathbf{v}_d(\mathbf{x})\,\mathbf{v}_d(\mathbf{x})^T$.

Or, again equivalently, if $r = \deg u$, then $\mathbf{M}_d(u\,\mathbf{y}) = \sum_{\alpha \in \mathbb{N}_{2d+r}^n} \mathbf{C}_\alpha\, y_\alpha$, where the real symmetric matrices $\mathbf{C}_\alpha$, $\alpha \in \mathbb{N}_{2d+r}^n$ are defined from:

$$\mathbf{x} \mapsto u(\mathbf{x})\,\mathbf{v}_d(\mathbf{x})\,\mathbf{v}_d(\mathbf{x})^T = \sum_{\alpha \in \mathbb{N}_{2d+r}^n} \mathbf{C}_\alpha\, \mathbf{x}^\alpha, \qquad \mathbf{x} \in \mathbb{R}^n$$

(see Section 2.4.2). For instance, when $n = 2$, $d = 1$, and with $\mathbf{x} \mapsto u(\mathbf{x}) := a - x_1^2 - x_2^2$,

$$\mathbf{M}_1(u\,\mathbf{y}) = \begin{bmatrix} ay_{00} - y_{20} - y_{02} & ay_{10} - y_{30} - y_{12} & ay_{01} - y_{21} - y_{03} \\ ay_{10} - y_{30} - y_{12} & ay_{20} - y_{40} - y_{22} & ay_{11} - y_{31} - y_{13} \\ ay_{01} - y_{21} - y_{03} & ay_{11} - y_{31} - y_{13} & ay_{02} - y_{22} - y_{04} \end{bmatrix}.$$

And similarly to (2.34),

$$\langle \mathbf{p}, \mathbf{M}_d(u\,\mathbf{y})\,\mathbf{q} \rangle = L_\mathbf{y}(u\,pq),$$

for all polynomials $p, q \in \mathbb{R}[\mathbf{x}]_d$ with coefficient vectors $\mathbf{p}, \mathbf{q} \in \mathbb{R}^{s(d)}$. In particular, if $\mathbf{y}$ has a representing measure μ, then

$$\langle \mathbf{q}, \mathbf{M}_d(u\,\mathbf{y})\,\mathbf{q} \rangle = L_\mathbf{y}(u\,q^2) = \int_{\mathbb{R}^n} u\, q^2\, d\mu, \tag{2.38}$$

for every polynomial $q \in \mathbb{R}[\mathbf{x}]_d$ with coefficient vector $\mathbf{q} \in \mathbb{R}^{s(d)}$. Therefore, $\mathbf{M}_d(u\,\mathbf{y}) \succeq 0$ whenever μ has its support contained in the set $\{\mathbf{x} \in \mathbb{R}^n : u(\mathbf{x}) \geq 0\}$. It is also immediate to check that if the polynomial $u\,q^2$ is expanded as $u(\mathbf{x})\,q(\mathbf{x})^2 = \sum_{\alpha \in \mathbb{N}^n} h_\alpha \mathbf{x}^\alpha$ then

$$\langle \mathbf{q}, \mathbf{M}_d(u\,\mathbf{y})\,\mathbf{q} \rangle \;=\; \sum_{\alpha \in \mathbb{N}^n} h_\alpha\, y_\alpha \;=\; L_{\mathbf{y}}(u\,q^2). \tag{2.39}$$

2.7.2 Positive and flat extensions of moment matrices

We next discuss the important notion of positive extension for moment matrices.

Definition 2.39 Given a finite sequence $\mathbf{y} = (y_\alpha)$, $\alpha \in \mathbb{N}^n_{2d}$, with $\mathbf{M}_d(\mathbf{y}) \succeq 0$, the *moment extension problem* is defined as follows: extend the sequence $\mathbf{y}$ with new scalars y_β, $2d < |\beta| \leq 2(d+1)$, so as to obtain a new finite sequence (y_α), $\alpha \in \mathbb{N}^n_{2d+2}$, such that $\mathbf{M}_{d+1}(\mathbf{y}) \succeq 0$.

If such an extension $\mathbf{M}_{d+1}(\mathbf{y})$ is possible, it is called a *positive extension* of $\mathbf{M}_d(\mathbf{y})$. If in addition, rank $\mathbf{M}_{d+1}(\mathbf{y}) = $ rank $\mathbf{M}_d(\mathbf{y})$, then $\mathbf{M}_{d+1}(\mathbf{y})$ is called a *flat extension* of $\mathbf{M}_d(\mathbf{y})$.

For truncated moment problems, flat extensions play an important role.

Theorem 2.40 (Flat extension) *Let* $\mathbf{y} = (y_\alpha)$, $\alpha \in \mathbb{N}^n_{2d}$. *Then the sequence* $\mathbf{y}$ *admits a* rank $\mathbf{M}_d(\mathbf{y})$*-atomic representing measure* μ *on* $\mathbb{R}^n$ *if and only if* $\mathbf{M}_d(\mathbf{y}) \succeq 0$ *and* $\mathbf{M}_d(\mathbf{y})$ *admits a flat extension* $\mathbf{M}_{d+1}(\mathbf{y}) \succeq 0$.

Theorem 2.40 is useful as it provides a simple numerical means to check whether a finite sequence has a representing measure.

Example 2.41 Let μ be the measure on $\mathbb{R}$ defined to be $\mu = \delta_0 + \delta_1$, that is, μ is the sum of two Dirac measures at the points $\{0\}$ and $\{1\}$. Then

$$\mathbf{M}_1(\mathbf{y}) = \begin{bmatrix} 2 & 1 \\ 1 & 1 \end{bmatrix}, \quad \mathbf{M}_2(\mathbf{y}) = \begin{bmatrix} 2 & 1 & 1 \\ 1 & 1 & 1 \\ 1 & 1 & 1 \end{bmatrix},$$

and obviously, rank $\mathbf{M}_2(\mathbf{y}) = $ rank $\mathbf{M}_1(\mathbf{y}) = 2$.

2.7.3 The K-moment problem

For $\mathbf{K} \subset \mathbb{R}^n$ and given $d \in \mathbb{N}$, let $C_d(\mathbf{K}) \subset \mathbb{R}[\mathbf{x}]_d$ be the convex cone of polynomials of degree at most d that are nonnegative on $\mathbf{K}$, and denote by $C_d(\mathbf{K})^*$ its dual cone which can be viewed as a convex cone in $\mathbb{R}^{s(d)}$ where $s(d) = \binom{n+d}{n}$. Of course one has the inclusion

$$C_d(\mathbf{K})^* = \{\, \mathbf{y} = (y_\alpha),\ \alpha \in \mathbb{N}^n_d\ :\quad L_{\mathbf{y}}(g) \geq 0, \quad \forall g \in C_d(\mathbf{K})\,\}$$

$$\supseteq \left\{\, \mathbf{y} = (y_\alpha),\ \alpha \in \mathbb{N}^n_d\ :\ \exists\, \mu\ \text{s.t.}\ y_\alpha = \int_{\mathbf{K}} \mathbf{x}^\alpha\, d\mu,\ \forall \alpha \in \mathbb{N}^n_d \,\right\},$$

and equality holds when $\mathbf{K}$ is compact. (See more details in Section 4.4.)

The (real) **K**-moment problem identifies those sequences **y** that are moment sequences of a measure μ whose support supp μ is contained in a set $\mathbf{K} \subset \mathbb{R}^n$. The next two results are a truncated version of Haviland's Theorem 2.34.

Theorem 2.42 *Let* $\mathbf{K} \subset \mathbb{R}^n$ *be closed and let* $\mathbf{y} \in C_{2d}(\mathbf{K})^*$. *Then the subsequence* $\mathbf{y} = (y_\alpha)$, $\alpha \in \mathbb{N}^n_{2d-1}$, *has a representing measure supported on* **K**.

Theorem 2.43 *Let* $\mathbf{K} \subset \mathbb{R}^n$ *be closed. A sequence* $\mathbf{y} = (y_\alpha)$, $\alpha \in \mathbb{N}^n_{2d}$, *has a representing measure on* **K** *if and only if* **y** *has an extension* $\tilde{\mathbf{y}} = (\tilde{y}_\alpha)$, $\alpha \in \mathbb{N}^n_{2d+2}$, *which belongs to* $C_{2d+2}(\mathbf{K})^*$.

Observe that Theorem 2.42 and Theorem 2.43 provide a theoretical solution to the **K**-moment problem but, as for Haviland's Theorem 2.34, they are of little practical use because one does not have a tractable characterization of the convex cone $C_{2d}(\mathbf{K})^*$.

Therefore to obtain a more concrete solution to the **K**-moment problem we need to consider more specific subsets $\mathbf{K} \subset \mathbb{R}^n$, namely basic semi-algebraic sets

$$\mathbf{K} := \{\, \mathbf{x} \in \mathbb{R}^n : g_j(\mathbf{x}) \geq 0, \quad j = 1, \ldots, m \,\}, \tag{2.40}$$

for some given polynomials $g_j \in \mathbb{R}[\mathbf{x}]$, $j = 1, \ldots, m$.

Conditions based on linear matrix inequalities For notational convenience, we also define $g_0 \in \mathbb{R}[\mathbf{x}]$ to be the constant polynomial with value 1 (i.e., $g_0 = 1$).

Recall that given a family $(g_j)_{j=1}^m \subset \mathbb{R}[\mathbf{x}]$, we denote by g_J, $J \subseteq \{1, \ldots, m\}$, the polynomial $\mathbf{x} \mapsto g_J(\mathbf{x}) := \prod_{j \in J} g_j(\mathbf{x})$. In particular, when $J = \emptyset$, $g_\emptyset = 1$.

Let $\mathbf{y} = (y_\alpha)$, $\alpha \in \mathbb{N}^n$, be a given infinite sequence. For every $d \in \mathbb{N}$ and every $J \subseteq \{1, \ldots, m\}$, let $\mathbf{M}_d(g_J \, \mathbf{y})$ be the localizing matrix of order d with respect to the polynomial $g_J := \prod_{j \in J} g_j$ (so that with $J := \emptyset$, $\mathbf{M}_d(g_\emptyset \, \mathbf{y}) = \mathbf{M}_d(\mathbf{y})$ is the moment matrix (of order d) associated with **y**).

As we have already seen, there is a duality between the theory of moments and the representation of positive polynomials. The following important theorem is the dual facet of Theorem 2.13 and Theorem 2.15, and makes this duality statement more precise.

Theorem 2.44 (Dual facet of Schmüdgen and Putinar Positivstellensätze) *Let* $\mathbf{y} = (y_\alpha) \subset \mathbb{R}$, $\alpha \in \mathbb{N}^n$, *be a given infinite sequence in* $\mathbb{R}$, $L_\mathbf{y}$:

$\mathbb{R}[\mathbf{x}] \to \mathbb{R}$ *be the Riesz functional introduced in (2.31), and let* **K** *in (2.40) be compact.*

(a) *The sequence* **y** *has a finite Borel representing measure with support contained in* **K**, *if and only if*

$$L_{\mathbf{y}}(f^2 g_J) \geq 0, \qquad \forall J \subseteq \{1, \ldots, m\}, \quad \forall f \in \mathbb{R}[\mathbf{x}], \qquad (2.41)$$

or, equivalently, if and only if

$$M_d(g_J \, \mathbf{y}) \succeq \mathbf{0}, \qquad \forall J \subseteq \{1, \ldots, m\}, \quad \forall d \in \mathbb{N}. \qquad (2.42)$$

(b) *If Assumption 2.14 holds true then* **y** *has a finite Borel representing measure with support contained in* **K**, *if and only if*

$$L_{\mathbf{y}}(f^2 g_j) \geq 0, \qquad \forall j = 0, 1, \ldots, m, \quad \forall f \in \mathbb{R}[\mathbf{x}], \qquad (2.43)$$

or, equivalently, if and only if

$$M_d(g_j \, \mathbf{y}) \succeq \mathbf{0}, \qquad \forall j = 0, 1, \ldots, m, \quad \forall d \in \mathbb{N}. \qquad (2.44)$$

Proof (a) For every $J \subseteq \{1, \ldots, m\}$ and $f \in \mathbb{R}[\mathbf{x}]_d$, the polynomial $f^2 g_J$ is nonnegative on **K**. Therefore, if **y** is the sequence of moments of a measure μ supported on **K**, then $\int f^2 g_J \, d\mu \geq 0$. Equivalently, $L_{\mathbf{y}}(f^2 g_J) \geq 0$, or, in view of (2.39), $M_d(g_J \, \mathbf{y}) \succeq \mathbf{0}$. Hence (2.41)–(2.42) hold.

Conversely, assume that (2.41), or equivalently, (2.42) holds. As **K** is compact, by Theorem 2.34, **y** is the moment sequence of a measure with support contained in **K** if and only if $\sum_{\alpha \in \mathbb{N}^n} f_\alpha y_\alpha \geq 0$ for all polynomials $f \geq 0$ on **K**.

Let $f > 0$ on **K**, so that by Theorem 2.13

$$f = \sum_{J \subseteq \{1, \ldots, m\}} p_J \, g_J, \qquad (2.45)$$

for some polynomials $\{p_J\} \subset \mathbb{R}[\mathbf{x}]$, all sums of squares. Hence, since $p_J \in \Sigma[\mathbf{x}]$, from (2.41) and from the linearity of $L_{\mathbf{y}}$, we have $L_{\mathbf{y}}(f) \geq 0$. Hence, for all polynomials $f > 0$ on **K**, we have $L_{\mathbf{y}}(f) = \sum_\alpha f_\alpha y_\alpha \geq 0$. Next, let $f \in \mathbb{R}[\mathbf{x}]$ be nonnegative on **K**. Then for arbitrary $\epsilon > 0$, $f + \epsilon > 0$ on **K**, and thus, $L_{\mathbf{y}}(f + \epsilon) = L_{\mathbf{y}}(f) + \epsilon y_0 \geq 0$. As $\epsilon > 0$ was arbitrary, $L_{\mathbf{y}}(f) \geq 0$ follows. Therefore, $L_{\mathbf{y}}(f) \geq 0$ for all $f \in \mathbb{R}[\mathbf{x}]$, nonnegative on **K**, which by Theorem 2.34, implies that **y** is the moment sequence of some measure with support contained in **K**.

(b) The proof is similar to part (a) and is left as an exercise. □

Note that the conditions (2.42) and (2.44) of Theorem 2.44 are stated in terms of positive semidefiniteness of the localizing matrices associated with the sequence $\mathbf{y}$ and the polynomials g_J and g_j involved in the definition (2.40) of the compact set $\mathbf{K}$.

Conditions based on linear inequalities Recall the notation $\widehat{\mathbf{g}}$ and $(1 - \widehat{\mathbf{g}})$ in (2.16). Alternatively, we also have the following.

Theorem 2.45 *Let $\mathbf{y} = (y_\alpha) \subset \mathbb{R}$, $\alpha \in \mathbb{N}^n$, be a given infinite sequence, let $L_{\mathbf{y}} : \mathbb{R}[\mathbf{x}] \rightarrow \mathbb{R}$ be the Riesz functional introduced in (2.31), and let $\mathbf{K}$ be as in (2.40), assumed to be compact. Let $C_G \subset \mathbb{R}[\mathbf{x}]$ be the convex cone defined in (2.16), and let Assumption 2.21 hold. Then, $\mathbf{y}$ has a representing measure μ with support contained in $\mathbf{K}$ if and only if*

$$L_{\mathbf{y}}(f) \geq 0, \qquad \forall\, f \in C_G, \qquad (2.46)$$

or, equivalently,

$$L_{\mathbf{y}}(\widehat{\mathbf{g}}^\alpha\, (1 - \widehat{\mathbf{g}})^\beta) \geq 0, \qquad \forall\, \alpha, \beta \in \mathbb{N}^m \qquad (2.47)$$

with $\widehat{\mathbf{g}}$ as in (2.15).

Proof If $\mathbf{y}$ is the moment sequence of some measure with support contained in $\mathbf{K}$, then (2.47) follows directly from Theorem 2.34, because $\widehat{g}_j \geq 0$ and $1 - \widehat{g}_j \geq 0$ on $\mathbf{K}$, for all $j = 1, \ldots, m$.

Conversely, let (2.47) (and so (2.46)) hold, and let $f \in \mathbb{R}[\mathbf{x}]$, with $f > 0$ on $\mathbf{K}$. By Theorem 2.22, $f \in C_G$, and so f can be written as in (2.16), from which $L_{\mathbf{y}}(f) \geq 0$ follows. Finally, let $f \geq 0$ on $\mathbf{K}$, so that $f + \epsilon > 0$ on $\mathbf{K}$ for every $\epsilon > 0$. Therefore, $0 \leq L_{\mathbf{y}}(f + \epsilon) = L_{\mathbf{y}}(f) + \epsilon y_0$ because $f + \epsilon \in C_G$. As $\epsilon > 0$ was arbitrary, we obtain $L_{\mathbf{y}}(f) \geq 0$. Therefore, $L_{\mathbf{y}}(f) \geq 0$ for all $f \in \mathbb{R}[\mathbf{x}]$, nonnegative on $\mathbf{K}$, which by Theorem 2.34, implies that $\mathbf{y}$ is the moment sequence of some measure with support contained in $\mathbf{K}$. $\qquad \square$

Exactly as Theorem 2.44 was the dual facet of Theorem 2.13 and 2.15, Theorem 2.45 is the dual facet of Theorem 2.22.

Note that (2.47) reduces to countably many *linear* conditions on the sequence $\mathbf{y}$. Indeed, for fixed $\alpha, \beta \in \mathbb{N}^m$, write

$$\mathbf{x} \mapsto \quad \widehat{\mathbf{g}}(\mathbf{x})^\alpha\, (1 - \widehat{\mathbf{g}}(\mathbf{x}))^\beta := \sum_{\gamma \in \mathbb{N}^m} q_\gamma(\alpha, \beta)\, \mathbf{x}^\gamma,$$

for finitely many real coefficients $(q_\gamma(\alpha, \beta))$. Then, (2.47) becomes

$$\sum_{\gamma \in \mathbb{N}^m} q_\gamma(\alpha, \beta) \, y_\gamma \geq 0, \qquad \forall \, \alpha, \beta \in \mathbb{N}^m. \tag{2.48}$$

The linear conditions (2.48) are to be contrasted with the positive semidefiniteness conditions (2.43) of Theorem 2.44.

In the case where all the g_j in (2.40) are affine (so that $\mathbf{K}$ is a convex polytope), we also have a specialized version of Theorem 2.45.

Theorem 2.46   *Let $\mathbf{y} = (y_\alpha) \subset \mathbb{R}$, $\alpha \in \mathbb{N}^n$, be a given infinite sequence, and let $L_{\mathbf{y}} : \mathbb{R}[\mathbf{x}] \to \mathbb{R}$ be the Riesz functional introduced in (2.31). Assume that $\mathbf{K}$ is compact with nonempty interior, and all the g_j in (2.40) are affine (so that $\mathbf{K}$ is a convex polytope). Then, $\mathbf{y}$ has a finite Borel representing measure with support contained in $\mathbf{K}$ if and only if*

$$L_{\mathbf{y}}(\mathbf{g}^\alpha) \geq 0, \qquad \forall \, \alpha \in \mathbb{N}^m. \tag{2.49}$$

A sufficient condition for the truncated K-moment problem   Finally, we present a very important sufficient condition for the truncated $\mathbf{K}$-moment problem. That is, we provide a condition on a finite sequence $\mathbf{y} = (y_\alpha)$ to admit a finite Borel representing measure supported on $\mathbf{K}$. Moreover, this condition can be checked numerically by standard techniques from linear algebra.

Theorem 2.47   *Let $\mathbf{K} \subset \mathbb{R}^n$ be the basic semi-algebraic set*

$$\mathbf{K} := \{\, \mathbf{x} \in \mathbb{R}^n : g_j(\mathbf{x}) \geq 0, \ j = 1, \ldots, m \,\},$$

for some polynomials $g_j \in \mathbb{R}[\mathbf{x}]$ of degree $2v_j$ or $2v_j - 1$, for all $j = 1, \ldots, m$. Let $\mathbf{y} = (y_\alpha)$, $\alpha \in \mathbb{N}^n_{2d}$, and let $v := \max_j v_j$. Then $\mathbf{y}$ has a rank $\mathbf{M}_{d-v}(\mathbf{y})$ atomic representing measure μ with support contained in $\mathbf{K}$ if and only if:

  (a) $\mathbf{M}_d(\mathbf{y}) \succeq \mathbf{0}$, $\mathbf{M}_{d-v}(g_j \mathbf{y}) \succeq \mathbf{0}$, $j = 1, \ldots, m$, *and*
  (b) $\operatorname{rank} \mathbf{M}_d(\mathbf{y}) = \operatorname{rank} \mathbf{M}_{d-v}(\mathbf{y})$.

In addition, μ has $\operatorname{rank} \mathbf{M}_d(\mathbf{y}) - \operatorname{rank} \mathbf{M}_{d-v}(g_j \mathbf{y})$ atoms $\mathbf{x} \in \mathbb{R}^n$ that satisfy $g_j(\mathbf{x}) = 0$, for all $j = 1, \ldots, m$.

Note that in Theorem 2.47, the set $\mathbf{K}$ is *not* required to be compact. The rank-condition can be checked by standard techniques from numerical linear algebra. However, it is also important to remember that computing the rank is sensitive to numerical imprecisions.

When the conditions (a) and (b) of Theorem 2.47 hold, we describe in Section 6.1.2 a numerical linear algebra procedure for extracting s ($s = \operatorname{rank} \mathbf{M}_d(\mathbf{y})$) points in $\mathbf{K}$ which form the support of an atomic measure whose moments up to order $2d$ are the y_α, for all $\alpha \in \mathbb{N}^n_{2d}$.

Example 2.48   Let $n = 2$ and let

$$\mathbf{K} = \{ \mathbf{x} \in \mathbb{R}^2 : 1 - (x_1 - 1)^2 \geq 0; \; 1 - (x_1 - x_2)^2 \geq 0; \; 1 - (x_2 - 3)^2 \geq 0 \}.$$

Let $\mathbf{y} = (y_\alpha)$, $\alpha \in \mathbb{N}^2_4$, with

$$\mathbf{M}_2(\mathbf{y}) = \begin{bmatrix} 1.0000 & 1.5868 & 2.2477 & 2.7603 & 3.6690 & 5.2387 \\ 1.5868 & 2.7603 & 3.6690 & 5.1073 & 6.5115 & 8.8245 \\ 2.2477 & 3.6690 & 5.2387 & 6.5115 & 8.8245 & 12.7072 \\ 2.7603 & 5.1073 & 6.5115 & 9.8013 & 12.1965 & 15.9960 \\ 3.6690 & 6.5115 & 8.8245 & 12.1965 & 15.9960 & 22.1084 \\ 5.2387 & 8.8245 & 12.7072 & 15.9960 & 22.1084 & 32.1036 \end{bmatrix}.$$

One may check that $\operatorname{rank} \mathbf{M}_1(\mathbf{y}) = \operatorname{rank} \mathbf{M}_2(\mathbf{y}) = 3$ (up to numerical imprecisions). Three eigenvalues of $\mathbf{M}_2(\mathbf{y})$ are below 6×10^{-5} whereas the three other eigenvalues are well above 0.3. Moreover, $\mathbf{M}_1(g_j\,\mathbf{y}) \succeq 0$ for $j = 1, 2, 3$. Hence, since $v = 1$, the conditions (a) and (b) of Theorem 2.47 are satisfied, which implies that $\mathbf{y}$ has a representing measure on $\mathbf{K}$. In fact, the atomic measure

$$v = 0.4131\,\delta_{(1,2)} + 0.3391\,\delta_{(2,2)} + 0.2477\,\delta_{(2,3)},$$

supported on the three points $(1, 2)$, $(2, 2)$ and $(2, 3)$, all in $\mathbf{K}$, has all its moments up to order 4 which match the sequence $\mathbf{y}$. More details on how to extract these three points are given in Section 6.1.2 and Example 6.10.

2.7.4 From measures to atomic measures

In the previous section we have seen conditions for a sequence $\mathbf{y} = (y_\alpha)$, $\alpha \in \mathbb{N}^n$, to have a representing measure μ supported on a set $\mathbf{K} \subset \mathbb{R}^n$. Another interesting issue is concerned with how to obtain a representing measure μ, especially when only finitely many moments are known. Theorem 2.47 is a result of this type. The next results prove some other insights. Recall that for a Borel measure μ its support, denoted $\operatorname{supp}(\mu)$, is the smallest closed set $S \subset \mathbb{R}^n$ such that $\mu(\mathbb{R}^n \setminus S) = 0$.

Theorem 2.49   *If $\mathbf{y} = (y_\alpha)$, $\alpha \in \mathbb{N}^n_d$, has a representing measure μ then there exists an atomic representing measure v supported on at most $\binom{n+d}{n}$ atoms, and such that $\operatorname{supp}(v) \subset \operatorname{supp}(\mu)$.*

Theorem 2.49 provides the basis for the existence of *cubatures rules*. Indeed, express the measure ν in Theorem 2.49 as:

$$\nu := \sum_{k=1}^{s} \lambda_k \, \delta_{\mathbf{x}_k}, \quad \lambda_k > 0, \quad k = 1, \ldots, s$$

where $s \leq \binom{n+d}{n}$ and $\mathbf{x}_i \in \operatorname{supp}(\mu)$, $i = 1, \ldots, s$, are the atoms of ν. Then by Theorem 2.49,

$$\int_{\mathbb{R}^n} p(\mathbf{x}) \, d\mu(\mathbf{x}) = \sum_{k=1}^{s} \lambda_k \, p(\mathbf{x}_k), \quad \forall \, p \in \mathbb{R}[\mathbf{x}]_d,$$

and so for every μ-integrable function $f : \mathbb{R}^n \to \mathbb{R}$, the approximation

$$\int_{\mathbb{R}^n} f(\mathbf{x}) \, d\mu(\mathbf{x}) \approx \int f(\mathbf{x}) \, d\nu(\mathbf{x}) = \sum_{k=1}^{s} \lambda_k \, f(\mathbf{x}_k),$$

is called a cubature rule with *nodes* $(\mathbf{x}_1, \ldots, \mathbf{x}_s)$ and positive *weights* $(\lambda_1, \ldots, \lambda_s)$. This cubature rule is *exact* for all polynomials of degree at most d.

There is also another related and more general result which does not necessarily use monomial moments $\mathbf{x}^\alpha$.

Theorem 2.50   *Let $f_1, \ldots, f_N$ be real-valued Borel measurable functions on a measurable space Ω and let μ be a probability measure on Ω such that each f_i, $i = 1, \ldots, N$, is integrable with respect to μ. Then there exists an atomic probability measure ν with $\operatorname{supp}(\nu) \subset \Omega$ and such that:*

$$\int_{\Omega} f_i(\mathbf{x}) \, \mu(d\mathbf{x}) = \int_{\Omega} f_i(\mathbf{x}) \, \nu(d\mathbf{x}), \quad i = 1, \ldots, N.$$

One can even determine that the support of ν has at most $N + 1$ points.

In fact if $M(\Omega)$ denotes the space of probability measures on Ω, then the *moment space*

$$Y_N := \left\{ \mathbf{y} = \left(\int_{\Omega} f_k(\mathbf{x}) d\mu(\mathbf{x}) \right), \ k = 1, \ldots, N : \quad \mu \in M(\Omega) \right\}$$

is the convex hull of the set $f(\Omega) = \{(f_1(\mathbf{x}), \ldots, f_N(\mathbf{x})) : \mathbf{x} \in \Omega\}$ and each point $\mathbf{y} \in Y_N$ can be represented as the convex hull of at most $N + 1$ points $f(\mathbf{x}_i)$, $i = 1, \ldots, N + 1$. (See e.g. Section 3, p. 29 in Kemperman (1968).)

Observe that in Theorem 2.50 the functions f_i, $i = 1, \ldots, N$, are only required to be integrable with respect to μ, whereas Theorem 2.49 is stated for classical monomial moments. In the particular case where $y_0 = 1$ and the f_i are the monomials $\mathbf{x}^\alpha$, $\alpha \in \mathbb{N}_d^n$, then one retrieves Theorem 2.49.

2.8 Exercises

Exercise 2.1   (Goursat's Lemma) With $f \in \mathbb{R}[x]$ and $\deg f = m$, the polynomial $\hat{f}$

$$x \mapsto \hat{f}(x) := (1+x)^m f\left(\frac{1-x}{1+x}\right),$$

is called the Goursat transform of f.

(a) What is the Goursat transform of $\hat{f}$?

(b) Prove that $f > 0$ on $[-1, 1]$ if and only if $\hat{f} > 0$ on $[0, \infty)$ and degree $\hat{f} = m$.

(c) Prove that $f \geq 0$ on $[-1, 1]$ if and only if $\hat{f} \geq 0$ on $[0, \infty)$ and degree $\hat{f} \leq m$.

Exercise 2.2   Show that a Farkas certificate of $\emptyset = \{\mathbf{x} : \mathbf{Ax} = \mathbf{b}; \mathbf{x} \geq 0\}$ is a particularly simple Stengle certificate (2.6).

Exercise 2.3   Let $f \in \mathbb{R}[x]$ be the univariate polynomial $x \mapsto f(x) := x - x^2$.

(a) Show that f can be written in the form

$$f(x) = \sigma_0(x) + \sigma_1(x)\,x + \sigma_2(x)\,(1-x),$$

for some SOS polynomials σ_i, $i = 0, 1, 2$, of degree at most 2. On which set $\mathbf{K} \subset \mathbb{R}$ is it a certificate of positivity?

(b) Can we write f above in the form

$$f(x) = \sum_{i,j} c_{ij}\, x^i\,(1-x)^j,$$

for finitely many positive coefficients c_{ij}?

(c) Same question as in (a) and (b) but now for the univariate $x \mapsto f(x) := x^2 - x + 1/4$.

Exercise 2.4   Let $f, h, s_0, s_1, s_2 \in \mathbb{R}[\mathbf{x}]$ be defined by $\mathbf{x} \mapsto f(\mathbf{x}) := x_1 - x_2^2 + 3$ and $\mathbf{x} \mapsto h(\mathbf{x}) := x_2 + x_1^2 + 2$, and

$$s_0(\mathbf{x}) = \frac{1}{3} + 2\left(x_2 + \frac{3}{2}\right)^2 + 6\left(x_1 - \frac{1}{6}\right)^2, \quad s_1(\mathbf{x}) = 2; \quad s_2(\mathbf{x}) = -6, \quad \mathbf{x} \in \mathbb{R}^2.$$

(a) Describe the preordering generated by f and the ideal generated by h.

(b) Develop the polynomial $s_0 + s_1 f + s_2 h$.

(c) Consider the semi-algebraic set $\mathbf{K} := \{\mathbf{x} : f(\mathbf{x}) \geq 0; h(\mathbf{x}) = 0\}$. What can we conclude? (*Hint:* Use Stengle's certificate.)

Exercise 2.5   Let $n = 1$. Write the moment matrices $\mathbf{M}_1(\mathbf{y})$ and $\mathbf{M}_2(\mathbf{y})$ associated with a measure supported on the variety $\{x \in \mathbb{R} : x\,(x-1) = 0\}$ and with moment sequence $\mathbf{y} = (y_k)$, $k \in \mathbb{N}$. What do we observe? What can be said about $\mathbf{M}_k(\mathbf{y})$ for every $k > 2$?

Exercise 2.6   The spaces $\mathbb{R}[\mathbf{x}]$ and the space $\mathbb{R}^\infty$ of real sequences $\mathbf{y} = (y_\alpha)$, $\alpha \in \mathbb{N}^n$, are two vector spaces in duality via the Riesz functional

$$\langle p, \mathbf{y}\rangle = \sum_\alpha p_\alpha\, y_\alpha = L_\mathbf{y}(p), \quad \forall p \in \mathbb{R}[\mathbf{x}],\ \mathbf{y} \in \mathbb{R}^\infty.$$

The dual cone C^* of a convex cone $C \subset \mathbb{R}[\mathbf{x}]$ is defined by $C^* = \{\,\mathbf{y} \in \mathbb{R}^\infty :$ $\langle p, \mathbf{y}\rangle \geq 0,\ \forall p \in C\,\}$. Similarly, the dual cone Θ^* of a convex cone $\Theta \subset \mathbb{R}^\infty$ is defined by $\Theta^* = \{\,p \in \mathbb{R}[\mathbf{x}] : \langle p, \mathbf{y}\rangle \geq 0,\ \forall \mathbf{y} \in \Theta\,\}$.

Let $\mathcal{P} \subset \mathbb{R}[\mathbf{x}]$ (respectively $\Sigma[\mathbf{x}] \subset \mathbb{R}[\mathbf{x}]$) be the convex cone of nonnegative polynomials (respectively SOS polynomials) and let $\mathcal{M} \subset \mathbb{R}^\infty$ (respectively $\mathcal{C} \subset \mathbb{R}^\infty$) be the convex cone of sequences that have a representing measure on $\mathbb{R}^n$ (respectively sequences $\mathbf{y}$ such that $\mathbf{M}_d(\mathbf{y}) \succeq 0$ for all $d \in \mathbb{N}$). Show that:

(a) $\mathcal{P}^* = \mathcal{M}$ and $\mathcal{M}^* = \mathcal{P}$,
(b) $\Sigma[\mathbf{x}]^* = \mathcal{C}$ and $\mathcal{C}^* = \Sigma[\mathbf{x}]$.

2.9 Notes and sources

2.1 For a nice exposition on degree bounds for Hilbert's 17th problem see Schmid (1998) and the recent Lombardi et al. (2014). For a nice discussion on historical aspects on Hilbert's 17th problem the reader is referred to Reznick (2000). Blekherman (2006, 2012) has shown that the cone of nonnegative polynomials is much larger than that of SOS polynomials and the gap between the two cones increases with the number of variables. However, a denseness result of Berg (1987) states that any (fixed) nonnegative polynomial can be approximated as closely as desired (for the ℓ_1-norm) by SOS polynomials of increasing degree, and Lasserre and Netzer (2007) have provided an explicit construction of such an approximating sequence of SOS; see also Lasserre (2006c) for more details. Finally, Lasserre (2007) also provides sufficient conditions for a polynomial to be SOS, directly in terms of linear inequalities on its coefficients, later extended in Fidalgo and Kovacec (2011) and Ghasemi and Marshall (2010, 2012).

2.2 Most of the material for the one-dimensional case is taken from Powers and Reznick (2000). Theorem 2.4(a) is due to Fekete (1935) whereas

Theorem 2.4(b) is attributed to F. Luckàcs. Theorem 2.5 is from Pólya and Szegö (1976). Theorem 2.7 is due to Hausdorff (1915) and Bernstein (1921).

2.3 The important Theorem 2.8 is due to Pólya (1974) whereas Theorem 2.9 is due to Reznick (1995). The Positivstellensatz (Theorem 2.12) is credited to Stengle (1974) but was proved earlier by Krivine (1964a). For a complete and recent exposition of representation of positive polynomials, the reader is referred to Prestel and Delzell (2001); see also Kuhlmann et al. (2005) and the more recent Helton and Putinar (2007), Scheiderer (2008) and Marshall (2008).

2.4 Theorem 2.13 is due to Schmüdgen (1991). The machinery uses the spectral theory of self-adjoint operators in Hilbert spaces. The important Theorem 2.15 is due to Putinar (1993) and Jacobi and Prestel (2001), whereas Theorem 2.16 is due to Schmüdgen (1991). Concerning degree bounds for SOS terms that appear in those representation results, Theorem 2.17 is from Nie and Schweighofer (2007); see also Marshall (2009) and Schweighofer (2005). For the noncompact case, negative results are provided in Scheiderer (2008). However, some nice representation results have been obtained in some specific cases, see for example Marshall (2010). Theorem 2.19, the matrix-polynomial version of Putinar's Positivstellensatz, was first proved in Scherer and Hol (2004) and in Kojima and Maramatsu (2007) independently (with no degree bound). The version with degree bound is from Helton and Nie (2010). Theorem 2.23 for the case of polytopes is due to Cassier (1984) and Handelman (1988) and Theorem 2.22 for compact semi-algebraic sets follows from a result due to Krivine (1964a, 1964b), and restated later in Becker and Schwartz (1983), and Vasilescu (2003).

2.5 Theorem 2.25 is from Lasserre (2005) whereas Theorem 2.26 can be deduced from Theorem 6.15 in Laurent (2008) which improves results by Parrilo (2002), themselves extensions of previous results in Lasserre (2002b) for the grid case.

2.6 Theorem 2.29 was first proved in Lasserre (2006a) under the assumption that the feasible set **K** has a nonempty interior. This assumption was later removed in Kojima and Maramatsu (2009). For extensions of Theorem 2.29 to some noncompact cases, see the work of Kuhlmann and Putinar (2007, 2009). Interestingly, the running intersection property (2.27) is well known in graph theory, graphical models, and Bayesian networks. In the latter the nodes are random variables with an attached conditional probability distribution, and arcs provide statistical dependence between nodes. When the running intersection property is true, exact inference can be done efficiently. For more details see for example Jordan (2004).

2.7 Moment problems have a long and rich history. For historical remarks and details on various approaches for the moment problem, the interested reader is referred to Landau (1987); see also Akhiezer (1965), Curto and Fialkow (2000), and Simon (1998). Example 2.33 is from Feller (1966, p. 227) whereas Example 2.35 is from Laurent (2008). Theorem 2.34 was first proved by M. Riesz for closed sets $\mathbf{K} \subset \mathbb{R}$, and subsequently generalized to closed sets $\mathbf{K} \subset \mathbb{R}^n$ by Haviland (1935, 1936).

Theorem 2.36 (originally stated in an homogeneous setting) is from Blekherman (2014). The multivariate condition in Proposition 2.37 that generalizes an earlier result of Carleman (1926) in one dimension, is stated in Berg (1987), and was proved by Nussbaum (1966). Proposition 2.38 is taken from Lasserre (2007). The infinite and truncated moment matrices (and in particular their kernel) have a lot of very interesting properties. For more details, the interested reader is referred to Laurent (2008). The localizing matrix was introduced in Curto and Fialkow (2000) and Berg (1987).

Concerning the solution of the $\mathbf{K}$-moment problem, Theorems 2.42 and 2.43 and are from Curto and Fialkow (2008). For compact semi-algebraic sets, Theorem 2.44(a) was proved by Schmüdgen (1991) with a nice interplay between real algebraic geometry and functional analysis. Indeed, the proof uses Stengle's Positivstellensatz (Theorem 2.12) and the spectral theory of self-adjoint operators in Hilbert spaces. Its refinement (b) is due to Putinar (1993), and Jacobi and Prestel (2001). Incidently, in Schmüdgen (1991), the Positivstellensatz Theorem 2.13 appears as a Corollary of Theorem 2.44(a). Theorem 2.23 is due to Cassier (1984) and Handelman (1988), and appears prior to the more general Theorem 2.45 due to Vasilescu (2003). Theorems 2.40 and 2.47 are due to Curto and Fialkow (1991, 1996, 1998) and the results are stated for the the the complex plane $\mathbb{C}$, but generalize to $\mathbb{C}^n$ and $\mathbb{R}^n$. An alternative proof of some of these results can be found in Laurent (2005); for instance Theorem 2.47 follows from (Laurent, 2005, Theorem 5.23). When $\mathbf{K}$ is the closure of a (not necessarily bounded) open set, Blekherman and Lasserre (2012) have characterized the interior of the convex cone of finite moment sequences that have a representing measure supported on $\mathbf{K}$; it is the domain of the Legendre–Fenchel transform associated with a certain convex function. Theorem 2.49 is from Bayer and Teichmann (2006) and is a refinement of Tchalakoff's Theorem stated in Tchakaloff (1957) for the compact case and of Putinar (2000). Theorem 2.50 is from Mulholland and

Rogers (1958) who mention the extension to the multi-dimensional case, and is also stated in Kemperman (1968, 1987) and Anastassiou (1993); see also Richter (1957) and Rogosinsky (1958).

Finally, let us mention the **K**-moment problem with bounded density, i.e., the existence of a representing finite Borel measure absolutely continuous with respect to some reference measure μ and with a bounded density $f \in L_\infty(\mathbf{K}, \mu)$. It was initially studied by Markov on the interval [0, 1] with μ the Lebesgue measure, a refinement of the Hausdorff moment problem where one only asks for existence of some finite Borel representing measure ν on [0, 1]. For an interesting discussion with historical details, the reader is referred to Diaconis and Freedman (2006) where, in particular, the authors have proposed a simplified proof as well as conditions for existence of density in $L_p([0, 1], \mu)$ with a similar flavor. SOS conditions for existence of a bounded density are given in Lasserre (2009c). In the above cited works the conditions include the unknown bounding parameter whereas in Lasserre (2013a) the author provides conditions for existence of a density in $\cap_{p \geq 1} L_p(\mathbf{K}, \mu)$ with no a priori bounding parameter.

3

Another look at nonnegativity

In this chapter we look at nonnegativity on a closed set $\mathbf{K} \subset \mathbb{R}^n$ from a different point of view. We now assume that our knowledge on $\mathbf{K}$ is *not* based on the polynomials g_j that define its boundary as in (2.7), but rather on the knowledge of a finite Borel measure μ whose support is exactly $\mathbf{K}$. In other words, we now look at $\mathbf{K}$ from "inside" rather than from "outside."

As we next see, this new (and dual) point of view enables us to obtain a characterization of polynomials nonnegative on $\mathbf{K}$, different from those described in Chapter 2. In particular, we will see in Chapter 4 that it yields *outer* approximations of the cone of polynomials nonnegative on $\mathbf{K}$ whereas the characterizations of Chapter 2 yield *inner* approximations.

3.1 Nonnegativity on closed sets

Recall that if $\mathbf{X}$ is a separable metric space with Borel σ-field $\mathcal{B}$, the support supp μ of a Borel measure μ on $\mathbf{X}$ is the (unique) smallest closed set $B \in \mathcal{B}$ such that $\mu(X \backslash B) = 0$. Next, given a Borel measure μ on $\mathbb{R}^n$ and a measurable function $f : \mathbb{R}^n \to \mathbb{R}$, the mapping $B \mapsto \nu(B) := \int_B f d\mu$, $B \in \mathcal{B}$, defines a set function on $\mathcal{B}$. If f is nonnegative then ν is a Borel measure (which is finite if f is μ-integrable). If f is not nonnegative then setting $B_1 := \{\mathbf{x} : f(\mathbf{x}) \geq 0\}$ and $B_2 := \{\mathbf{x} : f(\mathbf{x}) < 0\}$, the set function ν can be written as the difference

$$\nu(B) = \nu_1(B) - \nu_2(B), \qquad B \in \mathcal{B}, \tag{3.1}$$

of the two positive Borel measures ν_1, ν_2 defined by

$$\nu_1(B) = \int_{B_1 \cap B} f d\mu, \quad \nu_2(B) = -\int_{B_2 \cap B} f d\mu, \quad \forall B \in \mathcal{B}. \tag{3.2}$$

Then ν is a *signed* Borel measure provided that either $\nu_1(B_1)$ or $\nu_2(B_2)$ is finite; see for example Royden (1988, p. 271). We first provide the following auxiliary result which is also of independent interest.

Lemma 3.1 *Let* $\mathbf{X}$ *be a separable metric space,* $\mathbf{K} \subseteq \mathbf{X}$ *a closed set, and* μ *a finite Borel measure on* $\mathbf{X}$ *with* supp $\mu = \mathbf{K}$. *A continuous function* $f : \mathbf{X} \rightarrow \mathbb{R}$ *is nonnegative on* $\mathbf{K}$ *if and only if the set function* $B \mapsto \nu(B) = \int_{\mathbf{K} \cap B} f \, d\mu$, $B \in \mathcal{B}$, *is a positive measure.*

Proof The *only if* part is straightforward. For the *if* part, if ν is a positive measure then $f(\mathbf{x}) \geq 0$ for μ-almost all $\mathbf{x} \in \mathbf{K}$. That is, there is a Borel set $\mathbf{G} \subset \mathbf{K}$ such that $\mu(\mathbf{G}) = 0$ and $f(\mathbf{x}) \geq 0$ on $\mathbf{K} \backslash \mathbf{G}$. Indeed, otherwise suppose that there exists a Borel set B_0 with $\mu(B_0) > 0$ and $f < 0$ on B_0; then one would get the contradiction that ν is not positive because $\nu(B_0) = \int_{B_0} f \, d\mu <$ 0. In fact, f is called the Radon–Nikodym derivative of ν with respect to μ; see Royden (1988, Theorem 23, p. 276).

Next, observe that $\overline{\mathbf{K} \backslash \mathbf{G}} \subset \mathbf{K}$ and $\mu(\overline{\mathbf{K} \backslash \mathbf{G}}) = \mu(\mathbf{K})$. Therefore $\overline{\mathbf{K} \backslash \mathbf{G}} = \mathbf{K}$ because supp μ (supp $\mu = \mathbf{K}$) is the unique smallest closed set such that $\mu(\mathbf{X} \backslash \mathbf{K}) = 0$. Hence, let $\mathbf{x} \in \mathbf{K}$ be fixed, arbitrary. As $\mathbf{K} = \overline{\mathbf{K} \backslash \mathbf{G}}$, there is a sequence $(\mathbf{x}_k) \subset \mathbf{K} \backslash \mathbf{G}, k \in \mathbb{N}$, with $\mathbf{x}_k \rightarrow \mathbf{x}$ as $k \rightarrow \infty$. But since f is continuous and $f(\mathbf{x}_k) \geq 0$ for every $k \in \mathbb{N}$, we obtain the desired result $f(\mathbf{x}) \geq 0$. $\qquad\square$

Lemma 3.1 itself provides a characterization of nonnegativity on $\mathbf{K}$ for a continuous function f on $\mathbf{X}$. However, a practical and more concrete characterization is highly desirable, especially for computational purposes. To this end, we first consider the case of a compact set $\mathbf{K} \subset \mathbb{R}^n$.

3.2 The compact case

Let $\mathbf{K}$ be a compact subset of $\mathbb{R}^n$. For simplicity, and with no loss of generality, we may and will assume that $\mathbf{K} \subseteq [-1, 1]^n$ (possibly after a change of variable).

Theorem 3.2 *Let* $\mathbf{K} \subseteq [-1, 1]^n$ *be compact and let* μ *be an arbitrary, fixed, finite Borel measure on* $\mathbf{K}$ *with* supp $\mu = \mathbf{K}$, *and with vector of moment* $\mathbf{y} = (y_\alpha)$, $\alpha \in \mathbb{N}^n$. *Let* f *be a continuous function on* $\mathbb{R}^n$.

(a) f *is nonnegative on* **K** *if and only if*

$$\int_{\mathbf{K}} g^2 \, f \, d\mu \ge 0, \qquad \forall g \in \mathbb{R}[\mathbf{x}], \qquad (3.3)$$

or, equivalently, if and only if

$$\mathbf{M}_d(\mathbf{z}) \succeq 0, \qquad d = 0, 1, \dots \qquad (3.4)$$

where $\mathbf{z} = (z_\alpha)$, $\alpha \in \mathbb{N}^n$, *with* $z_\alpha = \int \mathbf{x}^\alpha f(\mathbf{x}) d\mu(\mathbf{x})$, *and with* $\mathbf{M}_d(\mathbf{z})$ *being the moment matrix defined in (2.33).*

If in addition $f \in \mathbb{R}[\mathbf{x}]$ *then (3.4) reads* $\mathbf{M}_d(f\,\mathbf{y}) \succeq 0$, $d = 0, 1, \dots$, *where* $\mathbf{M}_d(f\,\mathbf{y})$ *is the localizing matrix defined in (2.37).*

(b) *If in addition to being continuous,* f *is also concave on* **K***, then* f *is nonnegative on the convex hull* co(**K**) *of* **K** *if and only if (3.3) holds.*

Proof The *only if* part is straightforward. Indeed, if $f \ge 0$ on **K** then $\mathbf{K} \subseteq \{\mathbf{x} : f(\mathbf{x}) \ge 0\}$ and so for any finite Borel measure μ on **K**, $\int_{\mathbf{K}} g^2 f d\mu \ge 0$ for every $g \in \mathbb{R}[\mathbf{x}]$. Next, if f is concave and $f \ge 0$ on co(**K**) then $f \ge 0$ on **K** and so the only if part of (b) also follows.

For the *If* part, the set function $\nu(B) = \int_B f d\mu$, $B \in \mathcal{B}$, can be written as the difference $\nu = \nu_1 - \nu_2$ of the two positive finite Borel measures ν_1, ν_2 described in (3.1)–(3.2), where $B_1 := \{\mathbf{x} \in \mathbf{K} : f(\mathbf{x}) \ge 0\}$ and $B_2 := \{\mathbf{x} \in \mathbf{K} : f(\mathbf{x}) < 0\}$. As **K** is compact and f is continuous, both ν_1 and ν_2 are finite, and so ν is a finite signed Borel measure; see Royden (1988, p. 271). In view of Lemma 3.1 it suffices to prove that in fact ν is a finite and positive Borel measure. So let $\mathbf{z} = (z_\alpha)$, $\alpha \in \mathbb{N}^n$, be the sequence defined by:

$$z_\alpha = \int_{\mathbf{K}} \mathbf{x}^\alpha \, d\nu(\mathbf{x}) := \int_{\mathbf{K}} \mathbf{x}^\alpha f(\mathbf{x}) \, d\mu(\mathbf{x}), \qquad \forall \alpha \in \mathbb{N}^n. \qquad (3.5)$$

Every z_α, $\alpha \in \mathbb{N}^n$, is finite because **K** is compact and f is continuous. So the condition

$$\int_{\mathbf{K}} g(\mathbf{x})^2 f(\mathbf{x}) \, d\mu(\mathbf{x}) \ge 0, \qquad \forall f \in \mathbb{R}[\mathbf{x}]_d,$$

reads $\langle \mathbf{g}, \mathbf{M}_d(\mathbf{z})\mathbf{g} \rangle \ge 0$ for all $\mathbf{g} \in \mathbb{R}^{s(d)}$, that is, $\mathbf{M}_d(\mathbf{z}) \succeq 0$, where $\mathbf{M}_d(\mathbf{z})$ is the moment matrix defined in (2.33). And so (3.3) implies $\mathbf{M}_d(\mathbf{z}) \succeq 0$ for every $d \in \mathbb{N}$. Moreover, as $\mathbf{K} \subseteq [-1, 1]^n$,

$$|z_\alpha| \le c := \int_{\mathbf{K}} |f| d\mu, \qquad \forall \alpha \in \mathbb{N}^n.$$

Hence, by Proposition 2.37(b), $\mathbf{z}$ is the moment sequence of a finite (positive) Borel measure ψ on $[-1, 1]^n$, that is, as supp $\nu \subseteq \mathbf{K} \subseteq [-1, 1]^n$,

$$\int_{[-1,1]^n} \mathbf{x}^\alpha \, dv(\mathbf{x}) = \int_{[-1,1]^n} \mathbf{x}^\alpha \, d\psi(\mathbf{x}), \qquad \forall \alpha \in \mathbb{N}^n. \qquad (3.6)$$

But then using (3.1) and (3.6) yields

$$\int_{[-1,1]^n} \mathbf{x}^\alpha \, dv_1(\mathbf{x}) = \int_{[-1,1]^n} \mathbf{x}^\alpha \, d(v_2 + \psi)(\mathbf{x}), \qquad \forall \alpha \in \mathbb{N}^n,$$

which in turn implies $v_1 = v_2 + \psi$ because measures on a compact set are determinate. Next, this implies $\psi = v_1 - v_2$ (equal to v) and so v is a positive Borel measure on $\mathbf{K}$. Hence by Lemma 3.1, $f(\mathbf{x}) \geq 0$ on $\mathbf{K}$.

If in addition $f \in \mathbb{R}[\mathbf{x}]$, the sequence $\mathbf{z} = (z_\alpha)$ is obtained as a linear combination of (y_α). Indeed if $f(\mathbf{x}) = \sum_\beta f_\beta \mathbf{x}^\beta$ then

$$z_\alpha = \sum_{\beta \in \mathbb{N}^n} f_\beta \, y_{\alpha+\beta}, \qquad \forall \alpha \in \mathbb{N}^n,$$

and so in (3.4), $\mathbf{M}_d(\mathbf{z})$ is nothing less than the *localizing* matrix $\mathbf{M}_d(f \, \mathbf{y})$ associated with $\mathbf{y} = (y_\alpha)$ and $f \in \mathbb{R}[\mathbf{x}]$, defined in (2.37), and (3.4) reads $\mathbf{M}_d(f \, \mathbf{y}) \succeq 0$ for all $d = 0, 1, \ldots$.

Finally, if f is concave then $f \geq 0$ on $\mathbf{K}$ implies $f \geq 0$ on co($\mathbf{K}$), and so the *if* part of (b) also follows.   □

Therefore to check whether a polynomial $f \in \mathbb{R}[\mathbf{x}]$ is nonnegative on $\mathbf{K}$, it suffices to check whether every element of the countable family of real symmetric matrices $(\mathbf{M}_d(f \, \mathbf{y}))$, $d \in \mathbb{N}$, is positive semidefinite. Examples of sets where all moments $\mathbf{y} = (y_\alpha)$, $\alpha \in \mathbb{N}^n$, of a Borel measure μ (with supp $\mu = \mathbf{K}$) can be computed easily include in particular hyper-rectangles, simplices, ellipsoids and hypercubes.

Example 3.3   For illustrative purposes, let $n = 2$ and let $\mathbf{K} \subset \mathbb{R}^2$ be the simplex $\{\mathbf{x} \geq 0 : x_1 + x_2 \leq 1\}$. Let $f \in \mathbb{R}[\mathbf{x}]_1$ be the affine polynomial $\mathbf{x} \mapsto f(\mathbf{x}) = f_{00} + f_{10} x_1 + f_{01} x_2$ and let μ be the Lebesgue measure on $\mathbf{K}$ with moments $\mathbf{y} = (y_\alpha)$ up to order 3

$$\mathbf{y} = (1/2, 1/6, 1/6, 1/12, 1/24, 1/12, 1/20, 1/60, 1/60, 1/20).$$

With $d = 1$, the condition $\mathbf{M}_1(f \, \mathbf{y}) \succeq 0$ in Theorem 3.2(a) reads:

$$\begin{bmatrix} 60 f_0 + 20 f_{10} + 20 f_{01} & 20 f_0 + 10 f_{10} + 5 f_{01} & 20 f_0 + 5 f_{10} + 10 f_{01} \\ 20 f_0 + 10 f_{10} + 2 f_{01} & 10 f_0 + 6 f_{10} + 2 f_{01} & 5 f_0 + 2 f_{10} + 2 f_{01} \\ 20 f_0 + 5 f_{10} + 10 f_{01} & 5 f_0 + 2 f_{10} + 2 f_{01} & 10 f_0 + 2 f_{10} + 6 f_{01} \end{bmatrix} \succeq 0.$$

So for instance, the vector $\mathbf{f} = (1, -1.2, -1.2)$ violates the condition since the matrix

$$\begin{bmatrix} 12 & 2 & 2 \\ 2 & 0.4 & 0.2 \\ 2 & 0.2 & 0.4 \end{bmatrix}$$

is not positive semidefinite. Therefore the affine polynomial $\mathbf{x} \mapsto f(\mathbf{x}) = 1 - 1.2\,x_1 - 1.2\,x_2$ is not nonnegative on $\mathbf{K}$ (also confirmed by evaluation at the point $\mathbf{x} = (0.5, 0.5) \in \mathbf{K}$).

3.3 The noncompact case

We now consider the more delicate case where $\mathbf{K}$ is a closed set of $\mathbb{R}^n$, not necessarily compact. To handle arbitrary noncompact sets $\mathbf{K}$ and arbitrary continuous functions f, we need a reference measure μ with $\operatorname{supp}\mu = \mathbf{K}$ and with nice properties so that integrals such as $\int g^2 f d\mu$, $g \in \mathbb{R}[\mathbf{x}]$, are well behaved.

So, let φ be an arbitrary finite Borel measure on $\mathbb{R}^n$ whose support is exactly $\mathbf{K}$, and let μ be the finite Borel measure defined by:

$$\mu(B) := \int_B \exp\left(-\sum_{i=1}^n |x_i|\right) d\varphi(\mathbf{x}), \qquad \forall B \in \mathcal{B}(\mathbb{R}^n). \qquad (3.7)$$

Observe that $\operatorname{supp}\mu = \mathbf{K}$ and μ satisfies Carleman's condition (2.35). Indeed, let $\mathbf{z} = (z_\alpha)$, $\alpha \in \mathbb{N}^n$, be the sequence of moments of μ. Then for every $i = 1, \ldots, n$, and every $k = 0, 1, \ldots$, using $x_i^{2k} \leq (2k)!\exp|x_i|$,

$$L_{\mathbf{z}}(x_i^{2k}) = \int_{\mathbf{K}} x_i^{2k}\, d\mu(\mathbf{x}) \leq (2k)! \int_{\mathbf{K}} \exp(|x_i|)\, d\mu(\mathbf{x})$$

$$\leq (2k)!\, \varphi(\mathbf{K}) =: (2k)!\, M. \qquad (3.8)$$

Therefore for every $i = 1, \ldots, n$, using $(2k)! \leq (2k)^{2k}$ for every k, yields

$$\sum_{k=1}^\infty L_{\mathbf{z}}(x_i^{2k})^{-1/2k} \geq \sum_{k=1}^\infty M^{-1/2k}\left((2k)!\right)^{-1/2k} \geq \sum_{k=1}^\infty \frac{M^{-1/2k}}{2k} = +\infty,$$

i.e., (2.35) holds. Notice also that all the moments of μ (defined in (3.7)) are finite, and so every polynomial is μ-integrable.

Theorem 3.4 *Let* $\mathbf{K} \subseteq \mathbb{R}^n$ *be closed and let* φ *be an arbitrary finite Borel measure whose support is exactly* $\mathbf{K}$. *Let* f *be a continuous function on* $\mathbb{R}^n$. *If* $f \in \mathbb{R}[\mathbf{x}]$ *(i.e., f is a polynomial) let* μ *be as in (3.7) whereas if f is not a polynomial let* μ *be defined by*

$$\mu(B) := \int_B \frac{\exp\left(-\sum_{i=1}^n |x_i|\right)}{1 + f(\mathbf{x})^2} \, d\varphi(\mathbf{x}), \qquad \forall \, B \in \mathcal{B}(\mathbb{R}^n). \qquad (3.9)$$

Then (a) and (b) of Theorem 3.2 hold.

It is important to emphasize that in Theorems 3.2 and 3.4, the set $\mathbf{K}$ is an *arbitrary* closed set of $\mathbb{R}^n$, and to the best of our knowledge, the characterization of nonnegativity of f in terms of positive definiteness of the moment matrices $\mathbf{M}_d(\mathbf{z})$ is new. But of course, this characterization becomes even more interesting when one knows how to compute the moment sequence $\mathbf{z} = (z_\alpha)$, $\alpha \in \mathbb{N}^n$, which is possible in a few special cases only.

Important particular cases of such nice sets $\mathbf{K}$ include boxes, hyperrectangles, ellipsoids, and simplices in the compact case, and the positive orthant or the whole space $\mathbb{R}^n$ in the noncompact case. For instance, for the whole space $\mathbf{K} = \mathbb{R}^n$ one may choose for μ in (3.7) the multivariate Gaussian (or normal) probability measure

$$\mu(B) := (2\pi)^{-n/2} \int_B \exp\left(-\frac{1}{2}\|\mathbf{x}\|^2\right) d\mathbf{x}, \qquad B \in \mathcal{B}(\mathbb{R}^n),$$

which is the n-times product of the one-dimensional normal distribution

$$\mu_i(B) := \frac{1}{\sqrt{2\pi}} \int_B \exp(-x_i^2/2) \, dx_i, \qquad B \in \mathcal{B}(\mathbb{R}),$$

whose moments are all easily available in closed form. In Theorem 3.4 this corresponds to the choice

$$\varphi(B) = (2\pi)^{-n/2} \int_B \frac{\exp(-\|\mathbf{x}\|^2/2)}{\exp(-\sum_{i=1}^n |x_i|)} \, d\mathbf{x}, \qquad B \in \mathcal{B}(\mathbb{R}^n). \qquad (3.10)$$

When $\mathbf{K}$ is the positive orthant $\mathbb{R}_+^n$ one may choose for μ the exponential probability measure

$$\mu(B) := \int_B \exp\left(-\sum_{i=1}^n x_i\right) d\mathbf{x}, \qquad B \in \mathcal{B}(\mathbb{R}_+^n), \qquad (3.11)$$

which is the n-times product of the one-dimensional exponential distribution

$$\mu_i(B) := \int_B \exp(-x_i) \, dx_i, \qquad B \in \mathcal{B}(\mathbb{R}_+),$$

whose moments are also easily available in closed form. In Theorem 3.4 this corresponds to the choice

$$\varphi(B) = 2^n \int_B \exp\left(-\sum_{i=1}^n x_i\right) dx, \qquad B \in \mathcal{B}(\mathbb{R}_+^n).$$

3.4 A symmetric duality principle

Chapters 1 and 2 illustrate the duality between positive polynomials and moments. For instance, this "algebra–functional analysis" duality is nicely expressed in Theorem 2.15 and Theorem 2.44(b), two dual facets of the same problem.

We show here that this duality is also expressed in a nice and elegant manner via the localizing matrix $\mathbf{M}_d(f\,\mathbf{y})$ introduced in (2.37) and defined for two objects, namely,

- a polynomial $f \in \mathbb{R}[\mathbf{x}]$, and
- a sequence $\mathbf{y} = (y_\alpha)$, $\alpha \in \mathbb{N}^n$,

which are in duality via the Riesz functional $L_\mathbf{y} : \mathbb{R}[\mathbf{x}] \to \mathbb{R}$ defined in (2.31).

Theorem 3.5 *Let* $\mathbf{y} = (y_\alpha)$, $\alpha \in \mathbb{N}^n$, *be such that* $\sup_{i,k} L_\mathbf{y}(x_i^{2k}/(2k!)) < M$ *for some* $M > 0$, *and* $\mathbf{M}_d(\mathbf{y}) \succeq 0$, *for all* $d = 0, 1, \ldots$. *Then* $\mathbf{y}$ *has a determinate representing measure* μ *and for every* $f \in \mathbb{R}[\mathbf{x}]$,

$$\operatorname{supp}\mu \subseteq \{\mathbf{x} : f(\mathbf{x}) \geq 0\} \quad \Longleftrightarrow \quad \mathbf{M}_d(f\,\mathbf{y}) \succeq 0, \quad d = 0, 1, \ldots. \tag{3.12}$$

Proof Since $\sup_{i,k} L_\mathbf{y}(x_i^{2k}/(2k!)) < M$ for some $M > 0$, Carleman's condition (2.35) holds for the sequence $\mathbf{y}$ and so by Proposition 2.37, $\mathbf{y}$ has a determinate representing measure μ. Next, the proof of "⇒" is straightforward whereas the proof of "⇐" follows from the proof of Theorem 3.4 in Lasserre (2011) where it is proved that the sequence $\mathbf{z} = (z_\alpha)$ with $z_\alpha = L_\mathbf{y}(f\,\mathbf{x}^\alpha)$, $\alpha \in \mathbb{N}^n$, satisfies Carleman's condition (2.35). Combining this with $\mathbf{M}_d(f\,\mathbf{y}) \succeq 0$ for all $d \in \mathbb{N}$, we deduce that $\mathbf{z}$ has a representing measure ψ on $\mathbb{R}^n$ and ψ is determinate. From this one may conclude that the set function $B \mapsto \nu(B) := \int_B f d\mu$ is a positive measure and by Lemma 3.1, f is nonnegative on $\operatorname{supp}\mu$. $\square$

We call (3.12) a *symmetric duality principle* because it gives the polynomial $f \in \mathbb{R}[\mathbf{x}]$ and the linear functional $\mathbf{y} \in \mathbb{R}[\mathbf{x}]^*$ symmetric roles. Indeed, we have the following.

- If $\mathbf{y}$ is fixed and known then for each $d \in \mathbb{N}$, the conditions $\mathbf{M}_d(f\,\mathbf{y}) \succeq 0$, $d = 0, \ldots$, are linear matrix inequalities in the coefficients (f_α) of $f \in \mathbb{R}[\mathbf{x}]$, and provide necessary and sufficient conditions for f to be nonnegative on $\mathbf{S} := \operatorname{supp} \mu$.
- Conversely, if $f \in \mathbb{R}[\mathbf{x}]$ is fixed and known and $\mathbf{y}$ satisfies Carleman's condition with $\mathbf{M}_d(\mathbf{y}) \succeq 0$, $d = 0, \ldots$, then the conditions $\mathbf{M}_d(f\,\mathbf{y}) \succeq 0$, $d = 0, \ldots$, are linear matrix inequalities in the variables (y_α) and provide necessary and sufficient conditions for $\mathbf{y}$ to have a representing measure supported on $\mathbf{S} := \{\mathbf{x} : f(\mathbf{x}) \geq 0\}$.

Finally, observe that if the level set $\{\mathbf{x} : f(\mathbf{x}) \geq 0\}$ is compact then the condition (3.12) is Theorem 2.44(b).

3.5 Exercises

Exercise 3.1 Let $\mathbf{K} \subset \mathbb{R}$ be the interval $[0, 1]$ and let μ be the probability measure uniformly distributed on $\mathbf{K}$ with associated sequence of moments $\mathbf{y} = (y_k)$, $k \in \mathbb{N}$. Let $f \in \mathbb{R}[x]_1$ be of the form $x \mapsto f(x) = -x + f_0$ for some $f_0 \in \mathbb{R}$.

(a) Check whether $\mathbf{M}_1(f\,\mathbf{y}) \succeq 0$ and $\mathbf{M}_2(f\,\mathbf{y}) \succeq 0$ if $f_0 = 0.78, 0.8$ and 0.9.

(b) Characterize those $f_0 \in \mathbb{R}$ such that $\mathbf{M}_1(f\,\mathbf{y}) \succeq 0$.

(c) Compare with the set of $f_0 \in \mathbb{R}$ such that $f \geq 0$ on $\mathbf{K}$.

Exercise 3.2 Let $\mathbf{K}$ be the whole space $\mathbb{R}^2$, and let μ be the measure with density $\mathbf{x} \mapsto \exp(-\|\mathbf{x}\|^2/2)$ with respect to the Lebesgue measure and with associated sequence of moments $\mathbf{y} = (y_\alpha)$, $\alpha \in \mathbb{N}^2$. Let $f \in \mathbb{R}[\mathbf{x}]_6$ be the polynomial $\mathbf{x} \mapsto f(\mathbf{x}) = x_1^2 x_2^2 (x_1^2 + x_2^2 - 1) - f_0$, for some scalar $f_0 \in \mathbb{R}$.

(a) Check whether $\mathbf{M}_1(f\,\mathbf{y}) \succeq 0$ and $\mathbf{M}_2(f\,\mathbf{y}) \succeq 0$ for $f_0 = -1/27$, and $f_0 = 5$.

(b) Characterize those $f_0 \in \mathbb{R}$ such that $\mathbf{M}_1(f\,\mathbf{y}) \succeq 0$.

3.6 Notes and sources

This chapter is mainly from Lasserre (2011, 2013d).

4

The cone of polynomials nonnegative on **K**

4.1 Introduction

Recall from Chapter 2 that the convex cone $C_d \subset \mathbb{R}[\mathbf{x}]_{2d}$ of nonnegative polynomials of degree at most $2d$ (a nonnegative polynomial has necessarily even degree) is much harder to characterize than its subcone $\Sigma[\mathbf{x}]_d$ of sums of squares (SOS). Indeed, while we have seen that the latter has a simple semidefinite representation in a higher dimensional space, so far there is no such simple representation for the former. In addition, when d is fixed Blekherman (2006) has shown that after proper normalization, the "gap" between C_d and $\Sigma[\mathbf{x}]_d$ increases unboundedly with the number of variables.

Similarly, for a subset $\mathbf{K} \subset \mathbb{R}^n$, the (finite-dimensional) convex cone $C_d(\mathbf{K})$ of polynomials of degree at most d and nonnegative on $\mathbf{K}$ does not have a simple and tractable representation. This is why when $f \in \mathbb{R}[\mathbf{x}]_d$ the optimization problem $f^* = \inf \{ f(\mathbf{x}) : \mathbf{x} \in \mathbf{K} \}$, which is equivalent to $f^* = \sup \{ \lambda : f(\mathbf{x}) - \lambda \geq 0, \ \forall \mathbf{x} \in \mathbf{K} \}$, is very difficult to solve in general even though it is also the finite-dimensional convex optimization problem $\sup \{ \lambda : f - \lambda \in C_d(\mathbf{K}) \}$. Finite-dimensional convex optimization problems are not always tractable!

However, we next show that the results of Chapter 2 and Chapter 3 provide us with tractable *inner* and *outer* approximations of $C_d(\mathbf{K})$, respectively. Those approximations will be very useful to approximate as closely as desired the global optimum f^* of the optimization problem **P** in (1.1).

4.2 Inner semidefinite approximations when **K** is compact

In this section we assume that $\mathbf{K} \subset \mathbb{R}^n$ is again the basic semi-algebraic set

$$\mathbf{K} = \{ \mathbf{x} \in \mathbb{R}^n : g_j(\mathbf{x}) \geq 0, \quad j = 1, \ldots, m \}, \tag{4.1}$$

for some polynomials $g_j \in \mathbb{R}[\mathbf{x}]$, $j = 1, \ldots, m$. Recall that for every $J \subseteq \{1, \ldots, m\}$, $g_J := \prod_{j \in J} g_j$ and $g_\emptyset = 1$. With $k \in \mathbb{N}$, and $g_0 = g_\emptyset$, let

$$P_k(g) = \left\{ \sum_{J \subseteq \{1,2,\ldots,m\}} \sigma_J \, g_J \; : \; \sigma_J \in \Sigma[\mathbf{x}], \quad \deg \sigma_J \, g_J \leq 2k \right\} \qquad (4.2)$$

$$Q_k(g) = \left\{ \sum_{j=0}^{m} \sigma_j \, g_j \; : \; \sigma_j \in \Sigma[\mathbf{x}], \quad \deg \sigma_j \, g_j \leq 2k \right\}. \qquad (4.3)$$

Both sets P_k and Q_k are truncated versions of the preordering $P(g)$ (respectively quadratic module $Q(g)$) already encountered in Chapter 2. These sets are convex cones contained in $\mathbb{R}[\mathbf{x}]_{2k}$ and can be viewed as convex cones in $\mathbb{R}^{s(2k)}$ (recall that $s(k) = \binom{n+k}{n}$).

Theorem 4.1   *Let* **K** $\subset \mathbb{R}^n$ *be the basic semi-algebraic set defined in (4.1) and let $C_d(\mathbf{K}) \subset \mathbb{R}[\mathbf{x}]_d$ be the convex cone of polynomials of degree at most d that are nonnegative on* **K**. *If* **K** *is compact then*

$$\overline{\bigcup_{k=0}^{\infty} (P_k(g) \cap \mathbb{R}[\mathbf{x}]_d)} = C_d(\mathbf{K}).$$

If, in addition, Assumption 2.14 holds true then

$$\overline{\bigcup_{k=0}^{\infty} (Q_k(g) \cap \mathbb{R}[\mathbf{x}]_d)} = C_d(\mathbf{K}).$$

Proof   Of course we have the inclusion

$$\overline{\bigcup_{k=0}^{\infty} (P_k(g) \cap \mathbb{R}[\mathbf{x}]_d)} \subseteq C_d(\mathbf{K}).$$

Next, if $f \in C_d(\mathbf{K})$ then for every $\ell \in \mathbb{N} \setminus \{0\}$, $f + 1/\ell$ is strictly positive on **K** and by Theorem 2.13, $f + 1/\ell \in P(g)$, i.e., there is some $k_\ell \in \mathbb{N}$ such that $f + 1/\ell \in P_k(g) \cap \mathbb{R}[\mathbf{x}]_d$ for all $k \geq k_\ell$. The result follows by letting $\ell \to \infty$. Finally, under Assumption 2.14 the same result holds by invoking Theorem 2.15.   □

Hence as $k \to \infty$, the convex cones $P_k(g) \cap \mathbb{R}[\mathbf{x}]_d$ (respectively $Q_k(g) \cap \mathbb{R}[\mathbf{x}]_d$) provide a convergent hierarchy of nested *inner* approximations of $C_d(\mathbf{K})$. If $\mathbf{K} = \mathbb{R}^n$ then $\Sigma[\mathbf{x}]_d$ is an obvious inner approximation of $C_d(\mathbb{R}^n)$ but we do not have a hierarchy of inner approximations.

Semidefinite representation Importantly, the convex cones $P_k(g) \cap \mathbb{R}[\mathbf{x}]_d$ (respectively $Q_k(g) \cap \mathbb{R}[\mathbf{x}]_d$) have a semidefinite representation and so can be used for practical computation as explained in Part II of this book.

Recall that $\mathbf{v}_d(\mathbf{x})$ is the vector $(\mathbf{x}^\alpha)$, $\alpha \in \mathbb{N}^n_d$, of the canonical basis of monomials of $\mathbb{R}[\mathbf{x}]_d$, and with every $f \in \mathbb{R}[\mathbf{x}]_d$ is associated its vector of coefficients $\mathbf{f} = (f_\alpha)$, $\alpha \in \mathbb{N}^n_d$, in the basis $\mathbf{v}_d(\mathbf{x})$. Conversely, with every $\mathbf{f} \in \mathbb{R}^{s(d)}$ is associated a polynomial $f \in \mathbb{R}[\mathbf{x}]_d$ with vector of coefficients $\mathbf{f} = (f_\alpha)$ in the canonical basis $\mathbf{v}_d(\mathbf{x})$. So we may and will identify $Q_k(g) \cap \mathbb{R}[\mathbf{x}]_d$ (a subset of $\mathbb{R}[\mathbf{x}]_d$) with a subset of $\mathbb{R}^{s(d)}$.

For every $j = 0, \ldots, m$ let $d_j := \lceil (\deg g_j)/2 \rceil$, and write the matrix-polynomial $g_j(\mathbf{x}) \, \mathbf{v}_{d-d_j}(\mathbf{x}) \mathbf{v}_{d-d_j}(\mathbf{x})^T$ in the form:

$$g_j(\mathbf{x}) \, \mathbf{v}_{d-d_j}(\mathbf{x}) \, \mathbf{v}_{d-d_j}(\mathbf{x})^T = \sum_{\alpha \in \mathbb{N}^n_{2k}} \mathbf{C}^j_\alpha \mathbf{x}^\alpha,$$

for real symmetric matrices $\mathbf{C}^j_\alpha$ of dimension $s(d - d_j)$, $\alpha \in \mathbb{N}^n_{2k}$. And so for instance, as we did in Section 2.4.2, for every $k \geq \lceil d/2 \rceil$,

$$Q_k(g) \cap \mathbb{R}[\mathbf{x}]_d = \Big\{ \mathbf{f} = (f_\alpha) : \exists \mathbf{X}_j \, (= \mathbf{X}^T_j) \succeq 0, \; j = 0, \ldots, m,$$

$$\text{s.t. } f_\alpha = \sum_{j=0}^m \langle \mathbf{X}_j, \mathbf{C}^j_\alpha \rangle, \quad |\alpha| \leq d$$

$$0 = \sum_{j=0}^m \langle \mathbf{X}_j, \mathbf{C}^j_\alpha \rangle, \quad d < |\alpha| \leq 2k \Big\}. \quad (4.4)$$

As we will see in Part II of this book, the semidefinite representation (4.4) is particularly useful to define a convergent hierarchy of semidefinite relaxations for polynomial optimization.

4.3 Outer semidefinite approximations

We next provide a convergent hierarchy of *outer* semidefinite approximations $(C^k_d(\mathbf{K}))$, $k \in \mathbb{N}$, of $C_d(\mathbf{K})$ where each $C^k_d(\mathbf{K})$ has a semidefinite representation with *no* lifting, i.e., no projection is needed and $C^k_d(\mathbf{K})$ is a spectrahedron[1] of $\mathbb{R}^{s(d)}$. We start with the important special case $\mathbf{K} = \mathbb{R}^n$ and then proceed with the general case.

---

[1] A spectrahedron is the intersection of the cone of positive semidefinite matrices with a linear affine subspace. The feasible set of a semidefinite program is a spectrahedron.

A hierarchy of outer approximations of $C_d(\mathbb{R}^n)$ Let $C_d := C_d(\mathbb{R}^n)$ denote
the convex cone of polynomials of degree at most d that are nonnegative on
the whole $\mathbb{R}^n$ (and so necessarily d is even). For every $k = 1, \ldots,$ let

$$
\gamma_p := \frac{1}{\sqrt{2\pi}} \int_{-\infty}^{\infty} x^p \exp(-x^2/2)\, dx = \begin{cases} 0 & \text{if } p = 2k+1, \\ \prod_{j=1}^{k}(2j-1) & \text{if } p = 2k, \end{cases}
$$

as $\gamma_{2k} = (2k-1)\gamma_{2(k-1)}$ for every $k \geq 1$.

Corollary 4.2   *Let μ be the probability measure on $\mathbb{R}^n$ which is the n-times
product of the normal distribution on $\mathbb{R}$, and so with moments $\mathbf{y} = (y_{\boldsymbol{\alpha}})$,
$\boldsymbol{\alpha} \in \mathbb{N}^n$,*

$$
y_{\boldsymbol{\alpha}} = \int_{\mathbb{R}^n} \mathbf{x}^{\boldsymbol{\alpha}}\, d\mu = \prod_{i=1}^{n} \left(\frac{1}{\sqrt{2\pi}} \int_{-\infty}^{\infty} x^{\alpha_i} \exp(-x^2/2) dx \right),
$$

$$
= \prod_{i=1}^{n} \gamma_{\alpha_i}, \qquad \forall\, \boldsymbol{\alpha} \in \mathbb{R}^n. \tag{4.5}
$$

For every $k \in \mathbb{N}$, let

$$
C_d^k := \{ \mathbf{f} \in \mathbb{R}^{s(d)} : \mathbf{M}_k(f\, \mathbf{y}) \succeq 0 \}, \tag{4.6}
$$

*where $\mathbf{M}_k(f\, \mathbf{y})$ is the localizing matrix in (2.37) associated with $\mathbf{y}$ and $f \in
\mathbb{R}[\mathbf{x}]_d$. Each C_d^k is a closed convex cone and a spectrahedron of $\mathbb{R}^{s(d)}$.*
   *Then $C_d \subset C_d^{k+1} \subset C_d^k$ for all $k \in \mathbb{N}$, and $f \in C_d$ if and only if its vector
of coefficients $\mathbf{f} \in \mathbb{R}^{s(d)}$ satisfies $\mathbf{f} \in C_d^k$, for every $k = 0, 1, \ldots.$ Equivalently,*

$$
C_d = \bigcap_{k=0}^{\infty} C_d^k. \tag{4.7}
$$

Proof   From its definition (2.37), all entries of the localizing matrix $\mathbf{M}_k(f\, \mathbf{y})$
are linear in $\mathbf{f} \in \mathbb{R}^{s(d)}$ and so $\mathbf{M}_k(f\, \mathbf{y}) \succeq 0$ is a linear matrix inequality (LMI).
Therefore C_d^k is a spectrahedron and a closed convex cone. Next, let $\mathbf{K} := \mathbb{R}^n$
and let μ be as in Corollary 4.2 and so of the form (3.7) with φ as in (3.10).
Then μ satisfies Carleman's condition (2.35). Hence, by Theorem 3.4 with
$\mathbf{K} = \mathbb{R}^n$, f is nonnegative on $\mathbf{K}$ if and only if (3.4) holds, which is equivalent
to stating that $\mathbf{M}_k(f\, \mathbf{y}) \succeq 0$, $k = 0, 1, \ldots,$ which in turn is equivalent to
stating that $\mathbf{f} \in C_d^k$, $k = 0, 1, \ldots.$ Finally (4.7) follows because the set $\cap_{k=0}^{\infty} C_d^k$
is closed. $\square$

   So the nested sequence of convex cones $C_d^0 \supset C_d^k \cdots \supset C_d$ defines a hierarchy of arbitrarily close outer approximations of C_d. It is worth emphasizing
that each C_d^k is a spectrahedron with *no* lifting, that is, C_d^k is defined solely

in terms of the vector of coefficients **f** with no additional variable (i.e., no projection is needed).

For instance, the first approximation C_d^0 is just the set $\{\mathbf{f} \in \mathbb{R}^{s(d)} : \int f d\mu \geq 0\}$, which is a half-space of $\mathbb{R}^{s(d)}$. And with $n = 2$,

$$
C_d^1 = \left\{ \mathbf{f} \in \mathbb{R}^{s(d)} : \begin{bmatrix} \int f d\mu & \int x_1 f d\mu & \int x_2 f d\mu \\ \int x_1 f d\mu & \int x_1^2 f d\mu & \int x_1 x_2 f d\mu \\ \int x_2 f d\mu & \int x_1 x_2 f d\mu & \int x_2^2 f d\mu \end{bmatrix} \succeq 0 \right\},
$$

or, equivalently, C_d^1 is the convex basic semi-algebraic set:

$$
\left\{ \mathbf{f} \in \mathbb{R}^{s(d)} : \int f d\mu \geq 0; \int x_i^2 f d\mu \geq 0, \ i = 1, 2; \right.
$$

$$
\left(\int x_i^2 f d\mu \right) \left(\int f d\mu \right) \geq \left(\int x_i f d\mu \right)^2, \quad i = 1, 2;
$$

$$
\left(\int x_1^2 f d\mu \right) \left(\int x_2^2 f d\mu \right) \geq \left(\int x_1 x_2 f d\mu \right)^2;
$$

$$
\left(\int f d\mu \right) \left[\left(\int x_1^2 f d\mu \right) \left(\int x_2^2 f d\mu \right) - \left(\int x_1 x_2 f d\mu \right)^2 \right]
$$

$$
- \left(\int x_1 f d\mu \right)^2 \left(\int x_2^2 f d\mu \right) - \left(\int x_2 f d\mu \right)^2 \left(\int x_1^2 f d\mu \right)
$$

$$
\left. + 2 \left(\int x_1 f d\mu \right) \left(\int x_2 f d\mu \right) \left(\int x_1 x_2 f d\mu \right) \geq 0 \right\},
$$

where we have just expressed the nonnegativity of all principal minors of $\mathbf{M}_1(f\,\mathbf{y})$.

On the other hand, observe that we do not have a hierarchy of tractable inner approximations of C_d. The simplest inner approximation is $\Sigma[\mathbf{x}]_{d/2}$ (the convex cone of SOS polynomials of degree at most d), but as proved in Blekherman (2006) the "gap" between $\Sigma[\mathbf{x}]_{d/2}$ and C_d increases unboundedly with the number of variables.

A hierarchy of outer approximations of $C_d(\mathbf{K})$ A very similar result holds for the convex cone $C_d(\mathbf{K})$ of polynomials of degree at most d, nonnegative on a closed set $\mathbf{K} \subset \mathbb{R}^n$.

If $\mathbf{K} \subset \mathbb{R}^n$ is a closed set, let μ be the Borel measure defined in (3.7) where φ is an arbitrary finite Borel measure whose support is exactly $\mathbf{K}$. Let $\mathbf{y} = (y_\alpha)$, $\alpha \in \mathbb{N}^n$, be its vector of moments, i.e.,

$$y_\alpha = \int_{\mathbf{K}} \mathbf{x}^\alpha \, d\mu(\mathbf{x}), \qquad \alpha \in \mathbb{N}^n. \tag{4.8}$$

Corollary 4.3 *Let* $\mathbf{K} \subset \mathbb{R}^n$ *be closed,* $\mathbf{y} = (y_\alpha)$ *be as in (4.8) and for every* $k \in \mathbb{N}$, *let*

$$C_d^k(\mathbf{K}) := \{ \mathbf{f} \in \mathbb{R}^{s(d)} : \mathbf{M}_k(f\,\mathbf{y}) \succeq 0 \}, \tag{4.9}$$

where $\mathbf{M}_k(f\,\mathbf{y})$ *is the localizing matrix in (2.37) associated with* $\mathbf{y}$ *and* $f \in \mathbb{R}[\mathbf{x}]_d$.

Each $C_d^k(\mathbf{K})$ *is a closed convex cone and a spectrahedron of* $\mathbb{R}^{s(d)}$. *Moreover,* $C_d(\mathbf{K}) \subset C_d^{k+1}(\mathbf{K}) \subset C_d^k(\mathbf{K})$ *for all* $k \in \mathbb{N}$, *and* $f \in C_d(\mathbf{K})$ *if and only if its vector of coefficients* $\mathbf{f} \in \mathbb{R}^{s(d)}$ *satisfies* $\mathbf{f} \in C_d^k(\mathbf{K})$, *for every* $k = 0, 1, \ldots$. *Equivalently,*

$$C_d(\mathbf{K}) = \bigcap_{k=0}^{\infty} C_d^k(\mathbf{K}). \tag{4.10}$$

The proof, which mimicks that of Corollary 4.2, is omitted. Of course, for practical computation one is restricted to sets $\mathbf{K}$ for which one may compute *effectively* moments of the measure μ. Examples of such noncompact sets $\mathbf{K}$ are $\mathbb{R}^n$ and the positive orthant; for the latter one may choose the measure μ in (3.11) for which all moments are explicitly available. For compact sets $\mathbf{K}$ let us mention balls, boxes, ellipsoids, and simplices. But again, *any* closed set $\mathbf{K}$ for which one knows how to compute all moments of some measure whose support is exactly $\mathbf{K}$, is fine.

Example 4.4 For illustrative purposes, let $n = 2$ and consider again Example 3.3 where $\mathbf{K} \subset \mathbb{R}^2$ is the simplex $\{ \mathbf{x} \geq 0 : x_1 + x_2 \leq 1 \}$. Clearly, $C_1(\mathbf{K})$, i.e., the cone of affine polynomials $\mathbf{x} \mapsto f_{00} + f_{10}x_1 + f_{01}x_2$ nonnegative on $\mathbf{K}$ is the set

$$C_1(\mathbf{K}) = \{ \mathbf{f} \in \mathbb{R}^3 : f_{00} \geq 0; \ f_{10} + f_{00} \geq 0; \ f_{01} + f_{00} \geq 0 \}.$$

Let μ be the Lebesgue measure on $\mathbf{K}$ with moments $\mathbf{y} = (y_\alpha)$ up to order 3

$$\mathbf{y} = (1/2, 1/6, 1/6, 1/12, 1/24, 1/12, 1/20, 1/60, 1/60, 1/20).$$

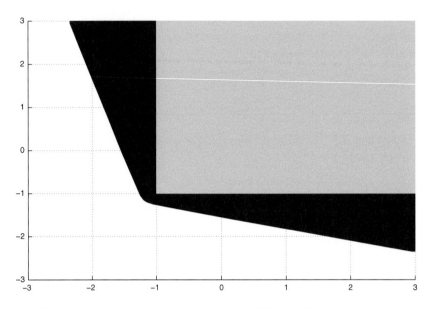

Figure 4.1 Example 4.4: outer approximation $C_1^1(\mathbf{K})$ of $C_1(\mathbf{K})$.

The first outer approximation of $C_1(\mathbf{K})$ is the convex cone $C_1^1(\mathbf{K})$ defined by the linear matrix inequalities:

$$\begin{bmatrix} 60f_0 + 20f_{10} + 20f_{01} & 20f_0 + 10f_{10} + 5f_{01} & 20f_0 + 5f_{10} + 10f_{01} \\ 20f_0 + 10f_{10} + 2f_{01} & 10f_0 + 6f_{10} + 2f_{01} & 5f_0 + 2f_{10} + 2f_{01} \\ 20f_0 + 5f_{10} + 10f_{01} & 5f_0 + 2f_{10} + 2f_{01} & 10f_0 + 2f_{10} + 6f_{01} \end{bmatrix} \succeq 0.$$

In fixing $f_{00} := 1$, the set $C_1^1(\mathbf{K})$ becomes the convex set displayed in Figure 4.1 to be compared with the set $\{ (f_{10}, f_{01}) : f_{10} \geq -1; f_{01} \geq -1 \}$. Observe that the first outer approximation $C_1^1(\mathbf{K})$ is already a relatively good and nontrivial approximation of $C_1(\mathbf{K})$.

Comparison with Stengle's Nichtnegativstellensatz The outer approximation (4.10) of the cone $C_d(\mathbf{K})$ in Corollary 4.3 holds for any closed set $\mathbf{K} \subset \mathbb{R}^n$. On the other hand, for the basic semi-algebraic set

$$\mathbf{K} = \{ \mathbf{x} \in \mathbb{R}^n : g_j(\mathbf{x}) \geq 0, \ j = 1, \dots, m \}. \tag{4.11}$$

Stengle's Nichtnegativstellensatz in Theorem 2.12(a) states that $f \in \mathbb{R}[\mathbf{x}]$ is nonnegative on $\mathbf{K}$ if and only if

$$p f = f^{2s} + q, \tag{4.12}$$

for some integer s and polynomials $p, q \in P(g)$, where $P(g)$ is the *pre-ordering* associated with the g_j; see Definition 2.10. In addition, there exist bounds on the integer s and the degree of the SOS weights in the definition of $p, q \in P(g)$, so that in principle, when f is *known*, checking whether $f \geq 0$ on **K** reduces to solving a single SDP to compute h, p in the nonnegativity certificate (4.12). However, the size of this SDP is potentially huge and makes it unpractical. Moreover, the representation of $C_d(\mathbf{K})$ in (4.12) is not convex in the vector of coefficients of f because it involves the power f^{2s} as well as the product pf.

Remark 4.5   If in Corollary 4.3 one replaces the finite-dimensional convex cone $C_d(\mathbf{K}) \subset \mathbb{R}[\mathbf{x}]_d$ with the infinite-dimensional convex cone $C(\mathbf{K}) \subset \mathbb{R}[\mathbf{x}]$ of *all* polynomials nonnegative on **K**, and $C_d^k(\mathbf{K}) \subset \mathbb{R}[\mathbf{x}]_d$ with $C^k(\mathbf{K}) = \{\mathbf{f} \in \mathbb{R}^{s(2k)} : \mathbf{M}_k(f\,\mathbf{y}) \succeq 0\}$, then the nested sequence of (increasing but finite-dimensional) convex cones $C^k(\mathbf{K})$, $k \in \mathbb{N}$, provides finite-dimensional approximations of $C(\mathbf{K})$.

4.4 Approximations of the dual cone

The dual cone of $C_d(\mathbf{K})$ is the convex cone $C_d(\mathbf{K})^* \subset \mathbb{R}[\mathbf{x}]_d^*$ defined by:

$$C_d(\mathbf{K})^* = \{\mathbf{z} \in \mathbb{R}^{s(d)} : \langle \mathbf{z}, \mathbf{f} \rangle \geq 0, \quad \forall f \in C_d(\mathbf{K})\},$$

where again $f \in \mathbb{R}[\mathbf{x}]_d$ is associated with its vector of coefficients $\mathbf{f} \in \mathbb{R}^{s(d)}$.

Lemma 4.6   *If* $\mathbf{K} \subset \mathbb{R}^n$ *has nonempty interior then the interior of* $C_d(\mathbf{K})^*$ *is nonempty.*

Proof   Since $C_d(\mathbf{K})$ is nonempty and closed then, by Faraut and Korányi (1994, Proposition I.1.4, p. 3)

$$\operatorname{int}(C_d(\mathbf{K})^*) = \{\mathbf{y} : \langle \mathbf{y}, \mathbf{g} \rangle > 0, \quad \forall g \in C_d(\mathbf{K}) \setminus \{0\}\},$$

where $\mathbf{g} \in \mathbb{R}^{s(d)}$ is the coefficient vector of $g \in C_d(\mathbf{K})$, and

$$\operatorname{int}(C_d(\mathbf{K})^*) \neq \emptyset \iff C_d(\mathbf{K}) \cap (-C_d(\mathbf{K})) = \{0\}.$$

But $g \in C_d(\mathbf{K}) \cap (-C_d(\mathbf{K}))$ implies $g \geq 0$ and $g \leq 0$ on **K**, which in turn implies $g = 0$ because **K** has nonempty interior. $\qquad\square$

We next provide a characterization of $C_d(\mathbf{K})^*$ when **K** is compact. Recall that $\mathcal{M}(\mathbf{K})$ denotes the space of finite signed Borel measures on **K** and $\mathcal{M}(\mathbf{K})_+ \subset \mathcal{M}(\mathbf{K})$ its subset of finite (nonnegative) Borel measures on **K**.

Lemma 4.7   *Let* $\mathbf{K} \subset \mathbb{R}^n$ *be compact. For every* $d \in \mathbb{N}$:

$$C_d(\mathbf{K})^* = \left\{ \left(\int_{\mathbf{K}} \mathbf{x}^\alpha \, d\phi \right), \; \alpha \in \mathbb{N}_d^n \; : \; \phi \in \mathcal{M}(\mathbf{K})_+ \right\}, \qquad (4.13)$$

i.e., the dual cone of $C_d(\mathbf{K})$ *is the convex cone of vectors of* $\mathbb{R}^{s(d)}$ *which have a finite representing measure with support contained in* **K**.

Proof   Let Δ_d be the convex cone in the right-handside of (4.13). For all $\mathbf{y} = (y_\alpha) \in \Delta_d$ and $f \in C_d(\mathbf{K})$,

$$\langle \mathbf{y}, \mathbf{f} \rangle = \sum_{\alpha \in \mathbb{N}_d^n} f_\alpha \, y_\alpha = \sum_{\alpha \in \mathbb{N}_d^n} \int_{\mathbf{K}} f_\alpha \, \mathbf{x}^\alpha \, d\phi = \int_{\mathbf{K}} f \, d\phi \geq 0,$$

so that necessarily $\Delta_d \subseteq C_d(\mathbf{K})^*$ and similarly, $C_d(\mathbf{K}) \subseteq \Delta_d^*$. Next,

$$\Delta_d^* = \left\{ \mathbf{f} \in \mathbb{R}^{s(d)} \; : \; \langle \mathbf{f}, \mathbf{y} \rangle \geq 0, \quad \forall \mathbf{y} \in \Delta_d \right\}$$

$$= \left\{ f \in \mathbb{R}[\mathbf{x}]_d \; : \; \int_{\mathbf{K}} f \, d\phi \geq 0, \quad \forall \phi \in \mathcal{M}(\mathbf{K})_+ \right\}$$

$$\Rightarrow \Delta_d^* \subseteq C_d(\mathbf{K}),$$

and so $\Delta_d^* = C_d(\mathbf{K})$. Hence the result follows if one proves that Δ_d is closed, because then $C_d(\mathbf{K})^* = (\Delta_d^*)^* = \Delta_d$, the desired result. So let $(\mathbf{y}^k) \subset \Delta_d, k \in \mathbb{N}$, be such that $\mathbf{y}^k \to \mathbf{y}$ as $k \to \infty$. Equivalently, $\int_{\mathbf{K}} \mathbf{x}^\alpha \, d\phi_k \to y_\alpha$ for all $\alpha \in \mathbb{N}_d^n$. In particular, the convergence $y_0^k \to y_0$ implies that the sequence of measures (ϕ_k), $k \in \mathbb{N}$, is bounded, that is, $\sup_k \phi_k(\mathbf{K}) < M$ for some $M > 0$. As **K** is compact, the unit ball of $\mathcal{M}(\mathbf{K})$ is sequentially compact in the weak $\star$ topology[2] $\sigma(\mathcal{M}(\mathbf{K}), C(\mathbf{K}))$ where $C(\mathbf{K})$ is the space of continuous functions on **K**.

Hence there is a finite Borel measure $\phi \in \mathcal{M}(\mathbf{K})_+$ and a subsequence (k_i) such that $\int_{\mathbf{K}} g \, d\phi_{k_i} \to \int_{\mathbf{K}} g \, d\phi$ as $i \to \infty$, for all $g \in C(\mathbf{K})$. In particular, for every $\alpha \in \mathbb{N}_d^n$,

$$y_\alpha = \lim_{k \to \infty} y_\alpha^k = \lim_{i \to \infty} y_\alpha^{k_i} = \lim_{i \to \infty} \int_{\mathbf{K}} \mathbf{x}^\alpha \, d\phi_{k_i} = \int_{\mathbf{K}} \mathbf{x}^\alpha \, d\phi,$$

which shows that $\mathbf{y} \in \Delta_d$, and so Δ_d is closed. $\square$

Of course, since $(C_d^k(\mathbf{K}))$, $k \in \mathbb{N}$, is a hierarchy of outer approximations of $C_d(\mathbf{K})$ then $(C_d^k(\mathbf{K})^*)$, $k \in \mathbb{N}$, provides a hierarchy of inner approximations of

---

[2] In the weak $\star$ topology, a bounded sequence $(\mu_k) \subset \mathcal{M}(\mathbf{K}), k \in \mathbb{N}$, converges to $\mu \in \mathcal{M}(\mathbf{K})$ if and only if $\int_{\mathbf{K}} h \, d\mu_n \to \int_{\mathbf{K}} h \, d\mu$ for all $h \in C(\mathbf{K})$.

the dual cone $C_d(\mathbf{K})^*$. More precisely, one has the following dual version of Corollary 4.3.

Corollary 4.8 *Let $\mathbf{K} \subset \mathbb{R}^n$ be closed and let $\mathbf{y} = (y_\alpha)$, $\alpha \in \mathbb{N}^n$, be as in (4.8) (where μ is defined in (3.7)). Then for every $k \in \mathbb{N}$:*

$$C_d^k(\mathbf{K})^* = \overline{\left\{ \langle \mathbf{X}, \mathbf{M}_k(\mathbf{x}^\alpha\, \mathbf{y}) \rangle, \ \alpha \in \mathbb{N}_d^n : \ \ \mathbf{X} \succeq 0 \right\}}$$

$$= \overline{\left\{ \left(\underbrace{\int_{\mathbf{K}} \mathbf{x}^\alpha\, \sigma\, d\mu}_{d\mu_\sigma} \right), \ \alpha \in \mathbb{N}_d^n : \sigma \in \Sigma[\mathbf{x}]_k \right\}}, \qquad (4.14)$$

and $C_d(\mathbf{K})^* = \overline{\displaystyle\bigcup_{k=0}^{\infty} C_d^k(\mathbf{K})^*}.$

Proof Let $\Delta_k = \{ \langle \mathbf{X}, \mathbf{M}_d(\mathbf{x}^\alpha\, \mathbf{y}) \rangle, \ \alpha \in \mathbb{N}_d^n : \mathbf{X} \succeq 0 \}$. Then

$$\mathbf{f} \in \Delta_k^* \Leftrightarrow \sum_{\alpha \in \mathbb{N}_d^n} f_\alpha \, \langle \mathbf{X}, \mathbf{M}_d(\mathbf{x}^\alpha\, \mathbf{y}) \rangle = \langle \mathbf{X}, \mathbf{M}_d(f\, \mathbf{y}) \rangle \geq 0, \quad \forall \mathbf{X} \Leftrightarrow \mathbf{M}_k(f\, \mathbf{y}) \succeq 0,$$

and so $\Delta_k^* = C_d^k(\mathbf{K})$. Therefore $C_d^k(\mathbf{K})^* = (\Delta_k^*)^* = \overline{\Delta_k}$, the desired result. The last statement follows from

$$C_d(\mathbf{K})^* = \left(\bigcap_{k=0}^{\infty} C_d^k(\mathbf{K}) \right)^* = \overline{\mathrm{conv}\left(\bigcup_{k=0}^{\infty} C_d^k(\mathbf{K})^* \right)} = \overline{\bigcup_{k=0}^{\infty} C_d^k(\mathbf{K})^*},$$

where we have used Rockafellar (1970, §16). □

From (4.14), the convex cone $C_d^k(\mathbf{K})^*$ has a simple interpretation. It consists of all vectors of moments up to order d, generated by the finite Borel measures $\{\mu_\sigma \,(= \sigma d\mu) : \sigma \in \Sigma[\mathbf{x}]_k\}$ absolutely continuous with respect to μ and whose density is an SOS polynomial σ of degree at most $2k$.

4.5 The cone of copositive matrices and its dual

In addition to being of independent interest, the convex cone $\mathcal{C}$ of *copositive* matrices and its dual cone $\mathcal{C}^*$ of *completely positive* matrices have attracted a lot of attention, in part because several interesting NP-hard optimization problems can be modeled as convex conic optimization problems over these cones.

Let $\mathcal{S}$ denote the cone of real symmetric $n \times n$ matrices. With $\mathbf{A} = (a_{ij}) \in \mathcal{S}$, we denote by $f_{\mathbf{A}} \in \mathbb{R}[\mathbf{x}]_2$ the homogeneous polynomial associated with the quadratic form $\mathbf{x} \mapsto \mathbf{x}^T \mathbf{A} \mathbf{x}$, and let μ be the exponential probability measure supported on $\mathbb{R}^n_+$, and with moments $\mathbf{y} = (y_\alpha)$, $\alpha \in \mathbb{N}^n$ given by:

$$y_\alpha = \int_{\mathbb{R}^n_+} \mathbf{x}^\alpha \exp\left(-\sum_{i=1}^n x_i\right) d\mu(\mathbf{x}) = \prod_{i=1}^n \alpha_i!, \qquad \forall \alpha \in \mathbb{N}^n. \quad (4.15)$$

Recall that a matrix $\mathbf{A} \in \mathcal{S}$ is *copositive* if $f_{\mathbf{A}}(\mathbf{x}) \geq 0$ for all $\mathbf{x} \in \mathbb{R}^n_+$, and denote by $\mathcal{C} \subset \mathcal{S}$ the cone of copositive matrices, i.e.,

$$\mathcal{C} := \{\, \mathbf{A} \in \mathcal{S} \,:\, f_{\mathbf{A}}(\mathbf{x}) \geq 0, \quad \forall \mathbf{x} \geq 0 \,\}. \quad (4.16)$$

Notice that $\mathcal{C}$ is also the convex cone of quadratic forms that are nonnegative on $\mathbf{K} = \mathbb{R}^n_+$. Its dual cone is the closed convex cone of *completely positive* matrices, i.e., matrices of $\mathcal{S}$ that can be written as the sum of finitely many rank-one matrices $\mathbf{x}\mathbf{x}^T$, with $\mathbf{x} \in \mathbb{R}^n_+$,

$$\mathcal{C}^* = \operatorname{conv}\{\, \mathbf{x}\mathbf{x}^T \,:\, \mathbf{x} \in \mathbb{R}^n_+ \,\}. \quad (4.17)$$

4.5.1 Inner approximations of $\mathcal{C}$

Let $\mathcal{S}_+$ be the convex cone of real $n \times n$ positive semidefinite matrices and let $\mathcal{N}$ be the convex set of real symmetric matrices all of whose entries are nonnegative. The elements of $\mathcal{S}_+ \cap \mathcal{N}$ are called *doubly nonnegative* matrices, and $(\mathcal{S}_+ \cap \mathcal{N})^* = \mathcal{N}^* + \mathcal{S}_+^* = \mathcal{N} + \mathcal{S}_+$. Moreover,

$$\mathcal{C}^* \subset \mathcal{N} \cap \mathcal{S}_+ \subset \mathcal{S}_+ \subset \mathcal{N} + \mathcal{S}_+ \subset \mathcal{C},$$

and so $\mathcal{N} + \mathcal{S}_+$ provides a simple inner approximation of $\mathcal{C}$. In addition, it turns out that for $n \leq 4$, one has $\mathcal{C}^* = \mathcal{N} \cap \mathcal{S}_+$ and $\mathcal{C} = \mathcal{N} + \mathcal{S}_+$. For $n = 5$ a famous counterexample is provided by the so-called Horn matrix

$$\mathbf{H} = \begin{bmatrix} 1 & -1 & 1 & 1 & -1 \\ -1 & 1 & -1 & 1 & 1 \\ 1 & -1 & 1 & -1 & 1 \\ 1 & 1 & -1 & 1 & -1 \\ -1 & 1 & 1 & -1 & 1 \end{bmatrix}$$

which is copositive but is not a member of $\mathcal{N} + \mathcal{S}_+$.

A hierarchy of inner approximations Let $\mathcal{N} + \mathcal{S}_+ =: \mathcal{K}_0 \subset \mathcal{C}$ and for $d \geq 1$ and $S = (S_{ij}) \in \mathcal{S}$, let

$$\mathbf{x} \mapsto p_d^S(\mathbf{x}) := \left(\sum_{i=1}^n x_i^2\right)^d \sum_{i,j} x_i^2 x_j^2 S_{ij}.$$

Then one may define the following two hierarchies of convex cones:

$$\mathcal{K}_d^1 = \{S \in \mathcal{S} \: : \: p_d^S \text{ has nonnegative coefficients}\} \qquad (4.18)$$

$$\mathcal{K}_d^2 = \{S \in \mathcal{S} \: : \: p_d^S \text{ is SOS}\}. \qquad (4.19)$$

For each d, the convex cone $\mathcal{K}_d^1$ (respectively $\mathcal{K}_d^2$) can be described by linear inequalities (respectively linear matrix inequalities) on the entries of S. Moreover, we have the following result.

Proposition 4.9   *Let $\mathcal{K}_d^i, \; i = 1, 2,$ be as in (4.18)–(4.19). Then for $i = 1, 2,$*

$$\mathcal{K}_d^i \subset \mathcal{K}_{d+1}^i \subset \mathcal{C}, \quad \forall d \in \mathbb{N}; \quad \mathcal{C}^* \subset (\mathcal{K}_{d+1}^i)^* \subset (\mathcal{K}_d^i)^*, \quad \forall d \in \mathbb{N},$$

and

$$\overline{\left(\bigcup_{d=0}^{\infty} \mathcal{K}_d^i \right)} = \mathcal{C} \quad and \quad \bigcap_{d=0}^{\infty} \left(\mathcal{K}_d^i \right)^* = \mathcal{C}^*. \qquad (4.20)$$

But of course, the description of the cones $\mathcal{K}_d^i$ (and their duals) is not polynomially bounded as d increases.

4.5.2 Outer approximations of $\mathcal{C}$

We next provide a hierarchy of outer approximations of $\mathcal{C}$ which complements the hierarchy of inner approximations $(\mathcal{K}_d^i)$, $i = 1, 2$ and $d = 0, 1, \ldots$, obtained in Proposition 4.9. Similarly, by duality, we also obtain a hierarchy of inner approximations of $\mathcal{C}^*$ which complements the hierarchy of outer approximations $(\mathcal{K}_d^i)^*$ obtained in (4.20).

Theorem 4.10   *Let $\mathbf{y}$ be as in (4.15) and let $\mathcal{C}_d \subset \mathcal{S}$ be the closed convex cone*

$$\mathcal{C}_d := \{\, \mathbf{A} \in \mathcal{S} \: : \: \mathbf{M}_d(f_{\mathbf{A}} \, \mathbf{y}) \succeq 0 \,\}, \qquad d = 0, 1, \ldots \qquad (4.21)$$

where $\mathbf{M}_d(f_{\mathbf{A}} \, \mathbf{y})$ is the localizing matrix associated with $f_{\mathbf{A}}$ and $\mathbf{y}$, defined in (2.37).

Then $\mathcal{C}_{d+1} \subset \mathcal{C}_d$ for all $d \in \mathbb{N}$ and $\mathcal{C} = \bigcap_{d=0}^{\infty} \mathcal{C}_d.$

The proof is a direct consequence of Theorem 3.4 with $\mathbf{K} = \mathbb{R}_+^n$ and $f = f_{\mathbf{A}}$. Since $f_{\mathbf{A}}$ is homogeneous, alternatively one may use the probability measure ν uniformly supported on the n-dimensional simplex $\mathbf{K} = \{\mathbf{x} \in \mathbb{R}_+^n : \sum_i x_i \leq 1\}$ and invoke Theorem 3.2.

Observe that in view of the definition (2.37) of the localizing matrix, the entries of the matrix $\mathbf{M}_d(f_{\mathbf{A}} \, \mathbf{y})$ are linear in $\mathbf{A}$. Therefore, each convex cone $\mathcal{C}_d \subset \mathcal{S}$ is a spectrahedron in $\mathbb{R}^{n(n+1)/2}$ defined solely in terms of the entries (a_{ij}) of $\mathbf{A} \in \mathcal{S}$, and the hierarchy of spectrahedra $(\mathcal{C}_d)$, $d \in \mathbb{N}$, provides a nested sequence of outer approximations of $\mathcal{C}$. It is a special case of Corollary 4.3 and $\mathcal{C} \subset \mathcal{C}_2(\mathbb{R}_+^n)$ is the convex cone of quadratic forms that are nonnegative on $\mathbf{K} = \mathbb{R}_+^n$.

Also, the set-membership problem "$\mathbf{A} \in \mathcal{C}_d$," i.e., testing whether a given matrix $\mathbf{A} \in \mathcal{S}$ belongs to $\mathcal{C}_d$, is an eigenvalue problem as one has to check whether the smallest eigenvalue of $\mathbf{M}_d(f_{\mathbf{A}}, \mathbf{y})$ is nonnegative. Therefore, instead of using standard software packages for linear matrix inequalities, one may use powerful specialized softwares for eigenvalues.

We next describe an inner approximation of the convex cone $\mathcal{C}^*$ via the hierarchy of convex cones $(\mathcal{C}_d^*)$, $d \in \mathbb{N}$, where each $\mathcal{C}_d^*$ is the dual cone of $\mathcal{C}_d$ in Theorem 4.10.

Recall that $\Sigma[\mathbf{x}]_d$ is the space of polynomials that are sums of squares of polynomials of degree at most d. A matrix $\mathbf{A} \in \mathcal{S}$ is also identified with a vector $\mathbf{a} \in \mathbb{R}^{n(n+1)/2}$ in the obvious way, and conversely, with any vector $\mathbf{a} \in \mathbb{R}^{n(n+1)/2}$ is associated a matrix $\mathbf{A} \in \mathcal{S}$. For instance, with $n = 2$,

$$\mathbf{A} = \begin{bmatrix} a & b \\ b & c \end{bmatrix} \quad \leftrightarrow \quad \mathbf{a} = \begin{bmatrix} a \\ 2b \\ c \end{bmatrix}. \tag{4.22}$$

So we will not distinguish between a convex cone in $\mathbb{R}^{n(n+1)/2}$ and the corresponding cone in $\mathcal{S}$. For a nonnegative polynomial $\sigma \in \mathbb{R}[\mathbf{x}]$, denote by μ_σ the finite Borel measure such that $\mu_\sigma(B) = \int_B \sigma \, d\mu$ for all Borel sets B.

Theorem 4.11 *Let $\mathcal{C}_d \subset \mathcal{S}$ be the convex cone defined in (4.21). Then*

$$\mathcal{C}_d^* = \mathrm{cl} \left\{ \left(\langle \mathbf{X}, \mathbf{M}_d(x_i x_j \, \mathbf{y}) \rangle \right)_{1 \leq i \leq j \leq n} : \mathbf{X} \in \mathcal{S}_+^{s(d)} \right\}. \tag{4.23}$$

Equivalently:

$$\mathcal{C}_d^* = \mathrm{cl} \left(\int_{\mathbb{R}_+^n} \mathbf{x}\mathbf{x}^T \underbrace{\sigma(\mathbf{x}) \, d\mu(\mathbf{x})}_{d\mu_\sigma(\mathbf{x})} : \sigma \in \Sigma[\mathbf{x}]_d \right)$$

$$= \mathrm{cl} \left\{ \int_{\mathbb{R}_+^n} \mathbf{x}\mathbf{x}^T \, d\mu_\sigma(\mathbf{x}) : \sigma \in \Sigma[\mathbf{x}]_d \right\} \tag{4.24}$$

(to compare with $\mathcal{C}^* = \mathrm{conv} \{ \mathbf{x}\mathbf{x}^T : \mathbf{x} \in \mathbb{R}_+^n \}$).

Proof   Let

$$\Delta_d := \left\{ \left(\langle \mathbf{X}, \mathbf{M}_d(x_i x_j\, \mathbf{y}) \rangle \right)_{1 \le i \le j \le n} : \mathbf{X} \in \mathcal{S}_+^{s(d)} \right\},$$

so that

$$\Delta_d^* = \left\{ \mathbf{a} \in \mathbb{R}^{n(n+1)/2} : \sum_{1 \le i \le j \le n} a_{ij} \langle \mathbf{X}, \mathbf{M}_d(x_i x_j\, \mathbf{y}) \ge 0, \quad \forall \mathbf{X} \in \mathcal{S}_+^{s(d)} \right\},$$

$$= \left\{ \mathbf{a} \in \mathbb{R}^{n(n+1)/2} : \left\langle \mathbf{X}, \mathbf{M}_d\left(\left(\sum_{1 \le i \le j \le n} a_{ij} x_i x_j \right) \mathbf{y} \right) \right\rangle \ge 0, \quad \forall \mathbf{X} \in \mathcal{S}_+^{s(d)} \right\},$$

$$= \{ \mathbf{A} \in \mathcal{S} : \langle \mathbf{X}, \mathbf{M}_d(f_{\mathbf{A}}\, \mathbf{y}) \rangle \ge 0, \quad \forall \mathbf{X} \in \mathcal{S}_+^{s(d)} \} \quad \text{[with } \mathbf{A}, \mathbf{a} \text{ as in (4.22)]}$$

$$= \{ \mathbf{A} \in \mathcal{S} : \mathbf{M}_d(f_{\mathbf{A}}\, \mathbf{y}) \succeq 0 \} = \mathcal{C}_d.$$

And so we obtain the desired result $\mathcal{C}_d^* = (\Delta_d^*)^* = \overline{\Delta_d}$. Next, writing the singular decomposition of $\mathbf{X}$ as $\sum_{\ell=0}^{s} \mathbf{q}_\ell \mathbf{q}_\ell^T$ for some $s \in \mathbb{N}$ and some vectors $(\mathbf{q}_\ell) \subset \mathbb{R}^{s(d)}$, one obtains that for every $1 \le i \le j \le n$,

$$\langle \mathbf{X}, \mathbf{M}_d(x_i x_j\, \mathbf{y}) \rangle = \sum_{\ell=0}^{s} \langle \mathbf{q}_\ell \mathbf{q}_\ell^T, \mathbf{M}_d(x_i x_j\, \mathbf{y}) \rangle = \sum_{\ell=0}^{s} \langle \mathbf{q}_\ell, \mathbf{M}_d(x_i x_j\, \mathbf{y}) \mathbf{q}_\ell \rangle$$

$$= \sum_{\ell=0}^{s} \int x_i x_j\, q_\ell(\mathbf{x})^2\, d\mu(\mathbf{x}) \quad \text{[by (2.38)]}$$

$$= \int x_i x_j\, \underbrace{\sigma(\mathbf{x})\, d\mu(\mathbf{x})}_{d\mu_\sigma(\mathbf{x})}.$$

$\square$

So Theorem 4.11 states that $\mathcal{C}_d^*$ is the closure of the convex cone generated by the matrix of second-order moments of measures $d\mu_\sigma = \sigma d\mu$, absolutely continuous with respect to μ (hence with support on $\mathbb{R}_+^n$) and with density being an SOS polynomial σ of degree at most $2d$. Of course we immediately have the following.

Corollary 4.12   *Let* $\mathcal{C}_d^*$, $d \in \mathbb{N}$, *be as in (4.24). Then*

$$\mathcal{C}_d^* \subset \mathcal{C}_{d+1}^*, \quad \forall d \in \mathbb{N} \quad and \quad \mathcal{C}^* = \overline{\left(\bigcup_{d=0}^{\infty} \mathcal{C}_d^* \right)}.$$

Proof   As $C_d^* \subset C_{d+1}^*$ for all $d \in \mathbb{N}$, the result follows from

$$C^* = \left(\bigcap_{d=0}^{\infty} C_d \right)^* = \overline{\mathrm{conv} \left(\bigcup_{d=0}^{\infty} C_d^{**} \right)} = \overline{\left(\bigcup_{d=0}^{\infty} C_d^* \right)},$$

see for example Rockafellar (1970, §16).   □

In other words, C_d^* approximates $C^* = \mathrm{conv} \{\mathbf{x}\mathbf{x}^T : \mathbf{x} \in \mathbb{R}_+^n\}$ (i.e., the convex hull of second-order moments of Dirac measures with support in $\mathbb{R}_+^n$) from inside by second-order moments of measures $\mu_\sigma \ll \mu$ whose density is an SOS polynomial σ of degree at most $2d$, and better and better approximations are obtained by letting d increase.

In fact Theorem 4.11 and Corollary 4.12 are a specialization of Corollary 4.8 to the convex cone of quadratic forms nonnegative on $\mathbf{K} = \mathbb{R}_+^n$.

Example 4.13   For instance, with $n = 2$, and $\mathbf{A} = \begin{bmatrix} a & b \\ b & c \end{bmatrix}$, it is known that

$$\mathbf{A} \text{ is copositive if and only if } a, c \geq 0 \text{ and } b + \sqrt{ac} \geq 0.$$

With $f_\mathbf{A}(\mathbf{x}) := ax_1^2 + 2bx_1x_2 + cx_2^2$ and $d = 1$, the condition $\mathbf{M}_d(f_\mathbf{A}\, \mathbf{y}) \succeq 0$ which reads

$$\begin{bmatrix} a+b+c & 3a+2b+c & a+2b+3c \\ 3a+2b+c & 12a+6b+2c & 3a+4b+3c \\ a+2b+3c & 3a+4b+3c & 2a+6b+12c \end{bmatrix} \succeq 0,$$

defines the convex cone $C_1 \subset \mathbb{R}^3$. It is a connected component of the basic semi-algebraic set $\{(a, b, c) : \det(\mathbf{M}_1(f_\mathbf{A}\, \mathbf{y})) \geq 0\}$, that is, elements (a, b, c) such that:

$$3a^3 + 15a^2b + 29a^2c + 16ab^2 + 50abc + 29ac^2 + 4b^3 + 16b^2c + 15bc^2 + 3c^3 \geq 0.$$

Figure 4.2 displays the projection on the (a, b)-plane of the sets C_1 and C intersected with the unit ball.

4.6 Exercises

Exercise 4.1   Let $\mathbf{K} \subset \mathbb{R}$ be the interval $[0, 1]$ and let μ be the probability measure uniformly distributed on $\mathbf{K}$ with associated sequence of moments $\mathbf{y} = (y_k)$, $k \in \mathbb{N}$. Let $C_1(\mathbf{K}) \subset \mathbb{R}[x]_1$ be the convex cone of polynomials of the form $x \mapsto f_0 + f_1x$, $(f_0, f_1) \in \mathbb{R}^2$, that are nonnegative on $\mathbf{K}$.

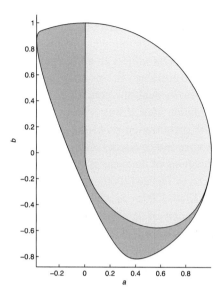

Figure 4.2 $n = 2$. Projection on the (a, b)-plane of $\mathcal{C}$ versus $\mathcal{C}_1$, both intersected with the unit ball.

(a) Characterize $C_1(\mathbf{K})$ in terms of conditions on $\mathbf{f} = (f_0, f_1) \in \mathbb{R}^2$.

(b) Characterize $C_1^1(\mathbf{K}) = \{\mathbf{f} \in \mathbb{R}^2 : \mathbf{M}_1(f \, \mathbf{y}) \succeq 0\}$ and compare (graphically) $C_1^1(\mathbf{K})$ with $C_1(\mathbf{K})$.

(c) Same thing as in (b) but now with $C_1^2(\mathbf{K}) = \{\mathbf{f} \in \mathbb{R}^2 : \mathbf{M}_2(f \, \mathbf{y}) \succeq 0\}$.

(d) Retrieve the results of Exercise 3.1.

Exercise 4.2 Let $\mathbf{K}$ be the whole space $\mathbb{R}^2$, and let μ be the measure with density $\mathbf{x} \mapsto \exp(-\|\mathbf{x}\|^2/2)$ with respect to the Lebesgue measure and with associated sequence of moments $\mathbf{y} = (y_\alpha)$, $\alpha \in \mathbb{N}^2$. Let $\Theta \subset \mathbb{R}[\mathbf{x}]_6$ be the set of polynomials of the form $\mathbf{x} \mapsto f(\mathbf{x}) = x_1^2 x_2^2 \, (x_1^2 + x_2^2 - f_{22}) - f_0$, for some vector $\mathbf{f} = (f_0, f_{22}) \in \mathbb{R}^2$.

(a) Show that Θ is a convex set and characterize those $\mathbf{f} = (f_0, f_{22})$ such that $\mathbf{M}_1(f \, \mathbf{y}) \succeq 0$ and $\mathbf{M}_2(f \, \mathbf{y}) \succeq 0$.

(b) Retrieve the results of Exercise 3.2.

4.7 Notes and sources

4.2 and 4.3 The results are essentially deduced from Chapter 2, Chapter 3 and Lasserre (2011).

4.5 *Copositive programming* (a term that was coined in Bomze et al. (2000)) is a special case of conic optimization, which consists of minimizing a linear function over a (convex) cone subject to additional (inhomogeneous) linear (inequality or equality) constraints. In this case the convex cone is the cone of copositive matrices or its dual, the cone of completely positive matrices. It attracted special attention after it was proved in Burer (2009) that several important NP-hard 0/1 optimization problems can be modeled as copositive programs; see also Bomze and de Klerk (2002).

The inner approximations of the cone of copositive matrices $\mathcal{C}$ in Proposition 4.9 were introduced by Parrilo (2000), and then by de Klerk and Pasechnik (2002), Bomze and de Klerk (2002), as well as Pena et al. (2007). For example, the first cone in the hierarchy of de Klerk and Pasechnik (2002) is $\mathcal{N}$, and the first cone is $\mathcal{N} + \mathcal{S}_+$ in the hierarchy of Parrilo (2000), whereas the hierarchy of Pena et al. (2007) is sandwiched between them. For a survey on copositive programming the interested reader is referred to, for example, Dürr (2010), Burer (2012), Bomze (2012) and Bomze et al. (2012). Proposition 4.9 is taken from Bomze (2012). Theorem 4.10 and its dual version 4.11 are consequences of Lasserre (2011) and are detailed in Lasserre (2014b). Let us also mention Gaddum's characterization of copositive matrices described in Gaddum (1958) for which the set-membership problem "$\mathbf{A} \in \mathcal{C}$" (with $\mathbf{A}$ known) reduces to solving a linear programming problem (whose size is not polynomially bounded in the input size of the matrix $\mathbf{A}$). However, Gaddum's characterization is *not* convex in the coefficients of the matrix $\mathbf{A}$, and so cannot be used to provide a hierarchy of outer approximations of $\mathcal{C}$. On the other hand, de Klerk and Pasechnik (2007) used Gaddum's characterization to help solve quadratic optimization problems on the simplex via solving a hierarchy of LP-relaxations of increasing size and with *finite* convergence of the process.

PART II

Polynomial and semi-algebraic optimization

5

The primal and dual points of view

Consider the following polynomial optimization problem:

$$f^* := \inf \{ f(\mathbf{x}) : \mathbf{x} \in \mathbf{K} \} \qquad (5.1)$$

with feasible set $\mathbf{K} \subset \mathbb{R}^n$ defined by

$$\mathbf{K} = \{ \mathbf{x} \in \mathbb{R}^n : g_j(\mathbf{x}) \geq 0, \ j = 1, \ldots, m \}, \qquad (5.2)$$

and where $f, g_j \in \mathbb{R}[\mathbf{x}]$ are real-valued polynomials, $j = 1, \ldots, m$.

In the real algebraic geometry terminology, such a set $\mathbf{K}$ defined by finitely many polynomial inequalities is called a basic closed semi-algebraic set.

Whenever $\mathbf{K} \neq \mathbb{R}^n$ and unless otherwise stated, we will assume that the set $\mathbf{K}$ is compact but we do not assume that $\mathbf{K}$ is convex or even connected. This is a rather rich modeling framework that includes linear, quadratic, 0/1, and mixed 0/1 optimization problems as special cases. In particular, constraints of the type $x_i \in \{0, 1\}$ can be written as $x_i^2 - x_i \geq 0$ and $x_i - x_i^2 \geq 0$, or as the single equality constraint $x_i^2 - x_i = 0$.

When $n = 1$, we have seen in Chapter 2 that a univariate polynomial nonnegative on $\mathbf{K} = \mathbb{R}$ is a sum of squares and that a univariate polynomial nonnegative on an interval $\mathbf{K} = (-\infty, b]$, $\mathbf{K} = [a, b]$ or $\mathbf{K} = [a, \infty)$, can be written in a specific form involving sums of squares whose degree is known. We will see that this naturally leads to reformulating problem (5.1) as a single semidefinite optimization problem for which efficient algorithms and software packages are available. Interestingly, this nonconvex problem can be reformulated as a tractable convex problem and underscores the importance of the representation theorems from Chapter 2.

On the other hand, the multivariate case differs radically from the univariate case because not every polynomial nonnegative on $\mathbf{K} = \mathbb{R}^n$ can be written as a sum of squares of polynomials. Moreover, problem (5.1) (with $\mathbf{K} = \mathbb{R}^n$) involving a polynomial f of degree greater than or equal to four on n variables is NP-hard.

Even though problem (5.1) is nonlinear, it can always be rephrased as a linear program (hence convex!). Of course there is a price to pay; namely this linear program is *infinite dimensional* and cannot be solved by usual LP software packages. It can also be restated as a finite-dimensional convex optimization problem (however, NP-hard in general!).

5.1 Polynomial optimization as an infinite-dimensional LP

To see how to transform (5.1) into a linear program, let $\mathcal{M}(\mathbf{K})$ be the vector space of finite signed Borel measures[1] on $\mathbf{K}$ and let $\mathcal{M}(\mathbf{K})_+ \subset \mathcal{M}(\mathbf{K})$ be the convex cone of nonnegative finite Borel measures on $\mathbf{K}$. Then:

$$f^* = \inf_{\mathbf{x}} \{ f(\mathbf{x}) \, : \, \mathbf{x} \in \mathbf{K} \} = \inf_{\mu \in \mathcal{M}(\mathbf{K})_+} \left\{ \int_{\mathbf{K}} f \, d\mu \, : \, \mu(\mathbf{K}) = 1 \right\}. \quad (5.3)$$

Indeed let ρ be the infimum in the right-hand side of (5.3). With every $\mathbf{x} \in \mathbf{K}$ one may associate the Borel probability measure $\mu := \delta_{\mathbf{x}} \in \mathcal{M}(\mathbf{K})_+$ which is the Dirac measure at the point $\mathbf{x} \in \mathbf{K}$. Then $\int_{\mathbf{K}} f \, d\mu = \int_{\mathbf{K}} f \, d\delta_{\mathbf{x}} = f(\mathbf{x})$ and therefore $\rho \leq f^*$. Conversely, if $f^* > -\infty$ and as $f(\mathbf{x}) \geq f^*$ for every $\mathbf{x} \in \mathbf{K}$, $\int_{\mathbf{K}} f \, d\mu \geq \int_{\mathbf{K}} f^* d\mu = f^*$ for every probability measure $\mu \in \mathcal{M}(\mathbf{K})_+$, i.e., $\rho \geq f^*$, and so $\rho = f^*$.

To write (5.3) as a linear program, and to avoid technicalities, assume that the basic semi-algebraic set $\mathbf{K}$ in (5.2) is compact. Let $\tilde{\mathcal{C}}(\mathbf{K})$ be the Banach space of real-valued continuous functions on $\mathbf{K}$, equipped with the sup-norm, and let $C(\mathbf{K}) \subset \tilde{\mathcal{C}}(\mathbf{K})$ be the convex cone of continuous functions nonnegative on $\mathbf{K}$. The topological dual of $\tilde{\mathcal{C}}(\mathbf{K})$ is precisely isometrically isomorphic to $\mathcal{M}(\mathbf{K})$, denoted $\mathcal{M}(\mathbf{K}) \simeq \tilde{\mathcal{C}}(\mathbf{K})^*$, and with duality bracket

$$\langle f, \mu \rangle = \int_{\mathbf{K}} f \, d\mu, \qquad \forall f \in \tilde{\mathcal{C}}(\mathbf{K}), \ \mu \in \mathcal{M}(\mathbf{K}).$$

Moreover, $\mathcal{M}(\mathbf{K})_+$ is the dual cone of $C(\mathbf{K})$, i.e.,

$$\mathcal{M}(\mathbf{K})_+ = C(\mathbf{K})^* = \{ \mu : \langle f, \mu \rangle \geq 0, \quad \forall f \in C(\mathbf{K}) \},$$

also written $\{ \mu \in \mathcal{M}(\mathbf{K}) \, : \, \mu \geq 0 \}$.

---

[1] Recall that a signed Borel measure on $\mathbf{K} \subset \mathbb{R}^n$ is a countably additive set function on the Borel σ-field of $\mathbf{K}$.

With the above notation, (5.3) can be rewritten as the infinite-dimensional linear program

$$f^* = \inf_{\mu \in \mathcal{M}(\mathbf{K})} \{ \langle f, \mu \rangle : \langle 1, \mu \rangle = 1, \quad \mu \geq 0 \}, \qquad (5.4)$$

whose dual is the infinite-dimensional linear program

$$\rho = \sup_{\lambda} \{ \lambda : f(\mathbf{x}) - \lambda \geq 0, \quad \forall \mathbf{x} \in \mathbf{K} \}, \qquad (5.5)$$

where λ is the dual variable associated to the constraint $\langle 1, \mu \rangle = 1$ of (5.4).

Observe that (5.5) is a linear program and even with a single variable! Indeed for each fixed $\mathbf{x} \in \mathbf{K}$, the constraint $f(\mathbf{x}) - \lambda$ is a linear constraint on λ; however there are uncountably many of them.

Moreover, recall that f is a polynomial (say of degree d) and so can be written

$$\mathbf{x} \mapsto f(\mathbf{x}) = \sum_{\alpha \in \mathbb{N}_d^n} f_\alpha \mathbf{x}^\alpha = \sum_{\alpha \in \mathbb{N}^n} f_\alpha x_1^{\alpha_1} \cdots x_n^{\alpha_n},$$

for finitely many real nonzero coefficients (f_α) in the canonical basis of monomials $(\mathbf{x}^\alpha)$, $\alpha \in \mathbb{N}_d^n$, of $\mathbb{R}[\mathbf{x}]$. And so (5.4) reads:

$$f^* = \inf_{\mathbf{y}} \left\{ \sum_{\alpha \in \mathbb{N}_d^n} f_\alpha y_\alpha : \exists \mu \in \mathcal{M}(\mathbf{K})_+ \text{ such that} \right.$$

$$\left. y_0 = 1; \; y_\alpha = \int_{\mathbf{K}} \mathbf{x}^\alpha d\mu, \forall \alpha \in \mathbb{N}_d^n \right\}, \qquad (5.6)$$

so that the unknown measure $\mu \in \mathcal{M}(\mathbf{K})_+$ only apppears through the scalars

$$y_\alpha = \int_{\mathbf{K}} \mathbf{x}^\alpha d\mu = \int_{\mathbf{K}} x_1^{\alpha_1} \cdots x_n^{\alpha_n} d\mu(\mathbf{x}), \qquad \alpha \in \mathbb{N}_d^n,$$

called the *moments* of μ (up to order d).

Hence in fact the two linear programs (5.4) and (5.5) express two dual points of view.

- The primal in (5.4) (or, equivalently, (5.6)) considers problem (5.1) from the point of view of *moments* as one searches for the sequence $\mathbf{y}$ of moments of a Borel probability measure μ supported on $\mathbf{K}$. Ideally, if there is a unique global minimizer $\mathbf{x}^* \in \mathbf{K}$, then one searches for the vector

$$\mathbf{y} = (1, x_1^*, \ldots, x_n^*, (x_1^*)^2, x_1^* x_2^*, \ldots, (x_n^*)^d)$$

of moments (up to order d) of the Dirac measure $\delta_{\mathbf{x}^*}$ at $\mathbf{x}^*$. This is why we call (5.4) the *primal* as the search is in a lifted space of the primal variables $\mathbf{x} \in \mathbb{R}^n$ (i.e., not only the n first-order moments $\mathbf{x} = (x_1, \ldots, x_n)$ but also other higher order moments of the Dirac measure at $\mathbf{x}$).

- The dual (5.5) considers problem (5.1) from the point of view of *positive polynomials* as one characterizes f^* as being the largest scalar λ such that $f - \lambda$ is nonnegative on $\mathbf{K}$.

It is important to emphasize that both points of view (5.4) and (5.5) characterize the global minimum f^* and are not valid for a local minimum.

5.2 Polynomial optimization as a finite-dimensional convex optimization problem

With $\mathbf{K}$ as in (5.2), the set $\Delta_d(\mathbf{K}) \subset \mathbb{R}^{s(d)}$ (with $s(d) = \binom{n+d}{n}$) defined by:

$$\Delta_d(\mathbf{K}) := \left\{ \mathbf{y} = (y_\alpha), \; \boldsymbol{\alpha} \in \mathbb{N}_d^n \; : \; \exists \mu \in \mathcal{M}(\mathbf{K})_+ \text{ such that} \right.$$

$$\left. y_0 = 1; \; y_\alpha = \int_{\mathbf{K}} \mathbf{x}^\alpha \, d\mu, \; \forall \alpha \in \mathbb{N}_d^n \right\},$$

is a finite-dimensional convex cone.

And so (5.6) also reads:

$$f^* = \inf_{\mathbf{y}} \left\{ \sum_{\alpha \in \mathbb{N}_d^n} f_\alpha \, y_\alpha \; : \; y_0 = 1; \quad \mathbf{y} \in \Delta_d(\mathbf{K}) \right\}, \qquad (5.7)$$

which is a finite-dimensional convex optimization problem.

Similarly, let $C_d(\mathbf{K})$ be the convex cone of polynomials of degree at most d that are nonnegative on $\mathbf{K}$, already encountered in Chapter 3 and Chapter 4. Recall that $f \in \mathbb{R}[\mathbf{x}]_d$ is associated with its vector of coefficients $\mathbf{f} = (f_\alpha) \in \mathbb{R}^{s(d)}$ and so the convex cone $C_d(\mathbf{K}) \subset \mathbb{R}[\mathbf{x}]_d$ can be identified with a finite-dimensional convex cone in $\mathbb{R}^{s(d)}$ which is given the same name $C_d(\mathbf{K})$ (by a slight abuse of notation). Therefore, if we denote by $\mathbf{e}_1 = (1, 0, \ldots, 0) \in \mathbb{R}^{s(d)}$ the vector of coefficients of the constant polynomial equal to 1 then we may alternatively write $f - \lambda \in C_d(\mathbf{K})$ or $\mathbf{f} - \lambda \mathbf{e}_1 \in C_d(\mathbf{K})$ with no ambiguity.

And so, if $f \in \mathbb{R}[\mathbf{x}]_d$ the dual problem (5.5) also reads:

$$f^* = \sup_{\lambda} \{ \lambda \; : \; \mathbf{f} - \lambda \, \mathbf{e}_1 \in C_d(\mathbf{K}) \}, \qquad (5.8)$$

a finite-dimensional convex optimization problem with a single variable!

Of course there is no magical trick in either (5.7) or (5.8) as there is no simple and tractable way to characterize the convex cones $\Delta_d(\mathbf{K})$ or $C_d(\mathbf{K})$, and so in general (5.7) (respectively (5.8)) is just a compact way of rephrasing the linear program (5.4) (respectively (5.5)). Note in passing that this contradicts the celebrated statement that *the frontier between easy and hard optimization problems is not between linear and nonlinear problems but between convex and nonconvex problems*.

There are convex problems that are not easy and (5.7)–(5.8) are among them! In fact, for efficiency purposes what counts is not the convexity of the feasible set itself but rather whether or not this convex set has a tractable characterization that can be exploited efficiently.

However, when $\mathbf{K}$ in (5.2) is compact we have seen in Chapter 4 that thanks to the powerful representation results of Chapter 2, one may define a hierarchy of inner approximations of $C_d(\mathbf{K})$. As we will see in the next chapters, these hierarchies can be exploited to solve the dual LP (5.5) (or at least to approximate f^* as closely as desired). By duality, it should not be a surprise that dual counterparts in Section 2.7 are also exploited to solve the primal LP (5.4).

In doing so,

- for $\mathbf{K} = \mathbb{R}^n$ we obtain a *single* semidefinite relaxation, which is exact if and only if the nonnegative polynomial $\mathbf{x} \mapsto f(\mathbf{x}) - f^*$ is a sum of squares (in short SOS);
- for $\mathbf{K}$ as in (5.2) and compact, we obtain a *hierarchy* of semidefinite relaxations whose size depends on the degree bound k on the SOS weights in Putinar's representation (2.10) of the polynomial $f - \lambda$ in (5.5) which is nonnegative on $\mathbf{K}$. This representation amounts to considering the *inner* and tractable approximations (4.3) of the convex cone $C_d(\mathbf{K})$. The larger is the degree k the better is the optimal value λ_k (but of course the larger is the size of the resulting semidefinite program to solve). As the degree k increases, asymptotic convergence $\lambda_k \to f^*$ is guaranteed by invoking (the powerful) Theorem 2.15.

Alternatively one may also consider the (not SOS based) representation (2.16) of the polynomial $f - \lambda$ to obtain a hierarchy of *linear* relaxations instead of semidefinite relaxations. In this case, the resulting asymptotic convergence $\lambda_k \to f^*$ is now guaranteed by invoking Theorem 2.22. It is rather unfortunate that such a hierarchy of linear programs suffers several serious drawbacks, for one would much prefer to solve linear programs than semidefinite programs. Indeed, even though both are convex conic programs, the latter are more difficult to solve and, in addition, the present status of semidefinite solvers is far from being comparable to that of linear program solvers which can handle problems with millions of variables!

Finally, one may and will also consider *outer* approximations of the convex cone $C_d(\mathbf{K})$ developed in Chapter 4 and in this case one obtains a hierarchy of semidefinite programs whose associated monotone sequence of optimal values (ρ_k), $k \in \mathbb{N}$, provide upper bounds on f^* with $\rho_k \to f^*$ as $k \to \infty$. In fact, in this case one ends up solving a generalized eigenvalue problem instead of a semidefinite program. This approach is possible only for sets $\mathbf{K}$ (not necessarily compact) for which one may compute all moments of some finite Borel measure whose support is exactly $\mathbf{K}$. Typical examples of such sets $\mathbf{K}$ include the whole space $\mathbb{R}^n$, the positive orthant $\mathbb{R}^n_+$, boxes or hyper-rectangles, ellipsoids, simplices and the hypercube $\{-1, 1\}^n$.

5.3 Exercises

Exercise 5.1 Show that the characterization of f^* in (5.4) and (5.5) is proper to the global minimum f^* and is not true for a local minimum.

6

Semidefinite relaxations for polynomial optimization

Consider the polynomial optimization problem **P** of Chapter 5, i.e.,

$$\mathbf{P}: \quad f^* := \inf \{ f(\mathbf{x}) \ : \ \mathbf{x} \in \mathbf{K} \} \qquad (6.1)$$

with feasible set $\mathbf{K} \subset \mathbb{R}^n$ defined by

$$\mathbf{K} = \{ \mathbf{x} \in \mathbb{R}^n \ : \ g_j(\mathbf{x}) \geq 0, \ j = 1, \dots, m \}, \qquad (6.2)$$

where $f, g_j \in \mathbb{R}[\mathbf{x}]$ are real-valued polynomials, $j = 1, \dots, m$.

We provide a hierarchy of semidefinite relaxations whose optimal values form a monotone nondecreasing sequence of lower bounds which converges to the global optimum f^*. The rationale behind this convergence is provided by the powerful Positivstellensatz of Putinar (Theorem 2.15). Moreover, finite convergence is generic!

6.1 Constrained polynomial optimization

In this section, we address problem (6.1) where $f \in \mathbb{R}[\mathbf{x}]$ and the set $\mathbf{K}$ defined in (6.2) is assumed to be a compact basic semi-algebraic subset of $\mathbb{R}^n$. We suppose that Assumption 2.14 holds, which, as we discussed in Section 2.4, is verified in many cases, for example if there is one polynomial g_k such that $\{ \mathbf{x} \in \mathbb{R}^n \ : \ g_k(\mathbf{x}) \geq 0 \}$ is compact. In particular, as $\mathbf{K}$ is compact there exists $M > 0$ such that the polynomial $\mathbf{x} \mapsto g_{m+1}(\mathbf{x}) := M - \|\mathbf{x}\|^2$ is nonnegative on $\mathbf{K}$. So if one knows such an M one can include the redundant quadratic constraint $g_{m+1}(\mathbf{x}) \geq 0$ in the definition (6.2) of $\mathbf{K}$ and Assumption 2.14 is satisfied; see the discussion following Theorem 2.15 and Theorem 2.16.

6.1.1 A hierarchy of semidefinite relaxations

With $g_0 = 1$ and $\mathbf{K}$ as in (6.2), let $v_j := \lceil (\deg g_j)/2 \rceil$, $j = 0, \ldots, m$. For $d \geq d_0 := \max[\lceil (\deg f)/2 \rceil, \max_{j=1,\ldots,m} v_j]$, consider the following semidefinite program:

$$
\begin{aligned}
\rho_d = \inf_{\mathbf{y}} \quad & L_{\mathbf{y}}(f) \\
\text{s.t.} \quad & \mathbf{M}_d(\mathbf{y}) \succeq \mathbf{0}, \\
& \mathbf{M}_{d-v_j}(g_j\,\mathbf{y}) \succeq \mathbf{0}, \quad j = 1, \ldots, m \\
& y_0 = 1,
\end{aligned}
\tag{6.3}
$$

where $L_{\mathbf{y}}$ is the Riesz linear functional and $\mathbf{M}_d(\mathbf{y})$ (respectively $\mathbf{M}_{d-v_j}(g_j\,\mathbf{y})$) is the moment matrix (respectively localizing matrix) associated with $\mathbf{y}$ and the polynomial g_j, all defined in Section 2.7.1.

Obviously, (6.3) is a relaxation of (6.1) (and so of (5.3)). Indeed, let $\mathbf{x} \in \mathbf{K}$ be a feasible solution and let $\mathbf{y} := \mathbf{v}_{2d}(\mathbf{x}) \ (= (\mathbf{x}^\alpha), \ \alpha \in \mathbb{N}^n_{2d})$. That is, $\mathbf{y}$ is the vector of moments (up to order $2d$) of the Dirac measure $\mu := \delta_{\mathbf{x}}$ at $\mathbf{x} \in \mathbf{K}$. Therefore $\mathbf{y}$ is a feasible solution of (6.3) with value $L_{\mathbf{y}}(f) = f(\mathbf{x})$, which proves that $\rho_d \leq f^*$ for every d. Moreover $\rho_{d+1} \geq \rho_d$ for all $d \geq d_0$ because the semidefinite program (6.3) with $d + 1$ is more constrained than with d.

Hence (6.3) defines a *hierarchy* of semidefinite relaxations of **P**.

The dual To obtain the dual of (6.3) write

$$
\mathbf{M}_d(\mathbf{y}) = \sum_{|\alpha| \leq 2d} y_\alpha\, \mathbf{C}^0_\alpha, \quad \mathbf{M}_{d-v_j}(g_j\,\mathbf{y}) = \sum_{|\alpha| \leq 2d} y_\alpha\, \mathbf{C}^j_\alpha, \quad j = 1, \ldots, m,
$$

for some appropriate real symmetric matrices $\mathbf{C}^j_\alpha$, $j = 0, \ldots, m$, $\alpha \in \mathbb{N}^n_{2d}$ already defined in Sections 2.4.2 and 2.7.1. Then the dual of (6.3) is the semidefinite program

$$
\begin{aligned}
\rho^*_d = \sup_{\lambda, \mathbf{X}_j} \quad & \lambda \\
\text{s.t.} \quad & f_\alpha - \lambda \mathbf{1}_{\alpha=0} = \sum_{j=0}^{m} \left\langle \mathbf{X}_j, \mathbf{C}^j_\alpha \right\rangle, \quad \forall \alpha \in \mathbb{N}^n_{2d} \\
& \mathbf{X}_j \succeq \mathbf{0}, \quad j = 0, \ldots, m,
\end{aligned}
$$

where $\mathbf{X}_j$ are all real symmetric matrices, $j = 0, \ldots, m$. To interpret the constraint in this semidefinite program, multiply each side of the above equality with $\mathbf{x}^\alpha$ and sum up to obtain

$$\sum_{\alpha \in \mathbb{N}_{2d}^n} f_\alpha \, \mathbf{x}^\alpha - \lambda = \sum_{j=0}^m \left\langle \mathbf{X}_j, \sum_\alpha \mathbf{C}_\alpha^j \, \mathbf{x}^\alpha \right\rangle, \qquad \forall \mathbf{x} \in \mathbb{R}^n,$$

which is the same as

$$f(\mathbf{x}) - \lambda = \sum_{j=0}^m g_j(\mathbf{x}) \left\langle \mathbf{X}_j, \mathbf{v}_{d-v_j}(\mathbf{x}) \mathbf{v}_{d-v_j}(\mathbf{x})^T \right\rangle$$

$$= \sum_{j=0}^m g_j(\mathbf{x}) \, \mathbf{v}_{d-v_j}(\mathbf{x})^T \mathbf{X}_j \mathbf{v}_{d-v_j}(\mathbf{x}), \qquad \forall \mathbf{x} \in \mathbb{R}^n.$$

Next, for every $j = 0, \ldots, m$, write $\mathbf{X}_j \succeq 0$ as $\sum_{\ell=1}^{t_j} \mathbf{q}_{j\ell} \mathbf{q}_{j\ell}^T$ for some $t_j \in \mathbb{N}$, to obtain

$$f(\mathbf{x}) - \lambda = \sum_{j=0}^m g_j(\mathbf{x}) \left(\sum_{\ell=1}^{t_j} \underbrace{(\mathbf{q}_{j\ell}^T \mathbf{v}_{d-v_j}(\mathbf{x}))^2}_{q_{j\ell}(\mathbf{x})^2} \right), \qquad \forall \mathbf{x} \in \mathbb{R}^n,$$

where $q_{j\ell} \in \mathbb{R}[\mathbf{x}]_{d-v_j}$ has coefficient vector $\mathbf{q}_{j\ell} \in \mathbb{R}^{s(d-v_j)}$, for all $\ell = 1, \ldots, t_j$, and all $j = 0, \ldots, m$. Equivalently,

$$f - \lambda = \sum_{j=0}^m \sigma_j \, g_j,$$

where $\sigma_j \in \Sigma[\mathbf{x}]_{d-v_j}$ is the SOS polynomial $\sum_{\ell=1}^{t_j} q_{j\ell}^2$, for every $j = 0, \ldots, m$. And so the dual of (6.3) is a semidefinite program which in compact form reads:

$$
\begin{aligned}
\rho_d^* = \sup_{\lambda, \{\sigma_j\}} \quad & \lambda \\
\text{s.t.} \quad & f - \lambda = \sum_{j=0}^m \sigma_j \, g_j, \quad \sigma_j \in \Sigma[\mathbf{x}]_{d-v_j}, \ j = 0, \ldots, m.
\end{aligned}
$$

$$(6.4)$$

Note that the semidefinite program (6.4) is a *strengthening* of the infinite-dimensional LP (5.5). Indeed, one has replaced the constraint $f - \lambda \geq 0$ on $\mathbf{K}$

with the more restrictive positivity certificate $f - \lambda = \sum_{j=0}^{m} \sigma_j g_j$ for some SOS polynomials σ_j of degree bounded by $2(d - v_j)$, $j = 0, \ldots, m$. This is consistent with the primal (6.3) being a relaxation of the primal infinite-dimensional LP (5.3). By duality in optimization, a relaxation of a primal optimization problem induces a strengthening of its dual.

If in the definition of **K** there is an equality constraint $g_j(\mathbf{x}) = 0$ (i.e., two opposite inequality constraints $g_j(\mathbf{x}) \geq 0$ and $-g_j(\mathbf{x}) \geq 0$), then one has the equality constraint $\mathbf{M}_{d-v_j}(g_j \, \mathbf{y}) = 0$ in (6.3) and accordingly, in (6.4), the weight $\sigma_j \in \mathbb{R}[\mathbf{x}]$ is not required to be an SOS polynomial. Moreover, in the case where $d_j = \deg(g_j)$ is odd then in (6.3) one may replace $\mathbf{M}_{d-v_j}(g_j \, \mathbf{y}) = 0$ with the slightly stronger constraint $L_{\mathbf{y}}(\mathbf{x}^\alpha g_j) = 0$ for all $\alpha \in \mathbb{N}^n_{2d-d_j}$.

The overall algorithm is as follows.

Algorithm 6.1   (Constrained polynomial optimization)

Input A polynomial $\mathbf{x} \mapsto f(\mathbf{x})$; a set $\mathbf{K} = \{\mathbf{x} \in \mathbb{R}^n : g_j(\mathbf{x}) \geq 0, \ j = 1, \ldots, m \}$, where the polynomials g_j are of degree $2v_j$ or $2v_j - 1$, $j = 1, \ldots, m$; a number k of the highest relaxation.

Output The value $f^* = \inf_{\mathbf{x} \in \mathbf{K}} f(\mathbf{x})$ and a list of global minimizers or a lower bound ρ_k on f^*.

Algorithm

1. Solve the semidefinite optimization problem (6.3) with optimal value ρ_d, and optimal solution $\mathbf{y}^*$ (if it exists).

2. If there is no optimal solution $\mathbf{y}^*$ then ρ_d only provides a lower bound $\rho_d \leq f^*$. If $d < k$ then increase d by one and go to Step 1; otherwise stop and output ρ_k.

3. If rank $\mathbf{M}_{d-v}(\mathbf{y}^*) = $ rank $\mathbf{M}_d(\mathbf{y}^*)$ (with $v := \max_j v_j$), then $\rho_d = f^*$ and there are at least rank $\mathbf{M}_d(\mathbf{y}^*)$ global minimizers which can be extracted.

4. If rank $\mathbf{M}_{d-v}(\mathbf{y}^*) \neq $ rank $\mathbf{M}_d(\mathbf{y}^*)$, and $d < k$, then increase d by one and go to Step 1; otherwise, stop and output ρ_k which only provides a lower bound $\rho_k \leq f^*$.

And we have the following convergence result.

Theorem 6.2   *Let Assumption 2.14 hold and consider the semidefinite relaxation (6.3) with optimal value ρ_d.*

(a) $\rho_d \uparrow f^*$ *as* $d \to \infty$.

(b) *Assume that (6.1) has a unique global minimizer* $\mathbf{x}^* \in \mathbf{K}$ *and let* $\mathbf{y}^d$ *be a nearly optimal solution of (6.3) with value* $L_{\mathbf{y}}(f) \leq \rho_d + 1/d$. *Then as* $d \to \infty$, $L_{\mathbf{y}^d}(x_j) \to x_j^*$ *for every* $j = 1, \ldots, n$.

Sketch of the proof As $\mathbf{K}$ is compact, $f^* > -\infty$, and with $\epsilon > 0$ arbitrary, the polynomial $f - (f^* - \epsilon)$ is strictly positive on $\mathbf{K}$. Therefore by Theorem 2.15

$$f - (f^* - \epsilon) = \sigma_{0\epsilon} + \sum_{j=1}^{m} \sigma_{j\epsilon}\, g_j,$$

for some SOS polynomials $(\sigma_{j\epsilon}) \subset \Sigma[\mathbf{x}]$. So $(f^* - \epsilon, (\sigma_{j\epsilon}))$ is a feasible solution of (6.4) as soon as d is large enough. Hence $\rho_d^* \geq f^* - \epsilon$. On the other hand, by weak duality one has $\rho_d^* \leq \rho_d$ for every d and we have seen that $\rho_d \leq f^*$, which yields $f^* - \epsilon \leq \rho_d^* \leq \rho_d \leq f^*$. Hence, as $\epsilon > 0$ was arbitrary, (a) follows.

For (b) one first proves that a subsequence $\mathbf{y}^{d_k}$ (considered as an infinite sequence by completing with zeros) converges pointwise to an infinite sequence $\mathbf{y}$, i.e., $\lim_{k \to \infty} y_{\boldsymbol{\alpha}}^{d_k} = y_{\boldsymbol{\alpha}}$ for every $\boldsymbol{\alpha} \in \mathbb{N}^n$. Then by invoking Theorem 2.44, the sequence $\mathbf{y}$ is shown to be the moment sequence of a measure μ supported on $\mathbf{K}$ and with value $L_{\mathbf{y}}(f) = \lim_{k \to \infty} L_{\mathbf{y}^{d_k}}(f) \leq f^*$. Hence necessarily μ is the Dirac measure $\delta_{\mathbf{x}^*}$ at the unique global minimizer $\mathbf{x}^* \in \mathbf{K}$. Therefore the whole sequence $\mathbf{y}^d$ converges pointwise to the vector $\mathbf{v}(\mathbf{x}^*)$ of moments of μ. In particular, for every $j = 1, \ldots, n$, $L_{\mathbf{y}^d}(x_j) \to x_j^*$ as $d \to \infty$. □

Finite convergence In the definition (6.2) of $\mathbf{K}$, an equality constraint $g_j(\mathbf{x}) = 0$ (if present) is taken into account via the two inequality constraints $g_j(\mathbf{x}) \geq 0$ and $-g_j(\mathbf{x}) \geq 0$. As a result, in (6.3) the associated semidefinite constraint $\mathbf{M}_{d-v_j}(g_j\,\mathbf{y}) \succeq 0$ simply becomes the equality constraint $\mathbf{M}_{d-v_j}(g_j\,\mathbf{y}) = 0$.

As we next see, finite convergence of the hierarchy of semidefinite relaxations (6.3) takes place under some classical conditions already encountered in nonlinear programming, but to detail such conditions we need to distinguish between equality constraints and inequality constraints. Therefore we rewrite the basic semi-algebraic set $\mathbf{K} \subset \mathbb{R}^n$ in the general form where equality and inequality constraints are identified explicitly:

$$\mathbf{K} = \{\mathbf{x} \in \mathbb{R}^n : g_j(\mathbf{x}) \geq 0,\ j = 1, \ldots, m_1;\ h_\ell(\mathbf{x}) = 0,\ \ell = 1, \ldots, m_2\},$$
(6.5)

for some integers m_1, m_2 and some polynomials $g_j, h_\ell \in \mathbb{R}[\mathbf{x}]$. With $\tau_\ell = \lceil (\deg h_\ell)/2 \rceil$, $\ell = 1, \ldots, m_2$, the semidefinite relaxation (6.3) now reads:

$$
\begin{aligned}
\rho_d = \inf_{\mathbf{y}} \quad & L_{\mathbf{y}}(f) \\
\text{s.t.} \quad & \mathbf{M}_d(\mathbf{y}) \succeq \mathbf{0}, \\
& \mathbf{M}_{d-v_j}(g_j\, \mathbf{y}) \succeq \mathbf{0}, \quad j = 1, \ldots, m_1 \\
& \mathbf{M}_{d-\tau_\ell}(h_\ell\, \mathbf{y}) = \mathbf{0}, \quad \ell = 1, \ldots, m_2 \\
& y_{\mathbf{0}} = 1,
\end{aligned}
\tag{6.6}
$$

and its dual is the semidefinite program

$$
\begin{aligned}
\rho_d^* = \sup_{\lambda, \{\sigma_j\}} \quad & \lambda \\
\text{s.t.} \quad & f - \lambda = \sum_{j=0}^{m_1} \sigma_j\, g_j + \sum_{\ell=1}^{m_2} \psi_\ell\, h_\ell \\
& \sigma_j \in \Sigma[\mathbf{x}]_{d-v_j}, \quad \psi_\ell \in \mathbb{R}[\mathbf{x}]_{d-\tau_\ell} \\
& j = 0, \ldots, m_1, \quad \ell = 1, \ldots, m_2.
\end{aligned}
\tag{6.7}
$$

If $\rho_d = f^*$ then we say the semidefinite relaxation (6.6) is *exact*.

Let $J(\mathbf{x}^*) := \{ j : g_j(\mathbf{x}^*) = 0 \}$, i.e., $J(\mathbf{x}^*)$ is the set of inequality constraints that are active at $\mathbf{x}^* \in \mathbf{K}$.

Definition 6.3 A triplet $(\mathbf{x}^*, \boldsymbol{\lambda}^*, \mathbf{v}^*) \in \mathbf{K} \times \mathbb{R}^{m_1} \times \mathbb{R}^{m_2}$ is called a Karush–Kuhn–Tucker (KKT) triplet if it satisfies:

$$
\begin{aligned}
\nabla f(\mathbf{x}^*) &= \sum_{j=1}^{m_1} \lambda_j^* \nabla g_j(\mathbf{x}^*) + \sum_{\ell=1}^{m_2} v_\ell^* \nabla h_\ell(\mathbf{x}^*), \\
\lambda_j^* g_j(\mathbf{x}^*) &= 0, \quad j = 1, \ldots, m_1 \quad \text{[Complementarity]} \\
\lambda_j^* &\geq 0, \quad j = 1, \ldots, m_1.
\end{aligned}
\tag{6.8}
$$

The polynomial $L(\cdot, \boldsymbol{\lambda}^*, \mathbf{v}^*) : \mathbb{R}^n \to \mathbb{R}$ defined by:

$$
\mathbf{x} \mapsto L(\mathbf{x}, \boldsymbol{\lambda}^*, \mathbf{v}^*) := f(\mathbf{x}) - f^* - \sum_{j=1}^{m_1} \lambda_j^* g_j(\mathbf{x}) - \sum_{\ell=1}^{m_2} v_\ell^* h_\ell(\mathbf{x}),
$$

is called the *Lagrangian* associated with the triplet $(\mathbf{x}^*, \boldsymbol{\lambda}^*, \mathbf{v}^*)$.

Note that if $\mathbf{x}^*$ is a local minimizer and the gradients $(\nabla g_j(\mathbf{x}^*), \nabla h_\ell(\mathbf{x}^*))$, $j \in J(\mathbf{x}^*)$, $\ell = 1, \ldots, m_2$, are linearly independent then such a KKT triplet

$(\mathbf{x}^*, \boldsymbol{\lambda}^*, \mathbf{v}^*)$ exists. Moreover, if $\lambda_j^* > 0$ for all $j \in J(\mathbf{x}^*)$ one says that *strict complementarity* holds.

When $\mathbf{K}$ is described in the general form (6.5) then (the Archimedean) Assumption 2.14 becomes as follows.

Assumption 6.4 *There exists $M > 0$ such that the quadratic polynomial $\mathbf{x} \mapsto M - \|\mathbf{x}\|^2$ can be rewritten as:*

$$M - \|\mathbf{x}\|^2 = \sum_{j=0}^{m_1} \sigma_j \, g_j + \sum_{\ell=1}^{m_2} \psi_\ell \, h_\ell, \qquad (6.9)$$

for some polynomials $\psi_\ell \in \mathbb{R}[\mathbf{x}]$ and some SOS polynomials $\sigma_j \in \Sigma[\mathbf{x}]$.

Recall that associated with an ideal $J \subset \mathbb{R}[\mathbf{x}]$ is the real variety $V_{\mathbb{R}}(J) := \{ \mathbf{x} \in \mathbb{R}^n : h(\mathbf{x}) = 0, \; \forall h \in J \}$. Similarly, associated with a set $V \subset \mathbb{R}^n$ is the ideal $I(V) := \{ p \in \mathbb{R}[\mathbf{x}] : p(\mathbf{x}) = 0, \; \forall \mathbf{x} \in V \}$, called the *vanishing* ideal. Then an ideal $J \subset \mathbb{R}[\mathbf{x}]$ is *real* (or real radical) if $J = I(V_{\mathbb{R}}(J))$.

Theorem 6.5 *Let $\mathbf{K}$ be as in (6.5) and let Assumption 6.4 hold. Let ρ_d, $d \in \mathbb{N}$, be the optimal value of the semidefinite relaxation (6.6).*

Then $\rho_d = f^*$ for some $d \in \mathbb{N}$, i.e., finite convergence takes place and (6.6) has an optimal solution, if at every global minimizer $\mathbf{x}^* \in \mathbf{K}$ of problem (6.1) with $f^* = f(\mathbf{x}^*)$:

- *the gradients $(\nabla g_j(\mathbf{x}^*), \nabla h_\ell(\mathbf{x}^*))$, $j \in J(\mathbf{x}^*)$, $\ell = 1, \ldots, m_2$, are linearly independent (and so a KKT triplet $(\mathbf{x}^*, \boldsymbol{\lambda}^*, \mathbf{v}^*)$ exists),*
- *strict complementarity holds,*
- *second-order sufficiency holds, i.e., $\mathbf{u}^T \nabla_{\mathbf{x}}^2 L(\mathbf{x}^*, \boldsymbol{\lambda}^*, \mathbf{v}^*) \mathbf{u} > 0$ for all $0 \neq \mathbf{u}$ such that $\nabla g_j(\mathbf{x}^*)^T \mathbf{u} = 0$ and $\nabla h_\ell(\mathbf{x}^*)^T \mathbf{u} = 0$, $j \in J(\mathbf{x}^*)$, $\ell = 1, \ldots, m_2$.*

If in addition the ideal $\langle h \rangle = \langle h_1, \ldots, h_{m_2} \rangle$ is real then there exists $d \in \mathbb{N}$, polynomials $\psi_\ell \in \mathbb{R}[\mathbf{x}]_{d-\tau_\ell}$, $\ell = 1, \ldots, m_2$, and SOS polynomials $\sigma_j \in \Sigma[\mathbf{x}]_{d-v_j}$, $j = 0, \ldots, m_1$, such that

$$f - f^* = \sum_{j=0}^{m_1} \sigma_j \, g_j + \sum_{\ell=1}^{m} \psi_\ell \, h_\ell, \qquad (6.10)$$

i.e., $\rho_d^* = f^*$ and (6.7) has an optimal solution.

Observe that linear independence of gradients, strict complementarity and second-order sufficiency condition at a point $\mathbf{x}^* \in \mathbf{K}$ are well-known standard sufficient conditions in nonlinear programming for $\mathbf{x}^*$ to be a local minimizer. Theorem 6.5 states that if such conditions hold at every global minimizer $\mathbf{x}^* \in \mathbf{K}$ then one also obtains a certificate of global optimality, either by finite convergence of the semidefinite relaxations (6.6) or by the Putinar certificate of positivity (6.10) (at least when the ideal $\langle h \rangle$ is real).

Moreover, in Chapter 7 we will see that the conditions in Theorem 6.5 are *generic* (in a rigorous sense) and so is finite convergence of the hierarchy of SDP-relaxations (6.6).

6.1.2 Obtaining global minimizers

After solving the semidefinite relaxation (6.3) (or (6.6) if equality constraints are explicit) for some value of $d \in \mathbb{N}$, we are left with two issues.

1. How may we decide whether $\rho_d < f^*$ or $\rho_d = f^*$?
2. If $\rho_d = f^*$ can we obtain at least one global minimizer $\mathbf{x}^* \in \mathbf{K}$?

Again an easy case is when (6.3) has an optimal solution $\mathbf{y}^*$ which satisfies rank $\mathbf{M}_d(\mathbf{y}^*) = 1$, and so, necessarily, $\mathbf{M}_d(\mathbf{y}^*) = \mathbf{v}_d(\mathbf{x}^*)\mathbf{v}_d(\mathbf{x}^*)^T$ for some $\mathbf{x}^* \in \mathbb{R}^n$. In addition, the constraints $\mathbf{M}_{d-v_j}(g_j \, \mathbf{y}^*) \succeq 0$ imply that $\mathbf{x}^* \in \mathbf{K}$. That is, $\mathbf{y}^*$ is the vector of moments up to order $2d$ of the Dirac measure at $\mathbf{x}^* \in \mathbf{K}$, and one reads the optimal solution $\mathbf{x}^*$, from the subvector of first "moments" y_α^* with $|\alpha| = 1$.

For the case of multiple global minimizers we have the following sufficient condition which is implemented at Step 3 of Algorithm 6.1.

Theorem 6.6 *Let $f \in \mathbb{R}[\mathbf{x}]$, and suppose that the optimal value ρ_d of problem (6.3) is attained at some optimal solution $\mathbf{y}^*$. Let $v := \max_{j=1,\ldots,m} v_j$. If rank $\mathbf{M}_{d-v}(\mathbf{y}^*) = \text{rank } \mathbf{M}_d(\mathbf{y}^*)$, then $\rho_d = f^*$ and there exist at least $s := \text{rank } \mathbf{M}_d(\mathbf{y}^*)$ global minimizers which can be extracted by Algorithm 6.9.*

Proof We already know that $\rho_d \leq f^*$ for all d. If $s = \text{rank } \mathbf{M}_{d-v}(\mathbf{y}^*) = \text{rank } \mathbf{M}_d(\mathbf{y}^*)$, then by Theorem 2.47, $\mathbf{y}^*$ is the moment vector of some s-atomic probability measure μ^* on $\mathbf{K}$. We then argue that each of the s atoms of μ^* is a global minimizer of f on $\mathbf{K}$. Indeed, being s-atomic, there is a family $(\mathbf{x}(k))_{k=1}^s \subset \mathbf{K}$ and a family $(\lambda_k)_{k=1}^s \subset \mathbb{R}$, such that

$$\mu^* = \sum_{k=1}^s \lambda_k \, \delta_{\mathbf{x}(k)}, \quad \lambda_k > 0, \quad \forall k = 1, \ldots, s; \quad \sum_{k=1}^s \lambda_k = 1.$$

Hence,

$$f^* \geq \rho_d = L_{\mathbf{y}^*}(f) = \int_{\mathbf{K}} f \, d\mu^* = \sum_{k=1}^s \lambda_k \, f(\mathbf{x}(k)) \geq f^*,$$

which clearly implies $f^* = \rho_d$, and $f(\mathbf{x}(k)) = f^*$, for all $k = 1, \ldots, s$, the desired result. □

If the rank test in Theorem 6.6 is satisfied then rank $\mathbf{M}_d(\mathbf{y}^*)$ global minimizers are encoded in the optimal solution $\mathbf{y}^*$ of (6.3). In order to extract these solutions from $\mathbf{y}^*$, we apply Algorithm 6.9 described below and implemented in the GloptiPoly software described in Appendix B.

Example 6.7 Let $n = 2$ and consider the optimization problem

$$f^* = \inf \quad x_1^2 x_2^2 (x_1^2 + x_2^2 - 1)$$
$$\text{s.t.} \quad x_1^2 + x_2^2 \leq 4,$$

which is the same as in Example 6.21, but with the additional ball constraint $\|\mathbf{x}\|^2 \leq 4$. Therefore the optimal value is $f^* = -1/27$ with global minimizers $\mathbf{x}^* = (\pm \sqrt{3}/3, \pm \sqrt{3}/3)$. Applying Algorithm 6.1 for $k = 4$ and using GloptyPoly, the optimal value is obtained with $\rho_4 \approx -1/27$ and the four optimal solutions $(\pm 0.5774, \pm 0.5774)^T \approx \mathbf{x}^*$ are extracted.

When we add the additional nonconvex constraint $x_1 x_2 \geq 1$, we find that the optimal value is obtained with $\rho_3 \approx 1$ and the two optimal solutions $(x_1^*, x_2^*) = (-1, -1)$ and $(x_1^*, x_2^*) = (1, 1)$ are extracted. In both cases, the rank test is satisfied for an optimal solution $\mathbf{y}^*$ and a global optimality certificate is provided due to Theorem 6.6.

Example 6.8 Let $n = 2$ and $f \in \mathbb{R}[\mathbf{x}]$ be the concave[1] polynomial

$$f(\mathbf{x}) := -(x_1 - 1)^2 - (x_1 - x_2)^2 - (x_2 - 3)^2$$

and let $\mathbf{K} \subset \mathbb{R}^2$ be the set:

$$\mathbf{K} := \{\mathbf{x} \in \mathbb{R}^2 : 1 - (x_1 - 1)^2 \geq 0, \ 1 - (x_1 - x_2)^2 \geq 0, \ 1 - (x_2 - 3)^2 \geq 0\}.$$

The optimal value is $f^* = -2$. Solving problem (6.3) for $d = 1$ yields $\rho_1 = -3$ instead of the desired value -2. On the other hand, solving problem (6.3) for $d = 2$ yields $\rho_2 \approx -2$ and the three optimal solutions $(x_1^*, x_2^*) = (1, 2), (2, 2), (2, 3)$ are extracted. Hence, with polynomials of degree 4 instead of 2, we obtain (a good approximation of) the correct value. Note that there exist scalars $\lambda_j = 1 \geq 0$ such that

$$f(\mathbf{x}) + 3 = 0 + \sum_{j=1}^{3} \lambda_j g_j(\mathbf{x}),$$

but $f(\mathbf{x}) - f^*$ (equal to $f(\mathbf{x}) + 2$) cannot be written in this way (otherwise ρ_1 would be the optimal value -2).

---

[1] A real function $f : \mathbb{R}^n \rightarrow \mathbb{R}$ is concave if $-f$ is convex, that is, $f(\lambda \mathbf{x} + (1 - \lambda)\mathbf{y}) \geq \lambda f(\mathbf{x}) + (1 - \lambda)f(\mathbf{y})$ for all $\mathbf{x}, \mathbf{y} \in \mathbb{R}^n$ and all $\lambda \in [0, 1]$.

For quadratically constrained nonconvex quadratic problems, the semidefinite program (6.3) with $d = 1$ is a well-known relaxation. But ρ_1 which sometimes provides directly the exact global minimum value, is only a lower bound in general.

An algorithm for extraction of global minimizers

We now describe Algorithm 6.9, a procedure to *extract* global minimizers of problem (6.1) when the semidefinite relaxation (6.3) is exact at some step s of the hierarchy, i.e., with $\rho_s = f^*$, and the rank condition

$$\operatorname{rank} \mathbf{M}_{s-v}(\mathbf{y}^*) = \operatorname{rank} \mathbf{M}_s(\mathbf{y}^*) \qquad (6.11)$$

in Theorem 6.6 holds at an optimal solution $\mathbf{y}^*$ of (6.3). The main steps of the extraction algorithm can be sketched as follows.

- **Cholesky factorization** As condition (6.11) holds, $\mathbf{y}^*$ is the vector of a rank $\mathbf{M}_s(\mathbf{y}^*)$-atomic Borel probability measure μ supported on $\mathbf{K}$. That is, there are r ($r = \operatorname{rank} \mathbf{M}_s(\mathbf{y}^*)$) points $(\mathbf{x}(k))_{k=1}^r \subset \mathbf{K}$ such that

$$\mu = \sum_{j=1}^{r} \kappa_j^2 \, \delta_{\mathbf{x}(j)}, \quad \kappa_j \neq 0, \; \forall j; \quad \sum_{j=1}^{r} \kappa_j^2 = y_0 = 1, \qquad (6.12)$$

with $\delta_{\mathbf{x}}$ being the Dirac measure at the point $\mathbf{x}$. Hence, by construction of $\mathbf{M}_s(\mathbf{y}^*)$,

$$\mathbf{M}_s(\mathbf{y}^*) = \sum_{j=1}^{r} \kappa_j^2 \, \mathbf{v}_s(\mathbf{x}^*(j))(\mathbf{v}_s(\mathbf{x}^*(j)))' = \mathbf{V}^* \mathbf{D}(\mathbf{V}^*)^T \qquad (6.13)$$

where $\mathbf{V}^*$ is written columnwise as

$$\mathbf{V}^* = \begin{bmatrix} \mathbf{v}_s(\mathbf{x}^*(1)) & \mathbf{v}_s(\mathbf{x}^*(2)) & \cdots & \mathbf{v}_s(\mathbf{x}^*(r)) \end{bmatrix}$$

with $\mathbf{v}_s(\mathbf{x})$ as in (2.1), and $\mathbf{D}$ is an $r \times r$ diagonal matrix with entries $D(i,i) = \kappa_i^2, i = 1, \ldots, r$.

In fact, the weights $(\kappa_j)_{j=1}^r$ do not play any role in the sequel. As long as $\kappa_j \neq 0$ for all j, the rank of the moment matrix $\mathbf{M}_s(\mathbf{y}^*)$ associated with the Borel measure μ defined in (6.12) does *not* depend on the weights κ_j. The extraction procedure with another matrix $\mathbf{M}_s(\tilde{\mathbf{y}})$ written as in (6.13) but with different weights $\tilde{\kappa}_j$, would yield the same global minimizers $(\mathbf{x}^*(j))_{j=1}^r$. Of course, the new associated vector $\tilde{\mathbf{y}}$ would also be an optimal solution of the semidefinite relaxation with value $\rho_s = f^*$.

Extract a Cholesky factor $\mathbf{V}$ of the positive semidefinite moment matrix $\mathbf{M}_s(\mathbf{y}^*)$, i.e., a matrix $\mathbf{V}$ with r columns satisfying

$$\mathbf{M}_s(\mathbf{y}^*) = \mathbf{V}\mathbf{V}^T. \tag{6.14}$$

Such a Cholesky factor[2] can be obtained via singular value decomposition, or any cheaper alternative; see for example Golub and Loan (1996).

The matrices $\mathbf{V}$ and $\mathbf{V}^*$ span the same linear subspace, so the solution extraction algorithm consists of transforming $\mathbf{V}$ into $\mathbf{V}^*$ by suitable column operations. This is described in the sequel.

- **Column echelon form** Reduce $\mathbf{V}$ to column echelon form

$$\mathbf{U} = \begin{bmatrix} 1 & & & & \\ \star & & & & \\ 0 & 1 & & & \\ 0 & 0 & 1 & & \\ \star & \star & \star & & \\ \vdots & & & \ddots & \\ 0 & 0 & 0 & \cdots & 1 \\ \star & \star & \star & \cdots & \star \\ \vdots & & & & \vdots \\ \star & \star & \star & \cdots & \star \end{bmatrix}$$

by Gaussian elimination with column pivoting. By construction of the moment matrix, each row in $\mathbf{U}$ is indexed by a monomial $\mathbf{x}^\alpha$ in the basis $\mathbf{v}_s(\mathbf{x})$. Pivot elements in $\mathbf{U}$ (i.e., the first nonzero elements in each column) correspond to monomials $\mathbf{x}^{\beta_j}$, $j = 1, 2, \ldots, r$, of the basis generating the r solutions. In other words, if

$$\mathbf{w}(\mathbf{x}) = \begin{bmatrix} \mathbf{x}^{\beta_1} & \mathbf{x}^{\beta_2} & \ldots & \mathbf{x}^{\beta_r} \end{bmatrix}^T \tag{6.15}$$

denotes this generating basis, then

$$\mathbf{v}_s(\mathbf{x}) = \mathbf{U}\mathbf{w}(\mathbf{x}) \tag{6.16}$$

for all solutions $\mathbf{x} = \mathbf{x}^*(j)$, $j = 1, 2, \ldots, r$. In summary, extracting the solutions amounts to solving the system of polynomial equations (6.16).

- **Solving the system of polynomial equations (6.16)** Once a generating monomial basis $\mathbf{w}(\mathbf{x})$ is available, it turns out that extracting solutions of the system of polynomial equations (6.16) reduces to solving the linear algebra problem described below.

---

[2] *Cholesky factor* is a slight abuse of terminology since rigorously it should apply for an invertible matrix, in which case $\mathbf{V}$ can be made triangular by a change of basis.

1. **Multiplication matrices** For each degree one monomial $x_i, i = 1, 2, \ldots, n$, extract from **U** the so-called $r \times r$ *multiplication* (by x_i) matrix $\mathbf{N}_i$ containing the coefficients of monomials $x_i \, \mathbf{x}^{\beta_j}, j = 1, 2, \ldots, r$, in the generating basis (6.15), i.e., such that

$$\mathbf{N}_i \, \mathbf{w}(\mathbf{x}) = x_i \, \mathbf{w}(\mathbf{x}), \quad i = 1, 2, \ldots, n. \qquad (6.17)$$

   The entries of global minimizers $\mathbf{x}^*(j), j = 1, 2, \ldots, r$ are all eigenvalues of multiplication matrices $\mathbf{N}_i, i = 1, 2, \ldots, n$. That is,

$$\mathbf{N}_i \, \mathbf{w}(\mathbf{x}^*(j)) = x_i^*(j) \, \mathbf{w}(\mathbf{x}^*(j)), \quad i = 1, \ldots, n, \quad j = 1, \ldots, r.$$

   But how to reconstruct the solutions $(\mathbf{x}^*(j))$ from knowledge of the eigenvalues of the $\mathbf{N}_i$? Indeed, from the n r-tuples of eigenvalues one could build up to r^n possible vectors of $\mathbb{R}^n$ whereas we are looking for only r of them.

2. **Common eigenspaces** Observe that for every $j = 1, \ldots, r$, the vector $\mathbf{w}(\mathbf{x}^*(j))$ is an eigenvector common to *all* matrices $\mathbf{N}_i, i = 1, \ldots, n$. Therefore, in order to compute $(\mathbf{x}^*(j))$, one builds up a random combination

$$\mathbf{N} = \sum_{i=1}^{n} \lambda_i \, \mathbf{N}_i$$

   of the multiplication matrices $\mathbf{N}_i$, where $\lambda_i, i = 1, 2, \ldots, n$ are nonnegative real numbers summing up to one. Then with probability 1, the eigenvalues of $\mathbf{N}$ are all distinct and so $\mathbf{N}$ is nonderogatory, i.e., all its eigenspaces are one dimensional (and spanned by the vectors $\mathbf{w}(\mathbf{x}^*(j)), j = 1, \ldots, r$).

   Then, compute the ordered Schur decomposition

$$\mathbf{N} = \mathbf{Q}\mathbf{T}\mathbf{Q}^T \qquad (6.18)$$

   where $\mathbf{Q} = [\ \mathbf{q}_1 \quad \mathbf{q}_2 \quad \cdots \quad \mathbf{q}_r\ ]$ is an orthogonal matrix (i.e., $\mathbf{q}_i^T \mathbf{q}_i = 1$ and $\mathbf{q}_i^T \mathbf{q}_j = 0$ for $i \neq j$) and **T** is upper-triangular with eigenvalues of **N** sorted in increasing order along the diagonal. Finally, the ith entry $x_i^*(j)$ of $\mathbf{x}^*(j) \in \mathbb{R}^n$ is given by

$$x_i^*(j) = \mathbf{q}_j' \mathbf{N}_i \mathbf{q}_j, \quad i = 1, 2, \ldots, n, \quad j = 1, 2, \ldots, r. \qquad (6.19)$$

   In summary the extraction algorithm reads as follows.

Algorithm 6.9 (The extraction algorithm)

Input The moment matrix $\mathbf{M}_s(\mathbf{y}^*)$ of rank r.

Output The r points $\mathbf{x}^*(i) \in \mathbf{K}, i = 1, \ldots, r$, support of an optimal solution of the moment problem.

Algorithm

1. Find the **Cholesky factorization** $\mathbf{V}\,\mathbf{V}^T$ of $\mathbf{M}_s(\mathbf{y}^*)$.

2. Reduce $\mathbf{V}$ to an **echelon form U**.

3. Extract from $\mathbf{U}$ the multiplication matrices $\mathbf{N}_i$, $i = 1, \ldots, n$.

4. Compute $\mathbf{N} := \sum_{i=1}^{n} \lambda_i \mathbf{N}_i$ with randomly generated coefficients λ_i, and the Schur decomposition $\mathbf{N} = \mathbf{Q}\mathbf{T}\mathbf{Q}^T$. Compute $\mathbf{Q} = \begin{bmatrix} \mathbf{q}_1 & \mathbf{q}_2 & \cdots & \mathbf{q}_r \end{bmatrix}$ and $x_i^*(j) = \mathbf{q}_j' \mathbf{N}_i \mathbf{q}_j$, for every $i = 1, 2, \ldots, n$, and every $j = 1, 2, \ldots, r$.

Example 6.10   Consider the bivariate optimization problem with data:

$$\mathbf{x} \mapsto f(\mathbf{x}) = -(x_1 - 1)^2 - (x_1 - x_2)^2 - (x_2 - 3)^2,$$

$$\mathbf{K} = \{\mathbf{x} \in \mathbb{R}^2 : 1 - (x_1 - 1)^2 \geq 0, \ 1 - (x_1 - x_2)^2 \geq 0, \ 1 - (x_2 - 3)^2 \geq 0\}.$$

The first $(d = 1)$ semidefinite relaxation yields $\rho_1 = -3$ and rank $\mathbf{M}_1(\mathbf{y}^*) = 3$, whereas the second $(d = 2)$ semidefinite relaxation yields $\rho_2 = -2$ and rank $\mathbf{M}_1(\mathbf{y}^*) = \text{rank}\,\mathbf{M}_2(\mathbf{y}^*) = 3$. Hence, the rank condition (6.11) is satisfied, which implies that $-2 = \rho_2 = f^*$. The moment matrix $\mathbf{M}_2(\mathbf{y}^*)$ reads

$$\mathbf{M}_2(\mathbf{y}^*) = \begin{bmatrix} 1.0000 & 1.5868 & 2.2477 & 2.7603 & 3.6690 & 5.2387 \\ 1.5868 & 2.7603 & 3.6690 & 5.1073 & 6.5115 & 8.8245 \\ 2.2477 & 3.6690 & 5.2387 & 6.5115 & 8.8245 & 12.7072 \\ 2.7603 & 5.1073 & 6.5115 & 9.8013 & 12.1965 & 15.9960 \\ 3.6690 & 6.5115 & 8.8245 & 12.1965 & 15.9960 & 22.1084 \\ 5.2387 & 8.8245 & 12.7072 & 15.9960 & 22.1084 & 32.1036 \end{bmatrix}$$

and the monomial basis is

$$\mathbf{v}_2(\mathbf{x}) = \begin{bmatrix} 1 & x_1 & x_2 & x_1^2 & x_1 x_2 & x_2^2 \end{bmatrix}^T.$$

The Cholesky factor (6.14) of $\mathbf{M}_2(\mathbf{y}^*)$ is given by

$$\mathbf{V} = \begin{bmatrix} -0.9384 & -0.0247 & 0.3447 \\ -1.6188 & 0.3036 & 0.2182 \\ -2.2486 & -0.1822 & 0.3864 \\ -2.9796 & 0.9603 & -0.0348 \\ -3.9813 & 0.3417 & -0.1697 \\ -5.6128 & -0.7627 & -0.1365 \end{bmatrix}$$

with column echelon form which reads (after rounding)

$$
U = \begin{bmatrix}
1 & & \\
0 & 1 & \\
0 & 0 & 1 \\
-2 & 3 & 0 \\
-4 & 2 & 2 \\
-6 & 0 & 5
\end{bmatrix}.
$$

Pivot entries correspond to the following generating basis (6.15)

$$
\mathbf{w}(\mathbf{x}) = \begin{bmatrix} 1 & x_1 & x_2 \end{bmatrix}^T.
$$

From the subsequent rows in matrix U we deduce from (6.16) that all solutions $\mathbf{x}$ must satisfy the three polynomial equations

$$
\begin{aligned}
x_1^2 &= -2 + 3x_1 \\
x_1 x_2 &= -4 + 2x_1 + 2x_2 \\
x_2^2 &= -6 + 5x_2.
\end{aligned}
$$

Multiplication matrices (6.17) (by x_1 and x_2) in the basis $\mathbf{w}(\mathbf{x})$ are readily extracted from rows in U:

$$
N_1 = \begin{bmatrix}
0 & 1 & 0 \\
-2 & 3 & 0 \\
-4 & 2 & 2
\end{bmatrix}, \quad
N_2 = \begin{bmatrix}
0 & 0 & 1 \\
-4 & 2 & 2 \\
-6 & 0 & 5
\end{bmatrix}.
$$

A randomly chosen convex combination of N_1 and N_2 yields

$$
N = 0.6909\, N_1 + 0.3091\, N_2 = \begin{bmatrix}
0 & 0.6909 & 0.3091 \\
-2.6183 & 2.6909 & 0.6183 \\
-4.6183 & 1.3817 & 2.9274
\end{bmatrix}
$$

with orthogonal matrix in Schur decomposition (6.18) given by

$$
Q = \begin{bmatrix}
0.4082 & 0.1826 & -0.8944 \\
0.4082 & -0.9129 & -0.0000 \\
0.8165 & 0.3651 & 0.4472
\end{bmatrix}.
$$

From equation (6.19), we obtain the three optimal solutions

$$
\mathbf{x}^*(1) = \begin{bmatrix} 1 \\ 2 \end{bmatrix}, \quad
\mathbf{x}^*(2) = \begin{bmatrix} 2 \\ 2 \end{bmatrix}, \quad
\mathbf{x}^*(3) = \begin{bmatrix} 2 \\ 3 \end{bmatrix}.
$$

6.1.3 The univariate case

If we consider the univariate case $n = 1$ with $\mathbf{K} = [a, b]$ or $\mathbf{K} = [0, \infty)$, the corresponding hierarchy of semidefinite relaxations (6.3) simply reduces to a *single* relaxation. In other words, the minimization of a univariate polynomial on an interval of the real line (bounded or not) is a convex optimization problem and reduces to solving a semidefinite program.

Indeed, consider for instance the case where f has degree $2d$ and $\mathbf{K} \subset \mathbb{R}$ is the interval $[-1, 1]$. If $f^* = \inf\{ f(x) : x \in [-1, 1]\}$ then $f - f^* \geq 0$ on $[-1, 1]$ and by Theorem 2.4(b)

$$f(x) - f^* = f_0(x) + f_3(x)(1 - x^2), \qquad x \in \mathbb{R}, \qquad (6.20)$$

for some SOS polynomials $f_0, f_3 \in \Sigma[x]$ such that the degree of each summand is less than $2d$.

Theorem 6.11   *Let $f \in \mathbb{R}[x]$ be a univariate polynomial of degree $2d$. The semidefinite relaxation (6.3) of the problem*

$$f^* = \inf_x \{ f(x) : 1 - x^2 \geq 0 \},$$

is exact, i.e., $\rho_d = f^*$. In addition $\rho_d^* = f^*$ and both (6.3) and (6.4) have an optimal solution.

Proof   From (6.20), (f^*, f_0, f_3) is a feasible solution of (6.4) and so optimal because $\rho_d^* \leq f^*$ for all d. Therefore, from $\rho_d^* \leq \rho_d \leq f^*$, we also obtain $\rho_d = f^*$, which in turn implies that $\mathbf{y}^* := \mathbf{v}_{2d}(x^*)$ is an optimal solution, for any global minimizer $x^* \in \mathbf{K}$.   □

A similar argument holds if $f \in \mathbb{R}[x]$ has odd degree $2d - 1$, in which case

$$f(x) - f^* = f_1(x)(1 + x) + f_2(x)(1 - x), \qquad x \in \mathbb{R},$$

for some SOS polynomials $f_1, f_2 \in \Sigma[x]$ such that the degree of each summand is less than $2d - 1$. Again, for the problem

$$f^* = \inf_x \{ f(x) : 1 - x \geq 0, \ 1 + x \geq 0 \},$$

the relaxation (6.3) is exact.

Example 6.12   Let $d = 2$ and $x \mapsto f(x) = 1 + 0.1x + 5x^2/2 - 3x^4$ and $\mathbf{K} = [-1, 1]$ displayed in Figure 6.1. Then computing the global optimum f^* on $\mathbf{K}$ reduces to solving the single semidefinite program:

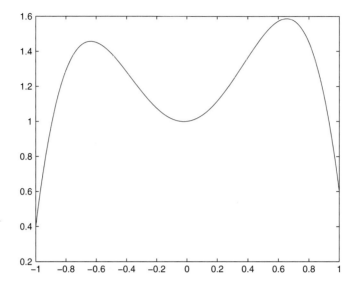

Figure 6.1 Example 6.12: $f(x) = 1 + 0.1x + 5x^2/2 - 3x^4$.

$$\rho_2 = \inf_{\mathbf{y}} \left\{ y_0 + 0.1y_1 + 5y_2/2 - 3y_4 \right.$$

$$\text{s.t.} \quad \begin{bmatrix} y_0 & y_1 & y_2 \\ y_1 & y_2 & y_3 \\ y_2 & y_3 & y_4 \end{bmatrix} \succeq 0; \quad \begin{bmatrix} y_0 - y_2 & y_1 - y_3 \\ y_1 - y_3 & y_2 - y_4 \end{bmatrix} \succeq 0; \quad \left. y_0 = 1 \right\},$$

in which case $f^* = \rho_2 = 0.4$ and $x^* = -1$.

So finding the global minimum of a univariate polynomial on an interval (finite or not) is a convex optimization problem as it reduces to solving the single semidefinite program (6.3) (and not a hierarchy of semidefinite programs as in the multivariate case).

6.1.4 Numerical experiments

In this section, we report on the performance of Algorithm 6.1 using the software package GloptiPoly (described in Appendix B) on a series of benchmarks for nonconvex continuous optimization problems. It is worth noticing that Algorithm 6.9 is implemented in GloptiPoly to extract global minimizers.

In Table 6.1, we record the problem name, the source of the problem, the number of decision variables (var), the number of inequality or equality constraints (cstr), and the maximum degree arising in the polynomial expressions (deg), the CPU time in seconds (CPU) and the order of the relaxation (order).

Table 6.1 *Continuous optimization problems, CPU times and semidefinite relaxation orders required to reach global optimality*

Problem	var	cstr	deg	CPU	order
Lasserre (2001, Ex. 1)	2	0	4	0.13	2
Lasserre (2001, Ex. 2)	2	0	4	0.13	2
Lasserre (2001, Ex. 3)	2	0	6	1.13	8
Lasserre (2001, Ex. 5)	2	3	2	0.22	2
Floudas et al. (1999, Pb. 2.2)	5	11	2	11.8	3
Floudas et al. (1999, Pb. 2.3)	6	13	2	1.86	2
Floudas et al. (1999, Pb. 2.4)	13	35	2	1012	2
Floudas et al. (1999, Pb. 2.5)	6	15	2	1.58	2
Floudas et al. (1999, Pb. 2.6)	10	31	2	67.7	2
Floudas et al. (1999, Pb. 2.7)	10	25	2	75.3	2
Floudas et al. (1999, Pb. 2.8)	20	10	2	–	dim
Floudas et al. (1999, Pb. 2.9)	24	10	2	–	dim
Floudas et al. (1999, Pb. 2.10)	10	11	2	45.3	2
Floudas et al. (1999, Pb. 2.11)	20	10	2	–	dim
Floudas et al. (1999, Pb. 3.2)	8	22	2	3032	3
Floudas et al. (1999, Pb. 3.3)	5	16	2	1.20	2
Floudas et al. (1999, Pb. 3.4)	6	16	2	1.50	2
Floudas et al. (1999, Pb. 3.5)	3	8	2	2.42	4
Floudas et al. (1999, Pb. 4.2)	1	2	6	0.17	3
Floudas et al. (1999, Pb. 4.3)	1	2	50	0.94	25
Floudas et al. (1999, Pb. 4.4)	1	2	5	0.25	3
Floudas et al. (1999, Pb. 4.5)	1	2	4	0.14	2
Floudas et al. (1999, Pb. 4.6)	2	2	6	0.41	3
Floudas et al. (1999, Pb. 4.7)	1	2	6	0.20	3
Floudas et al. (1999, Pb. 4.8)	1	2	4	0.16	2
Floudas et al. (1999, Pb. 4.9)	2	5	4	0.31	2
Floudas et al. (1999, Pb. 4.10)	2	6	4	0.58	4

At the time of the experiment, GloptiPoly was using the semidefinite optimization solver SeDuMi; see Sturm (1999). As indicated by the label **dim** in the rightmost column, quadratic problems 2.8, 2.9 and 2.11 in Floudas et al. (1999) involve more than 19 variables and could not be handled by the current version of GloptiPoly. Except for problems 2.4 and 3.2, the computational load is moderate. In almost all reported instances the global optimum was reached exactly by a semidefinite relaxation of small order.

6.2 Discrete optimization

In this section, we consider problem (6.1) with $\mathbf{K}$ being a (finite) real variety. More precisely, in the definition (5.2) of $\mathbf{K}$, the g_i are such that $\mathbf{K}$ can be rewritten in the general form (6.5) with equality and inequality constraints, i.e.,

$$\mathbf{K} = \{ \mathbf{x} \in \mathbb{R}^n : g_j(\mathbf{x}) \geq 0, \ j = 1, \ldots, m_1; \ h_\ell(\mathbf{x}) = 0, \ \ell = 1, \ldots, m_2 \},$$
(6.21)

for some integers m_1, m_2 and some polynomials $g_j, h_\ell \in \mathbb{R}[\mathbf{x}]$. Indeed, (6.21) is a particular case of (5.2) where some inequality constraints $g_i(\mathbf{x}) \geq 0$ and $-g_i(\mathbf{x}) \geq 0$ are present.

The polynomials $(h_\ell)_{\ell=1}^{m_2} \subset \mathbb{R}[\mathbf{x}]$ define an ideal $J := \langle h_1, \ldots, h_{m_2} \rangle \subset \mathbb{R}[\mathbf{x}]$, and we will consider the case where J is a *zero-dimensional* ideal, that is, the algebraic variety

$$V_{\mathbb{C}}(J) := \{ \mathbf{x} \in \mathbb{C}^n : h_\ell(\mathbf{x}) = 0, \ \ell = 1, \ldots, m_2 \},$$

is a finite set; see Section 2.5.

This is an important special case as it covers 0/1 and discrete optimization problems. For instance, if we let $h_\ell(\mathbf{x}) := x_\ell^2 - x_\ell$, for all $\ell = 1, \ldots, n$, then we recover 0/1 optimization, in which case, $V_{\mathbb{C}}(J) = \{0, 1\}^n$ and the ideal $J = \langle x_1^2 - x_1, \ldots, x_n^2 - x_n \rangle$ is even radical (and real). Similarly, given $(r_\ell)_{\ell=1}^n \subset \mathbb{N}$, and a finite set of points $(x_{\ell j})_{j=1}^{r_\ell} \subset \mathbb{R}, \ \ell = 1, \ldots, n$, let

$$h_\ell(\mathbf{x}) := \prod_{j=1}^{r_\ell} (x_\ell - x_{\ell j}), \quad \ell = 1, \ldots, n.$$

Then we recover (bounded) integer optimization problems, in which case $V_{\mathbb{C}}(J)$ is the *grid* $\{(x_{1j_1}, x_{2j_2}, \cdots, x_{nj_n})\}$, where $j_\ell \in \{1, \ldots, r_\ell\}$ for all $\ell = 1, \ldots, n$.

Theorem 6.13   *Let $f \in \mathbb{R}[\mathbf{x}]$, $\mathbf{K} \subset \mathbb{R}^n$ be as in (6.21) and let the ideal $J = \langle h_1, \ldots, h_{m_2} \rangle$ be zero dimensional. Let ρ_d and ρ_d^* be the respective optimal values of the semidefinite relaxation (6.6) and its dual (6.7). Then, there is some $d_0 \in \mathbb{N}$ such that $\rho_d = \rho_d^* = f^*$ for all $d \geq d_0$. In addition both (6.6) and its dual have an optimal solution.*

Proof   The polynomial $f - f^*$ is nonnegative on $\mathbf{K}$. Therefore, by Theorem 2.26 there exist polynomials $(\psi_\ell)_{\ell=1}^{m_2} \subset \mathbb{R}[\mathbf{x}]$, and SOS polynomials $(\sigma_j)_{j=0}^{m_1} \subset \Sigma[\mathbf{x}]$, such that:

$$f - f^* = \sum_{j=0}^{m_1} \sigma_j \, g_j + \sum_{\ell=1}^{m_2} \psi_\ell \, h_\ell$$

(recall that $g_0(\mathbf{x}) = 1$ for all $\mathbf{x}$). Let d_1, d_2 be the maximum degree of the polynomials $(\sigma_j \, g_j)_{j=0}^{m_1}$ and $(\psi_\ell \, h_\ell)_{\ell=1}^{m_2}$ respectively, and let $2d_0 \geq \max[d_1, d_2]$.

Then, $(f^*, (\sigma_j), (\psi_\ell))$ is a feasible solution of the semidefinite relaxation (6.7) for $d = d_0$, and with value f^*, so that $\rho^*_{d_0} \geq f^*$. As we also have $\rho^*_d \leq \rho_d \leq f^*$ whenever the semidefinite relaxations are well defined, we conclude that $\rho^*_d = \rho_d = f^*$ for all $d \geq d_0$. Finally, let μ be the Dirac probability measure at some global minimizer $\mathbf{x}^* \in \mathbf{K}$ of problem (6.1), and let $\mathbf{y} \subset \mathbb{R}$ be the vector of its moments. Then $\mathbf{y}$ is feasible for all the semidefinite relaxations (6.6), with value f^*, which completes the proof. $\qquad\square$

In fact, one may refine Theorem 6.13 and remove the assumption that the ideal is zero dimensional but the real variety $V_{\mathbb{R}}(J)$ is finite.

Theorem 6.14 *Let $f \in \mathbb{R}[\mathbf{x}]$, $\mathbf{K} \subset \mathbb{R}^n$ be as in (6.21) and assume that the ideal $J = \langle h_1, \ldots, h_{m_2} \rangle$ is such that $V_{\mathbb{R}}(J) = \{ \mathbf{x} \in \mathbb{R}^n : h_\ell(\mathbf{x}) = 0, \ell = 1, \ldots, m_2 \}$ is finite. Let ρ_d be the optimal value of the semidefinite relaxation (6.6). Then, there is some $d_0 \in \mathbb{N}$ such that (6.6) has an optimal solution and $\rho_d = f^*$ for all $d \geq d_0$.*

Notice that in contrast with Theorem 6.13, in Theorem 6.14 the semidefinite program (6.7) (dual of (6.6)) may not have an optimal solution for $d \geq d_0$.

6.2.1 0/1 optimization

It is worth noting that in the semidefinite relaxations (6.6), the constraints $\mathbf{M}_d(h_\ell\, \mathbf{y}) = 0$ translate into simplifications via elimination of variables in the moment matrix $\mathbf{M}_d(\mathbf{y})$ and the localizing matrices $\mathbf{M}_d(g_j\, \mathbf{y})$. Indeed, consider for instance the Boolean optimization case, i.e., when $h_\ell(\mathbf{x}) = x_\ell^2 - x_\ell$ for all $\ell = 1, \ldots, n$. Then the constraints $\mathbf{M}_d(h_\ell\, \mathbf{y}) = \mathbf{0}$ for all $\ell = 1, \ldots, n$, simply state that whenever $|\alpha| \leq 2d$, one replaces every variable y_α with the variable y_β, where for $k = 1, \ldots, n$:

$$\beta_k = \begin{cases} 0 & \text{if } \alpha_k = 0, \\ 1 & \text{otherwise.} \end{cases}$$

Indeed, with $x_i^2 = x_i$ for all $i = 1, \ldots, n$, one has $\mathbf{x}^\alpha = \mathbf{x}^\beta$, with β as above. For instance, with $n = 2$, we obtain

$$\mathbf{M}_2(\mathbf{y}) = \begin{bmatrix} y_{00} & y_{10} & y_{01} & y_{10} & y_{11} & y_{01} \\ y_{10} & y_{10} & y_{11} & y_{10} & y_{11} & y_{11} \\ y_{01} & y_{11} & y_{01} & y_{11} & y_{11} & y_{01} \\ y_{10} & y_{10} & y_{11} & y_{10} & y_{11} & y_{11} \\ y_{11} & y_{11} & y_{11} & y_{11} & y_{11} & y_{11} \\ y_{01} & y_{11} & y_{01} & y_{11} & y_{11} & y_{01} \end{bmatrix}.$$

In addition, every column (row) of $\mathbf{M}_d(\mathbf{y})$ corresponding to a monomial $\mathbf{x}^\alpha$, with $\alpha_k > 1$ for some $k \in \{1, \ldots, n\}$, is identical to the column corresponding to the monomial $\mathbf{x}^\beta$, with β as above. Hence, the constraint $\mathbf{M}_d(\mathbf{y}) \succeq \mathbf{0}$ reduces to the new constraint $\widehat{\mathbf{M}}_d(\mathbf{y}) \succeq \mathbf{0}$, with the new simplified moment matrix

$$
\widehat{\mathbf{M}}_2(\mathbf{y}) =
\begin{bmatrix}
y_{00} & y_{10} & y_{01} & y_{11} \\
y_{10} & y_{10} & y_{11} & y_{11} \\
y_{01} & y_{11} & y_{01} & y_{11} \\
y_{11} & y_{11} & y_{11} & y_{11}
\end{bmatrix}.
$$

Theorem 6.13 has little practical value. For instance, in the case of Boolean optimization, we may easily show that $\rho_d = \rho_{d_0}$ for all $d \geq d_0 := n$. This is because every column associated with $\alpha \in \mathbb{N}^n$, $|\alpha| > n$, is identical to another column associated with some $\beta \in \mathbb{N}^n$, $|\beta| \leq n$. Equivalently, rank $\mathbf{M}_d(\mathbf{y}) = $ rank $\mathbf{M}_n(\mathbf{y})$ for all $d > n$. But in this case the simplified matrix $\widehat{\mathbf{M}}_n(\mathbf{y})$ has size $2^n \times 2^n$ and solving problem (6.1) by simple enumeration would be as efficient! However, in general one expects to obtain the exact global optimum f^* at some (much) earlier relaxation $d \ll d_0$.

6.2.2 Numerical experiments

The performance of GloptiPoly has also been reported for a series of small-size combinatorial optimization problems (in particular, the MAXCUT problem). In Table 6.2 we first let GloptiPoly converge to the global optimum, in general extracting several solutions. The number of extracted solutions is reported in the column labelled sol; "dim" indicates that the problem could not be solved because of excessive memory requirement.

Then, we slightly perturbed the criterion to be optimized in order to destroy the problem symmetry. Proceeding this way, the optimum solution is generically unique and convergence to the global optimum is easier.

Of course, the size of the combinatorial problems is relatively small and GloptiPoly cannot compete with ad hoc heuristics which may solve problems with many more variables. But these numerical experiments are reported only to show the potential of the method, as in most cases the global optimum is reached at the second semidefinite relaxation in the hierarchy.

6.3 Unconstrained polynomial optimization

If $\mathbf{K} = \mathbb{R}^n$, the only interesting case is when deg f is even, otherwise necessarily $f^* = -\infty$. So let $2d$ be the degree of $f \in \mathbb{R}[\mathbf{x}]$, let $\ell \geq d$, and consider the semidefinite program:

Table 6.2 *Discrete optimization problems, CPU times and semidefinite relaxation order required to reach f^* and extract several solutions*

Problem	var	cstr	deg	CPU	order	sol
QP Floudas et al. (1999, Pb. 13.2.1.1)	4	4	2	0.10	1	1
QP Floudas et al. (1999, Pb. 13.2.1.2)	10	0	2	3.61	2	1
P_1 Floudas et al. (1999)	10	0	2	38.1	3	10
P_2 Floudas et al. (1999)	10	0	2	2.7	2	2
P_3 Floudas et al. (1999)	10	0	2	2.6	2	2
P_4 Floudas et al. (1999)	10	0	2	2.6	2	2
P_5 Floudas et al. (1999)	10	0	2	–	4	dim
P_6 Floudas et al. (1999)	10	0	2	2.6	2	2
P_7 Floudas et al. (1999)	10	0	2	44.3	3	4
P_8 Floudas et al. (1999)	10	0	2	2.6	2	2
P_9 Floudas et al. (1999)	10	0	2	49.3	3	6
cycle C_5 Anjos (2001)	5	0	2	0.19	3	10
complete K_5 Anjos (2001)	5	0	2	0.19	4	20
5-node Anjos (2001)	5	0	2	0.24	3	6
antiweb AW_9^2 Anjos (2001)	9	0	2	–	4	dim
10-node Petersen Anjos (2001)	10	0	2	39.6	3	10
12-node Anjos (2001)	12	0	2	–	3	dim

$$\rho_\ell = \inf_{\mathbf{y}} \{ \, L_{\mathbf{y}}(f) \, : \, \mathbf{M}_\ell(\mathbf{y}) \succeq \mathbf{0}, \ y_0 = 1 \, \}, \tag{6.22}$$

where $\mathbf{M}_\ell(\mathbf{y})$ is the moment matrix of order ℓ associated with the sequence $\mathbf{y}$, already defined in Section 2.7.1.

Again writing $\mathbf{M}_\ell(\mathbf{y}) = \sum_{|\alpha| \le 2\ell} y_\alpha \mathbf{C}_\alpha^0$ for appropriate symmetric matrices $(\mathbf{C}_\alpha^0)$, the dual of (6.22) is the semidefinite program:

$$\rho_\ell^* = \sup_{\lambda, \mathbf{X}} \{ \, \lambda \, : \, \mathbf{X} \succeq \mathbf{0}; \quad \langle \mathbf{X}, \mathbf{C}_\alpha^0 \rangle = f_\alpha - \lambda \delta_{\alpha=0}, \quad \forall \, |\alpha| \le 2\ell \, \} \tag{6.23}$$

where $\delta_{\alpha=0}$ is the Kronecker symbol at $\alpha = 0$. Recall the definition of the vector $\mathbf{v}_d(\mathbf{x})$ in Section 2.1. Mutliplying each side of the constraints by $\mathbf{x}^\alpha$ and summing up yields:

$$\sum_{\alpha \in \mathbb{N}^n} f_\alpha \mathbf{x}^\alpha - \lambda = \left\langle \mathbf{X}, \sum_{\alpha \in \mathbb{N}^n} \mathbf{C}_\alpha^0 \mathbf{x}^\alpha \right\rangle = \langle \mathbf{X}, \mathbf{v}_\ell(\mathbf{x}) \mathbf{v}_\ell(\mathbf{x})^T \rangle$$

$$= \sum_{k=1}^s \langle \mathbf{q}_k \mathbf{q}_k^T, \mathbf{v}_\ell(\mathbf{x}) \mathbf{v}_\ell(\mathbf{x})^T \rangle = \sum_{k=1}^s (\mathbf{q}_k^T \mathbf{v}_\ell(\mathbf{x}))^2 = \sum_{k=1}^s q_k(\mathbf{x})^2,$$

where we have used the spectral decomposition $\mathbf{X} = \sum_k \mathbf{q}_k \mathbf{q}_k^T$ of the positive semidefinite matrix $\mathbf{X}$, and interpreted $\mathbf{q}_k$ as the vector of coefficients of the polynomial $q_k \in \mathbb{R}[\mathbf{x}]_\ell$. Hence, (6.23) has the equivalent compact formulation:

$$\rho_\ell^* = \sup_\lambda \{\lambda : f - \lambda \in \Sigma(\mathbf{x})_\ell\}. \qquad (6.24)$$

Therefore, $\rho_\ell^* = \rho_d^*$ for every $\ell \geq d$ because obviously, if f has degree $2d$ then $f - \lambda$ cannot be a sum of squares of polynomials with degree larger than d.

Proposition 6.15 *There is no duality gap, that is, $\rho_d = \rho_d^*$. Moreover, if $\rho_d > -\infty$ then (6.23) has an optimal solution.*

Proof The result follows from standard duality in conic optimization if we can prove that there is a strictly feasible solution $\mathbf{y}$ of problem (6.22), i.e., such that $\mathbf{M}_d(\mathbf{y}) \succ \mathbf{0}$ (Slater condition). So let μ be a probability measure on $\mathbb{R}^n$ with a strictly positive density f with respect to the Lebesgue measure and with all its moments finite; that is, μ is such that

$$y_\alpha = \int_{\mathbb{R}^n} \mathbf{x}^\alpha \, d\mu = \int_{\mathbb{R}^n} \mathbf{x}^\alpha f(\mathbf{x}) \, d\mathbf{x} < \infty, \quad \forall \alpha \in \mathbb{N}^n.$$

Then $\mathbf{M}_d(\mathbf{y})$, with $\mathbf{y}$ as above, is such that $\mathbf{M}_d(\mathbf{y}) \succ \mathbf{0}$. To see this, recall that for every polynomial $q \in \mathbb{R}[\mathbf{x}]_d$ of degree at most d, and vector of coefficients $\mathbf{q} \in \mathbb{R}^{s(d)}$,

$$\begin{aligned}
\langle q, q \rangle_{\mathbf{y}} = \langle \mathbf{q}, \mathbf{M}_d(\mathbf{y})\mathbf{q} \rangle &= \int_{\mathbb{R}^n} q^2 \, d\mu && \text{[from (2.34)]} \\
&= \int_{\mathbb{R}^n} q(\mathbf{x})^2 f(\mathbf{x}) \, d\mathbf{x} \\
&> 0, && \text{whenever } q \neq 0 \quad (\text{as } f > 0).
\end{aligned}$$

Therefore, $\mathbf{y}$ is strictly feasible for problem (6.22), i.e., $\mathbf{M}_d(\mathbf{y}) \succ \mathbf{0}$, the desired result.   □

We next prove the main result of this section.

Theorem 6.16 *Let $f \in \mathbb{R}[\mathbf{x}]$ be a $2d$-degree polynomial with global minimum f^* on $\mathbf{K} = \mathbb{R}^n$.*

(a) *If the nonnegative polynomial $f - f^*$ is SOS, then problem (6.1) is equivalent to the semidefinite optimization problem (6.22), i.e., $f^* = \rho_d^* = \rho_d$ and if $\mathbf{x}^* \in \mathbb{R}^n$ is a global minimizer of (6.1), then the moment vector*

$$\mathbf{y}^* := (x_1^*, \ldots, x_n^*, (x_1^*)^2, x_1^* x_2^*, \ldots, (x_1^*)^{2d}, \ldots, (x_n^*)^{2d}) \qquad (6.25)$$

is a minimizer of problem (6.22).

(b) *If problem (6.23) has a feasible solution and $f^* = \rho_d^*$, then $f - f^*$ is SOS.*

Proof   (a) Let $f - f^*$ be SOS, that is,

$$f(\mathbf{x}) - f^* = \sum_{\ell=1}^{k} q_\ell(\mathbf{x})^2, \qquad \mathbf{x} \in \mathbb{R}^n, \qquad (6.26)$$

for some polynomials $\{q_\ell\}_{\ell=1}^{k} \subset \mathbb{R}[\mathbf{x}]_d$ with coefficient vectors $\mathbf{q}_\ell \in \mathbb{R}^{s(d)}$, $\ell = 1, 2, \ldots, k$, with $s(d) = \binom{n+d}{n}$. Equivalently, with $\mathbf{v}_d(\mathbf{x})$ as in (2.1),

$$f(\mathbf{x}) - f^* = \langle \mathbf{X}, \mathbf{M}_d(\mathbf{y}) \rangle, \qquad \mathbf{x} \in \mathbb{R}^n, \qquad (6.27)$$

with $\mathbf{X} = \sum_{\ell=1}^{k} \mathbf{q}_\ell \mathbf{q}_\ell^T \succeq 0$ and $\mathbf{y} = \mathbf{v}_{2d}(\mathbf{x})$. From (6.27) it follows that

$$\langle \mathbf{X}, \mathbf{C}_0^0 \rangle = f_0 - f^*, \quad \langle \mathbf{X}, \mathbf{C}_\alpha^0 \rangle = f_\alpha, \text{ for all } 0 \neq \alpha, \ |\alpha| \leq 2d,$$

so that (as $\mathbf{X} \succeq \mathbf{0}$) $\mathbf{X}$ is feasible for problem (6.23) with value $\lambda = f^*$. Since $\mathbf{y}^*$ in (6.25) is feasible for problem (6.22) with value f^* and $\rho_d^* = \rho_d$, it follows that $\mathbf{y}^*$ and $\mathbf{X}$ are optimal solutions to problems (6.22) and (6.23), respectively.

(b) Suppose that problem (6.23) has a feasible solution and $f^* = \rho_d^*$. Then, by Proposition 6.15, problem (6.23) has an optimal solution $(\mathbf{X}^*, f^*)$, with $\mathbf{X}^* \succeq \mathbf{0}$, and there is no duality gap, that is, $\rho_d = \rho_d^*$. As $\mathbf{X}^* \succeq \mathbf{0}$, we use its spectral decomposition to write $\mathbf{X}^* = \sum_{\ell=1}^{k} \mathbf{q}_\ell \mathbf{q}_\ell^T$ for some vectors $(\mathbf{q}_\ell) \subset \mathbb{R}^{s(d)}$. Using the feasibility of $(\mathbf{X}^*, f^*)$ in (6.23), we obtain

$$\left\langle \mathbf{X}^*, \sum_{\alpha} \mathbf{C}_\alpha^0 \mathbf{x}^\alpha \right\rangle = f(\mathbf{x}) - f^*,$$

which, using $\mathbf{X}^* = \sum_{\ell=1}^{k} \mathbf{q}_\ell \mathbf{q}_\ell^T$ and

$$\sum_{\alpha \in \mathbb{N}_{2d}^n} \mathbf{C}_\alpha^0 \mathbf{x}^\alpha = \mathbf{M}_d(\mathbf{v}_{2d}(\mathbf{x})) = \mathbf{v}_d(\mathbf{x}) \, \mathbf{v}_d(\mathbf{x})^T,$$

yields the desired sum of squares

$$f(\mathbf{x}) - f^* = \sum_{\ell=1}^{k} \langle \mathbf{q}_\ell \mathbf{q}_\ell^T, \mathbf{v}_d(\mathbf{x}) \mathbf{v}_d(\mathbf{x})^T \rangle = \sum_{\ell=1}^{k} \langle \mathbf{q}_\ell, \mathbf{v}_d(\mathbf{x}) \rangle^2 = \sum_{\ell=1}^{k} q_\ell(\mathbf{x})^2,$$

with the polynomials $\mathbf{x} \mapsto q_\ell(\mathbf{x}) := \langle \mathbf{q}_\ell, \mathbf{v}_d(\mathbf{x}) \rangle$, for all $\ell = 1, \ldots, k$. $\square$

From the proof of Theorem 6.16, it is obvious that if $f^* = \rho_d^*$, then any global minimizer $\mathbf{x}^*$ of f is a zero of each polynomial q_ℓ, where $\mathbf{X}^* = \sum_{\ell=1}^{k} \mathbf{q}_\ell \mathbf{q}_\ell^T$ at an optimal solution $\mathbf{X}^*$ of problem (6.23). When $f - f^*$ is SOS, solving problem (6.23) provides the polynomials q_ℓ of such a decomposition. As a corollary, we obtain the following.

Corollary 6.17   *Let $f \in \mathbb{R}[\mathbf{x}]$ be of degree 2d. Assume that problem (6.23) has a feasible solution. Then,*

$$f(\mathbf{x}) - f^* = \sum_{\ell=1}^{k} q_\ell(\mathbf{x})^2 - [f^* - \rho_d^*], \quad \mathbf{x} \in \mathbb{R}^n, \tag{6.28}$$

for some real-valued polynomials $q_\ell \in \mathbb{R}[\mathbf{x}]_d$, $\ell = 1, 2, \ldots, k$.

The proof is the same as that of Theorem 6.16(b), except now we may not have $f^* = \rho_d^*$, but instead $\rho_d^* \leq f^*$. Hence, ρ_d^* always provides a lower bound on f^*.

Corollary 6.17 states that, up to some constant, one may always write $f - f^*$ as an SOS whenever problem (6.23) has a feasible solution. The previous development leads to the following algorithm either for solving problem (6.1) or for providing a lower bound on its optimal value f^*.

Algorithm 6.18   (Unconstrained polynomial optimization)

Input A polynomial $\mathbf{x} \mapsto f(\mathbf{x}) = \sum_{\alpha \in \mathbb{N}_{2d}^n} f_\alpha \mathbf{x}^\alpha$ of degree $2d$.

Output The value $f^* = \inf_{\mathbf{x} \in \mathbb{R}^n} f(\mathbf{x})$ or a lower bound ρ_d on f^*.
Algorithm

1. Solve the semidefinite optimization problem (6.22) with optimal value ρ_d and optimal solution $\mathbf{y}^*$ (if $\mathbf{y}^*$ exists).
2. (a) If $\operatorname{rank} \mathbf{M}_d(\mathbf{y}^*) \leq 3d - 3$ if $d \geq 3$ or $\operatorname{rank} \mathbf{M}_d(\mathbf{y}^*) \leq 6$ if $d = 2$, then $f^* = \rho_d$, and there are at least $\operatorname{rank} \mathbf{M}_d(\mathbf{y}^*)$ global minimizers.
   (b) If $\operatorname{rank} \mathbf{M}_{d-1}(\mathbf{y}^*) = \operatorname{rank} \mathbf{M}_d(\mathbf{y}^*)$, then $f^* = \rho_d$ and there are at least $\operatorname{rank} \mathbf{M}_d(\mathbf{y}^*)$ global minimizers which can be extracted by Algorithm 6.9.
3. Otherwise ρ_d is only a lower bound of f^*.

We next show that Algorithm 6.18 correctly determines whether ρ_d is the exact solution value f^* or a lower bound.

Theorem 6.19   *Let $f \in \mathbb{R}[\mathbf{x}]$ with degree 2d, and suppose that the optimal value ρ_d of problem (6.22) is attained at some optimal solution $\mathbf{y}^*$.*

(a) If $\operatorname{rank} \mathbf{M}_d(\mathbf{y}^*) \leq 3d - 3$ if $d \geq 3$ or $\operatorname{rank} \mathbf{M}_d(\mathbf{y}^*) \leq 6$ if $d = 2$, then $f^* = \rho_d$, and there are at least $\operatorname{rank} \mathbf{M}_d(\mathbf{y}^*)$ global minimizers.
(b) If $\operatorname{rank} \mathbf{M}_{d-1}(\mathbf{y}^*) = \operatorname{rank} \mathbf{M}_d(\mathbf{y}^*)$, then $f^* = \rho_d$, and there are at least $\operatorname{rank} \mathbf{M}_d(\mathbf{y}^*)$ global minimizers which can be extracted by Algorithm 6.9.

Proof   (a) We have already shown that $\rho_d \leq f^*$. Next if $s := \text{rank}\, \mathbf{M}_d(\mathbf{y}^*) \leq 3d - 3$ (or ≤ 6 if $d = 2$) then by Theorem 2.36, $\mathbf{y}^*$ is coming from an atomic measure μ^* supported on s points of $\mathbb{R}^n$. Therefore, $\rho_d = L_{\mathbf{y}}(f) = \int f \, d\mu^*$, which proves that μ^* is an optimal solution of (5.3), and $\rho_d = f^*$ because we always have $\rho_d \leq f^*$. We next show that each of the s atoms of μ^* is a global minimizer of f. Indeed, being s-atomic, there is a family $(\mathbf{x}(k))_{k=1}^s \subset \mathbb{R}^n$ and a family $(\beta_k)_{k=1}^s \subset \mathbb{R}$, such that

$$\mu^* = \sum_{k=1}^s \beta_k \, \delta_{\mathbf{x}(k)}, \quad \beta_k > 0, \quad \forall k = 1, \ldots, s; \quad \sum_{k=1}^s \beta_k = 1.$$

Hence,

$$f^* = \rho_d = \int_{\mathbb{R}^n} f \, d\mu^* = \sum_{k=1}^s \beta_k \, f(\mathbf{x}(k)),$$

which, in view of $f(\mathbf{x}(k)) \geq f^*$, for all $k = 1, \ldots, s$, implies the desired result $f(\mathbf{x}(k)) = f^*$, for all $k = 1, \ldots, s$.

(b) If $\text{rank}\, \mathbf{M}_{d-1}(\mathbf{y}^*) = \text{rank}\, \mathbf{M}_d(\mathbf{y}^*) =: s$ then $\mathbf{M}_d(\mathbf{y}^*)$ is a flat extension of $\mathbf{M}_{d-1}(\mathbf{y}^*)$, and so, by Theorem 2.40, $\mathbf{y}^*$ is the vector of moments up to order $2d$, of some s-atomic probability measure μ^* on $\mathbb{R}^n$. The rest of the proof is the same as for (a). In addition one can extract the s global minimizers by Algorithm 6.9.   □

As an illustration, suppose that $\mathbf{y}^*$ satisfies $\text{rank}\, \mathbf{M}_d(\mathbf{y}^*) = 1$. Therefore, $\mathbf{M}_d(\mathbf{y}^*) = v_d(\mathbf{x}^*)v_d(\mathbf{x}^*)^T$ for some $\mathbf{x}^* \in \mathbb{R}^n$, that is, $\mathbf{y}^*$ is the vector of moments up to order $2d$ of the Dirac measure at $\mathbf{x}^*$, and one **reads** an optimal solution $\mathbf{x}^*$ from the subvector of first "moments" y_α^* with $|\alpha| = 1$. Note also that when $n = 1$, $f - f^*$ is always an SOS polynomial, so that we expect that $\rho_d = f^*$, where $2d$ is the degree of f.

In other words, the global unconstrained minimization of a univariate polynomial is a convex optimization problem, the semidefinite program (6.22).

Therefore, in view of Section 2.3, $\rho_d = f^*$ for quadratic polynomials, and bivariate polynomials of degree 4. Let us illustrate these properties with an example.

Example 6.20   We consider the bivariate polynomial f:

$$\mathbf{x} \mapsto f(\mathbf{x}) = (x_1^2 + 1)^2 + (x_2^2 + 1)^2 + (x_1 + x_2 + 1)^2.$$

Note that in this case $d = 2$ and $f - f^*$ is a bivariate polynomial of degree 4, and therefore it is a sum of squares. We thus expect that $\rho_2 = f^*$. Solving problem (6.22) for $\ell = 2$ yields a minimum value of $\rho_2 = -0.4926$. In this case, it turns out that $\mathbf{M}_2(\mathbf{y}^*)$ has rank one, and from the optimal solution $\mathbf{y}^*$, we check that

$$\mathbf{y} = (1, x_1^*, x_2^*, (x_1^*)^2, x_1^* x_2^*, (x_2^*)^2, \ldots, (x_1^*)^4, \ldots, (x_2^*)^4),$$

with $x_1^* = x_2^* \approx -0.2428$. We observe that the solution $\mathbf{x}^*$ is a good approximation of a global minimizer of problem (6.1), since the gradient vector verifies

$$\left.\frac{\partial f}{\partial x_1}\right|_{\mathbf{x}=\mathbf{x}^*} = \left.\frac{\partial f}{\partial x_2}\right|_{\mathbf{x}=\mathbf{x}^*} = 4 \times 10^{-9}.$$

In this example, it follows that the semidefinite relaxation is exact. The reason we have $\nabla f(\mathbf{x}^*) \approx \mathbf{0}$, but not exactly $\mathbf{0}$, is due to unavoidable numerical inaccuracies when using an SDP solver.

The next example illustrates the effect of perturbations.

Example 6.21 Consider the bivariate polynomial $\mathbf{x} \mapsto f(\mathbf{x}) = x_1^2 x_2^2 (x_1^2 + x_2^2 - 1)$. It turns out that $f + 1$ is positive, but not a sum of squares. Note that the global optimum is $f^* = -1/27$ and there are four global minimizers $\mathbf{x}^* = (\pm\sqrt{3}/3, \pm\sqrt{3}/3)$. If we consider the problem $f^* = \inf f(\mathbf{x})$ and apply Algorithm 6.18 for $\ell = 3$, we obtain that $\rho_3 = -\infty$, that is, the bound is uninformative.

However, consider now the perturbed problem of minimizing the polynomial $f_\epsilon(\mathbf{x}) := f(\mathbf{x}) + \epsilon(x_1^{10} + x_2^{10})$ with $\epsilon = 0.001$. Applying Algorithm 6.18 using the GloptiPoly software we find that $\rho_5 \approx f^*$ and four optimal solutions $(x_1, x_2) \approx \mathbf{x}^*$ are extracted. In other words, while the approach fails when applied to the original polynomial f, it succeeds when applied to the perturbed polynomial f_ϵ.

Reformulation as a constrained optimization problem

We have seen in Section 6.3 that for the unconstrained optimization problem (6.1), the semidefinite relaxations (6.22) reduce to a single one, and with a 0/1 answer, depending on whether or not the polynomial $f - f^*$ is a sum of squares. Therefore, in general, according to Theorem 6.16, the SDP (6.22) provides only a lower bound on f^*.

However, if one knows a priori some bound M on the Euclidean norm $\|\mathbf{x}^*\|$ of a global minimizer $\mathbf{x}^* \in \mathbb{R}^n$, then it suffices to replace the original unconstrained problem (6.1) (where $\mathbf{K} = \mathbb{R}^n$) with a constrained problem where $\mathbf{K}$ is the basic semi-algebraic set $\{\mathbf{x} \in \mathbb{R}^n : M^2 - \|\mathbf{x}\|^2 \geq 0\}$. It is immediate to verify that Assumption 2.14 holds, and therefore the machinery described in Section 6.1 applies, and the semidefinite relaxations (6.3) with $\mathbf{K}$ as above converge to f^*.

Another approach which avoids the a priori knowledge of this bound M consists of taking

$$\mathbf{K} := \{ \mathbf{x} \in \mathbb{R}^n \: : \: \nabla f(\mathbf{x}) = 0 \}, \qquad (6.29)$$

since if a global minimizer $\mathbf{x}^* \in \mathbb{R}^n$ exists, then necessarily $\nabla f(\mathbf{x}^*) = 0$, and in addition $\mathbf{x}^*$ is also a global minimizer of f on $\mathbf{K}$ defined in (6.29). However, convergence of the hierarchy of semidefinite relaxations (6.3) has been proved for compact basic semi-algebraic sets only. Fortunately, the set $\mathbf{K}$ in (6.29) has nice properties.

Proposition 6.22 *For almost all polynomials $f \in \mathbb{R}[\mathbf{x}]_d$, the gradient ideal $J_f := \langle \frac{\partial f}{\partial x_1}, \ldots, \frac{\partial f}{\partial x_n} \rangle$ is zero dimensional and radical.*[3]

And so we have the following result.

Theorem 6.23 *With $f \in \mathbb{R}[\mathbf{x}]_t$, and $\mathbf{K} \subset \mathbb{R}^n$ as in (6.29), consider the semidefinite relaxation defined in (6.3) with optimal value ρ_d. Let*

$$\mathcal{F}_t := \{ f \in \mathbb{R}[\mathbf{x}]_t \: : \: \exists \mathbf{x}^* \in \mathbb{R}^n \ s.t. \ f(\mathbf{x}^*) = f^* = \inf\{ f(\mathbf{x}) \: : \: \mathbf{x} \in \mathbb{R}^n \} \}.$$

Then for almost all $f \in \mathcal{F}_t$, $\rho_d = f^*$ for some index d, i.e., finite convergence takes place.

Theorem 6.23 is a direct consequence of Proposition 6.22 and Theorem 6.13.

6.4 Exercises

Exercise 6.1 Let $n = 1$ and consider the univariate optimization problem $f^* = \inf\{ x \: : \: -x^2 \geq 0 \} = 0$, with obvious optimal solution $x^* = 0$.

(a) Show that the first semidefinite relaxation of the hierarchy (6.3) is exact, that is, $\rho_1 = f^*$. What about the dual semidefinite program? Does it have an optimal solution? Explain why.

(b) What can be said about the Karush–Kuhn–Tucker optimality conditions associated with $x^* \in \mathbf{K}$.

(c) Consider now the problem $f_\epsilon^* = \inf\{ x \: : \: -x^2 \geq -\epsilon \}$ for some given positive scalar $\epsilon > 0$. What is an optimal solution? What can be said about the first semidefinite relaxation (6.3) and its dual?

---

[3] Recall that given $S \subset \mathbb{C}^n$, the set $I(S) := \{ f \in \mathbb{C}[\mathbf{x}] \: : \: f(\mathbf{z}) = 0, \ \forall \mathbf{z} \in S \}$ is the vanishing ideal associated with S. An ideal $J \subset \mathbb{C}[\mathbf{x}]$ is zero dimensional if the algebraic variety $V_{\mathbb{C}}(J) = \{ \mathbf{z} \in \mathbb{C}^n \: : \: g(\mathbf{z}) = 0, \ \forall g \in J \}$ is finite and J is radical if $J = I(V_{\mathbb{C}}(J))$.

Exercise 6.2   Let $f \in \mathbb{R}[\mathbf{x}]_d$ and $\mathbf{K} = [-1, 1]^n$.

(a) Show that there exists $M > 0$ such that

$$f(\mathbf{x}) + M = \sigma_0(\mathbf{x}) + \sum_{i=1}^{n} \sigma_i(\mathbf{x})(1 - x_i^2), \qquad \forall \mathbf{x} \in \mathbb{R}^n,$$

for some SOS polynomials σ_i such that $\deg \sigma_0 \leq 2\lceil d/2 \rceil$ and $\deg \sigma_i \leq 2\lceil d/2 \rceil - 2$ for every $i = 1, \ldots, n$. (*Hint:* Consider the first semidefinite relaxation of the hierarchy (6.3) associated with the optimization problem $\inf \{ f(\mathbf{x}) : \mathbf{x} \in \mathbf{K} \}$. Show that its feasible set is compact and Slater's condition holds.)

(b) Show the same result as in (a) with $\mathbf{K} = [0, 1]^n$ (and with $x_i(1 - x_i)$ in lieu of $(1 - x_i^2)$, $i = 1, \ldots, n$).

Exercise 6.3   Consider the MAXCUT problem with equal weights:

$$f^* = \inf \left\{ \sum_{1 \leq i < j \leq n} x_i x_j : \mathbf{x} \in \{-1, 1\}^n \right\}.$$

(a) Write the problem as a polynomial optimization problem.

(b) Show that the optimal value ρ_1 of the first semidefinite relaxation (6.3) satisfies $\rho_1 \geq -n/2$ and discuss the cases n even and n odd. (*Hint:* Express the polynomial $\sum_{1 \leq i < j \leq n} x_i x_j + n/2$ in terms of the polynomials $(\sum_i x_i)^2$ and $x_i^2 - 1$, $i = 1, \ldots, n$.)

6.5 Notes and sources

For a survey on semidefinite programming and its multiple applications, the interested reader is referred to Vandenberghe and Boyd (1996).

6.1–6.3 Most of the material in this chapter is from Lasserre (2000, 2001, 2002a, 2002b, 2002c, 2004, 2006b). Shor (1987, 1998) was the first to prove that the global minimization of a univariate polynomial is a convex optimization problem. Later, Nesterov (2000) defined exact semidefinite formulations for the univariate case, while converging semidefinite relaxations for the general multivariate case were treated in Lasserre (2000, 2001, 2002a, 2002b) and Parrilo (2000, 2003). de Klerk et al. (2006) provided a polynomial time approximation scheme (PTAS) for minimizing polynomials of fixed degree on the simplex. The finite convergence in Theorem 6.5 is due to Nie (2014) who used previous results by Marshall (2006, 2009). In particular Nie (2014) showed that the usual linear independence of gradients, strict complementarity and the standard second-order sufficient condition of nonlinear programming imply that the crucial boundary Hessian condition introduced in Marshall (2006) holds. The procedure for extraction of solutions in Section 6.1.2 is from Henrion and Lasserre (2005).

6.2 For discrete 0/1 optimization problems a hierarchy of semidefinite relaxations for 0/1 problems was first proposed by Lovász and Schrijver (1991) who also proved finite convergence. For 0/1 problems, Laurent (2003) compared the linear relaxations of Sherali and Adams (1990) and Lovász and Schrijver (1991), and the semidefinite relaxations of Lovász and Schrijver (1991) and Lasserre (2002a) within the common framework of the moment matrix, and proved that the latter semidefinite relaxations are the strongest. This has motivated research on integrality gaps for difficult combinatorial optimization problems. (The integrality gap is the ratio between the optimal value of the relaxation and that of the problem to solve. It provides a measure of the quality of the approximation.) In particular, Chlamtac (2007) and Chlamtac and Singh (2008) showed that the hierarchy of semidefinite relaxations provides improved approximation algorithms for finding independent sets in graphs, and for coloring problems. See also the related work of Schoenebeck (2008). For a recent analysis and comparison of various hierarchies of convex relaxations for hard combinatorial optimization problems, the interested reader is referred to Chlamtac and Tulsiani (2012). Theorem 6.13 and Theorem 6.14 can be deduced from Theorem 6.15 in Laurent (2008) which extends previous results by Parrilo (2003), themselves extensions to the general setting (6.21) of the grid case studied in Lasserre (2002a, 2002b). See also Laurent (2007a) for additional results.

6.3 Recent approaches to unconstrained optimization via optimizing on the gradient ideal appear in Hanzon and Jibetean (2003) with matrix methods in Jibetean and Laurent (2005) with semidefinite programming. In both approaches one slightly perturbates p (of degree say $2d$) by adding monomials $\{x_i^{2d+2}\}$ with a small coefficient ϵ and obtains a sequence of polynomials f_ϵ with the property that $V = \{\mathbf{x} : \nabla f_\epsilon(\mathbf{x}) = 0\}$ is finite and the minima f_ϵ^* converge to f^*. In particular $\{\partial f_\epsilon/\partial x_i\}$ form a Gröbner basis of the ideal they generate. On the other hand, Proposition 6.22 is Proposition 1 in Nie et al. (2006). To handle the case where no global minimizer exists, Schweighofer (2006) used SOS and the concept of gradient tentacles and Vui and Son (2008) used the truncated tangency variety.

7

Global optimality certificates

In this chapter, we derive global optimality conditions for polynomial optimization which generalize the local first-order optimality conditions due to Fritz John and Karush–Kuhn–Tucker (KKT) for nonlinear optimization. We also show how those global optimality conditions recognize the constraints that are *important* even though they may not be active at a global minimizer.

7.1 Putinar versus Karush–Kuhn–Tucker

So with $f \in \mathbb{R}[\mathbf{x}]_{d_0}$ consider the polynomial optimization problem

$$\mathbf{P}: \qquad f^* = \inf_{\mathbf{x}} \{ f(\mathbf{x}) : \mathbf{x} \in \mathbf{K} \}, \qquad (7.1)$$

whose feasible set $\mathbf{K}$ is basic semi-algebraic and is in the form:

$$\mathbf{K} = \{ \mathbf{x} \in \mathbb{R}^n : g_j(\mathbf{x}) \geq 0, \ j = 1, \dots, m_1; \ h_\ell(\mathbf{x}) = 0, \ \ell = 1, \dots, m_2 \}, \qquad (7.2)$$

for some given polynomials $g_j, h_\ell \in \mathbb{R}[\mathbf{x}]$ with respective degrees d_j, $j = 1, \dots, m_1$, and $d'_\ell, \ell = 1, \dots, m_2$.

Definition 7.1 A point $\mathbf{x} \in \mathbf{K}$ is a local minimizer of $\mathbf{P}$ if there exists $\epsilon > 0$ and a neighborhood $\mathbf{B}(\mathbf{x}, \epsilon)$ ($:= \{ \mathbf{z} : \|\mathbf{z} - \mathbf{x}\| < \epsilon \}$) such that $f(\mathbf{x}) \leq f(\mathbf{z})$ for all $\mathbf{z} \in \mathbf{K} \cap \mathbf{B}(\mathbf{x}, \epsilon)$. It is a strict local minimizer if $f(\mathbf{x}) < f(\mathbf{z})$ for all $\mathbf{z} \in \mathbf{K} \cap \mathbf{B}(\mathbf{x}, \epsilon)$.

First-order KKT optimality conditions Given $\mathbf{x} \in \mathbf{K}$, let $J(\mathbf{x})$ be the set of inequality constraints that are active (or saturated) at $\mathbf{x}$, i.e., $J(\mathbf{x}) := \{ j \in \{1, \dots, m_1\} : g_j(\mathbf{x}) = 0 \}$.

Theorem 7.2 (First-order KKT necessary conditions) *Let* $\mathbf{x}^*$ *be a local minimum of problem* **P** *and assume that the vectors* $(\nabla g_j(\mathbf{x}^*), \nabla h_\ell(\mathbf{x}^*))$, $j \in J(\mathbf{x}^*)$, $\ell = 1, \ldots, m_2$, *are linearly independent. Then there exist* $\lambda \in \mathbb{R}_+^{m_1}$ *and* $\mathbf{v}^* \in \mathbb{R}^{m_2}$ *such that:*

$$
\begin{aligned}
\nabla f(\mathbf{x}^*) &= \sum_{j=1}^{m_1} \lambda_j^* \nabla g_j(\mathbf{x}^*) + \sum_{\ell=1}^{m_2} v_\ell^* \nabla h_\ell(\mathbf{x}^*), \\
\lambda_j^* g_j(\mathbf{x}^*) &= 0, \quad j = 1, \ldots, m_1, \\
\lambda_j^* &\geq 0, \quad j = 1, \ldots, m_1.
\end{aligned}
\tag{7.3}
$$

A triplet $(\mathbf{x}^*, \lambda^*, \mathbf{v}^*) \in \mathbf{K} \times \mathbb{R}^{m_1} \times \mathbb{R}^{m_2}$ which satisfies the first-order (necessary) KKT optimality conditions (7.3) associated with problem **P** is called a KKT triplet, and the dual variables $(\lambda^*, \mathbf{v}^*) \in \mathbb{R}^{m_1} \times \mathbb{R}^{m_2}$ are called Lagrange Karush–Kuhn–Tucker (or Lagrange KKT) multipliers. We have already encountered such a triplet in Chapter 6 when analyzing the finite convergence of the semidefinite relaxations (6.6). In Theorem 7.2, the linear independence condition

(QC) The family $(\nabla g_j(\mathbf{x}^*), \nabla h_\ell(\mathbf{x}^*))$, $j \in J(\mathbf{x}^*)$, $\ell = 1, \ldots, m_2$,

   is linearly idependent
$$\tag{7.4}$$

on the gradient vectors $\nabla g_j(\mathbf{x}^*), \nabla h_\ell(\mathbf{x}^*)$, $j \in J(\mathbf{x}^*)$, $\ell = 1, \ldots, m_2$, at $\mathbf{x}^* \in \mathbf{K}$, is called a *constraint qualification* (QC). In fact there exist other QCs under which Theorem 7.2 holds, in particular the Mangasarian–Fromovitz QC and the Slater QC.

The well-known Mangasarian–Fromovitz QC states that:

the gradients $\nabla h_\ell(\mathbf{x}^*)$, $\ell = 1, \ldots, m_2$, are linearly independent and
$\exists \mathbf{u} \in \mathbb{R}^n$ s.t. $\langle \nabla g_j(\mathbf{x}^*), \mathbf{u} \rangle > 0$ and $\langle \nabla h_\ell(\mathbf{x}^*), \mathbf{u} \rangle = 0$, $\tag{7.5}$
$\forall j \in J(\mathbf{x}^*)$, $\ell = 1, \ldots, m_2$.

When f and $-g_j$ are convex, $j = 1, \ldots, m_1$, and h_ℓ are all affine (i.e., $h_\ell(\mathbf{x}) = \mathbf{a}_\ell^T \mathbf{x} + c_\ell$ for some $(\mathbf{a}_\ell, c_\ell) \in \mathbb{R}^n \times \mathbb{R}$, $\ell = 1, \ldots, m_2$), the well-known Slater condition (or Slater QC) states that the $\mathbf{a}_\ell$ are linearly independent and there exists $\mathbf{x}_0 \in \mathbf{K}$ such that $g_j(\mathbf{x}_0) > 0$ for all $j = 1, \ldots, m_1$.

- In fact, most optimization algorithms try to find a KKT triplet $(\mathbf{x}^*, \boldsymbol{\lambda}^*, \mathbf{v}^*) \in \mathbb{R}^n \times \mathbb{R}_+^{m_1} \times \mathbb{R}^{m_2}$ that satisfies (7.3).
- From the first condition in (7.3), $\mathbf{x}^*$ is a *stationary point* of the *Lagrangian polynomial*

$$\mathbf{x} \mapsto \quad L(\mathbf{x}, \boldsymbol{\lambda}^*, \mathbf{v}^*) := f(\mathbf{x}) - f^* - \sum_{j=1}^{m_1} \lambda_j^* g_j(\mathbf{x}) - \sum_{\ell=1}^{m_2} v_\ell^* h_\ell(\mathbf{x}), \quad (7.6)$$

but in general, $\mathbf{x}^*$ is *not* a global minimizer of L (and may not even be a local minimizer).

- However, if f and $-g_j$ are all convex, the h_ℓ are all affine and Slater's condition holds, then (7.3) are necessary and sufficient optimality conditions for $\mathbf{x}^*$ to be an optimal solution of **P**. Moreover, $\mathbf{x}^*$ is a global minimizer of the Lagrangian polynomial L which is nonnegative on $\mathbb{R}^n$, with $L(\mathbf{x}^*, \boldsymbol{\lambda}^*, \mathbf{v}^*) = 0$. In fact, the triplet $(\mathbf{x}^*, \boldsymbol{\lambda}^*, \mathbf{v}^*) \in \mathbf{K} \times \mathbb{R}_+^{m_1} \times \mathbb{R}^{m_2}$ is a *saddle point* of the Lagrangian L, i.e.,

$$L(\mathbf{x}^*, \boldsymbol{\lambda}, \mathbf{v}) \leq L(\mathbf{x}^*, \boldsymbol{\lambda}^*, \mathbf{v}^*) \leq L(\mathbf{x}, \boldsymbol{\lambda}^*, \mathbf{v}^*), \quad \forall (\mathbf{x}, \boldsymbol{\lambda}, \mathbf{v}) \in \mathbb{R}^n \times \mathbb{R}_+^{m_1} \times \mathbb{R}^{m_2}.$$

In this case one may state that $(\boldsymbol{\lambda}^*, \mathbf{v}^*) \in \mathbb{R}^{m_1} \times \mathbb{R}^{m_2}$ provides a *certificate of global optimality* for $\mathbf{x}^* \in \mathbf{K}$. From $L(\mathbf{x}, \boldsymbol{\lambda}^*, \mathbf{v}^*) \geq L(\mathbf{x}^*, \boldsymbol{\lambda}^*, \mathbf{v}^*) = 0$ for all $\mathbf{x} \in \mathbb{R}^n$ we obtain $f(\mathbf{x}) \geq f^*$ for all $\mathbf{x} \in \mathbf{K}$, and $f(\mathbf{x}^*) - f^* = L(\mathbf{x}^*, \boldsymbol{\lambda}^*, \mathbf{v}^*) = 0$ so that $\mathbf{x}^*$ is a global minimizer of f on $\mathbf{K}$.

Second-order sufficient conditions When $f, -g_j$ are convex, $j = 1, \ldots, m_1$, and the h_ℓ are affine, we have seen that if Slater's condition holds then the first-order KKT conditions are sufficient for $\mathbf{x}^* \in \mathbf{K}$ to be a minimizer. In the general nonconvex case there also exist sufficient conditions which involve second-order information at the triplet $(\mathbf{x}^*, \boldsymbol{\lambda}^*, \mathbf{v}^*)$.

Namely, let $(\mathbf{x}^*, \boldsymbol{\lambda}^*, \mathbf{v}^*) \in \mathbf{K} \times \mathbb{R}_+^{m_1} \times \mathbb{R}^{m_2}$ be a KKT triplet and let L be the Lagrangian defined in (7.6).

The *second-order sufficient condition* holds at $(\mathbf{x}^*, \boldsymbol{\lambda}^*, \mathbf{v}^*)$ if

$$\mathbf{u}^T \nabla_{\mathbf{x}}^2 L(\mathbf{x}^*, \boldsymbol{\lambda}^*, \mathbf{v}^*) \mathbf{u} > 0, \quad \text{for all } 0 \neq \mathbf{u} \text{ such that}$$

$$\nabla g_j(\mathbf{x}^*)^T \mathbf{u} = 0, \quad \nabla h_\ell(\mathbf{x}^*)^T \mathbf{u} = 0 \quad \forall j \in J(\mathbf{x}^*), \ \ell = 1, \ldots, m_2. \quad (7.7)$$

Moreover *strict complementarity* holds if $\lambda_j^* > 0$ for all $j \in J(\mathbf{x}^*)$.

And we have the following result.

Theorem 7.3 (Second-order sufficient conditions) *Let* $(\mathbf{x}^*, \boldsymbol{\lambda}^*, \mathbf{v}^*) \in \mathbf{K} \times \mathbb{R}^{m_1} \times \mathbb{R}^{m_2}$ *be a KKT triplet, i.e.,* $(\mathbf{x}^*, \boldsymbol{\lambda}^*, \mathbf{v}^*)$ *satisfies (7.3), and let L be the Lagrangian defined in (7.6). If the strict complementarity and second-order sufficient conditions (7.7) hold at* $(\mathbf{x}^*, \boldsymbol{\lambda}^*, \mathbf{v}^*)$ *then* $\mathbf{x}^*$ *is a (strict) local minimizer of* **P.**

7.1.1 Putinar versus KKT

Notice that if $\mathbf{x}^* \in \mathbf{K}$ is a global minimizer of **P** then the polynomial $\mathbf{x} \mapsto f(\mathbf{x}) - f(\mathbf{x}^*)$ is nonnegative on **K**. And so if Assumption 6.4 holds then by Theorem 2.15, for every $\epsilon > 0$,

$$f(\mathbf{x}) - f(\mathbf{x}^*) + \epsilon = \sigma_0(\mathbf{x}) + \sum_{j=1}^{m_1} \sigma_j(\mathbf{x}) \, g_j(\mathbf{x}) + \sum_{\ell=1}^{m_2} \psi_\ell(\mathbf{x}) \, h_\ell(\mathbf{x}),$$

for some polynomials $(\psi_\ell) \subset \mathbb{R}[\mathbf{x}]$ and some SOS polynomials $(\sigma_j) \subset \Sigma[\mathbf{x}]$. In many cases it happens that this SOS representation also holds even with $\epsilon = 0$, which naturally leads us to the following global optimality conditions.

Theorem 7.4 *Let* $\mathbf{x}^* \in \mathbf{K}$ *be a global minimizer for problem* **P** *in (7.1), with global optimum* f^*, *and assume that* $f - f^*$ *has the representation*

$$f(\mathbf{x}) - f^* = \sigma_0(\mathbf{x}) + \sum_{j=1}^{m_1} \sigma_j(\mathbf{x}) \, g_j(\mathbf{x}) + \sum_{\ell=1}^{m_2} \psi_\ell(\mathbf{x}) \, h_\ell(\mathbf{x}), \quad \mathbf{x} \in \mathbb{R}^n,$$

$$(7.8)$$

for some polynomials $(\psi_\ell)_{\ell=1}^{m_2} \subset \mathbb{R}[\mathbf{x}]$ *and some SOS polynomials* $(\sigma_j)_{j=0}^{m} \subset \Sigma[\mathbf{x}]$. *Then:*

(a) $\sigma_j(\mathbf{x}^*) \geq 0$ *and* $g_j(\mathbf{x}^*) \geq 0$, *for all* $j = 1, \ldots, m_1$,

(b) $\sigma_j(\mathbf{x}^*) g_j(\mathbf{x}^*) = 0$, *for all* $j = 1, \ldots, m_1$,

(c) $\nabla f(\mathbf{x}^*) = \sum_{j=1}^{m_1} \sigma_j(\mathbf{x}^*) \nabla g_j(\mathbf{x}^*) + \sum_{\ell=1}^{m_2} \psi_\ell(\mathbf{x}^*) \nabla h_\ell(\mathbf{x}^*)$, *that is,* $(\mathbf{x}^*, \boldsymbol{\lambda}^*, \mathbf{v}^*)$ *is a KKT triplet, with* $\lambda_j^* := \sigma_j(\mathbf{x}^*)$ *and* $v_\ell^* = \psi_\ell(\mathbf{x}^*)$ *for all* $j = 1, \ldots, m_1, \ell = 1, \ldots, m_2$,

(d) $\mathbf{x}^*$ *is a global minimizer of the (generalized) Lagrangian polynomial*

$$\mathbf{x} \mapsto \mathcal{L}(\mathbf{x}, \sigma, \boldsymbol{\psi}) := f(\mathbf{x}) - f^* - \sum_{j=1}^{m_1} \sigma_j(\mathbf{x}) \, g_j(\mathbf{x}) - \sum_{\ell=1}^{m_2} \psi_\ell(\mathbf{x}) \, h_\ell(\mathbf{x}).$$

Proof   (a) Part (a) is obvious from $\mathbf{x}^* \in \mathbf{K}$, and the polynomials σ_j are SOS.

(b) From (7.8) and the fact that $\mathbf{x}^*$ is a global minimizer, we obtain

$$f(\mathbf{x}^*) - f^* = 0 = \sigma_0(\mathbf{x}^*) + \sum_{j=1}^{m_1} \sigma_j(\mathbf{x}^*) g_j(\mathbf{x}^*),$$

which in turn implies part (b) because $g_j(\mathbf{x}^*) \geq 0$ for all $j = 1, \ldots, m_1$, and the polynomials σ_j are all SOS, hence nonnegative. This also implies $\sigma_0(\mathbf{x}^*) = 0$.

(c) Differentiating and using the fact that the polynomials σ_j are SOS, and using part (b), yields part (c).

(d) From (7.8) we obtain,

$$\mathcal{L}(\mathbf{x}, \boldsymbol{\sigma}, \boldsymbol{\psi}) = f(\mathbf{x}) - f^* - \sum_{j=1}^{m_1} \sigma_j(\mathbf{x}) \, g_j(\mathbf{x}) - \sum_{\ell=1}^{m_2} \psi_\ell(\mathbf{x}) \, h_\ell(\mathbf{x}) = \sigma_0(\mathbf{x}) \geq 0,$$

for all $\mathbf{x}$, because $\sigma_0 \in \mathbb{R}[\mathbf{x}]$ is SOS, and using property (b),

$$\mathcal{L}(\mathbf{x}^*, \boldsymbol{\sigma}, \boldsymbol{\psi}) = f(\mathbf{x}^*) - f^* - \sum_{j=1}^{m_1} \sigma_j(\mathbf{x}^*) \, g_j(\mathbf{x}^*) = 0,$$

which shows that $\mathbf{x}^*$ is a global minimizer of $\mathcal{L}(\cdot, \boldsymbol{\sigma}, \boldsymbol{\psi})$. $\square$

So if $f - f^*$ has the representation (7.8) then Theorem 7.4 implies the following.

1. (7.8) should be interpreted as a *global optimality condition*.
2. The polynomial $\mathcal{L}(\cdot, \boldsymbol{\sigma}, \boldsymbol{\psi}) = f - f^* - \sum_j \sigma_j g_j - \sum_\ell \psi_\ell h_\ell$ is a *generalized (polynomial) Lagrangian*, with generalized Lagrange KKT (polynomial) multipliers $(\boldsymbol{\sigma}, \boldsymbol{\psi}) \in \Sigma[\mathbf{x}]^{m_1} \times \mathbb{R}[\mathbf{x}]^{m_2}$ instead of scalar multipliers $(\boldsymbol{\lambda}, \mathbf{v}) \in \mathbb{R}_+^{m_1} \times \mathbb{R}^{m_2}$. It is SOS (hence nonnegative on $\mathbb{R}^n$), vanishes at every global minimizer $\mathbf{x}^* \in \mathbf{K}$, and so $\mathbf{x}^*$ is also a global minimizer of the generalized Lagrangian.
3. The generalized Lagrange KKT multipliers $(\boldsymbol{\sigma}, \boldsymbol{\psi}) \in \Sigma[\mathbf{x}]^{m_1} \times \mathbb{R}[\mathbf{x}]^{m_2}$ provide a *certificate of global optimality* for $\mathbf{x}^* \in \mathbf{K}$ in the nonconvex case exactly as the Lagrange KKT multipliers $(\boldsymbol{\lambda}, \mathbf{v}) \in \mathbb{R}_+^{m_1} \times \mathbb{R}^{m_2}$ provide a certificate in the *convex* case.

Of course, in view of the power of Theorem 7.4, one immediately wonders when condition (7.8) holds and how restrictive is this condition. And in fact the answer is that Theorem 7.4 holds whenever the constraint qualification (QC) in (7.4), strict complementarity and second-order sufficiency condition hold.

Theorem 7.5 *Let* $\mathbf{K} \subset \mathbb{R}^n$ *be as in (7.2), let Assumption 6.4 hold and assume that the ideal* $J := \langle h_1, \ldots, h_{m_2} \rangle \subset \mathbb{R}[\mathbf{x}]$ *is real radical (in particular if there is no equality constraint then* J *(equal to* $\langle 0 \rangle$*) is real radical).*

Let $\mathbf{x}^* \in \mathbf{K}$ *be a global minimizer for problem* $\mathbf{P}$ *in (7.1) with global optimum* f^*, *and assume that the constraint qualification (QC) in (7.4) holds (so that there exists a KKT triplet* $(\mathbf{x}^*, \boldsymbol{\lambda}^*, \mathbf{v}^*) \in \mathbf{K} \times \mathbb{R}_+^{m_1} \times \mathbb{R}^{m_2})$.

If the strict complementarity and second-order sufficient condition hold at $(\mathbf{x}^*, \boldsymbol{\lambda}^*, \mathbf{v}^*)$ *then (7.8) holds, i.e.,*

$$f(\mathbf{x}) - f^* = \sigma_0(\mathbf{x}) + \sum_{j=1}^{m_1} \sigma_j(\mathbf{x})\, g_j(\mathbf{x}) + \sum_{\ell=1}^{m_2} \psi_\ell(\mathbf{x})\, h_\ell(\mathbf{x}), \quad \mathbf{x} \in \mathbb{R}^n,$$

for some polynomials $(\psi_\ell)_{\ell=1}^{m_2} \subset \mathbb{R}[\mathbf{x}]$ *and some SOS polynomials* $(\sigma_j)_{j=0}^m \subset \Sigma[\mathbf{x}]$. *Moreover, Theorem 7.4 applies.*

If there is no equality constraint then J ($J = \langle 0 \rangle$) is real radical as $V_{\mathbb{R}}(J) = \mathbb{R}^n$. If the ideal $J = \langle h_1, \ldots, h_{m_2} \rangle$ is not real radical then one may replace the equality constraints $h_\ell(\mathbf{x}) = 0$, $\ell = 1, \ldots, m_2$, with the equality constraints $p_k(\mathbf{x}) = 0$, $k = 1, \ldots, s$, where the polynomials (p_k) generate the vanishing ideal $I(V_{\mathbb{R}}(J))$ (which by definition is real radical). Then Theorem 7.5 applies. However, computing the generators (p_k) of the vanishing ideal is difficult in general.

So the same second-order sufficient condition (7.7) that allows the characterization of a strict local minimum via Theorem 7.3 also allows a characterization of global optimality via Theorem 7.4. Furthermore, these conditions hold generically as stated in the next theorem.

Theorem 7.6 *Let* $d_0, \ldots, d_{m_1}, d'_1, \ldots, d'_{m_2} \in \mathbb{N}$ *be fixed, and consider the polynomial optimization problem* $\mathbf{P} : \inf\{\, f(\mathbf{x}) : \mathbf{x} \in \mathbf{K} \,\}$ *with* $\mathbf{K}$ *as in (7.2).*

There exist finitely many real polynomials $\varphi_1, \ldots, \varphi_L$ *in the coefficients of the polynomials* $(f, g_1, \ldots, g_{m_1}, h_1, \ldots, h_{m_2}) \subset \mathbb{R}[\mathbf{x}]$ *such that if* $\varphi_1, \ldots, \varphi_L$ *do not vanish at the input polynomials* $(f, g_1, \ldots, g_{m_1}, h_1, \ldots, h_{m_2})$ *then the constraint qualification (QC) in (7.4), strict complementarity and second-order sufficiency condition (7.7) hold at every local minimizer* $\mathbf{x}^* \in \mathbf{K}$ *of* $\mathbf{P}$.

So Theorem 7.5 and Theorem 7.6 have two important consequences.

- In polynomial optimization, the constraint qualification (QC) (7.4), strict complementarity and second-order sufficient condition (7.7) hold generically at every local minimizer of $\mathbf{P}$ (and hence at every global minimizer as well).
- The characterization of global optimality in Theorem 7.4 is also generic for problems $\mathbf{P}$ with compact set $\mathbf{K}$ and no equality constraint. (One may always add a strictly redundant quadratic constraint $M - \|\mathbf{x}\|^2 \geq 0$ in the definition of $\mathbf{K}$ to ensure that (the Archimedean) Assumption 2.14 holds true.)

  This characterization of global optimality is an extension of the KKT optimality conditions for a local optimum, under the same sufficient conditions. In particular, generically, solving a polynomial optimization problem $\mathbf{P}$ in (7.1) with compact set $\mathbf{K}$ and no equality constraint, reduces to solving a single semidefinite program (whose size is not known in advance).

7.1.2 On constraints not active at a global minimum

In this section we focus on an important property associated with the SOS multipliers σ_j in Theorem 7.4, which is *not* valid in the usual KKT optimality conditions (7.3) of nonlinear programming. This property is a key difference between the local KKT optimality conditions and the global optimality conditions.

For every $\mathbf{x} \in \mathbf{K}$, let $J(\mathbf{x}) \subseteq \{1, \ldots, m_1\}$ be the set of indices such that $j \in J(\mathbf{x})$ if and only if $g_j(\mathbf{x}) = 0$, i.e., if and only if the constraint $g_j(\mathbf{x}) \geq 0$ is active at $\mathbf{x}$.

In the local KKT optimality conditions (7.3), only the constraints $g_j(\mathbf{x}) \geq 0$ that are active at $\mathbf{x}^*$ have a possibly nontrivial associated Lagrange (scalar) multiplier λ_j^*. Hence the nonactive constraints do *not* appear in the Lagrangian L defined in (7.6). In contrast, in the global optimality condition (7.8), every constraint $g_j(\mathbf{x}) \geq 0$ has a possibly nontrivial SOS polynomial Lagrange multiplier $\mathbf{x} \mapsto \sigma_j(\mathbf{x})$. But if $g_j(\mathbf{x}^*) > 0$ then necessarily $\sigma_j(\mathbf{x}^*) = 0 = \lambda_j^*$, as in the local KKT optimality conditions. Of course if f and $-g_j$ are convex, $j = 1, \ldots, m_1$, and all h_ℓ are affine, then the nonactive constraints are indeed not important (or are redundant) because if one now considers the optimization problem

$$\mathbf{P}' : \quad \inf_{\mathbf{x}} \{ f(\mathbf{x}) : g_j(\mathbf{x}) \geq 0, \ j \in J(\mathbf{x}^*), \ h_\ell(\mathbf{x}) = 0, \ \ell = 1, \ldots, m_2 \},$$

then $\mathbf{x}^*$ is still a global minimizer of $\mathbf{P}'$.

But in nonconvex optimization, a constraint $g_j(\mathbf{x}) \geq 0$ which is not active at a global minimizer $\mathbf{x}^* \in \mathbf{K}$ is not necessarily redundant and may still be important. That is, if it is removed from the definition of $\mathbf{K}$, then the global minimum f^* may decrease strictly. In this case, and in contrast to the local KKT optimality conditions (7.3), g_j is necessarily involved in the representation (7.8) of $f - f^*$ (when the latter exists), hence with a nontrivial SOS multiplier σ_j which vanishes at $\mathbf{x}^*$.

Proposition 7.7 *Let $\mathbf{x}^* \in \mathbf{K}$ be a global minimizer of* **P**. *Assume that (7.8) holds and that*

$$f^* > \rho_k^* = \inf_{\mathbf{x}} \{ f(\mathbf{x}) : g_j(\mathbf{x}) \geq 0, \ \forall j \neq k; \ h_\ell = 0, \ \ell = 1, \ldots, m_2 \},$$

(7.9)

for some $k \notin J(\mathbf{x}^*)$. Then $\sigma_k \neq 0$ in (7.8).

Proof Assume that (7.9) holds for some $k \notin J(\mathbf{x}^*)$ and suppose that $\sigma_k = 0$ in (7.8). Since $f^* > \rho_k^*$ then necessarily there exists $\mathbf{x}_0 \in \mathbf{K}' := \{ \mathbf{x} : g_j(\mathbf{x}) \geq 0, \ \forall j \neq k; \ h_\ell(\mathbf{x}) = 0, \ \forall \ell \}$ such that $f(\mathbf{x}_0) < f^*$. But then from (7.8) one obtains the contradiction:

$$0 > f(\mathbf{x}_0) - f^* = \sigma_0(\mathbf{x}_0) + \sum_{j=1}^{m_1} \sigma_j(\mathbf{x}_0)\, g_j(\mathbf{x}_0)$$

$$= \sigma_0(\mathbf{x}_0) + \underbrace{\sum_{j \neq k} \sigma_j(\mathbf{x}_0)\, g_j(\mathbf{x}_0)}_{\geq 0} \quad [\text{as } \sigma_k = 0]$$

$$\geq 0.$$

$\square$

In other words, since the polynomial $f - f^*$ is nonnegative on $\mathbf{K}$, when its representation (7.8) holds then it must necessarily involve all constraints $g_j(\mathbf{x}) \geq 0$ that are important and each such g_j has a nontrivial associated SOS generalized Lagrange KKT multiplier σ_j even though the constraint may not be active at $\mathbf{x}^*$.

Equivalently, and in contrast to the usual first-order KKT optimality conditions, the presence of a trivial multiplier σ_j in (7.8) certifies that the constraint $g_j(\mathbf{x}) \geq 0$ is not important for problem **P**.

We next illustrate this property in the following trivial example.

Example 7.8 Let $n = 1$ and consider the following problem:

$$f^* = \inf_{x} \{ -x : 1/2 - x \geq 0, \ x^2 - 1 = 0 \},$$

with optimal value $f^* = 1$, and global minimizer $x^* = -1$. The constraint $1/2 - x \geq 0$ is not active at $x^* = -1$, but if removed, the global minimum jumps to -1 with new global minimizer $x^* = 1$. In fact, we have the representation

$$f(x) - f^* = -(x + 1) = (1/2 - x)(x + 1)^2 + (x^2 - 1)(x + 3/2),$$

which shows the important role of the constraint $1/2 - x \geq 0$ in the representation of $f - f^*$, via its nontrivial multiplier $x \mapsto \sigma_1(x) := (x + 1)^2$ (which vanishes at $x^* = -1$). Note also that $\sigma_1(x^*) = 0 = \lambda_1$ and $\sigma_2(x^*) = x^* + 3/2 = -1/2 = \lambda_2$ are the Lagrange KKT multipliers $(\lambda_1, \lambda_2) \in \mathbb{R}_+ \times \mathbb{R}$ in the local optimality conditions (7.3). The Lagrange multiplier λ_2 is not constrained in sign because it is associated with an equality constraint (and not an inequality constraint as for λ_1); similarly, σ_2 is not required to be an SOS.

7.2 Krivine–Stengle versus Fritz John

First-order local optimality conditions for **P** due to Fritz John and weaker than the first-order KKT optimality conditions (7.3), state that with **K** as in (7.2):

---

if $\mathbf{x}^* \in \mathbf{K}$ is a local minimum then

$$
\begin{aligned}
\lambda_0^* \nabla f(\mathbf{x}^*) &= \sum_{j=1}^{m_1} \lambda_j^* \nabla g_j(\mathbf{x}^*) + \sum_{\ell=1}^{m_2} v_\ell^* \nabla h_\ell(\mathbf{x}^*), \\
\lambda_j^* g_j(\mathbf{x}^*) &= 0, \quad j = 1, \ldots, m_1, \\
\lambda_j^* &\geq 0, \quad j = 0, 1, \ldots, m_1,
\end{aligned}
\tag{7.10}
$$

for some nonnegative vector $\boldsymbol{\lambda}^* \in \mathbb{R}_+^{m_1+1}$ and some vector $\mathbf{v}^* \in \mathbb{R}^{m_2}$.

In contrast to the first-order KKT optimality conditions (7.3), no constraint qualification (QC) is needed for (7.10) to hold; observe that by homogeneity, if $\lambda_0^* > 0$ then (7.10) reduces to the KKT optimality conditions (7.3).

Notice also that if $\lambda_0^* = 0$ then (7.10) is not very informative on $\mathbf{x}^*$ as a minimizer of f on **K**! So this case should be viewed as a degenerate case and a QC is precisely a sufficient condition to ensure that $\lambda_0^* \neq 0$ in (7.10).

In the first-order Fritz John optimality conditions (7.10) the point $\mathbf{x}^* \in \mathbf{K}$ is a stationary point of the Lagrangian

$$L(\mathbf{x}, \boldsymbol{\lambda}^*, \mathbf{v}^*) := \lambda_0^* (f - f^*) - \sum_{j=1}^{m_1} \lambda_j^* g_j(\mathbf{x}) - \sum_{\ell=1}^{m_2} v_\ell^* h_\ell(\mathbf{x}), \qquad \mathbf{x} \in \mathbb{R}^n$$

because $\nabla_{\mathbf{x}} L(\mathbf{x}^*, \boldsymbol{\lambda}^*, \mathbf{v}^*) = 0$, but in general $\mathbf{x}^*$ is not a global minimum of L on $\mathbb{R}^n$.

Theorem 7.9 *Let* $\mathbf{x}^* \in \mathbf{K}$ *be a global minimizer of* f *on* $\mathbf{K}$ *with* $f^* = f(\mathbf{x}^*)$ *and let* $J(\mathbf{x}^*) = \{ j \in \{1, \ldots, m_1\} : g_j(\mathbf{x}^*) = 0 \}$. *Then there exist polynomials* Λ_j, $j = 0, 1, \ldots, m_1$, *all nonnegative on* $\mathbf{K}$, *and polynomials* ψ_ℓ, $\ell = 1, \ldots, m_2$, *such that the generalized Lagrangian* $\mathbf{x} \mapsto \mathcal{L}(\mathbf{x}, \Lambda, \boldsymbol{\psi})$ *defined by:*

$$\mathcal{L}(\mathbf{x}, \Lambda, \boldsymbol{\psi}) := \Lambda_0(\mathbf{x}) \, (f(\mathbf{x}) - f^*) - \sum_{j=1}^{m_1} \Lambda_j(\mathbf{x}) \, g_j(\mathbf{x}) - \sum_{\ell=1}^{m_2} \psi_\ell(\mathbf{x}) \, h_\ell(\mathbf{x})$$

(7.11)

is an SOS which vanishes at $\mathbf{x}^*$ *(and so* $\mathbf{x}^*$ *is a global minimizer of* $\mathcal{L}$ *on* $\mathbb{R}^n$*) and is positive whenever* $f(\mathbf{x}) \neq f^*$. *Moreover,*

$$\begin{aligned} \lambda_0 \, \nabla f(\mathbf{x}^*) &= \sum_{j=1}^{m_1} \lambda_j^* \nabla g_j(\mathbf{x}^*) + \sum_{\ell=1}^{m_2} v_\ell^* \nabla h_\ell(\mathbf{x}^*) \\ \lambda_j^* \, g_j(\mathbf{x}^*) &= 0, \quad \forall j = 1, \ldots, m_1, \end{aligned}$$

(7.12)

for some vector $\mathbf{v}^* = (\psi_\ell(\mathbf{x}^*)) \in \mathbb{R}^{m_2}$, *and some nonnegative vector* $\boldsymbol{\lambda}^* = (\Lambda_j(\mathbf{x}^*)) \in \mathbb{R}_+^{m_1+1}$ *obtained from values at* $\mathbf{x}^*$ *of the polynomials* g_j *and* σ_j.

Proof $\mathbf{x}^* \in \mathbf{K}$ is a global minimizer of f on $\mathbf{K}$ if and only if $f(\mathbf{x}) - f^* \geq 0$ on $\mathbf{K}$, and by Stengle's Theorem 2.11, if and only if:

$$p \, (f - f^*) = (f - f^*)^{2s} + q + h,$$

(7.13)

for some integer s, some polynomials p, q in the preordering $P(g)$ generated by the g_j (and defined in (2.10)), and some polynomial h in the ideal $\langle h_1, \ldots, h_{m_2} \rangle$ generated by the h_ℓ, (i.e., $h = \sum_\ell \psi_\ell h_\ell$ for some polynomials $(\psi_\ell) \subset \mathbb{R}[\mathbf{x}]$).

Evaluating at $\mathbf{x}^* \in \mathbf{K}$ in (7.13) yields $q(\mathbf{x}^*) = 0$ because $q \in P(g)$, and so $\mathbf{x}^*$ is a global minimum of q on $\mathbf{K}$. Recall that

$$q = \sum_{J \subseteq \{1,\dots,m_1\}} \sigma_J \underbrace{\prod_{j \in J} g_j}_{g_J} = \sum_{J \subseteq \{1,\dots,m_1\}} \sigma_J \, g_J,$$

for some SOS polynomials $\sigma_J \in \Sigma[\mathbf{x}]$, and so $\sigma_J(\mathbf{x}^*)g_J(\mathbf{x}^*) = 0$ for every $J \subseteq \{1,\dots,m_1\}$. Moreover, $\sigma_J(\mathbf{x}^*) = 0$ whenever $J \cap J(\mathbf{x}^*) = \emptyset$, which in turn implies $\nabla \sigma_J(\mathbf{x}^*) = 0$ whenever $J \cap J(\mathbf{x}^*) = \emptyset$ because σ_J is SOS.

Letting $\Lambda_0 := p$, define the polynomial $\mathbf{x} \mapsto \mathcal{L}(\mathbf{x}, \Lambda_0, \boldsymbol{\sigma}, \boldsymbol{\psi}) \in \mathbb{R}[\mathbf{x}]$

$$\mathcal{L}(\mathbf{x}, \Lambda_0, \boldsymbol{\sigma}, \boldsymbol{\psi}) := \Lambda_0(\mathbf{x})\,(f(\mathbf{x}) - f^*) - \sum_{\emptyset \neq J \subseteq \{1,\dots,m_1\}} \sigma_J(\mathbf{x})\, g_J(\mathbf{x})$$

$$- \sum_{\ell=1}^{m_2} \psi_\ell(\mathbf{x})\, h_\ell(\mathbf{x}) = \underbrace{(f(\mathbf{x}) - f^*)^{2s} + \sigma_\emptyset(\mathbf{x})}_{\text{SOS}},$$

and so $\mathcal{L}(\mathbf{x}, \Lambda_0, \boldsymbol{\sigma}, \boldsymbol{\psi}) \geq 0$ for all $\mathbf{x} \in \mathbb{R}^n$. In fact $\mathcal{L}$ is even an SOS which is positive whenever $f(\mathbf{x}) \neq f^*$. In addition, $\mathcal{L}(\mathbf{x}^*, \Lambda_0, \boldsymbol{\sigma}, \boldsymbol{\psi}) = 0$ which implies that $\mathbf{x}^*$ is a global minimum of $\mathcal{L}$ and so $\nabla_{\mathbf{x}} \mathcal{L}(\mathbf{x}^*, \Lambda_0, \boldsymbol{\sigma}, \boldsymbol{\psi}) = 0$. Therefore one may rewrite $\mathcal{L}$ as:

$$\Lambda_0(f - f^*) - \sum_{j=1}^{m_1} g_j \underbrace{\left(\sum_{\emptyset \neq J \subseteq \{1,\dots,m_1\}; J \ni j} \sigma_J \prod_{k \in J; k \neq j} g_k \right)}_{\Lambda_j \geq 0 \text{ on } \mathbf{K}} - \sum_{\ell=1}^{m_2} \psi_\ell h_\ell,$$

which is of the form (7.11) for some polynomials Λ_j, all nonnegative on $\mathbf{K}$, $j = 0, 1, \dots, m_1$. In particular, one has $\Lambda_j(\mathbf{x}^*)g_j(\mathbf{x}^*) = 0$ for all $j = 1, \dots, m_1$. Next, observe that since $\sigma_J(\mathbf{x}^*)\, g_J(\mathbf{x}^*) = 0$ for all $J \subseteq \{1,\dots,m_1\}$ and σ_J is SOS, one must have $(\nabla \sigma_J(\mathbf{x}^*))\, g_J(\mathbf{x}^*) = 0$ for all $J \subseteq \{1,\dots,m_1\}$. Moreover,

$$\sigma_J(\mathbf{x}^*)\nabla g_J(\mathbf{x}^*) = \sigma_J(\mathbf{x}^*) \sum_{\ell \in J} \nabla g_\ell(\mathbf{x}^*) \left(\prod_{k \in J; k \neq \ell} g_k(\mathbf{x}^*) \right)$$

$$= \underbrace{\sigma_J(\mathbf{x}^*)g_J(\mathbf{x}^*) \sum_{\ell \in J; \ell \notin J(\mathbf{x}^*)} \nabla g_\ell(\mathbf{x}^*)/g_\ell(\mathbf{x}^*)}_{=0}$$

$$+ \sum_{\ell \in J \cap J(\mathbf{x}^*)} \nabla g_\ell(\mathbf{x}^*)\, \sigma_J(\mathbf{x}^*) \left(\prod_{k \in J; k \neq \ell} g_k(\mathbf{x}^*) \right).$$

Hence differentiating $\mathcal{L}$ at $\mathbf{x}^*$ yields:

$$0 = \nabla_{\mathbf{x}}\mathcal{L}(\mathbf{x}^*, \Lambda_0, \sigma, \boldsymbol{\psi})$$

$$= \Lambda_0(\mathbf{x}^*)\nabla f(\mathbf{x}^*) - \sum_{\emptyset \neq J \subseteq \{1,\dots,m_1\}} \sigma_J(\mathbf{x}^*)\nabla g_J(\mathbf{x}^*) - \sum_{\ell=1}^{m_2} \psi_\ell(\mathbf{x}^*)\nabla h_\ell(\mathbf{x}^*)$$

$$= \underbrace{\Lambda_0(\mathbf{x}^*)}_{\lambda_0^* \geq 0} \nabla f(\mathbf{x}^*) - \sum_{\ell \in J(\mathbf{x}^*)} \nabla g_\ell(\mathbf{x}^*) \underbrace{\left(\sum_{\emptyset \neq J \subseteq \{1,\dots,m_1\}; J \ni \ell} \sigma_J(\mathbf{x}^*) \prod_{k \in J; k \neq \ell} g_k(\mathbf{x}^*) \right)}_{\Lambda_\ell(\mathbf{x}^*) = \lambda_\ell^* \geq 0}$$

$$- \sum_{\ell=1}^{m_2} \underbrace{\psi_\ell(\mathbf{x}^*)}_{v_\ell^*} \nabla h_\ell(\mathbf{x}^*),$$

which yields (7.12). $\qquad\square$

Observe that (7.12) is a Fritz John optimality condition. But the nice additional feature when compared with the classical first-order Fritz John optimality conditions (7.10) is that Theorem 7.9 asserts the existence of the generalized Lagrangian $\mathcal{L}$ in (7.11) for which $\mathbf{x}^*$ is a global minimum on $\mathbb{R}^n$!

In other words, the polynomials Λ_j, $j = 0, \dots, m_1$, and ψ_ℓ, $\ell = 1, \dots, m_2$, in (7.11) provide a certificate of nonnegativity on $\mathbf{K}$ for the polynomial $f - f^*$, and so a certificate of global optimality for $\mathbf{x}^* \in \mathbf{K}$.

Indeed suppose that $f(\mathbf{z}) - f^* < 0$ for some $\mathbf{z} \in \mathbf{K}$. Then recalling that $\mathcal{L}$ is an SOS which is positive whenever $f(\mathbf{x}) \neq f^*$ and evaluating $\mathcal{L}(\cdot, \Lambda, \boldsymbol{\psi})$ at $\mathbf{z}$ yields the contradiction

$$\underbrace{\mathcal{L}(\mathbf{z}, \Lambda, \boldsymbol{\psi})}_{>0} = \underbrace{\Lambda_0(\mathbf{z})}_{\geq 0} \underbrace{(f(\mathbf{z}) - f^*)}_{<0} - \underbrace{\sum_{j=1}^{m_1} \Lambda_j(\mathbf{z})\, g_j(\mathbf{z}) - \sum_{\ell=1}^{m_2} \psi_\ell(\mathbf{z})\, h_\ell(\mathbf{z})}_{\leq 0}.$$

This result is in the same vein as for the generalized Lagrangian defined in Theorem 7.4(d) but now for an arbitrary closed basic semi-algebraic set $\mathbf{K} \subset \mathbb{R}^n$ of the form (7.2), not necessarily compact. Moreover, since the polynomial weights $\Lambda_j \in \mathbb{R}[\mathbf{x}]$, $j = 0, \dots, m_1$, are in the preordering $P(g)$, they are now only nonnegative on $\mathbf{K}$ and not SOS.

Example 7.10 Let us illustrate Theorem 7.9 on the following toy example. In the univariate case consider the problem $\mathbf{P} : f^* = \inf_x \{ x : x^3 \geq 0 \}$

for which $f^* = 0$ and $x^* = 0$. The first-order Fritz John optimality conditions (7.10) hold with $\lambda_0^* = 0$ and $\lambda_1^* > 0$ arbitrary. Notice that $x^* = 0$ is not a global minimum of the Lagrangian $x \mapsto L(x, \lambda) = \lambda_0^*(f - f^*) - \lambda_1^* x^3 = -\lambda_1^* x^3$.

In addition, the polynomial $x \mapsto f(x) - f^*$ *(equal to x)* cannot be written in the form $\sigma_0 + \sigma_1 x^3$ for some SOS univariate polynomials σ_0 and σ_1. On the other hand observe that

$$\underbrace{(x^3 + x^2)}_{\geq 0 \text{ on } \mathbf{K}} (f(x) - f^*) = x^4 + x^3 = \underbrace{x^4}_{\text{SOS}} + \underbrace{1}_{\text{SOS}} \cdot x^3,$$

and so with $x \mapsto \Lambda_0(x) := x^3 + x^2 \in P(g)$ and $x \mapsto \Lambda_1(x) := 1 \in P(g)$, the extended Lagrangian

$$x \mapsto \mathcal{L}(x, \Lambda) = \Lambda_0(x) x - \Lambda_1(x) x^3 = x^4,$$

is SOS and $x^* = 0$ is its global minimizer. Moreover, $\Lambda_0(x^*) = 0 = \lambda_0^*$ and $\Lambda_1(x^*) = 1 =: \lambda_1^* > 0$.

7.3 Exercises

Exercise 7.1 Consider Example 7.10.

(a) What can be said of the hierarchy of semidefinite relaxations (6.3) and their duals (6.4)?

(b) Replace the constraint "$x^3 \geq 0$" with the equivalent constraint "$x \geq 0$." Show that the first semidefinite relaxation of the hierarchy (6.3) is exact.

(c) Show that its dual (6.4) is also exact and has an optimal solution. Why? (*Hint:* See Section 2.2 about positivity certificates in the univariate case.) What can we conclude about the representation of the feasible set for optimization problems?

Exercise 7.2 Consider the optimization problem $f^* = \inf\{x : -x^2 \geq 0\}$.

(a) Show that the first semidefinite program of the hierarchy (6.3) is exact (and so has an optimal solution). Can its dual have an optimal solution? (*Hint:* Try to write the KKT optimality conditions.)

(b) Write the Fritz John optimality conditions. Exhibit polynomials $(\Lambda_j) \subset \mathbb{R}[x]$ as in (7.11) of Theorem 7.9.

7.4 Notes and sources

7.1–7.1.2 Most of the material is taken from Lasserre (2001) and Chapter 5 in Lasserre (2009c). Theorem 7.3 is taken from Bertsimas (1995, §3.3); see also Güler (2010, Chapter 9) in which the reader will find an interesting and detailed discussion on various constraint qualifications as well as a first-order sufficient condition of optimality. The important results in Theorems 7.5 and 7.6 are from Nie (2014). This shows in particular that in polynomial optimization, the usual constraint qualification (QC) (7.4), strict complementarity and second-order sufficient condition (7.7) hold generically! To prove Theorem 7.5, Nie (2014) shows that the so-called *boundary Hessian condition* (BHC) of Marshall (2009) holds, which in turn implies that (7.8) holds; see also Marshall (2006).

7.2 The "Fritz John optimality conditions" originally due to Karush (1939) were rediscovered in John (1948), and for this reason they are called Karush–John optimality conditions in Schichl and Neumaier (2006) where the authors also use Stengle's Positivstellensatz to provide qualification-free global optimality conditions for multi-objective optimization with polynomial data.

8

Exploiting sparsity or symmetry

In previous chapters we have seen that the semidefinite relaxations (6.3) of the moment-SOS approach described in Chapter 6 have very nice properties. But there is a price to pay. Their size grows rapidly with the dimension n of the original problem. Typically, the moment matrix $\mathbf{M}_d(\mathbf{y})$ is $s(d) \times s(d)$ with $s(d) = \binom{n+d}{n}$ (which grows as $O(n^d)$ with n if d is fixed), and there are $\binom{n+2d}{n} (= O(n^{2d}))$ variables (y_α). This makes the applicability of Algorithm 6.1 limited to problems of modest size only, especially in view of the present status of semidefinite programming software packages.

Fortunately, in many practical applications of large size optimization problems, some sparsity pattern and/or symmetry is often present and may be exploited. For instance, very often all polynomials involved in the description of the optimization problem $\mathbf{P}$ in $\mathbb{R}^n$ have low degree. Therefore even if n is large, any monomial contains at most d variables (if d is the maximum degree of the polynomials involved) which makes the coupling of n variables relatively weak. If in addition each constraint $g_j(\mathbf{x}) \geq 0$ is concerned with few variables only, then this weak coupling of variables can be exploited to define specialized semidefinite relaxations of $\mathbf{P}$ of much smaller size while keeping their convergence properties. In a similar spirit, if some symmetry is present it can also be exploited to obtain specialized semidefinite relaxations of much smaller size.

8.1 Exploiting sparsity

Consider the polynomial optimization problem:

$$f^* := \inf \{ f(\mathbf{x}) : \mathbf{x} \in \mathbf{K} \}, \tag{8.1}$$

where the feasible set $\mathbf{K} \subset \mathbb{R}^n$ is the basic semi-algebraic set defined by:

$$\mathbf{K} = \{\mathbf{x} \in \mathbb{R}^n : g_j(\mathbf{x}) \geq 0, \ j = 1, \ldots, m\}, \tag{8.2}$$

for some real-valued polynomials $f, g_j \in \mathbb{R}[\mathbf{x}], j = 1, \ldots, m$.

Suppose that there is no *coupling* between some subsets of variables in the polynomials (g_j) that define the set $\mathbf{K}$, and the polynomial objective function f. By no coupling between two sets of variables, we mean that there is *no* monomial involving some variables of such subsets in any of the polynomials $f, (g_j)$.

More precisely, recalling the notation of Section 2.6, let $I_0 := \{1, \ldots, n\}$ be the union $\cup_{k=1}^p I_k$ (with possible overlaps) of p subsets $I_k, k = 1, \ldots, p$, with cardinal denoted n_k. For an arbitrary $J \subseteq I_0$, let $\mathbb{R}[\mathbf{x}(J)]$ denote the ring of polynomials in the variables $\mathbf{x}(J) = \{x_i : i \in J\}$, and so $\mathbb{R}[\mathbf{x}(I_0)] = \mathbb{R}[\mathbf{x}]$.

As in Section 2.6 we make the following assumption.

Assumption 8.1 *Let $\mathbf{K} \subset \mathbb{R}^n$ be as in (8.2). A scalar $M > 0$ is known such that $\|\mathbf{x}\|_\infty < M$ for all $\mathbf{x} \in \mathbf{K}$.*

Note that under Assumption 8.1, we have $\sum_{i \in I_k} x_i^2 \leq n_k M^2, k = 1, \ldots, p$, and therefore, in the description (8.2) of $\mathbf{K}$, we may and will add the p redundant quadratic constraints

$$0 \leq g_{m+k}(\mathbf{x}) \ \left(:= n_k M^2 - \sum_{i \in I_k} x_i^2\right), \quad k = 1, \ldots, p, \tag{8.3}$$

and set $m' = m + p$, so that $\mathbf{K}$ is now described by:

$$\mathbf{K} := \{\mathbf{x} \in \mathbb{R}^n : g_j(\mathbf{x}) \geq 0, \quad j = 1, \ldots, m'\}. \tag{8.4}$$

Note that $g_{m+k} \in \mathbb{R}[\mathbf{x}(I_k)]$, for all $k = 1, \ldots, p$.

For convenience we restate Assumption 2.28 of Chapter 2 which gives the sparsity pattern $\{I_k\}$ some structure:

Assumption 8.2 *Let $\mathbf{K} \subset \mathbb{R}^n$ be as in (8.4). The index set $J = \{1, \ldots, m'\}$ is partitioned into p disjoint sets $J_k, k = 1, \ldots, p$, and the collections $\{I_k\}$ and $\{J_k\}$ satisfy the following.*

(a) For every $j \in J_k, g_j \in \mathbb{R}[\mathbf{x}(I_k)]$, that is, for every $j \in J_k$, the constraint $g_j(\mathbf{x}) \geq 0$ only involves the variables $\mathbf{x}(I_k) = \{x_i : i \in I_k\}$.

(b) *The objective function $f \in \mathbb{R}[\mathbf{x}]$ can be written as*

$$f = \sum_{k=1}^{p} f_k, \quad \text{with } f_k \in \mathbb{R}[\mathbf{x}(I_k)], \quad k = 1, \ldots, p. \tag{8.5}$$

Sparse semidefinite relaxations

We now describe how to specialize the semidefinite relaxations (6.3) to the present context of structured sparsity satisfying Assumption 8.2.

With $k \in \{1, \ldots, p\}$ fixed, and $g \in \mathbb{R}[\mathbf{x}(I_k)]$, let $\mathbf{M}_d(\mathbf{y}, I_k)$ (respectively $\mathbf{M}_d(g\,\mathbf{y}, I_k)$) be the moment (respectively localizing) submatrix obtained from $\mathbf{M}_d(\mathbf{y})$ (respectively $\mathbf{M}_d(g\,\mathbf{y})$) by retaining only those rows (and columns) $\alpha \in \mathbb{N}^n$ of $\mathbf{M}_d(\mathbf{y})$ (respectively $\mathbf{M}_d(g\,\mathbf{y})$) such that $\alpha_i = 0$ whenever $i \notin I_k$. What we call a *sparse* semidefinite relaxation is a semidefinite program of the following form:

$$\rho_d^{\text{sparse}} = \inf_{\mathbf{y}} L_{\mathbf{y}}(f)$$
$$\text{s.t.} \quad \mathbf{M}_d(\mathbf{y}, I_k) \succeq \mathbf{0}, \qquad k = 1, \ldots, p$$
$$\mathbf{M}_{d-v_j}(g_j\mathbf{y}, I_k) \succeq \mathbf{0}, \quad j \in J_k,\ k = 1, \ldots, p \tag{8.6}$$
$$y_\mathbf{0} = 1.$$

Its dual is the semidefinite program

$$(\rho_d^{\text{sparse}})^* = \sup_{\lambda, \sigma_{kj}} \lambda$$
$$\text{s.t.} \quad f - \lambda = \sum_{k=1}^{p} \left(\sigma_{k0} + \sum_{j \in J_k} \sigma_{kj}\, g_j \right) \tag{8.7}$$
$$\sigma_{k0}, \sigma_{kj} \in \Sigma[\mathbf{x}(I_k)], \quad k = 1, \ldots, p$$
$$\deg \sigma_{k0},\ \deg \sigma_{kj} g_j \leq 2d, \quad k = 1, \ldots, p.$$

As we next see, this specialized *sparse* version of the hierarchy of *dense* semidefinite relaxations (6.3) retains the convergence properties of the latter.

Theorem 8.3 *Let Assumptions 8.1 and 8.2 hold, and consider the sparse semidefinite relaxations (8.6) and (8.7).*

If the running intersection property (2.27) holds then $\rho_d^{\text{sparse}} \to f^*$ and $(\rho_d^{\text{sparse}})^* \to f^*$ as $d \to \infty$.

The proof is very similar to that of Theorem 6.2 except that one now uses the sparse version Theorem 2.29 of Theorem 2.15.

To see the gain in terms of number of variables and size of the moment and localizing matrices, let $\tau := \sup_k |I(k)|$. Then the semidefinite relaxation (8.6) has at most $O(p\tau^{2d})$ moment variables instead of $O(n^{2d})$, and p moment matrices of size at most $O(\tau^d)$ instead of a single one of size $O(n^d)$. Similarly, the localizing matrix $\mathbf{M}_{d-v_j}(g_j\mathbf{y}, I_k)$ is of size at most $O(\tau^{d-v_j})$ instead of $O(n^{d-v_j})$.

This yields big savings if τ is small compared to n. For instance (with the present status of semidefinite programming software packages) if τ is relatively small (say e.g. 6, 7) then one can solve optimization problems with n as large as 1000 variables, whereas with $n = 1000$ one cannot even implement the first standard semidefinite relaxation of the hierarchy (6.3) with $d = 1$!

Example 8.4 Let $I_0 = \{1, \ldots, n\}$ and $I_k := \{k, k+1\}$, for $k = 1, \ldots n - 1$ (i.e., $p = n - 1$). In this sparsity pattern all polynomials see only monomials of the form $x_k x_{k+1}, k = 1, \ldots, n - 1$.

And so if Assumption 8.2 holds for problem **P**, the semidefinite relaxation (8.6) contains p moment matrices $\mathbf{M}_d(y, I_k)$, each of size $(d + 2)(d + 1)/2$, instead of a single moment matrix of size $\binom{n+d}{n}$ in the dense semidefinite relaxation (6.3). Similarly, the semidefinite program (8.7) reads:

$$
(\rho_d^{\text{sparse}})^* = \sup_{\lambda, \sigma_{kj}} \lambda
$$
$$
\text{s.t.} \sum_{k=1}^p f_k(x_k, x_{k+1}) - \lambda = \sum_{k=1}^p \left(\sigma_{k0}(x_k, x_{k+1}) \right.
$$
$$
\left. + \sum_{j \in J_k} \sigma_{kj}(x_k, x_{k+1}) \, g_j(x_k, x_{k+1}) \right)
$$
$$
\deg \sigma_{k0}, \ \deg \sigma_{kj} g_j \leq 2d, \quad k = 1, \ldots, p,
$$

i.e., all SOS weights σ_{kj} are bivariate polynomials.

8.2 Exploiting symmetry

In this section we briefly describe how symmetry can be exploited to replace a semidefinite program invariant under the action of some group of permutations, with a much simpler one. In particular, it can be applied to the semidefinite relaxations (6.3) when $f \in \mathbb{R}[\mathbf{x}]$ and the polynomials $(g_j) \subset \mathbb{R}[\mathbf{x}]$ that define $\mathbf{K}$ are all invariant under some group of permutations.

Let S_n be the space of $n \times n$ real symmetric matrices and let Aut(S_n) be the group of automorphisms on S_n. Let $\mathcal{G}$ be a finite group acting on $\mathbb{R}^n$ via $\rho_0 : \mathcal{G} \to \text{GL}(\mathbb{R}^n)$, which in turn induces an action $\rho : \mathcal{G} \to \text{Aut}(S_n)$ on S_n by $\rho(g)(\mathbf{X}) = \rho_0(g)^T \mathbf{X} \rho_0(g)$ for every $g \in \mathcal{G}$, $\mathbf{X} \in S_n$.

Assume that $\rho_0(g)$ is orthonormal for every $g \in \mathcal{G}$. A matrix $\mathbf{X} \in \mathbb{R}^{n \times n}$ is said to be invariant under the action of $\mathcal{G}$ if $\rho(g)(\mathbf{X}) = \mathbf{X}$ for every $g \in \mathcal{G}$, and $\mathbf{X}$ is invariant if and only if $\mathbf{X}$ is an element of the *commutant algebra*

$$A^{\mathcal{G}} := \{ \mathbf{X} \in \mathbb{R}^{n \times n} : \rho_0(g)\,\mathbf{X} = \mathbf{X}\,\rho_0(g), \quad \forall g \in \mathcal{G} \}. \qquad (8.8)$$

Of particular interest is when $\mathcal{G}$ is a subgroup of the group $\mathcal{P}_n$ of permutations of $\{1, \ldots, n\}$, in which case $\rho_0(g)(\mathbf{x}) = (x_{g(i)})$ for every $\mathbf{x} \in \mathbb{R}^n$, and $\rho(g)(\mathbf{X})_{ij} = X_{g(i),g(j)}$ for every $1 \le i, j \le n$.

For every $(i, j) \in \{1, \ldots, n\} \times \{1, \ldots, n\}$, the orbit $O_{\mathcal{G}}(i, j)$ under action of $\mathcal{G}$, is the set of couples $\{(g(i), g(j)) : g \in \mathcal{G}\}$. With ω the number of orbits, and $1 \le \ell \le \omega$, define the $n \times n$ matrix $\tilde{\mathbf{D}}_\ell$ by $(\tilde{\mathbf{D}}_\ell)_{ij} := 1$ if (i, j) belongs to orbit ℓ, and 0 otherwise. Normalize to $\mathbf{D}_\ell := \tilde{\mathbf{D}}_\ell / \sqrt{\langle \tilde{\mathbf{D}}_\ell, \tilde{\mathbf{D}}_\ell \rangle}$, for every $1 \le \ell \le \omega$, and define:

- the multiplication table

$$\mathbf{D}_i\,\mathbf{D}_j = \sum_{\ell=1}^{\omega} \gamma_{ij}^\ell\,\mathbf{D}_\ell, \qquad i, j = 1, \ldots, \omega,$$

for some $(\gamma_{ij}^\ell) \subset \mathbb{R}$,

- the $\omega \times \omega$ matrices $\mathbf{L}_1, \ldots, \mathbf{L}_\omega$ by

$$(\mathbf{L}_k)_{ij} := \gamma_{kj}^i, \qquad i, j, k = 1, \ldots, \omega.$$

Then the commutant algebra (8.8) reads

$$A^{\mathcal{G}} = \left\{ \sum_{\ell=1}^{\omega} x_\ell\,\mathbf{D}_\ell : x_\ell \in \mathbb{R} \right\}$$

with dimension dim $A^{\mathcal{G}} = \omega$.

Exploiting symmetry in semidefinite programming is possible thanks to the following crucial property of the matrices $(\mathbf{D}_l)$.

Theorem 8.5 *The mapping $\mathbf{D}_\ell \to \mathbf{L}_\ell$ is a $\star$-isomorphism called the regular $\star$-representation of $A^{\mathcal{G}}$, and in particular:*

$$\sum_{\ell=1}^{\omega} x_\ell\,\mathbf{D}_\ell \succeq 0 \quad \Longleftrightarrow \quad \sum_{\ell=1}^{\omega} x_\ell\,\mathbf{L}_\ell \succeq 0. \qquad (8.9)$$

Application to semidefinite programming   Consider the semidefinite program

$$\sup_{\mathbf{X} \succeq 0} \{ \langle \mathbf{C}, \mathbf{X} \rangle \; : \; \langle \mathbf{A}_k, \mathbf{X} \rangle = b_k, \quad k = 1, \ldots, p \} \tag{8.10}$$

and assume it is invariant under the action of $\mathcal{G}$, that is, $\mathbf{C} \in A^{\mathcal{G}}$ and the feasible region is globally invariant, meaning that if $\mathbf{X}$ is feasible in (8.10) then so is $\rho(g)(\mathbf{X})$ for every $g \in \mathcal{G}$.

By convexity, for every feasible $\mathbf{X}$ of (8.10), the matrix $\mathbf{X}_0 := \frac{1}{|\mathcal{G}|} \sum_{g \in \mathcal{G}} \rho(g)(\mathbf{X})$ is feasible, invariant under the action of $\mathcal{G}$ and with the same objective value as $\mathbf{X}$. Therefore, we can include in the semidefinite program (8.10) the additional linear constraint $\mathbf{X} \in A^{\mathcal{G}}$ without affecting the optimal value.

Therefore, writing $\mathbf{X} = \sum_{l=1}^{\omega} x_l \mathbf{D}_l$ and setting

$$c_l := \langle \mathbf{C}, \mathbf{D}_l \rangle, \quad a_{kl} := \langle \mathbf{A}_k, \mathbf{D}_l \rangle, \qquad \forall l = 1, \ldots, \omega, \; k = 1, \ldots, p,$$

the semidefinite program (8.10) has the same optimal value as

$$\sup_{\mathbf{x} \in \mathbb{R}^{\omega}} \left\{ \mathbf{c}' \mathbf{x} \; : \; \mathbf{a}_k' \mathbf{x} = b_k, \quad k = 1, \ldots, p; \quad \sum_{l=1}^{\omega} x_l \mathbf{L}_l \succeq 0 \right\}. \tag{8.11}$$

Observe that in (8.10) we have n variables and an $n \times n$ positive semidefinite matrix $\mathbf{X}$, whereas in (8.11) we only have ω variables and an $\omega \times \omega$ positive semidefinite matrix.

8.3   Exercises

Exercise 8.1   Let $I_0 := \{1, \ldots, n\} = I_1 \cup I_2$ with some possible overlap between I_1 and I_2. Suppose that Assumption 8.2 holds. What can we conclude for the hierarchy of sparse semidefinite relaxations (8.6)?

Exercise 8.2   In typical discrete-time optimal control problems over T periods, with initial state $\mathbf{x}_0$, state variables $\mathbf{x}_t \in \mathbb{R}^n$ and control variables $\mathbf{u}_t \in \mathbb{R}^m$, the goal is to minimize a criterion of the form $J = \sum_{t=1}^{T} p_t(\mathbf{x}_t, \mathbf{u}_{t-1})$ under constraints of the form

$$\mathbf{x}_t = f(\mathbf{x}_{t-1}, \mathbf{u}_{t-1}), \quad h_\ell(\mathbf{x}_{t-1}, \mathbf{u}_{t-1}) \leq 0, \quad \ell = 1, \ldots, s, \; t = 1, \ldots, T.$$

The above equality constraints describe the dynamics of the system whereas the other constraints describe limitations on the state and/or control variables.

(a)   Show that such problems exhibit a natural sparsity pattern as in Assumption 8.1 and moreover this sparsity pattern satisfies the running intersection property (2.27).

(b) If f, h_ℓ and p_t are all polynomials what can we conclude for this class of optimal control problems?

8.4 Notes and sources

8.1 Polynomials satisfying sparsity patterns were investigated in Kojima et al. (2005) and the sparse semidefinite relaxations (8.6) were first proposed in Waki et al. (2006) as a heuristic to solve global optimization problems with a large number of variables and which satisfy some structured sparsity pattern (called a *correlative* sparsity pattern). Their convergence in Theorem 8.3 was proved in Lasserre (2006a) if the sparsity pattern satifies the running intersection property. The sparse relaxations have been implemented in the SparsePOP software of Waki et al. (2008) and numerical experiments show that one may then solve global optimization problems with $n = 1000$ variables for which even the first nonsparse semidefinite relaxation of the hierarchy (6.3) cannot be implemented. Kim et al. (2009) provides a nice application for sensor network localization.

8.2 This section is inspired by Laurent (2008) and Theorem 8.5 is from de Klerk et al. (2007). For exploiting symmetry in the context of sums of squares and semidefinite programming see also the work of Gaterman and Parrilo (2004) and Vallentin (2007). For instance, these ideas have been used successfully in coding theory for large error correcting codes based on computing the stability number of some related graph; see for example Laurent (2007b) and Schrijver (2005). Riener et al. (2013) describe how to derive explicitly a version of the semidefinite relaxations (6.3) adapted to the case where **P** is invariant under the action of some groups (in particular the symmetric group). In such a case, the moment and localizing matrices have a nice block-diagonal decomposition. For the symmetric group each block has almost constant (small) size as d increases in the hierarchy.

9

LP-relaxations for polynomial optimization

The semidefinite relaxations described in Chapter 6 are based on Putinar's certificate of positivity (Theorem 2.15) which is a weighted SOS representation of polynomials that are positive on $\mathbf{K} \subset \mathbb{R}^n$. But in Chapter 2 we have also seen other non SOS based representations. In this chapter we show how to use the certificate of positivity described in Theorem 2.22 to obtain a hierarchy of *linear programming* (LP) relaxations for $\mathbf{P}$ (as opposed to the hierarchy of semidefinite relaxations based on Putinar's certificate).

A well-known and common practice for solving difficult optimization problems (in particular 0/1 and combinatorial optimization problems) is to add redundant constraints in the definition of $\mathbf{P}$ to obtain an equivalent problem $\mathbf{P}_d$. The reason for doing so is because these additional constraints may not be redundant in certain relaxations of the modified problem $\mathbf{P}_d$. That is, even though $\mathbf{P}$ and $\mathbf{P}_d$ are equivalent, their respective relaxations are *not*, and in general the relaxation of $\mathbf{P}_d$ is "tighter" (i.e., more constrained) than that of $\mathbf{P}$. Hence adding more and more constraints may result in tighter and tighter lower bounds for the resulting relaxations. This is exactly the spirit of the so-called RLT-relaxations of Sherali and Adams (1990) for 0/1 programs, where redundant constraints are obtained by taking products of the original constraints. We will see that thanks to Theorem 2.22 we can provide a rigorous theoretical justification (in a relatively general framework) for doing so. Indeed, the sequence of optimal values of the corresponding LP-relaxations converges to the global optimum. Last but not least, we also show that the standard *dual* method of nonlinear programming (or Lagrangian relaxation), i.e., computing

$$\sup_{\mathbf{u} \geq 0} \left\{ \inf_{\mathbf{x}} \{ f(\mathbf{x}) - \sum_{j=1}^{m} u_j \, g_j(\mathbf{x}) \} \right\},$$

provides a lower bound as close as desired to the optimal value f^* of **P**, *provided that* it is applied to a redundant description of **K** (i.e., to problem $\mathbf{P}_d$) that includes sufficiently many products of the original constraints.

Of course, especially in view of the present relative status of linear programming and semidefinite solvers, one would much prefer to solve LP-relaxations rather than semidefinite relaxations. Indeed the former solvers can handle problems with millions of variables and constraints. However and unfortunately, we will see that the LP-relaxations suffer from serious drawbacks when compared with their semidefinite counterparts.

9.1 LP-relaxations

Again consider the optimization problem

$$\mathbf{P}: \qquad f^* = \inf_{\mathbf{x}} \{ f(\mathbf{x}) : \mathbf{x} \in \mathbf{K} \}, \tag{9.1}$$

where $\mathbf{K} = \{ \mathbf{x} \in \mathbb{R}^n : g_j(\mathbf{x}) \geq 0, \ j = 1, \ldots, m \}$ is a compact basic semi-algebraic set.

The case of a basic semi-algebraic set With **K** as in (5.2), let $\widehat{g}_j$ be the normalized version associated with g_j, $j = 1, \ldots, m$, defined in (2.15). Therefore $\widehat{g}_j(\mathbf{x}) \geq 0$ and $1 - \widehat{g}_j(\mathbf{x}) \geq 0$ for every $\mathbf{x} \in \mathbf{K}$. Also recall the vector notation of Section 2.4.3:

$$\widehat{\mathbf{g}} = \begin{pmatrix} \widehat{g}_1 \\ \vdots \\ \widehat{g}_m \end{pmatrix} \in \mathbb{R}[\mathbf{x}]^m, \quad 1 - \widehat{\mathbf{g}} = \begin{pmatrix} 1 - \widehat{g}_1 \\ \vdots \\ 1 - \widehat{g}_m \end{pmatrix} \in \mathbb{R}[\mathbf{x}]^m,$$

and $\mathbb{N}_d^{2m} = \{ (\boldsymbol{\alpha}, \boldsymbol{\beta}) \in \mathbb{N}^{2m} : |\boldsymbol{\alpha} + \boldsymbol{\beta}| \leq d \}$. Let $d \in \mathbb{N}$ be fixed and consider the following linear optimization problem:

$$L_d = \inf_{\mathbf{y}} \{ L_{\mathbf{y}}(f) : y_0 = 1; \ L_{\mathbf{y}}(\widehat{\mathbf{g}}^{\boldsymbol{\alpha}} (1 - \widehat{\mathbf{g}})^{\boldsymbol{\beta}}) \geq 0, \ \forall (\boldsymbol{\alpha}, \boldsymbol{\beta}) \in \mathbb{N}_d^{2m} \}, \tag{9.2}$$

where $L_{\mathbf{y}}$ is the Riesz linear functional defined in (2.31).

Indeed, problem (9.2) is a linear optimization problem, because since $|\boldsymbol{\alpha} + \boldsymbol{\beta}| \leq d$, expanding $\widehat{\mathbf{g}}^{\boldsymbol{\alpha}} (1 - \widehat{\mathbf{g}})^{\boldsymbol{\beta}}$ in the monomial basis $(\mathbf{x}^{\boldsymbol{\alpha}})$ and using linearity of $L_{\mathbf{y}}$ in the conditions $L_{\mathbf{y}}(\widehat{\mathbf{g}}^{\boldsymbol{\alpha}} (1 - \widehat{\mathbf{g}})^{\boldsymbol{\beta}}) \geq 0$, yields finitely many linear inequality constraints on finitely many coefficients $(y_{\boldsymbol{\alpha}})$ of the infinite sequence **y**.

The dual of (9.2) is the linear program:

$$L_d^* = \sup_{\mathbf{u} \geq 0, \lambda} \left\{ \lambda : f - \lambda = \sum_{(\boldsymbol{\alpha}, \boldsymbol{\beta}) \in \mathbb{N}_d^{2m}} u_{\boldsymbol{\alpha}\boldsymbol{\beta}} \, \widehat{\mathbf{g}}^{\boldsymbol{\alpha}} (1 - \widehat{\mathbf{g}})^{\boldsymbol{\beta}} \right\}. \tag{9.3}$$

Example 9.1 With $n = 1$, let $\mathbf{K} := [0, 1] = \{x : x \geq 0, (1 - x) \geq 0\} \subset \mathbb{R}$. With $(\alpha, \beta) \in \mathbb{N}_2^2$ one obtains:

$$L_\mathbf{y}(x^0(1 - x)^0) = y_0$$
$$L_\mathbf{y}(x^1(1 - x)^0) = y_1$$
$$L_\mathbf{y}(x^0(1 - x)^1) = y_0 - y_1$$
$$L_\mathbf{y}(x^2(1 - x)^0) = y_2$$
$$L_\mathbf{y}(x^1(1 - x)^1) = y_1 - y_2$$
$$L_\mathbf{y}(x^0(1 - x)^2) = y_0 - 2y_1 + y_2$$

and so if $f \in \mathbb{R}[x]_2$, the LP-relaxation (9.2) with $d = 2$ reads

$$L_2 = \inf_\mathbf{y} \sum_{k=0}^2 f_k\, y_k$$
$$\text{s.t.} \quad y_0 = 1, \ y_0 - y_1 \geq 0, \ y_1 - y_2 \geq 0, \ y_0 - 2y_1 + y_2 \geq 0, \ \mathbf{y} \geq 0,$$

which is a linear program. Its dual is the linear program:

$$\sup_{\mathbf{u} \geq 0, \lambda} \quad \lambda$$
$$\text{s.t.} \quad f_0 - \lambda = u_{00} + u_{01} + u_{02}$$
$$f_1 = u_{10} - u_{01} + u_{11} - 2u_{02}$$
$$f_2 = u_{20} - u_{11} + u_{02}.$$

As expected, multiplying both sides of the equality constraint "$f_i = \cdots$" by x^i and summing up yields

$$f(x) - \lambda = u_{00} + u_{10}\, x + u_{01}\, (1 - x) + u_{20}\, x^2 + u_{11}\, x\, (1 - x) + u_{02}\, (1 - x)^2,$$

for all x, which is the same as the constraint of the dual (9.3).

Of course each linear program (9.2), $d \in \mathbb{N}$, is a relaxation of the original optimization problem **P**. To see this, let $s := d \max_j [\deg g_j]$. Let $\mathbf{x} \in \mathbf{K}$ be fixed arbitrary, and let $\mathbf{y} = (\mathbf{x}^\gamma)$, $\gamma \in \mathbb{N}_s^n$ be the sequence of moments up to order s, of the Dirac measure at the point $\mathbf{x}$ (so that $y_0 = 1$). Then since $\widehat{\mathbf{g}}(\mathbf{x})^\alpha(1 - \widehat{\mathbf{g}}(\mathbf{x})^\beta) \geq 0$ it follows that $L_\mathbf{y}(\widehat{\mathbf{g}}^\alpha(1 - \widehat{\mathbf{g}}^\beta)) \geq 0$, for all $|\alpha + \beta| \leq d$. Hence $\mathbf{y}$ is feasible for (9.2) and $L_\mathbf{y}(f) = f(\mathbf{x})$. Therefore as $\mathbf{x} \in \mathbf{K}$ was arbitrary, $L_d \leq f^*$ for all d.

Theorem 9.2 *Assume that* $\mathbf{K}$ *in (9.1) is compact. The sequence* (L_d), $d \in \mathbb{N}$, *is monotone nondecreasing, and under Assumption 2.21,*

$$\lim_{d \to \infty} L_d^* = \lim_{d \to \infty} L_d = f^*. \tag{9.4}$$

The proof is along the same lines as that of Theorem 6.2 except that one now uses Theorem 2.22 instead of Theorem 2.15.

So the relaxations (9.2)–(9.3) are linear programs, good news because in principle, with current LP software packages, one is able to solve linear programs with millions of variables and constraints. However, as we will see in Section 9.3, the linear programs (9.2)–(9.3) suffer from some serious drawbacks.

The case of a convex polytope In the particular case where the basic semi-algebraic set $\mathbf{K} \subset \mathbb{R}^n$ is compact and all the polynomials (g_j) in the definition (5.2) of $\mathbf{K}$ are affine (and so $\mathbf{K}$ is a polytope), one may specialize the linear relaxation (9.2) to

$$L_d = \inf_{\mathbf{y}} \{ L_{\mathbf{y}}(f) : \ y_0 = 1; \ L_{\mathbf{y}}(\mathbf{g}^\alpha) \geq 0, \quad \forall \alpha \in \mathbb{N}_d^m \}, \qquad (9.5)$$

and its associated dual

$$L_d^* = \sup_{\lambda, \mathbf{u} \geq 0} \left\{ \lambda : \ f - \lambda = \sum_{\alpha \in \mathbb{N}_d^m} u_\alpha \, \mathbf{g}^\alpha \right\}. \qquad (9.6)$$

Theorem 9.3 *If all polynomials (g_j) in (5.2) are affine and $\mathbf{K}$ is a convex polytope with nonempty interior, then:*

$$\lim_{d \to \infty} L_d^* = \lim_{d \to \infty} L_d = f^*. \qquad (9.7)$$

The proof is similar to that of Theorem 9.2.

SDP- versus LP-relaxations The underlying principle behind the semidefinite relaxations (6.3) and the LP-relaxations (9.2) is the same but its technical implementation differs. Indeed in both cases it consists of relaxing the (hard) finite-dimensional conic optimization problem (5.7) by replacing the hard conic membership constraint $\mathbf{y} \in \Delta_d(\mathbf{K})$ (here d is the degree of f) with (only) necessary conditions for $\mathbf{y}$ to be a member of $\Delta_d(\mathbf{K})$.

- In the former case one uses the *semidefinite* constraints $L_{\mathbf{y}}(h^2 g_j) \geq 0$ for all $h \in \mathbb{R}[\mathbf{x}]_k$, (that is, $\mathbf{M}_k(\mathbf{y}), \mathbf{M}_k(\mathbf{y}\,g_j) \succeq 0, \ j = 1, \ldots, m$).
- In the latter case one uses the *linear* constraints $L_{\mathbf{y}}(\widehat{\mathbf{g}}^\alpha (1 - \widehat{\mathbf{g}}^\beta)) \geq 0$ for all $(\alpha, \beta) \in \mathbb{N}_k^{2m}$.

Equivalently, in both cases it consists of *strengthening* the dual conic optimization problem (5.8) by approximating from inside the convex cone $C_d(\mathbf{K})$ by using two different certificates of positivity on $\mathbf{K}$.

- In the former case one uses Putinar's positivity certificate

$$f - \lambda = \sum_{j=0}^{m} \sigma_j \, g_j, \quad \text{for some SOS } (\sigma_j) \subset \Sigma[\mathbf{x}]_k.$$

- In the latter case one uses Krivine–Vasilescu's positivity certificate:

$$f - \lambda = \sum_{(\alpha,\beta)\in\mathbb{N}_k^{2m}} u_{\alpha\beta}\, \mathbf{g}^\alpha (1 - \mathbf{g})^\beta \quad \text{for some scalars } u_{\alpha\beta} \geq 0.$$

9.2 Interpretation and the dual method in NLP

With $\mathbf{P}$ as in (9.1) and with $d \in \mathbb{N}$ fixed, consider the new problem

$$\mathbf{P}_d : \quad \inf \{ f(\mathbf{x}) : \mathbf{g}(\mathbf{x})^\alpha (1 - \mathbf{g}(\mathbf{x}))^\beta \geq 0, \quad \forall (\alpha, \beta) \in \mathbb{N}_d^{2m} \}.$$

Of course $\mathbf{P}_d$ is equivalent to $\mathbf{P}$ as we have only added redundant constraints to $\mathbf{P}$. Suppose that one uses a standard dual method in nonlinear programming to solve $\mathbf{P}_d$. That is, letting $s(d) = \binom{2m+d}{2m}$, solve

$$\theta_d = \sup_{\mathbf{u}\in\mathbb{R}^{s(d)}} \{ G_d(\mathbf{u}) : \mathbf{u} \geq 0 \}, \tag{9.8}$$

where $G_d : \mathbb{R}^{s(d)} \to \mathbb{R}$ is the function:

$$\mathbf{u} \mapsto G_d(\mathbf{u}) := \inf_{\mathbf{x}\in\mathbb{R}^n} \left\{ f(\mathbf{x}) - \sum_{(\alpha,\beta)\in\mathbb{N}_d^{2m}} u_{\alpha\beta}\, \mathbf{g}(\mathbf{x})^\alpha (1 - \mathbf{g}(\mathbf{x}))^\beta \right\}. \tag{9.9}$$

Of course $\theta_d \leq f^*$ for all d because $G_d(\mathbf{u}) \leq f^*$ for all $\mathbf{u} \geq 0$.

Theorem 9.4 *Let $\mathbf{K}$ in (9.1) be compact and let Assumption 2.21 hold. Let f^*, L_d^* and $G_d : \mathbb{R}^{s(d)} \to \mathbb{R}$ be as in (9.1), (9.3) and (9.9), respectively. Then*

$$L_d^* \leq \theta_d \leq f^* \quad \forall d \in \mathbb{N}, \quad \text{and} \quad \theta_d \to f^* \text{ as } d \to \infty.$$

Proof Let $d \in \mathbb{N}$ be fixed and define the set $\mathbf{U}_d \subset \mathbb{R}^{s(d)}$ by:

$$\mathbf{U}_d := \left\{ 0 \leq \mathbf{u} \in \mathbb{R}^{s(d)} : \exists \lambda \text{ s.t. } f - \lambda = \sum_{(\boldsymbol{\alpha},\boldsymbol{\beta}) \in \mathbb{N}_d^{2m}} u_{\alpha\beta} \, \widehat{\mathbf{g}}^{\boldsymbol{\alpha}} \, (1 - \widehat{\mathbf{g}})^{\boldsymbol{\beta}} \right\}.$$

(9.10)

Observe that $\mathbf{U}_d$ is a convex polyhedron, the projection on $\mathbb{R}^{s(d)}$ of the polyhedron

$$\left\{ (\mathbf{u}, \lambda) \in \mathbb{R}_+^{s(d)} \times \mathbb{R} : f(\mathbf{x}) - \lambda = \sum_{(\boldsymbol{\alpha},\boldsymbol{\beta}) \in \mathbb{N}_d^{2m}} u_{\alpha\beta} \, \widehat{\mathbf{g}}(\mathbf{x})^{\boldsymbol{\alpha}} \, (1 - \widehat{\mathbf{g}}(\mathbf{x}))^{\boldsymbol{\beta}}, \quad \forall \mathbf{x} \right\}.$$

(9.11)

Then recalling the definition of L_d^* in (9.3),

$$f^* \geq \theta_d = \sup\{G_d(\mathbf{u}) : \mathbf{u} \geq 0\} \geq \sup\{G_d(\mathbf{u}) : \mathbf{u} \in \mathbf{U}_d\} = L_d^*.$$

Next, by Theorem 9.2, $L_d^* \to f^*$ which in turn implies $\theta_d \to f^*$. □

In conclusion, Theorem 9.4 gives a rigorous theoretical justification for the well-known and common practice of adding redundant constraints in the definition of $\mathbf{P}$ (e.g. as in $\mathbf{P}_d$) to help solve $\mathbf{P}$ because they are no longer redundant in some relaxations of $\mathbf{P}_d$. In other words, even though $\mathbf{P}$ and $\mathbf{P}_d$ are equivalent, their relaxations are *not*. In general, the relaxation of $\mathbf{P}_d$ is "tighter" (i.e., more constrained) than that of $\mathbf{P}$. And so the more such constraints are added, the better is the lower bound obtained by solving the resulting relaxation of $\mathbf{P}_d$.

Of course computing the function $G_d(\mathbf{u})$ in (9.9) is very hard and in fact even evaluating $G_d(\mathbf{u})$ at any particular $\mathbf{u}$ is hard. Indeed, it requires computing the unconstrained global minimum of the function

$$\mathbf{x} \mapsto f(\mathbf{x}) - \sum_{(\boldsymbol{\alpha},\boldsymbol{\beta}) \in \mathbb{N}_d^{2m}} u_{\alpha\beta} \, \mathbf{g}(\mathbf{x})^{\boldsymbol{\alpha}} (1 - \mathbf{g}(\mathbf{x}))^{\boldsymbol{\beta}},$$

an NP-hard problem in general!

So in fact the LP-relaxations (9.3) can be viewed as a brute force simplification of the Lagrangian method (9.8) to make it tractable. This brute force simplification consists of maximizing $G_d(\mathbf{u})$ on the convex polyhedron $\mathbf{U}_d \subset \mathbb{R}_+^{s(d)}$ in (9.10) instead of maximizing $G_d(\mathbf{u})$ on $\mathbb{R}_+^{s(d)}$. Indeed, maximizing $G_d(\mathbf{u})$ on $\mathbf{U}_d$ reduces to solving an LP with variables $(\mathbf{u}, \lambda)$ and feasible set (9.11)!

In particular, the so-called RLT-relaxations of Sherali–Adams are using redundant constraint of **P** formed with products of the initial (g_j) in the definition (5.2), exactly as is done in $\mathbf{P}_d$.

Finally, the proof of Theorem 9.4 also shows a relatively surprising result. Namely, the standard dual method (9.8)–(9.9) of nonlinear programming provides a lower bound as close as desired to the optimal value f^* of the non convex problem **P**, *provided that* it is applied to a redundant description of **K** (i.e., to problem $\mathbf{P}_d$) that includes sufficiently many products of the original constraints. Usually, convergence of the Lagrangian relaxations is proved for convex problems only.

9.3 Contrasting semidefinite and LP-relaxations

Theorem 9.3 implies that we can approach the global optimal value f^* as closely as desired by solving the hierarchy of linear programs (9.2) of increasing size. This should be interesting because today's linear optimization software packages can solve problems of huge size, i.e., with millions of variables and constraints! In contrast, semidefinite optimization software packages are still in their infancy and the size limitation of problems that can be solved is far more severe. In addition, even though semidefinite optimization problems are convex, they are intrinsically more difficult to solve than linear programs. Therefore, solving very large size semidefinite problems (with no special structure) is likely to be impossible at least even in the near future.

Unfortunately we next show that in general the LP-relaxations (9.2) cannot be exact, that is, the convergence in (9.4) (or in (9.7) when **K** is a convex polytope) is only *asymptotic*, not finite.

With **K** compact as in (5.2) and $\mathbf{x} \in \mathbf{K}$, denote by $J(\mathbf{x})$ the set of active constraints at **x**, i.e., $J(\mathbf{x}) = \{j \in \{1, \ldots, m\} : \widehat{g}_j(\mathbf{x}) = 0\}$. Since $\widehat{g}$ is obtained from a normalization on **K**, we may and will assume that $(1 - \widehat{g}_j(\mathbf{x})) > 0$ for all $\mathbf{x} \in \mathbf{K}$ and all $j = 1, \ldots, m$.

Proposition 9.5 *Let* $\mathbf{K} \subset \mathbb{R}^n$ *be as in (5.2) and consider the LP-relaxations defined in (9.2). The finite convergence* $L_d = f^*$ *for some* $d \in \mathbb{N}$ *cannot take place if:*

(a) *there exists a global minimizer* $\mathbf{x}^* \in \mathbf{K}$ *such that* $\widehat{g}_j(\mathbf{x}^*) > 0$ *for all* $j = 1, \ldots, m$;

(b) *there exists* $\mathbf{x}_0 \in \mathbf{K}$ *with* $f(\mathbf{x}_0) > f^*$ *and* $J(\mathbf{x}_0) = J(\mathbf{x}^*)$ *for some global minimizer* $\mathbf{x}^* \in \mathbf{K}$.

Table 9.1 *The slow convergence of the linear relaxations (9.2)*

d	2	4	6	10	15
L_i	$-1/3$	$-1/3$	-0.3	-0.27	-0.2695

Proof (a) Assume that the convergence is finite, i.e., $L_d^* = f^*$ for some $d \in \mathbb{N}$. We would get the contradiction:

$$0 = f(\mathbf{x}^*) - L_d^* = \sum_{(\alpha,\beta)\in\mathbb{N}_d^{2m}} u_{\alpha\beta} \underbrace{\widehat{\mathbf{g}}(\mathbf{x}^*)^\alpha}_{>0} \underbrace{(1 - \widehat{\mathbf{g}}(\mathbf{x}^*))^\beta}_{>0} > 0.$$

(b) Similarly, suppose that $0 = f^* - L_d^* = f(\mathbf{x}^*) - L_d^*$. Then

$$u_{\alpha\beta} \neq 0 \Rightarrow \widehat{\mathbf{g}}^\alpha(\mathbf{x}^*) = 0 \Rightarrow \alpha_j > 0 \text{ for some } j \in J(\mathbf{x}^*) (= J(\mathbf{x}_0)).$$

But we now get the contradiction

$$0 < f(\mathbf{x}_0) - L_d^* = \sum_{(\alpha,\beta)\in\mathbb{N}_d^{2m}} u_{\alpha\beta} \underbrace{\widehat{\mathbf{g}}(\mathbf{x}_0)^\alpha}_{=0} (1 - \widehat{\mathbf{g}}(\mathbf{x}_0))^\beta,$$

because $\widehat{g}(\mathbf{x}_0)^\alpha = 0$ whenever $u_{\alpha\beta} > 0$ (since $J(\mathbf{x}_0) = J(\mathbf{x}^*)$).   □

Example 9.6   Consider the following toy example in the univariate case. Let $\mathbf{K} := \{x \in \mathbb{R} : 0 \leq x \leq 1\}$ and $x \mapsto f(x) := x(x-1)$. This is a convex optimization problem with global minimizer $x^* = 1/2$ in the interior of $\mathbf{K}$, and optimal value $f^* = -1/4$. The optimal values (L_d), $d \in \mathbb{N}$ of the linear relaxations (9.2) are reported in Table 9.1. The example shows the rather slow monotone convergence of $L_d \to -0.25$, despite the original problem being convex.

On the other hand, with $-f$ instead of f, the problem becomes a harder concave minimization problem. But this time the second relaxation is exact! Indeed, $f^* = 0$ and we have

$$f(x) - f^* = x - x^2 = x(1-x) = g_1(x)\,g_2(x),$$

with $g_1(x) = x$ and $g_2(x) = 1 - x$.

Example 9.6 illustrates that the convergence $L_d \uparrow f^*$ as $d \to \infty$, is only asymptotic in general, not finite; as underlined, if a global minimizer $\mathbf{x}^*$ is in the interior of $\mathbf{K}$ or if there is a nonoptimal solution $\mathbf{x}_0$ with same active constraints as $\mathbf{x}^*$, then convergence *cannot* be finite. This is typically the case in convex polynomial optimization (except for linear programs where, generically, an optimal solution $\mathbf{x}^*$ is a vertex of the convex polyhedron $\mathbf{K}$). This is clearly a drawback of the LP-relaxations since convex problems of the form (9.1) are considered easy and can be solved efficiently.

Interestingly, in computational complexity one is mainly concerned with analyzing the performance of an algorithm in the worst case situation (in a certain class $\mathcal{A}$ of problems to which the algorithm is applied) and usually one is not interested in the behavior of the algorithm when it is applied to a subclass $\mathcal{B} \subset \mathcal{A}$ of problems considered "easy." For instance, in the class $\mathcal{A}$ of polynomial optimization problems, convex optimization problems (9.1) form an interesting and important subclass $\mathcal{B} \subset \mathcal{A}$ of problems considered "easy." In our opinion, a general purpose algorithm (like the LP- or SDP-relaxations) should have the highly desirable feature of behaving efficiently on such a subclass $\mathcal{B}$ of easy problems. Indeed otherwise one might have reasonable doubts about its efficiency when applied to more difficult problems in $\mathcal{A}$. In this respect, the hierarchy of LP-relaxations (9.3) fails the test of efficiency on easy (convex) problems.

In addition, Example 9.6 exhibits an annoying paradox, namely that LP-relaxations may perform better for concave minimization problems than for a priori easier convex minimization problems. Finally, notice that for large values of d, the constraints of the LP-relaxations (9.3) should contain very large coefficients due to the presence of binomial coefficients when expanding the term $(1 - \widehat{\mathbf{g}})^\beta$ in the basis of monomials, a source of numerical instability and ill-conditioning.

9.4 An intermediate hierarchy of convex relaxations

So when comparing semidefinite and LP-relaxations we find the following.

- The hierarchy of semidefinite relaxations has very nice properties. In particular, its convergence to the global optimum f^* is generically finite (at least when $\mathbf{K}$ is compact with nonempty interior). However, as the size of the resulting semidefinite programs becomes rapidly prohibitive for state-of-the-art semidefinite solvers, application of the moment-SOS approach is currently limited to problems of modest size unless some sparsity and/or symmetries can be exploited.

- The hierarchy of LP-relaxations has been proved to be less powerful than the SDP hierarchy for solving hard 0/1 combinatorial problems (see e.g. Laurent (2003)) and cannot converge in finitely many steps for most convex problems. On the other hand, as each relaxation in the hierarchy is an LP, one can implement relaxations of relatively high order (and for problems of larger size) because LP software packages can handle problems with potentially millions of variables and constraints.

Therefore, in view of the respective drawbacks of the semidefinite and LP-relaxations one may wonder whether it is possible to define convex relaxations which try to combine the best of these two types of convex relaxations.

So let $k \in \mathbb{N}$ be *fixed* and consider the following hierarchy of semidefinite programs, indexed by $d \in \mathbb{N}$:

$$
\begin{aligned}
\rho_d = \inf_{\mathbf{y}} \quad & \{ L_{\mathbf{y}}(f) \\
\text{s.t.} \quad & y_0 = 1; \; \mathbf{M}_k(\mathbf{y}) \succeq 0; \\
& L_{\mathbf{y}}(\widehat{\mathbf{g}}^\alpha (1 - \widehat{\mathbf{g}})^\beta) \geq 0, \quad \forall (\alpha, \beta) \in \mathbb{N}_d^{2m} \}.
\end{aligned}
\tag{9.12}
$$

Observe that (9.12) has a single semidefinite constraint, namely $\mathbf{M}_k(\mathbf{y}) \succeq 0$. All the other constraints are linear inequality constraints. The dual of (9.12) is the semidefinite program:

$$
\begin{aligned}
\rho_d^* = \sup_{\lambda, \sigma, \mathbf{u}} \quad & \Big\{ \lambda : f - \lambda = \sigma + \sum_{(\alpha, \beta) \in \mathbb{N}_d^{2m}} u_{\alpha\beta} \, \widehat{\mathbf{g}}^\alpha (1 - \widehat{\mathbf{g}})^\beta, \\
& \sigma \in \Sigma[\mathbf{x}]_k, \; \mathbf{u} \geq 0 \Big\}.
\end{aligned}
\tag{9.13}
$$

Of course, with L_d as in (9.2) one has $f^* \geq \rho_d \geq L_d$ for every $d \in \mathbb{N}$, and so in view of Theorem 9.2, the convergence $\rho_d \to f^*$ as $d \to \infty$, takes place.

Recall that the LP-relaxations (9.3) can be viewed as a brute force simplification of the Lagrangian method (9.8) to make it tractable.

With k fixed, the duals of the semidefinite relaxations (9.12) are a (less brutal) simplification of the Lagrangian method (9.8). Instead of maximizing $G_d(\mathbf{u})$ on the positive orthant $\mathbf{u} \geq 0$, one maximizes $G_d(\mathbf{u})$ on the convex set of vectors $(\mathbf{u}, \lambda)$ such that the polynomial $f - \lambda - \sum_{\alpha, \beta} u_{\alpha\beta} \, \mathbf{g}^\alpha (1 - \widehat{\mathbf{g}})^\beta$ is an SOS of degree at most $2k$.

So one may say that the hierarchy of semidefinite relaxations (9.12) is a *level-k* strengthening of the LP hierarchy (9.2). On the other hand each relaxation (9.12) is a semidefinite program and not an LP anymore. But, and importantly, the only semidefinite constraint is $\mathbf{M}_k(\mathbf{y}) \succeq 0$ on the moment matrix

$\mathbf{M}_k(\mathbf{y})$ whose size $\binom{n+k}{n} \times \binom{n+k}{n}$ is independent of d, for *all* relaxations in the hierarchy. Recall that crucial for solving a semidefinite program is the size of the matrices that are constrained to be positive semidefinite. In fact one may even strengthen (9.12) by replacing the constraint $L_{\mathbf{y}}(\widehat{\mathbf{g}}^\alpha(1 - \widehat{\mathbf{g}})^\beta) \geq 0$ with the stronger $L_{\mathbf{y}}(h^2\widehat{\mathbf{g}}^\alpha(1 - \widehat{\mathbf{g}})^\beta) \geq 0$ for all $h \in \mathbb{R}[\mathbf{x}]_k$. Equivalently, in (9.13) the nonnegative scalar $u_{\alpha\beta}$ is replaced with an SOS of degree at most $2k$ (see Exercise 9.2).

Moreover, and in contrast with the LP hierarchy (9.2), finite convergence takes place for an important class of convex optimization problems. Recall that a polynomial f is SOS-convex if its Hessian $\nabla^2 f(\mathbf{x})$ factors as $\mathbf{M}(\mathbf{x})\mathbf{M}(\mathbf{x})^T$ for some polynomial matrix $\mathbf{M} \in \mathbb{R}[\mathbf{x}]^{n \times p}$ for some p.

Theorem 9.7   *With* $\mathbf{P}$ *as in (9.1) let* $f, -g_j$ *be convex,* $j = 1, \ldots, m$, *let Slater's condition hold and let* $f^* > -\infty$. *Then:*

(a) *if* $\max[\deg f, \deg g_j] \leq 2$ *then* $\rho_1 = f^*$, *i.e., the first relaxation of the hierarchy (9.12) parametrized by* $k = 1$, *is exact;*

(b) *if* $\max[\deg f, \deg g_j] \leq 2k$ *and* $f, -g_j$ *are all SOS-convex, then* $\rho_1 = f^*$, *i.e., the first relaxation of the hierarchy (9.12) parametrized by* k, *is exact.*

Proof   Under the assumptions of Theorem 9.7, $\mathbf{P}$ has a minimizer $\mathbf{x}^* \in \mathbf{K}$. Indeed as $f^* > -\infty$ and f and $-g_j$ are convex, $j = 1, \ldots, m$, then the minimum f^* is attained; see for example Belousov and Klatte (2002). Moreover, as Slater's condition holds, the KKT optimality conditions (7.3) hold at $(\mathbf{x}^*, \lambda^*) \in \mathbf{K} \times \mathbb{R}_+^m$ for some nonnegative $\lambda^* \in \mathbb{R}_+^m$. And so if $k = 1$, the Lagrangian polynomial $L(\cdot, \lambda^*)$ in (7.6), i.e.,

$$\mathbf{x} \mapsto L(\mathbf{x}, \lambda^*) := f(\mathbf{x}) - f^* - \sum_{j=1}^{m} \lambda_j^* g_j(\mathbf{x}),$$

is a nonnegative quadratic polynomial and so an SOS polynomial $\sigma^* \in \Sigma[\mathbf{x}]_1$. Therefore as $\rho_d \leq f^*$ for all d, the triplet $(\mathbf{u}^*, f^*, \sigma^*)$ (where $\lambda^* = (\mathbf{u}_{\alpha 0}^*)$, $|\alpha| = 1$, and $u_{\alpha\beta}^* = 0$ otherwise) is an optimal solution of (9.13) with $d = 1$ (and parameter $k = 1$), i.e., $\rho_1 = f^*$, which proves (a).

Next, if $k > 1$ and $f, -g_j$ are all SOS-convex then so is the Lagrangian polynomial $L(\cdot, \lambda^*)$. In addition, as $\nabla_{\mathbf{x}} L(\mathbf{x}^*, \lambda^*) = 0$ and $L(\mathbf{x}^*, \lambda^*) = 0$, the polynomial $L(\cdot, \lambda^*)$ is an SOS of degree at most $2k$; see for example Helton and Nie (2012, Lemma 4.2). Hence $L(\cdot, \lambda^*) = \sigma^*$ for some $\sigma^* \in \Sigma[\mathbf{x}]_k$, and again, the triplet $(\mathbf{u}^*, f^*, \sigma^*)$ (where $\lambda^* = (\mathbf{u}^*_{\alpha 0})$, $|\alpha| = 1$, and $u^*_{\alpha\beta} = 0$ otherwise) is an optimal solution of (9.13) with $d = 1$, which proves (b).   □

So Theorem 9.7 states that the SDP hierarchy (9.12)–(9.13) passes the test of efficiency on the subclass of "easy" SOS-convex problems. Finally, note that in Theorem 9.7, the convex set $\mathbf{K}$ may not be compact (and the polynomials $(1 - \widehat{\mathbf{g}}_j)$ may be ignored).

9.5 Exercises

Exercise 9.1   Consider Example 9.6 and with $k = 1$ show that the first semidefinite relaxation of the hierarchy (9.12) is exact, i.e., $\rho_1 = f^*$. What about its dual (9.13)?

Exercise 9.2   With $k \geq 1$ fixed, consider the hierarchy of semidefinite relaxations (9.12) and their duals (9.13). Next, define the new hierarchy of semidefinite programs:

$$v_d^* = \sup_{\lambda, \sigma} \left\{ \lambda : f - \lambda = \sigma_0 + \sum_{(\alpha, \beta) \in \mathbb{N}_d^{2m}} \sigma_{\alpha\beta}\, \widehat{\mathbf{g}}^\alpha (1 - \widehat{\mathbf{g}})^\beta, \right.$$

$$\left. \sigma_{\alpha\beta} \in \Sigma[\mathbf{x}]_k, \quad \forall\, (\alpha, \beta) \in \mathbb{N}_d^{2m} \right\}, \tag{9.14}$$

where the nonnegative scalar $u_{\alpha\beta}$ is replaced with the SOS polynomial $\sigma_{\alpha\beta}$.

(a)   Show that $\rho_d^* \leq v_d^*$ for all d and that $v_d^* \to f^*$ as $d \to \infty$, i.e., the semidefinite relaxations (9.14) are more powerful than (9.13).

(b)   Compare the dual of the semidefinite program (9.14) with (9.12).

(c)   What is the size of the semidefinite constraints in both (9.14) and its dual? Compare the respective computational complexities of the semidefinite programs (9.13) and (9.14) (and of their respective duals as well).

9.6 Notes and sources

9.1–9.2 Building on Shor's ideas, Sherali and Adams (1990, 1999) and Sherali et al. (1992), proposed the so-called RLT (Reformulation-Linearization Technique), the earliest hierarchy of LP relaxations for polynomial optimization, and proved finite convergence for 0/1 problems. Other linear relaxations for 0/1 programs have also been proposed by Balas et al. (1993) and Lovász and Schrijver (1991), while a hierarchy of semidefinite relaxations for 0/1 problems was first proposed by Lovász and Schrijver (1991) who also proved finite convergence. Section 9.2 is taken from Lasserre (2013e); to the best of our knowledge, the convergence of the standard dual method of nonlinear programming when applied to larger and larger redundant descriptions of **P**, is new.

9.3 A comparison between semidefinite and linear relaxations (in the spirit of Sherali and Adams) for general polynomial optimization problems is made in Lasserre (2002c), in the light of the results of Chapter 2 on the representation of positive polynomials. For 0/1 problems, Laurent (2003) compared the linear relaxations of Sherali and Adams (1990) and Lovász and Schrijver (1991), and the semidefinite relaxations of Lovász and Schrijver (1991) and Lasserre (2002a) within the common framework of the moment matrix, and proved that the latter semidefinite relaxations are the strongest. This has motivated research on integrality gaps for difficult combinatorial optimization problems. (The integrality gap measures the ratio between the optimal value of the relaxation and that of the problem to be solved.) In particular, Chlamtac (2007) and Chlamtac and Singh (2008) showed that the hierarchy of semidefinite relaxations provides improved approximation algorithms for finding independent sets in graphs, and for coloring problems. See also the related work of Schoenebeck (2008), Karlin et al. (2011) and the recent survey of Chlamtac and Tulsiani (2012).

9.4 For 0/1 combinatorial optimization problems, the intermediate hierarchy of semidefinite relaxation (9.12) with parameter $k = 1$ is called the "Sherali–Adams + SDP" hierarchy in for example Benabbas and Magen (2010) and Benabbas et al. (2012), in which the authors show that any (constant) level d of this hierarchy, viewed as a strengthening of the basic SDP-relaxation, does not make the integrality gap decrease. In fact, and in view of Section 9.2, the "Sherali–Adams + SDP" hierarchy is a *(level $k = 1$)* strengthening of the basic Sherali–Adams LP hierarchy rather than a strengthening of the basic SDP relaxations.

10

Minimization of rational functions

In this chapter we consider the minimization of a single rational function as well as the minimization of a sum of many rational functions on a basic semi-algebraic set $\mathbf{K} \subset \mathbb{R}^n$ of the form (5.2). Even though rational functions are a special case of semi-algebraic functions that we will consider in Chapter 11, the minimization of such functions on a basic semi-algebraic set deserves special attention.

In fact such a study was motivated by some important applications in *Computer Vision* where several typical problems (e.g. for camera calibration) require minimizing a sum of *many* rational functions. Even with few variables, such optimization problems can be quite challenging, and even more so when they involve many variables. Of course one can always reduce all rational fractions to the same denominator and end up with minimizing a single rational function for which the semidefinite relaxations (6.3) can be adapted easily (as described below). However, such a reduction would result in polynomials with significantly large degree and would prevent implementing these relaxations. Therefore to overcome this problem, we introduce a general methodology that avoids reducing to the same denominator.

10.1 Minimizing a rational function

We first consider the problem of minimizing a single rational function, i.e.,

$$\rho^* := \inf \left\{ \frac{p(\mathbf{x})}{q(\mathbf{x})} \ : \ \mathbf{x} \in \mathbf{K} \right\}, \tag{10.1}$$

where $p, q \in \mathbb{R}[\mathbf{x}]$ and $\mathbf{K}$ is defined as in (5.2). Note that if p, q have no common zero on $\mathbf{K}$, then $\rho^* > -\infty$ only if q does not change sign on $\mathbf{K}$.

Therefore, we will assume that q is strictly positive on $\mathbf{K}$. Recall the notation of Chapter 5 where $\mathcal{M}(\mathbf{K})_+$ denotes the space of finite Borel measures on $\mathbf{K} \subset \mathbb{R}^n$.

Proposition 10.1   *Asume that $q > 0$ on $\mathbf{K}$. The optimal value ρ^* of problem (10.1) is the same as the optimal value of the optimization problem:*

$$\rho = \inf_{\mu \in \mathcal{M}(\mathbf{K})_+} \left\{ \int_{\mathbf{K}} p\, d\mu \; : \; \int_{\mathbf{K}} q\, d\mu = 1 \right\}. \tag{10.2}$$

Proof   Assume first that $\rho^* > -\infty$, so that $p(\mathbf{x}) \geq \rho^* q(\mathbf{x})$ for all $\mathbf{x} \in \mathbf{K}$, and let $\mu \in \mathcal{M}(\mathbf{K})_+$ be a feasible measure for problem (10.2). Then $\int_{\mathbf{K}} p\, d\mu \geq \rho^* \int_{\mathbf{K}} q\, d\mu$, leading to $\rho \geq \rho^*$.

Conversely, let $\mathbf{x} \in \mathbf{K}$ be fixed arbitrary, and let $\mu \in \mathcal{M}(\mathbf{K})_+$ be the measure $q(\mathbf{x})^{-1}\delta_{\mathbf{x}}$, where $\delta_{\mathbf{x}}$ is the Dirac measure at the point $\mathbf{x} \in \mathbf{K}$. Then, $\int_{\mathbf{K}} q\, d\mu = 1$, so that μ is a feasible measure for problem (10.2). Moreover, its value satisfies $\int_{\mathbf{K}} p\, d\mu = p(\mathbf{x})/q(\mathbf{x})$. As $\mathbf{x} \in \mathbf{K}$ was arbitrary, $\rho \leq \rho^*$, and the result follows. Finally if $\rho^* = -\infty$ then from what precedes we also have $\rho = -\infty$.   □

Let $2v_j$ or $2v_j - 1$ be the degree of the polynomial $g_j \in \mathbb{R}[\mathbf{x}]$ in the definition (5.2) of $\mathbf{K}$, for all $j = 1, \dots, m$. Proceeding as in Section 6.1, we obtain the following hierarchy of semidefinite relaxations for $d \geq \max\{\deg p, \deg q, \max_j v_j\}$, which is very similar to the hierarchy (6.3):

$$\begin{aligned}
\rho_d = \inf_{\mathbf{y}} \quad & L_{\mathbf{y}}(p) \\
\text{s.t.} \quad & \mathbf{M}_d(\mathbf{y}),\ \mathbf{M}_{d-v_j}(g_j\, \mathbf{y}) \succeq \mathbf{0}, \quad j = 1, \dots, m \\
& L_{\mathbf{y}}(q) = 1.
\end{aligned} \tag{10.3}$$

Note that in contrast to (6.3) where $y_0 = 1$, in general $y_0 \neq 1$ in (10.3). In fact, the last constraint $L_{\mathbf{y}}(q) = 1$ in (10.3) yields $y_0 = 1$ whenever $q = 1$, that is, problem (10.2) reduces to problem (5.3). Again proceeding as in Section 6.1, the dual of problem (10.3) reads:

$$\begin{aligned}
\rho_d^* = \sup_{\sigma_j, \lambda} \quad & \lambda \\
\text{s.t.} \quad & p - \lambda q = \sum_{j=0}^{m} \sigma_j\, g_j \\
& \sigma_j \in \Sigma[\mathbf{x}], \quad \deg \sigma_j\, g_j \leq 2d, \ j = 0, 1, \dots, m
\end{aligned} \tag{10.4}$$

(with $g_0 = 1$ and $v_0 = 0$). Recall that $q > 0$ on $\mathbf{K}$.

Theorem 10.2 *Let* **K** *be as in (5.2), and let Assumption 2.14 hold. Consider the semidefinite relaxations (10.3) and (10.4). Then,*

(a) ρ^* *is finite and* $\rho_d^* \uparrow \rho^*$, $\rho_d \uparrow \rho^*$ *as* $d \to \infty$;

(b) *in addition, if* **K** $\subset \mathbb{R}^n$ *has nonempty interior then* $\rho_d^* = \rho_d$ *for every d;*

(c) *let* $\mathbf{x}^* \in \mathbf{K}$ *be a global minimizer of* p/q *on* **K**, *then if the polynomial* $p - \rho^* q \in \mathbb{R}[\mathbf{x}]$, *nonnegative on* **K**, *has Putinar's representation (2.10), both problems (10.3) and (10.4) have an optimal solution, and* $\rho_d^* = \rho_d = \rho^*$, *for all* $d \geq d_0$, *for some* $d_0 \in \mathbb{N}$;

(d) *let* $\mathbf{y}^*$ *be an optimal solution of (10.3) such that*

$$\operatorname{rank} \mathbf{M}_{d-v}(\mathbf{y}^*) = \operatorname{rank} \mathbf{M}_d(\mathbf{y}^*) \, (= s),$$

where $v = \max_j v_j$, *then there are s global minimizers and they can be extracted by Algorithm 6.9.*

Proof (a) As $q > 0$ on **K** and **K** is compact, the rational function p/q attains its minimum ρ^* on **K**, at some point $\mathbf{x}^* \in \mathbf{K}$. Let **y** be the vector of moments (up to order $2d$) of the measure $q(\mathbf{x}^*)^{-1}\delta_{\mathbf{x}^*}$. Obviously **y** is feasible for (10.3) with value $\rho_d = L_{\mathbf{y}}(p) = p(\mathbf{x}^*)/q(\mathbf{x}^*) = \rho^*$. Hence $\rho_d \leq \rho^*$ for every d, and by weak duality, $\rho_d^* \leq \rho_d \leq \rho^*$ for every d. On the other hand, observe that the polynomial $p - \rho^* q$ is nonnegative on **K**. So with $\epsilon > 0$ fixed arbitrary, $p - (\rho^* - \epsilon)q = p - \rho^* q + \epsilon q > 0$ on **K**. Therefore by Theorem 2.15,

$$p - (\rho^* - \epsilon)q = \sum_{j=0}^m \sigma_j \, g_j,$$

for some SOS polynomials $(\sigma_j) \subset \Sigma[\mathbf{x}]$. And so $(\rho^* - \epsilon, (\sigma_j))$ is a feasible solution of (10.4) provided that d is large enough. But then $\rho_d^* \geq \rho^* - \epsilon$, which combined with $\rho_d^* \leq \rho^*$ and the fact that $\epsilon > 0$ was arbitrary, yields the desired result.

(b) $\rho_d^* = \rho_d$ follows from Slater's condition which is satisfied for problem (10.3). Indeed, let $\mu \in \mathcal{M}(\mathbf{K})_+$ be a measure with uniform distribution on **K**, and scaled to ensure $\int q \, d\mu = 1$. (As **K** is compact, this is always possible.) Then, the vector **y** of its moments is a strictly feasible solution of problem (10.3) ($\mathbf{M}_d(\mathbf{y}) \succ \mathbf{0}$, and **K** having a nonempty interior implies $\mathbf{M}_{d-v_j}(g_j \, \mathbf{y}) \succ \mathbf{0}$ for all $j = 1, \ldots, m$). Thus, there is no duality gap, i.e., $\rho_d^* = \rho_d$, and the result follows from (a).

(c) If $p - \rho^* q$ has the representation (2.10), then as we did for part (a), we can find a feasible solution to problem (10.4) with value ρ^*, for all $d \geq d_0$, for some $d_0 \in \mathbb{N}$. Hence, $\rho_d^* = \rho_d = \rho^*$ for all $d \geq d_0$. Finally, (10.3) has an optimal solution **y**. It suffices to take for **y** the vector of moments up to order $2d$ of the measure $q(\mathbf{x}^*)^{-1}\delta_{\mathbf{x}^*}$, i.e., $y_\alpha = q(\mathbf{x}^*)^{-1}(\mathbf{x}^*)^\alpha$, for all $\alpha \in \mathbb{N}_{2d}^n$.

(d) Let $\mathbf{y}^*$ be an optimal solution of (10.3) such that the rank condition is satisfied. Then $\mathbf{y}^*$ is the vector of moments of the Borel measure μ on **K** supported on s points $\mathbf{x}(i) \in \mathbf{K}$, $i = 1, \ldots, s$, i.e., $\mu = \sum_{k=1}^s \alpha_i \delta_{\mathbf{x}(i)}$ for some positive weights $\alpha_i > 0$, $i = 1, \ldots, s$. In addition, the points $\mathbf{x}(i)$, $i = 1, \ldots, s$, can be extracted by running Algorithm 6.9. As $\mathbf{x}(i) \in \mathbf{K}$ one has $p(\mathbf{x}(i))/q(\mathbf{x}(i)) \geq f^*$ for all $i = 1, \ldots, s$. On the other hand,

$$\int_{\mathbf{K}} q \, d\mu = 1 \Rightarrow \sum_{i=1}^s \alpha_i \, q(\mathbf{x}(i)) = 1.$$

But then

$$f^* \geq \int_{\mathbf{K}} p \, d\mu = \sum_{i=1}^s \alpha_i \, p(\mathbf{x}(i)) = \sum_{i=1}^s \alpha_i \underbrace{\frac{p(\mathbf{x}(i))}{q(\mathbf{x}(i))}}_{\geq f^*} q(\mathbf{x}(i)) \geq f^*,$$

and so $p(\mathbf{x}(i))/q(\mathbf{x}(i)) = f^*$ for every $i = 1, \ldots, s$. □

Example 10.3 With $n = 2$ let $\mathbf{K} = [-1, 1]^2$, $\mathbf{x} \mapsto p(\mathbf{x}) = 2x_1 - x_2 - x_1^3 - x_2^3$, $\mathbf{x} \mapsto q(\mathbf{x}) = 1 + x_1^2 + 2x_2^2$. Modeling $[-1, 1]^2$ as $\{\mathbf{x} : 1 - x_1^2 \geq 0; \ 1 - x_2^2 \geq 0\}$, the first semidefinite relaxation of the hierarchy (10.3) with $d = 2$ reads:

$$\rho_2 = \inf_{\mathbf{y}} \quad 2y_{10} - y_{01} - y_{30} - y_{03}$$
$$\text{s.t.} \quad \mathbf{M}_2(\mathbf{y}) \succeq 0$$
$$y_{00} - y_{20} \geq 0, \quad y_{00} - y_{02} \geq 0$$
$$y_{00} + y_{20} + 2y_{02} = 1,$$

and one obtains the global minimum[1] $\rho_2 = \rho^* = 0.8887$ and the global minimizer $\mathbf{x}^* = (-0.5724, 1)$ is extracted.

10.2 Minimizing a sum of many rational functions

Consider now the following optimization problem:

$$\mathbf{P}: \qquad f^* := \inf_{\mathbf{x}} \left\{ \sum_{i=1}^s f_i(\mathbf{x}) : \mathbf{x} \in \mathbf{K} \right\}, \qquad (10.5)$$

---

[1] To solve (10.3) we have used the Gloptipoly software described in Appendix B.

where $s \in \mathbb{N}$ is some given number, $\mathbf{K}$ is the basic semi-algebraic set defined in (5.2), and for every $i = 1, \ldots, s$, f_i is a rational function p_i/q_i with $p_i, q_i \in \mathbb{R}[\mathbf{x}]$ and $q_i > 0$ on $\mathbf{K}$.

Of course, by reducing to the same denominator $q := \prod_{i=1}^{n} q_i$, problem $\mathbf{P}$ is the same as problem (10.1) and could be treated as in Section 10.1. However, if s is large and many q_i are nontrivial then the degree of q is too large and so even the first semidefinite relaxation of the hierarchy (10.3) cannot be implemented in practice!

To overcome this difficulty, we next present two strategies, namely the *epigraph* approach and the *multi-measures* approach. In both cases one avoids reducing to the same denominator (and thus the resulting degree explosion).

10.2.1 The epigraph approach

A first strategy is to introduce additional variables $\mathbf{r} = (r_i) \in \mathbb{R}^s$ (which we may call liftings) with associated constraints

$$\frac{p_i(\mathbf{x})}{q_i(\mathbf{x})} \le r_i, \qquad i = 1, \ldots, s,$$

and solve the equivalent problem:

$$f^* := \inf_{\mathbf{x}, \mathbf{r}} \left\{ \sum_{i=1}^{s} r_i \; : \; (\mathbf{x}, \mathbf{r}) \in \widehat{\mathbf{K}} \right\}, \tag{10.6}$$

which is now a *polynomial* optimization problem in the new variables $(\mathbf{x}, \mathbf{r}) \in \mathbb{R}^n \times \mathbb{R}^s$, and with new feasible set

$$\widehat{\mathbf{K}} = \{ (\mathbf{x}, \mathbf{r}) \in \mathbb{R}^n \times \mathbb{R}^s \; : \; \mathbf{x} \in \mathbf{K}; \quad r_i q_i(\mathbf{x}) - p_i(\mathbf{x}) \ge 0, \; i = 1, \ldots, s \}.$$

In doing so one introduces s additional variables and this may have an impact on the overall performance, especially if s is large. However, observe that some sparsity pattern is present in (10.6). Indeed, relabel the variables $\mathbf{r}$ as $(x_{n+1}, \ldots, x_{n+s})$, and let $I_k := \{1, \ldots, n\} \cup \{n+k\}$ be the index set associated with the $n + 1$ variables $(\mathbf{x}, r_k) = (\mathbf{x}, x_{n+k})$, $k = 1, \ldots, s$. Then

$$\left(\bigcup_{j=1}^{k-1} I_j \right) \cap I_k \subset I_1, \qquad \forall k = 2, \ldots, s,$$

that is, the *running intersection property* (2.27) holds.

Moreover, if $\mathbf{K}$ is compact one may in general obtain upper and lower bounds $\bar{r}_i, \underline{r}_i$ on the variables r_i so as to make $\widehat{\mathbf{K}}$ compact by adding the quadratic (redundant) constraints $(r_i - \underline{r}_i)(\bar{r}_i - r_i) \ge 0$, $i = 1, \ldots, s$.

Therefore as every g_j in the description of $\mathbf{K}$ is a polynomial in the variables (x_i), $i \in I_k$, for some k, one may implement the sparse semidefinite relaxations described in Section 8.1 whose convergence is proved in Theorem 8.3. Finally, if the original problem $\mathbf{P}$ already has some sparsity pattern, then it may be combined with the one above.

10.2.2 The multi-measures approach

In the approach described below one avoids introducing additional variables $\mathbf{r} \in \mathbb{R}^s$. Consider the infinite-dimensional linear problem:

$$
\hat{f} := \inf_{\mu_i \in \mathcal{M}(\mathbf{K})_+} \left\{ \sum_{i=1}^{s} \int_{\mathbf{K}} p_i \, d\mu_i : \int_{\mathbf{K}} q_1 d\mu_1 = 1 \right. \tag{10.7}
$$
$$
\left. \int_{\mathbf{K}} \mathbf{x}^\alpha q_i d\mu_i = \int_{\mathbf{K}} \mathbf{x}^\alpha q_1 d\mu_1, \quad \forall \alpha \in \mathbb{N}^n, \ i = 2, \ldots, s, \right\}
$$

where $\mathcal{M}(\mathbf{K})_+$ is the space of finite Borel measures supported on $\mathbf{K}$.

Lemma 10.4 *Let $\mathbf{K} \subset \mathbb{R}^n$ in (5.2) be compact and assume that $q_i > 0$ on $\mathbf{K}$, $i = 1, \ldots, s$. Then $\hat{f} = f^*$.*

Proof We first prove that $f^* \geq \hat{f}$. As $f = \sum_i p_i/q_i$ is continuous on $\mathbf{K}$, there exists a global minimizer $\mathbf{x}^* \in \mathbf{K}$ with $f(\mathbf{x}^*) = f^*$. Define $\mu_i := q_i(\mathbf{x}^*)^{-1}\delta_{\mathbf{x}^*}$, $i = 1, \ldots, s$, where $\delta_{\mathbf{x}^*}$ is the Dirac measure at $\mathbf{x}^*$. Then obviously, the measures (μ_i), $i = 1, \ldots, s$, are feasible for (10.7) with associated value

$$
\sum_{i=1}^{s} \int_{\mathbf{K}} p_i d\mu_i = \sum_{i=1}^{s} p_i(\mathbf{x}^*)/q_i(\mathbf{x}^*) = f(\mathbf{x}^*) = f^*.
$$

Conversely, let (μ_i) be a feasible solution of (10.7). For every $i = 1, \ldots, s$, let $d\nu_i$ be the measure $q_i d\mu_i$, i.e.,

$$
\nu_i(B) := \int_{\mathbf{K} \cap B} q_i(\mathbf{x}) d\mu_i(\mathbf{x})
$$

for all Borel subsets B of $\mathbb{R}^n$. Hence the support of ν_i is contained in $\mathbf{K}$. As measures on compact sets are moment determinate[2], the moment constraints in (10.7) imply that $\nu_i = \nu_1$, for every $i = 2, \ldots, s$, and from $\int_{\mathbf{K}} q_1 d\mu_1 = 1$ we also deduce that ν_1 is a probability measure on $\mathbf{K}$. But then

---

[2] Recall from Section 2.7 that a finite Borel measure is *determinate* if there is no other finite Borel measure with the same moments.

$$\sum_{i=1}^{s} \int_{\mathbf{K}} p_i d\mu_i = \sum_{i=1}^{s} \int_{\mathbf{K}} \frac{p_i}{q_i} q_i d\mu_i = \sum_{i=1}^{s} \int_{\mathbf{K}} \frac{p_i}{q_i} dv_1$$

$$= \int_{\mathbf{K}} \left(\sum_{i=1}^{s} \frac{p_i}{q_i} \right) dv_1 = \int_{\mathbf{K}} f dv_1 \geq \int_{\mathbf{K}} f^* dv_1 = f^*,$$

where we have used that $f \geq f^*$ on $\mathbf{K}$ and v_1 is a probability measure on $\mathbf{K}$.

□

A hierarchy of semidefinite relaxations For every $i = 1, \dots, s$, let $\mathbf{y}_i = (y_{i\alpha})$, $\alpha \in \mathbb{N}^n$, be a real sequence indexed in the canonical basis $(\mathbf{x}^\alpha)$ of $\mathbb{R}[\mathbf{x}]$, and recall that $\mathbb{N}_d^n := \{ \alpha \in \mathbb{N}^n : \sum_j \alpha_j \leq d \}$. Let $v_j := \lceil (\deg g_j)/2 \rceil$, $j = 1, \dots, m$, and $v := \max_j v_j$. Let $u_i := \deg q_i$, $i = 1, \dots, s$, and with no loss of generality assume that $u_1 \leq u_2 \leq \dots \leq u_s$. Consider the hierarchy of semidefinite programs indexed by $d \in \mathbb{N}$:

$$\begin{aligned} f_d^* = \inf_{\mathbf{y}_i} \quad & \sum_{i=1}^{s} L_{\mathbf{y}_i}(p_i) \\ \text{s.t.} \quad & \mathbf{M}_d(\mathbf{y}_i) \succeq 0, && i = 1, \dots, s \\ & \mathbf{M}_{d-v_j}(g_j \, \mathbf{y}_i) \succeq 0, && i = 1, \dots, s, \; j = 1, \dots, m \\ & L_{\mathbf{y}_1}(q_1) = 1 \\ & L_{\mathbf{y}_i}(\mathbf{x}^\alpha q_i) = L_{\mathbf{y}_1}(\mathbf{x}^\alpha q_1), && \forall \alpha \in \mathbb{N}_{2d-u_i}^n, \; i = 2, \dots, s, \end{aligned} \tag{10.8}$$

where $\mathbf{M}_d(\mathbf{y}_i)$ (respectively $\mathbf{M}_d(g_j \mathbf{y}_i)$) is the moment matrix (respectively localizing matrix) associated with the sequence $\mathbf{y}_i$ and the polynomial g_j.

It is straightforward to show that for each $d \in \mathbb{N}$, the semidefinite program (10.8) is a relaxation of the initial problem (10.2). Notice also that there is no degree explosion (as when reducing to the same denominator) and so if the degree of the g_j and q_j is modest then several relaxations of the hierarchy (10.8) can be implemented in practice.

Theorem 10.5 *Let Assumption 2.14 hold for the polynomials (g_j) that define $\mathbf{K}$ and consider the hierarchy of semidefinite relaxations (10.8).*

(a) *$f_d^* \uparrow f^*$ as $d \to \infty$.*
(b) *Moreover, let $(\mathbf{y}_i^d)$ be an optimal solution of (10.8) such that*

$$\operatorname{rank} \mathbf{M}_d(\mathbf{y}_i^d) = \operatorname{rank} \mathbf{M}_{d-v}(\mathbf{y}_i^d) =: R, \quad i = 1, \dots, s.$$

If $R > 1$ then for every $i = 1, \dots, s$, $\mathbf{y}_i$ is the truncated moment sequence of a measure μ_i supported on R points $\mathbf{x}^\ell(i) \in \mathbf{K}$,

$\ell = 1, \ldots, R$. *If* $\mathbf{x}^\ell(i) = \mathbf{x}^\ell(j)$ *for all* $i \neq j$ *and every* ℓ, *then* $f_d^* = f^*$ *and every* $\mathbf{x}^\ell(i)$ *is a global minimizer of* **P**. *If* $R = 1$ *then* $f_d^* = f^*$ *and* $\mathbf{x}(i) = \mathbf{x}^*$ *for all* $i = 1, \ldots, s$, *and for some* $\mathbf{x}^* \in \mathbf{K}$ *which is a global minimizer of* **P**.

Even though the epigraph and multi-measures strategies avoid degree explosion when reducing to the same denominator, the resulting semidefinite relaxations are only implementable for problems of modest size, i.e., with a relatively small number of variables. As we next see, for problems of larger size and with a sparsity pattern as described in Section 8.1, one may define "sparse" analogues of the dense semidefinite relaxations (10.8). (For the epigraph approach we have already indicated how to define sparse analogues of the epigraph semidefinite relaxations.)

10.2.3 The multi-measures approach with sparsity

We consider the same problem (10.5) but we now assume that n is typically large and a sparsity pattern (as described in Section 8.1) is present in the data of **P**. That is, every monomial contains at most d variables (if d is the maximum degree of the polynomials involved) which makes the coupling of n variables relatively weak, and in addition, each constraint is concerned with few variables.

So as in Section 8.1, let $I_0 := \{1, \ldots, n\} = \cup_{i=1}^s I_i$ with possible overlaps, and let $\mathbb{R}[x_k : k \in I_i]$ denote the ring of polynomials in the variables $\{x_k, \ k \in I_i\}$. Denote by n_i the cardinality of I_i.

We assume that $\mathbf{K} \subset \mathbb{R}^n$ in (5.2) is compact, and that we know some $M > 0$ such that $\mathbf{x} \in \mathbf{K} \Rightarrow M - \|\mathbf{x}\|^2 \geq 0$. For every $i \leq s$, introduce the quadratic polynomial $\mathbf{x} \mapsto g_{m+i}(\mathbf{x}) = M - \sum_{k \in I_i} x_k^2$.

The index set $\{1, \ldots, m + s\}$ has a partition $\cup_{i=1}^s J_i$ with $J_i \neq \emptyset$ for every $i = 1, \ldots, s$. In the sequel we assume that for every $i = 1, \ldots, s$, $p_i, q_i \in \mathbb{R}[x_k : k \in I_i]$ and for every $j \in J_i$, $g_j \in \mathbb{R}[x_k : k \in I_i]$. Next, for every $i = 1, \ldots, s$, let

$$\mathbf{K}_i := \{ \mathbf{z} \in \mathbb{R}^{n_i} : g_j(\mathbf{z}) \geq 0, \ j \in J_i \}$$

so that **K** in (5.2) has the equivalent characterization

$$\mathbf{K} = \{ \mathbf{x} \in \mathbb{R}^n : (x_k, k \in I_i) \in \mathbf{K}_i, \ i = 1, \ldots, s \}.$$

Similarly, for every $i, j \in \{1, \ldots, s\}$ such that $i \neq j$ and $I_i \cap I_j \neq \emptyset$,

$$\mathbf{K}_{ij} = \mathbf{K}_{ji} := \{ (x_k, k \in I_i \cap I_j) : (x_k, k \in I_i) \in \mathbf{K}_i; \ (x_k, k \in I_j) \in \mathbf{K}_j \}.$$

With $\mathcal{M}(\mathbf{K})_+$ being the space of finite Borel measures on $\mathbf{K}$, and for every $i = 1, \ldots, s$, let $\pi_i : \mathcal{M}(\mathbf{K})_+ \to \mathcal{M}(\mathbf{K}_i)_+$ denote the projection on $\mathbf{K}_i$, that is, for every $\mu \in \mathcal{M}(\mathbf{K})$:

$$\pi_i \mu(B) := \mu(\{ \mathbf{x} \in \mathbf{K} : (x_k, k \in I_i) \in B \}), \quad \forall B \in \mathcal{B}(\mathbf{K}_i),$$

where $\mathcal{B}(\mathbf{K}_i)$ is the usual Borel σ-algebra associated with $\mathbf{K}_i$.

For every $i, j \in \{1, \ldots, s\}$ such that $i \neq j$ and $I_i \cap I_j \neq \emptyset$, the projection $\pi_{ij} : \mathcal{M}(\mathbf{K}_i)_+ \to \mathcal{M}(\mathbf{K}_{ij})_+$ is also defined in an obvious similar manner. For every $i = 1, \ldots, s - 1$ define the set:

$$U_i := \{ j \in \{i + 1, \ldots, s\} : I_i \cap I_j \neq \emptyset \},$$

and consider the infinite-dimensional problem

$$\hat{f} := \inf_{\mu_i \in \mathcal{M}(\mathbf{K}_i)_+} \left\{ \sum_{i=1}^{s} \int_{\mathbf{K}_i} p_i \, d\mu_i : \int_{\mathbf{K}_i} q_i \, d\mu_i = 1, \ i = 1, \ldots, s \right. \tag{10.9}$$

$$\left. \pi_{ij}(q_i d\mu_i) = \pi_{ji}(q_j d\mu_j), \ \forall j \in U_i, \ i = 1, \ldots, s - 1 \right\}.$$

Theorem 10.6 *Let $\mathbf{K} \subset \mathbb{R}^n$ in (5.2) be compact. If the sparsity pattern $(I_i)_{i=1}^{s}$ satisfies the running intersection property (2.27) then $\hat{f} = f^*$.*

Proof That $\hat{f} \leq f^*$ is straightforward. As $\mathbf{K}$ is compact and $q_i > 0$ on $\mathbf{K}$ for every $i = 1, \ldots, s$, $f^* = \sum_{i=1}^{s} f_i(\mathbf{x}^*)$ for some $\mathbf{x}^* \in \mathbf{K}$. So let μ be the Dirac measure $\delta_{\mathbf{x}^*}$ at $\mathbf{x}^*$ and let ν_i be the projection $\pi_i \mu$ of μ on $\mathbf{K}_i$, $i = 1, \ldots, s$. That is $\nu_i = \delta_{(x_k^*, k \in I_i)}$, the Dirac measure at the point $(x_k^*, k \in I_i)$ of $\mathbf{K}_i$. Next, for every $i = 1, \ldots, s$, define the measure $d\mu_i := q_i(\mathbf{x}^*)^{-1} d\nu_i$. Obviously, (μ_i) is a feasible solution of (10.9) because $\mu_i \in \mathcal{M}(\mathbf{K}_i)_+$ and $\int q_i d\mu_i = 1$, for every $i = 1, \ldots, s$, and one also has:

$$\delta_{(x_k^*, k \in I_i \cap I_j)} = \pi_{ij} \mu_i = \pi_{ji} \mu_j, \quad \forall j \neq i \text{ such that } I_j \cap I_i \neq \emptyset.$$

Finally, its value satisfies

$$\sum_{i=1}^{s} \int_{\mathbf{K}_i} p_i d\mu_i = \sum_{i=1}^{s} p_i(\mathbf{x}^*)/q_i(\mathbf{x}^*) = f^*,$$

and so $\hat{f} \leq f^*$. We next prove the converse inequality $\hat{f} \geq f^*$. Let (μ_i) be an arbitrary feasible solution of (10.9), and for every $i = 1, \ldots, s$, denote by ν_i the probability measure on $\mathbf{K}_i$ with density q_i with respect to μ_i, that is,

$$\nu_i(B) := \int_{\mathbf{K}_i \cap B} q_i(\mathbf{x}) \, d\mu_i(\mathbf{x}), \quad \forall B \in \mathcal{B}(\mathbf{K}_i).$$

By definition of the linear program (10.9), $\pi_{ij} v_i = \pi_{ji} v_j$ for every couple $j \neq i$ such that $I_j \cap I_i \neq \emptyset$. Therefore, by Lasserre (2009c, Lemma B.13) there exists a probability measure v on $\mathbf{K}$ such that $\pi_i v = v_i$ for every $i = 1, \ldots, s$. But then

$$\sum_{i=1}^{s} \int_{\mathbf{K}_i} p_i \, d\mu_i = \sum_{i=1}^{s} \int_{\mathbf{K}_i} \frac{p_i}{q_i} \, dv_i = \sum_{i=1}^{s} \int_{\mathbf{K}_i} \frac{p_i}{q_i} \, dv = \int_{\mathbf{K}} \left(\sum_{i=1}^{s} \frac{p_i}{q_i} \right) dv \geq f^*,$$

and so $\hat{f} \geq f^*$.   □

A hierarchy of sparse SDP relaxations In a manner similar to what we did in Section 8.1 for polynomial optimization with a sparsity pattern, we now define a hierarchy of semidefinite relaxations for problem (10.9). For every $i = 1, \ldots, s$, let

$$\mathbb{N}^{(i)} := \{ \alpha \in \mathbb{N}^n : \alpha_k = 0 \text{ if } k \notin I_i \}, \quad \mathbb{N}_k^{(i)} := \left\{ \alpha \in \mathbb{N}^{(i)} : \sum_i \alpha_i \leq k \right\}.$$

An obvious similar definition of $\mathbb{N}^{(ij)}$ (equal to $\mathbb{N}^{(ji)}$) and $\mathbb{N}_k^{(ij)}$ (equal to $\mathbb{N}_k^{(ji)}$) applies when considering $I_j \cap I_i \neq \emptyset$.

For every $i = 1, \ldots, s$, let $\mathbf{y}_i = (y_{i\alpha})$ be a given sequence indexed in the canonical basis of $\mathbb{R}[\mathbf{x}]$. The sparse moment matrix $\mathbf{M}_d(\mathbf{y}_i, I_i)$ associated with $\mathbf{y}_i$ has its rows and columns indexed in the canonical basis $(\mathbf{x}^\alpha)$ of $\mathbb{R}[x_k : k \in I_i]$, and with entries:

$$\mathbf{M}_d(\mathbf{y}_i, I_i)(\alpha, \beta) = L_{\mathbf{y}_i}(\mathbf{x}^{\alpha+\beta}) = y_{i(\alpha+\beta)}, \qquad \forall \alpha, \beta \in \mathbb{N}_d^{(i)}.$$

Similarly, for a given polynomial $h \in \mathbb{R}[x_k : k \in I_i]$, the sparse localizing matrix $\mathbf{M}_d(h \, \mathbf{y}_i, I_i)$ associated with $\mathbf{y}$ and h, has its rows and columns indexed in the canonical basis $(\mathbf{x}^\alpha)$ of $\mathbb{R}[x_k : k \in I_i]$, and with entries:

$$\mathbf{M}_d(h \, \mathbf{y}_i, I_i)(\alpha, \beta) = L_{\mathbf{y}_i}(h \, \mathbf{x}^{\alpha+\beta}) = \sum_{\gamma \in \mathbb{N}^{(i)}} h_\gamma \, y_{i(\alpha+\beta+\gamma)}, \qquad \forall \alpha, \beta \in \mathbb{N}_d^{(i)}.$$

With $\mathbf{K} \subset \mathbb{R}^n$ defined in (5.2), let $v_j := \lceil (\deg g_j)/2 \rceil$, for every $j = 1, \ldots, m + s$. Consider the hierarchy of semidefinite programs indexed by $d \in \mathbb{N}$:

$$f_d^* = \inf_{\mathbf{y}_1, \ldots, \mathbf{y}_s} \sum_{i=1}^{s} L_{\mathbf{y}_i}(p_i)$$

$$\begin{aligned}
\text{s.t.} \quad & \mathbf{M}_d(\mathbf{y}_i, I_i) \succeq 0, & & i = 1, \ldots, s \\
& \mathbf{M}_{d-v_j}(g_j \mathbf{y}_i, I_i) \succeq 0, & & \forall j \in J_i, \ i = 1, \ldots, s \\
& L_{\mathbf{y}_i}(q_i) = 1, & & i = 1, \ldots, s \\
& L_{\mathbf{y}_i}(\mathbf{x}^\alpha q_i) = L_{\mathbf{y}_j}(\mathbf{x}^\alpha q_j), & & \forall \alpha \in \mathbb{N}^{(ij)}, \ \forall j \in U_i, \\
& i = 1, \ldots, s-1 \text{ and} \\
& |\alpha| + \max[\deg q_i, \deg q_j] \ \leq 2d,
\end{aligned}$$

(10.10)

which is the "sparse" analogue of the "dense" hierarchy (10.8). Again for each $d \in \mathbb{N}$, the semidefinite program (10.10) is a relaxation of the initial problem **P**, and so $f_d^* \leq f^*$ for all d. For every $i = 1, \ldots, s$, let $s_i := \max\{v_j : j \in J_i\}$.

Theorem 10.7 *Let* $\mathbf{K} \subset \mathbb{R}^n$ *in (5.2) be compact. Let the sparsity pattern* $(I_i)_{i=1}^s$ *satisfy the running intersection property (2.27), and consider the hierarchy of semidefinite relaxations defined in (10.10). Then*

 (a) $f_d^* \uparrow f^*$ *as* $d \to \infty$,
 (b) *if an optimal solution* $\mathbf{y}^* = (\mathbf{y}_1^*, \ldots, \mathbf{y}_s^*)$ *of (10.10) satisfies*

$$\mathrm{rank}\, \mathbf{M}_d(\mathbf{y}_i^*, I_i) = \mathrm{rank}\, \mathbf{M}_{d-s_i}(\mathbf{y}_i^*, I_i) =: R_i, \qquad \forall i = 1, \ldots, s,$$

and

$$\mathrm{rank}\, \mathbf{M}_d(\mathbf{y}_i^*, I_i \cap I_j) = 1, \qquad \forall j \in U_i, \; i = 1, \ldots, s-1,$$

then $f_d^* = f^*$ *and one may extract finitely many global minimizers.*

Example 10.8 We first illustrate the multi-measures approach on the modified Shekel foxhole minimization problem $\inf\{ f(\mathbf{x}) : \mathbf{x} \in \mathbf{K} \}$ where:

$$f(\mathbf{x}) := \sum_{i=1}^{s} \frac{1}{\sum_{j=1}^{n}(x_j - a_{ij})^2 + c_i}, \qquad \mathbf{K} := \left\{ \mathbf{x} : \sum_{i=1}^{n}(x_i - 5)^2 \leq 60 \right\},$$

for some given data $a_{ij}, c_i, \; i = 1, \ldots, s, \; j = 1, \ldots, n$. Such functions are designed to have multiple local minima and computing their global minimum can be quite challenging when s is not small. (An example of such a function for $n = 2$, with its level sets is displayed in Figure 10.1.) Here sparsity cannot be exploited as each rational function $f_i(\mathbf{x}) := \sum_j (x_j - a_{ij})^2 + c_i$ contains all variables. To avoid ill-conditioning of the moment-SOS approach, a suitable scaling $x_i \to 10x_i$ is applied. The approach has been tested with $s = 30$ rational functions on examples taken from Ali et al. (2005) for which the global minimum was known.

In the case $n = 5$ (respectively $n = 10$), one obtains a certificate of global optimality at order $d = 3$ in about 84 seconds (respectively at order $d = 2$ in about 284 seconds) on a standard PC. The extracted solution matches the known unique global minimizer up to four correct significant digits. For comparison, in both cases the BARON solver[3] provides an approximation of

---

[3] BARON is a general purpose solver for global optimization; for more details see for example Tawarmalani and Sahinidis (2005).

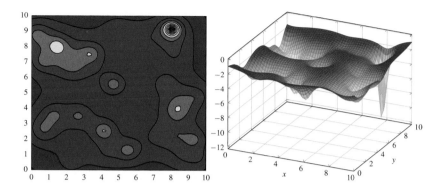

Figure 10.1   A bivariate Shekel function (right) and its level sets (left).

the unique global minimum with four correct significant digits in 1000 seconds (which at that time was the default usage limit on the NEOS server). Indeed, the solver interrupts because the limit resource is reached (i.e., with a solver status equal to 3). Consequently, the obtained approximation is not certified and is detected as a local minimum.

Example 10.9   We next illustrate the multi-measures approach with sparsity and consider the rational optimization problem:

$$f^* = \sup_{\mathbf{x} \in \mathbb{R}^n} \sum_{i=1}^{n-1} \frac{1}{100\,(x_{i+1} - x_i^2)^2 + (x_i - 1)^2 + 1} \tag{10.11}$$

which has the same critical points as the well-known Rosenbrock problem

$$\inf_{\mathbf{x} \in \mathbb{R}^n} \sum_{i=1}^{n-1} (100\,(x_{i+1} - x_i^2)^2 + (x_i - 1)^2) \tag{10.12}$$

whose geometry is troublesome for local optimization solvers. Notice that the obvious sparsity pattern

$$I_1 = \{1, 2\}, I_2 = \{2, 3\}, \dots, I_s = \{n - 1, n\},$$

(with $s = n - 1$) satisfies the running intersection property (2.27). It can easily be shown that the global maximum $f^* = n - 1$ of problem (10.11) is achieved at $x_i^* = 1, i = 1, \dots, n$. Experiments with local optimization algorithms reveal that standard quasi-Newton solvers or functions of the optimization toolbox for MATLAB®, called repeatedly with random initial guesses, typically yield local maxima quite far from the global maximum.

With the multi-measures approach and exploiting the structured sparsity pattern (10.12) (and adding bound constraints $x_i^2 \le 16$ for all $i = 1, \dots, n$),

Table 10.1 *Values of n for which $f^*_{BAR} \neq f^*$.*

n	f^*	f^*_{BAR}
240	239	235.248
260	259	252.753
280	279	252.753
340	339	336.403
420	419	414.606

one can solve problem (10.11) with a certificate of global optimality for n up to 1000 variables. A certificate of global optimality is obtained at order $d = 1$ with typical CPU times ranging from 10 seconds for $n = 100$ to 500 seconds for $n = 1000$.

By varying n from 20 to 1000 and submitting the corresponding problems to the BARON solver on the NEOS solver, we observe the following behavior. First, BARON provides a solution (i.e., with a termination status equal to 1) with typical CPU times ranging from less than 1 second to 500 seconds. Second, BARON finds the global maximum (denoted f^*_{BAR}) in most cases except for the values of n summarized in Table 10.1. Moreover, for $n \geqslant 640$ (except for $n = 860$), all solutions provided by BARON are wrong since all satisfy $x_1 = -0.995$ ($\neq 1$) and $x_k = 1$ for all $k \neq 1$ (the expected values of the global maximum); that is, problem (10.11) can create numerical instabilities.

10.3 Exercises

Exercise 10.1   Let $n = 1$ and $p, q \in \mathbb{R}[x]$ and let $I = [0, 1]$ or $I = [0, +\infty)$ or $I = \mathbb{R}$. With $q > 0$ on I, show that computing $f^* = \inf\{ p(x)/q(x) : x \in I \}$ reduces to solving a single semidefinite program. Write such a program and its dual.

Exercise 10.2   Let $\mathbf{K} \subset \mathbb{R}^n$ be compact and let $p, q \in \mathbb{R}[\mathbf{x}]$ be such that $q > 0$ on $\mathbf{K}$. Let $\mathcal{M}(\mathbf{K})_+$ denote the space of Borel (nonnegative) measures on $\mathbf{K}$ and $\mathcal{P}(\mathbf{K}) \subset \mathcal{M}(\mathbf{K})_+$ its subset of probability measures on $\mathbf{K}$.

(a) Show that

$$\inf_{\mathbf{x}} \left\{ \frac{p(\mathbf{x})}{q(\mathbf{x})} : \mathbf{x} \in \mathbf{K} \right\} = \inf_{\mu} \left\{ \int_{\mathbf{K}} \frac{p(\mathbf{x})}{q(\mathbf{x})} \, d\mu : \mu \in \mathcal{P}(\mathbf{K}) \right\}$$

$$= \inf_{\mu} \left\{ \frac{\int_{\mathbf{K}} p(\mathbf{x}) \, d\mu}{\int_{\mathbf{K}} q(\mathbf{x}) \, d\mu} : \mu \in \mathcal{P}(\mathbf{K}) \right\}$$

$$= \inf_{\mu} \left\{ \int_{\mathbf{K}} p(\mathbf{x}) \, d\mu : \int_{\mathbf{K}} q(\mathbf{x}) \, d\mu = 1; \ \mu \in \mathcal{M}(\mathbf{K})_+ \right\},$$

and the infimum is attained in all formulations.

(b) Show that two of the above formulations are an infinite-dimensional LP, but only one has polynomial data.

(c) With $\mathbf{K}$, p and q as above, consider the new optimization problem $\mathbf{P}'$: $\inf\{ z \ : \ zq(\mathbf{x}) - p(\mathbf{x}) = 0; \ (\mathbf{x}, z) \in \mathbf{K} \times \mathbb{R} \}$. Show how to obtain upper and lower bounds on z. Modify $\mathbf{P}'$ into an equivalent polynomial optimization problem over a compact basic semi-algebraic set. Try and compare formulations as in (b) and (c) on a simple example.

10.4 Notes and sources

10.1 The semidefinite relaxation (10.4) was first introduced in Jibetean and de Klerk (2006) where the authors also proved their convergence (Theorem 10.2(a)).

10.2 This section is from Bugarin et al. (2011) and was motivated by several important applications in computer vision.

11

Semidefinite relaxations for semi-algebraic optimization

11.1 Introduction

The power of the moment-SOS approach for polynomial optimization rests upon the fact that the powerful Putinar positivity certificate (2.10)

- is stated in terms of SOS polynomial weights, and
- is linear in the polynomials that describe the feasible set $\mathbf{K}$,

which allows a practical and efficient computational procedure to be defined, as described in Chapter 6. For instance, the stronger and more general Stengle positivity certificates of Theorem 2.12 do *not* have this property.

From an optimization (and practical) viewpoint there should be a tradeoff between the generality (or strength) of a positivity certificate and whether it can be implemented in some practical efficient numerical procedure. In this respect Putinar's Positivstellensatz in Theorem 2.15 achieves a good compromise since by being restricted to compact basic semi-algebraic sets $\mathbf{K}$ and polynomial objective functions, it still applies to a large class of important optimization problems while being implementable in an efficient practical procedure.

However, even though such polynomial optimization problems encompass many important applications of practical interest, one may wonder whether the moment-SOS approach described in Chapter 6 can be extended to a more general class of optimization problems that involve some algebra of functions larger than the polynomials.

In Chapter 10 we have already seen how to minimize a sum of rational functions on a compact basic semi-algebraic set. However, in several applications one has to deal with functions like absolute values, square roots, fractions with rational powers, etc., which appear in both the objective function to be minimized an the description of the feasible set $\mathbf{K}$. Again, to extend the moment-SOS approach to this new framework one needs an analogue of

Putinar's positivity certificate for such functions which also translates into a practical procedure for optimization purposes.

In this chapter we consider the algebra $\mathcal{A}$ which consists of functions on a basic semi-algebraic set $\mathbf{K} \subset \mathbb{R}^n$, generated by basic monadic and dyadic relations on polynomials. The monadic operations are the absolute value $|\cdot|$, and the fractional powers $(\cdot)^{1/p}$, $p \in \mathbb{N}$, while the dyadic operations are $(+, \times, \div, \vee, \wedge)$. Notice that this algebra, which is a subclass of semi-algebraic functions, contains highly nonlinear and nondifferentiable functions. However, it turns out that every element $f \in \mathcal{A}$ has the very nice property that its graph $\{(\mathbf{x}, f(\mathbf{x})) : \mathbf{x} \in \mathbf{K}\}$ has a *lifted* basic semi-algebraic representation. By this we mean that it is the projection onto some coordinates of a basic semi-algebraic set defined in a higher dimensional space $\mathbb{R}^{n+p}$ (for some p).

And so minimizing $f \in \mathcal{A}$ on $\mathbf{K}$ reduces to solving a polynomial optimization problem in $\mathbb{R}^{n+p}$ for which the moment-SOS approach described in Chapter 6 applies. In addition, the feasible set $\mathbf{K}$ may also be described with constraints of the form $g(\mathbf{x}) \geq 0$, where $g \in \mathcal{A}$. This enlarges significantly the range of applications of the moment-SOS approach.

11.2 Semi-algebraic functions

We consider an algebra of functions which forms an important subclass of semi-algebraic functions. Recall the notation

$$a \vee b = \max[a, b], \quad a \wedge b := \min[a, b], \quad |\mathbf{x}| = (|x_1|, \dots, |x_n|) \in \mathbb{R}^n.$$

As before let $\mathbf{K} \subset \mathbb{R}^n$ be the basic semi-algebraic set

$$\mathbf{K} := \{\mathbf{x} \in \mathbb{R}^n : g_j(\mathbf{x}) \geq 0, \; j = 1, \dots, m\}, \tag{11.1}$$

for some $(g_j)_{j=1}^m \subset \mathbb{R}[\mathbf{x}]$, and let $\mathcal{A}$ be the algebra of functions $f : \mathbf{K} \to \mathbb{R}$ generated by finitely many of the dyadic operations $\{+, \times, \div, \vee, \wedge\}$ and monadic operations $|\cdot|$ and $(\cdot)^{1/p}$ ($p \in \mathbb{N}$) on polynomials.

Example 11.1 For instance, with $a, \ell \in \mathbb{R}[\mathbf{x}]$ and $\mathbf{p} = (p_1, p_2) \in \mathbb{R}[\mathbf{x}]^2$, the function

$$\mathbf{x} \mapsto f(\mathbf{x}) := a(\mathbf{x}) \wedge \left(\sqrt{\|\mathbf{p}(\mathbf{x})\|_{2q+1} + \ell(\mathbf{x})} \right) \tag{11.2}$$

is a member of $\mathcal{A}$ (e.g. assuming that $\ell(\mathbf{x}) \geq 0$ for all $\mathbf{x} \in \mathbf{K}$).

Definition 11.2 Let $\mathbf{K}$ be a semi-algebraic set of $\mathbb{R}^n$. A function $f : \mathbf{K} \to \mathbb{R}$ is a *semi-algebraic* function if its graph $\Psi_f := \{(\mathbf{x}, f(\mathbf{x})) : \mathbf{x} \in \mathbf{K}\}$ is a semi-algebraic set of $\mathbb{R}^n \times \mathbb{R}$.

Motivated by the efficiency of the constructive aspects of nonpolynomial optimization, we adopt the following ad hoc terminology.

Definition 11.3 A function $f \in \mathcal{A}$ is said to have a *basic semi-algebraic lifting* (in short a b.s.a.l), or f is *basic semi-algebraic* (b.s.a.), if there exist $p, s \in \mathbb{N}$, polynomials $(h_k)_{k=1}^s \subset \mathbb{R}[\mathbf{x}, y_1, \dots, y_p]$ and a basic semi-algebraic set

$$\mathbf{K}_f := \{ (\mathbf{x}, \mathbf{y}) \in \mathbb{R}^{n+p} : \mathbf{x} \in \mathbf{K}; \; h_k(\mathbf{x}, \mathbf{y}) \geq 0, \quad k = 1, \dots, s \} \quad (11.3)$$

such that the graph of f (denoted Ψ_f) satisfies:

$$\Psi_f := \{ (\mathbf{x}, f(\mathbf{x})) : \mathbf{x} \in \mathbf{K} \} = \{ (\mathbf{x}, y_p) : (\mathbf{x}, \mathbf{y}) \in \mathbf{K}_f \}. \quad (11.4)$$

In other words, Ψ_f is an orthogonal projection of the basic semi-algebraic set $\mathbf{K}_f$ which lives in the lifted space $\mathbb{R}^{n+p}$.

Below are some basic examples of b.s.a.l. functions.

Example 11.4 With $\mathbf{K} \subseteq \mathbb{R}^n$ let $p, q \in \mathbb{R}[\mathbf{x}]$. Consider the following examples.

- $f := |p - q|$ is b.s.a. Indeed
$$\mathbf{K}_f = \{(\mathbf{x}, y) \in \mathbf{K} \times \mathbb{R} : y^2 - (p(\mathbf{x}) - q(\mathbf{x}))^2 = 0; \; y \geq 0\}.$$

- $f := p \wedge q$ is b.s.a. From $2(a \wedge b) = (a + b) - |a - b|$,
$$\mathbf{K}_f = \{(\mathbf{x}, \mathbf{y}) \in \mathbf{K} \times \mathbb{R}^2 : y_1^2 - (p(\mathbf{x}) - q(\mathbf{x}))^2 = 0; \; y_1 \geq 0;$$
$$2y_2 + y_1 - p(\mathbf{x}) - q(\mathbf{x}) = 0\}.$$

- $f := p \vee q$ is b.s.a. From $2(a \vee b) = (a + b) + |a - b|$,
$$\mathbf{K}_f = \{(\mathbf{x}, \mathbf{y}) \in \mathbf{K} \times \mathbb{R}^2 : y_1^2 - (p(\mathbf{x}) - q(\mathbf{x}))^2 = 0; \; y_1 \geq 0;$$
$$2y_2 - y_1 - p(\mathbf{x}) - q(\mathbf{x}) = 0\}.$$

- If $p \geq 0$ on $\mathbf{K}$ then $f := \sqrt{p}$ is b.s.a. Indeed
$$\mathbf{K}_f = \{(\mathbf{x}, y) \in \mathbf{K} \times \mathbb{R} : y^2 - p(\mathbf{x}) = 0; \; y \geq 0\}.$$

- If $p > 0$ on $\mathbf{K}$ then $f := p^{-1/(2\ell)}$ is b.s.a. Indeed
$$\mathbf{K}_f = \{(\mathbf{x}, y) \in \mathbf{K} \times \mathbb{R} : y^{2\ell} p(\mathbf{x}) = 1; \; y \geq 0\}.$$

And, by combining the above elementary b.s.a. functions one may easily construct sets $\mathbf{K}_f$ for much more complicated b.s.a. functions f.

Hence by the projection theorem of real algebraic geometry (Bochnak et al., 1998, §2.2) a b.s.a. function $f \in \mathcal{A}$ is semi-algebraic. This is because Ψ_f is the projection of the basic semi-algebraic set $\mathbf{K}_f$ onto some coordinates. As a matter of fact, *every* semi-algebraic function possesses a basic semi-algebraic lifting.[1] This can be deduced from the fact that every semi-algebraic set S is the projection of a closed, basic semi-algebraic set. Indeed, assume first that S is closed and $S = \cup_{i \in I} S_i$ where every S_i is closed, basic semi-algebraic and the set I is finite. Writing

$$S_i = \{\, \mathbf{x} \in \mathbb{R}^n : g_1(\mathbf{x}) \geq 0, \ldots, g_k(\mathbf{x}) \geq 0 \,\}$$

we can lift the inequalities to equalities as before:

$$S_i = \{\, \mathbf{x} \in \mathbb{R}^n : \exists\, \mathbf{y} \in \mathbb{R}^k, \ y_i^2 - g_i(\mathbf{x}) = 0, \ i = 1, \ldots, k \,\},$$

and remark that a union of real algebraic sets is real algebraic. The same procedure works for an open component of an arbitrary semi-algebraic set S:

$$G_j = \{\, \mathbf{x} \in \mathbb{R}^n : h_1(\mathbf{x}) > 0, \ldots, h_k(\mathbf{x}) > 0 \,\},$$

can be represented as

$$G_j = \{\, \mathbf{x} \in \mathbb{R}^n : \exists\, \mathbf{y} \in \mathbb{R}^k, \ 1 - y_i^2 h_i(\mathbf{x}) = 0, \quad i = 1, \ldots, k \,\}.$$

Lemma 11.5   *Let $\mathbf{K}$ be the basic semi-algebraic set in (11.1). Then every well-defined $f \in \mathcal{A}$ has a basic semi-algebraic lifting.*

Example 11.6   With $f \in \mathcal{A}$ as in Example 11.1, $\Psi_f = \{\, (\mathbf{x}, y_6) : (\mathbf{x}, \mathbf{y}) \in \mathbf{K}_f \,\}$, where $\mathbf{K}_f \subset \mathbb{R}^{n+6}$ is the basic semi-algebraic set:

$$\mathbf{K}_f = \Big\{ (\mathbf{x}, \mathbf{y}) \in \mathbb{R}^{n+6} : \quad g_j(\mathbf{x}) \geq 0, \quad j = 1, \ldots, m; $$
$$y_i^2 - p_i(\mathbf{x})^2 = 0, \ i = 1, 2; \quad y_3^{2q+1} - y_1^{2q+1} - y_2^{2q+1} = 0;$$
$$y_4^2 - \ell(\mathbf{x}) - y_3 = 0; \quad y_5^2 - (a(\mathbf{x}) - y_4)^2 = 0;$$
$$2y_6 - (a(\mathbf{x}) + y_4) + y_5 = 0; \quad y_1, y_2, y_4, y_5 \geq 0 \Big\}.$$

Observe that with each variable y_k of the lifting $\mathbf{y} = (y_1, \ldots, y_p)$ is associated a certain function $v_k \in \mathcal{A}$. For instance in Example 11.6:

$$y_i \to v_i(\mathbf{x}) := |p_i(\mathbf{x})|, \ i = 1, 2; \quad y_3 \to v_3(\mathbf{x}) := (v_1(\mathbf{x})^{2q+1} + v_2(\mathbf{x})^{2q+1})^{1/(2q+1)};$$

$$y_4 \to v_4(\mathbf{x}) := \sqrt{v_3(\mathbf{x}) + \ell(\mathbf{x})}; \quad y_5 \to v_5(\mathbf{x}) := |a(\mathbf{x}) - v_4(\mathbf{x})|;$$

$$y_6 \to v_6(\mathbf{x}) := f(\mathbf{x}).$$

---

[1] The author wishes to thank D. Plaumann for pointing out this fact.

Remark 11.7 In fact, we have just seen that for the algebra $\mathcal{A}$ that we are considering, the constraints $h_k(\mathbf{x}, \mathbf{y}) \geq 0, k = 1, \ldots, s$, in the definition (11.3) of $\mathbf{K}_f$, can be partitioned into:

- equality constraints $u_\ell(\mathbf{x}, \mathbf{y}) = 0$ that describe algebraic relationships, and
- inequality constraints of the form $y_j \geq 0, \; j \in J$ (for some index set J), needed for a complete description of the b.s.a. function f.

If $\mathbf{v} = (v_k) \subset \mathcal{A}$ denotes the intermediate b.s.a. functions associated with the lifting $\mathbf{y} = (y_k)$, then by construction $u_\ell(\mathbf{x}, \mathbf{v}(\mathbf{x})) = 0$ for all ℓ.

11.3 A Positivstellensatz for semi-algebraic functions

Next, with $\mathbf{K} \subset \mathbb{R}^n$ as in (11.1), consider the set $\mathbf{S} \subseteq \mathbf{K}$ defined by:

$$\mathbf{S} := \{\mathbf{x} \in \mathbf{K} : h_\ell(\mathbf{x}) \geq 0, \; \ell = 1, \ldots, s\} \tag{11.5}$$

for some finite family $(h_k)_{k=1}^{s} \subset \mathcal{A}$.

Lemma 11.8 *The set $\mathbf{S}$ in (11.5) is a semi-algebraic set which is the projection of a lifted basic semi-algebraic set. It can be written*

$$\mathbf{S} = \Big\{ \mathbf{x} \in \mathbb{R}^n : \exists \mathbf{y} \in \mathbb{R}^t \text{ s.t. } \mathbf{x} \in \mathbf{K}; \quad u_k(\mathbf{x}, \mathbf{y}) = 0, \quad k = 1, \ldots, r,$$
$$y_j \geq 0, \quad j \in J \Big\} \tag{11.6}$$

for some integers r, t, some polynomials $(u_k)_{k=1}^{r} \subset \mathbb{R}[\mathbf{x}, \mathbf{y}]$ and some index set $J \subseteq \{1, \ldots, t\}$. With the lifting $\mathbf{y} = (y_1, \ldots, y_t)$ is associated a vector of functions $\mathbf{v} = (v_1, \ldots, v_t) \in \mathcal{A}^t$.

So in (11.6) the equations $u_k(\mathbf{x}, \mathbf{y}) = 0$ are the algebraic relationships that link the b.s.a. functions $v_k \in \mathcal{A}$ with each other and with $\mathbf{x}$. See for instance the six equations that define the set $\mathbf{K}_f$ in Example 11.6. In particular, $u_k(\mathbf{x}, \mathbf{v}(\mathbf{x})) = 0$ for all $\mathbf{x} \in \mathbf{K}$. The additional inequalities $y_j \geq 0, j \in J$, completely characterize the functions $v_k \in \mathcal{A}$.

Theorem 11.9 *Let $\mathbf{K} \subset \mathbb{R}^n$ in (11.1) be compact and let $\mathbf{S}$ be as in (11.5) for some finite family $(h_\ell) \subset \mathcal{A}$. Let $f \in \mathcal{A}$ be such that*

$$\Psi_f = \Big\{ (\mathbf{x}, z_{m_f}) : \mathbf{x} \in \mathbf{K}; \; q_\ell(\mathbf{x}, \mathbf{z}) = 0, \; \ell = 1, \ldots, t_f; \; z_j \geq 0, \quad j \in J_f \Big\}$$

(where $\mathbf{z} = (z_1, \ldots, z_{m_f})$ and $J_f \subseteq \{1, \ldots, m_f\}$), and let $\mathbf{v} = (v_1, \ldots, v_t) \in \mathcal{A}^t$ (respectively $\mathbf{p} = (p_1, \ldots, p_{m_f}) \in \mathcal{A}^{m_f}$) be the vector of b.s.a. functions

associated with the lifting $\mathbf{y}$ *in* (11.6) *(respectively the lifting* $\mathbf{z}$). *If f is positive on* $\mathbf{S}$ *then*

$$f(\mathbf{x}) = \sigma_0(\mathbf{x}, \mathbf{v}(\mathbf{x}), \mathbf{p}(\mathbf{x})) + \sum_{j=1}^{m} \sigma_j(\mathbf{x}, \mathbf{v}(\mathbf{x}), \mathbf{p}(\mathbf{x})) \, g_j(\mathbf{x})$$

$$+ \sum_{k \in J} \psi_k(\mathbf{x}, \mathbf{v}(\mathbf{x}), \mathbf{p}(\mathbf{x})) \, v_k(\mathbf{x}) + \sum_{\ell \in J_f} \phi_\ell(\mathbf{x}, \mathbf{v}(\mathbf{x}), \mathbf{p}(\mathbf{x})) \, p_\ell(\mathbf{x})$$

$$+ \Delta(\mathbf{x}, \mathbf{v}(\mathbf{x}), \mathbf{p}(\mathbf{x})) \, (M - \|(\mathbf{x}, \mathbf{v}(\mathbf{x}), \mathbf{p}(\mathbf{x}))\|^2) \qquad (11.7)$$

for some SOS polynomials $(\sigma_j, \psi_k, \phi_\ell, \Delta) \subset \Sigma[\mathbf{x}, \mathbf{y}, \mathbf{z}]$ *and some sufficiently large* $M > 0$.

Proof   That f is positive on $\mathbf{S}$ is the same as stating that the polynomial $z_{mf} \in \mathbb{R}[\mathbf{x}, \mathbf{y}, \mathbf{z}]$ is positive on the basic semi-algebraic set:

$$\Omega = \{ (\mathbf{x}, \mathbf{y}, \mathbf{z}) : \begin{array}{llll} g_j(\mathbf{x}) & \geq 0, & j = 1, \ldots, m \\ u_k(\mathbf{x}, \mathbf{y}) & = 0, & k = 1, \ldots, r \\ q_\ell(\mathbf{x}, \mathbf{z}) & = 0, & \ell = 1, \ldots, t_f \\ M - \|(\mathbf{x}, \mathbf{y}, \mathbf{z})\|^2 & \geq 0 \\ y_k, z_j & \geq 0, & k \in J, \ j \in J_f \}, \end{array}$$

where $M > 0$ has been chosen sufficiently large so that the quadratic constraint $M - \|(\mathbf{x}, \mathbf{y}, \mathbf{z})\|^2 \geq 0$ is redundant. Indeed, since $\mathbf{K}$ is compact, all variables $\mathbf{x}, \mathbf{y}$ and $\mathbf{z}$ in the definition of Ω are bounded. Because of this constraint the quadratic module associated with the polynomials defining Ω is Archimedean, i.e., Assumption 2.14 holds. Next, if f is positive on $\mathbf{K}$ then so is z_{mf} on Ω and by Theorem 2.15 (with $g_0 = 1$),

$$z_{mf} = \sum_{j=0}^{m} \sigma_j(\mathbf{x}, \mathbf{y}, \mathbf{z}) g_j(\mathbf{x}) + \sum_{k=1}^{r} \theta_k(\mathbf{x}, \mathbf{y}, \mathbf{z}) u_k(\mathbf{x}, \mathbf{y}) + \sum_{\ell=1}^{t_f} \xi_\ell(\mathbf{x}, \mathbf{y}, \mathbf{z}) \, q_\ell(\mathbf{x}, \mathbf{z})$$

$$+ \Delta(\mathbf{x}, \mathbf{y}, \mathbf{z})(M - \|(\mathbf{x}, \mathbf{y}, \mathbf{z})\|^2) + \sum_{k \in J} \psi_k(\mathbf{x}, \mathbf{y}, \mathbf{z}) y_k + \sum_{j \in J_f} \phi_j(\mathbf{x}, \mathbf{y}, \mathbf{z}) z_j$$

for some SOS polynomials $\sigma_j, \Delta, \psi_k, \phi_j \in \Sigma[\mathbf{x}, \mathbf{y}, \mathbf{z}]$ and some polynomials $\theta_k, \xi_\ell \in \mathbb{R}[\mathbf{x}, \mathbf{y}, \mathbf{z}]$.

Finally one obtains (11.7) by replacing in the above identity

- y_k with $v_k(\mathbf{x})$ for every $k = 1, \ldots, t$,
- z_ℓ with $p_\ell(\mathbf{x})$ for every $\ell = 1, \ldots, m_f$,

and recalling (see Remark 11.7) that $u_k(\mathbf{x}, \mathbf{v}(\mathbf{x})) = 0$ and $q_\ell(\mathbf{x}, \mathbf{p}(\mathbf{x})) = 0$ for all $\mathbf{x}$ and all $k = 1, \ldots, r, \ell = 1, \ldots, t_f$. $\qquad \square$

11.4 Optimization of semi-algebraic functions

In the previous section we have seen that positivity of semi-algebraic functions $f \in \mathcal{A}$ on a set $\mathbf{S}$ of the form (11.6) reduces to positivity of polynomials on some basic semi-algebraic defined in a lifted space. And so, as one may expect, the apparatus available for polynomial optimization and described in Chapter 6 also works for these functions.

So let $\mathcal{A}$ be the algebra of functions considered in Section 11.2, let $\mathbf{K}, \mathbf{S} \subset \mathbb{R}^n$ be as in (11.1) and (11.5) respectively, and consider the global optimization problem:

$$\mathbf{P}: \quad f^* := \inf_{\mathbf{x}} \{ f(\mathbf{x}) \,:\, \mathbf{x} \in \mathbf{S} \} \tag{11.8}$$

where $f \in \mathcal{A}$.

Proposition 11.10   *Let $\mathbf{K} \subset \mathbb{R}^n$ be as in (11.1) and with $\mathbf{S}$ as in (11.5)–(11.6), let $f \in A$ be a b.s.a. function with*

$$\Psi_f = \left\{ (\mathbf{x}, z_f) \,:\, \mathbf{x} \in \mathbf{K}; \; q_\ell(\mathbf{x}, \mathbf{z}) = 0, \quad \ell = 1, \dots, t_f; \; z_j \ge 0, \quad j \in J_f \right\},$$

where $\mathbf{z} = (z_1, \dots, z_{m_f})$ and $J_f \subseteq \{1, \dots, m_f\}$. Then $\mathbf{P}$ is also the polynomial optimization problem:

$$f^* = \inf_{\mathbf{x}, \mathbf{y}, \mathbf{z}} \{ z_{m_f} \,:\, (\mathbf{x}, \mathbf{y}, \mathbf{z}) \in \mathbf{\Omega} \} \tag{11.9}$$

where $\mathbf{\Omega} \subset \mathbb{R}^{n+t+m_f}$ is the basic semi-algebraic set

$$\mathbf{\Omega} = \{ (\mathbf{x}, \mathbf{y}, \mathbf{z}) : \begin{array}{llll} g_j(\mathbf{x}) & \ge 0, & j = 1, \dots, m \\ u_k(\mathbf{x}, \mathbf{y}) & = 0, & k = 1, \dots, r \\ q_\ell(\mathbf{x}, \mathbf{z}) & = 0, & \ell = 1, \dots, t_f \\ y_k, z_j & \ge 0, & k \in J, \; j \in J_f \}. \end{array} \tag{11.10}$$

This follows from the fact that for every $(\mathbf{x}, \mathbf{y}, \mathbf{z}) \in \mathbf{\Omega}$, z_{m_f} is exactly equal to $f(\mathbf{x})$ when $\mathbf{x} \in \mathbf{S}$. Then one may apply the moment-SOS approach developed in Chapter 6 and build up the associated hierarchy of semidefinite relaxations (6.3). If in the definition of $\mathbf{\Omega}$ one adds the (redundant) Archimedean quadratic constraint $M - \|(\mathbf{x}, \mathbf{y}, \mathbf{z})\|^2 \ge 0$ (as in the proof of Theorem 11.9), then the sequence of optimal values (ρ_d), $d \in \mathbb{N}$, associated with the hierarchy of semidefinite relaxations (6.3) converges to the global optimum f^*. In addition, if a certain rank condition on the moment matrix is satisfied at an optimal solution of the dth semidefinite relaxation in the hierarchy, then $\rho_d = f^*$ (i.e., finite convergence takes place) and one may extract global minimizers; see Theorem 6.6.

A structured sparsity pattern   In the description of Ω in (11.10), a *lifting* $\mathbf{y}^\ell \in \mathbb{R}^{n_\ell}$ is associated with each nonpolynomial function $h_\ell \in \mathcal{A}$ that appears in the description of $\mathbf{P}$, as well as a lifting $\mathbf{z} \in \mathbb{R}^{m_f}$ for f (if f is not a polynomial). This can be rapidly penalizing. However, it is worth noting that there is no mixing between the lifting variables $\mathbf{z}$ and $\mathbf{y}^\ell$, as well as between the lifting variables $\mathbf{y}^\ell$ and $\mathbf{y}^k$, $k \neq \ell$. The coupling of all these variables is through the variables $\mathbf{x}$. Hence there is an obvious structured sparsity pattern if the set of all variables $(\mathbf{x}, \mathbf{y}, \mathbf{z})$ is written as the union (with overlaps)

$$\{\mathbf{x}, \mathbf{y}, \mathbf{z}\} = \{\mathbf{x}\} \cup \{\mathbf{x}, \mathbf{z}\} \cup \{\mathbf{x}, \mathbf{y}^1\} \cup \cdots \cup \{\mathbf{x}, \mathbf{y}^s\}$$

and this sparsity pattern obviously satisfies the so-called *running intersection property* defined in (2.27). Indeed, for each $1 \leq k < s$,

$$\{\mathbf{x}, \mathbf{y}^{k+1}\} \bigcap \left(\{\mathbf{x}\} \cup \{\mathbf{x}, \mathbf{z}\} \cup_{i=1}^{k} \{\mathbf{x}, \mathbf{y}^i\} \right) \subseteq \{\mathbf{x}\}.$$

This is good news because with such a sparsity pattern one may use the sparse semidefinite relaxations (8.6) defined in Chapter 8 whose associated sequence of optimal values still converges to f^*; see Theorem 8.3. Hence if on the one hand lifting is penalizing, on the other hand the special structure of the lifted polynomial optimization problem (11.9) permits the use of specialized "sparse" relaxations, which (partly) compensates for the increase in the number of variables.

Examples

For illustration purposes, consider the following two simple examples of semi-algebraic optimization problems $\mathbf{P}$.

Example 11.11   Let $n = 2$ and $\mathbf{P}$: $f^* = \sup_{\mathbf{x}} \{ x_1 \, |x_1 - 2x_2| : \|\mathbf{x}\|^2 = 1 \}$. Hence

$$\Omega = \{ (\mathbf{x}, z) : \|\mathbf{x}\|^2 - 1 = 0; \; z^2 - (x_1 - 2x_2)^2 = 0; \; z \geq 0 \},$$

and one may add the redundant Archimedean quadratic constraint $10 - \|(\mathbf{x}, z)\|^2 \geq 0$. At an optimal solution of the semidefinite relaxation (6.3) for $\mathbf{P}$ with $d = 2$, the rank condition of Theorem 6.6 is satisfied[2] so that $f^* = \rho_2 = 1.6180$ and one may extract the global minimizer $(\mathbf{x}^*, z^*) = (0.8507, -0.5257, 1.9021)$. (In fact, one even has $\rho_1 = f^*$.)

---

[2]   The semidefinite relaxation was solved with the GloptiPoly software described in Appendix B.

Example 11.12   Consider $n = 2$ and $\mathbf{P} : f^* = \sup_{\mathbf{x}} \{ f(\mathbf{x}) : \mathbf{x} \in \mathbf{K} = [0, 1]^2 \}$,
where f is the b.s.a. function of Example 11.6 with $q = 1$, $\mathbf{x} \mapsto p_1(\mathbf{x}) := 2x_1$,
$\mathbf{x} \mapsto p_2(\mathbf{x}) := x_2$, $\mathbf{x} \mapsto \ell(\mathbf{x}) := 0.1 + x_1^2 + x_2^2$, and $a(\mathbf{x}) = 2x_1 - x_2$. So with
$\mathbf{x} \mapsto g_i(\mathbf{x}) = x_i(1 - x_i)$, $i = 1, 2$, $\Omega \subset \mathbb{R}^8$ is the set:

$$
\Omega = \Big\{ (\mathbf{x}, \mathbf{z}) \in \mathbb{R}^8 : \mathbf{x} \in \mathbf{K}; \ z_i^2 - p_i(\mathbf{x})^2 = 0, \quad z_i \ge 0, \quad i = 1, 2;
$$

$$
z_3^{2q+1} - z_1^{2q+1} - z_2^{2q+1} = 0;
$$

$$
z_4^2 - \ell(\mathbf{x}) - z_3 = 0, \quad z_4 \ge 0;
$$

$$
z_5^2 - (a(\mathbf{x}) - z_4)^2 = 0, \quad z_5 \ge 0;
$$

$$
2z_6 - (a(\mathbf{x}) + z_4) + z_5 = 0 \Big\}.
$$

Solving the second semidefinite relaxation (i.e., with $d = 3$ and moments up
to order 6) provides the global minimum $\rho_3 = f^* = 3.553$ and the global
minimizer $\mathbf{x}^* = (1.0, 0.2106) \in \mathbf{K}$ is extracted.

11.5 Exercises

Exercise 11.1   Let $\mathbf{K} \subset \mathbb{R}^n$ as in (11.1) be compact and let $p_i, q_i \in \mathbb{R}[\mathbf{x}]$,
$i = 1, \ldots, 2$, with $q_i > 0$ on $\mathbf{K}$ for $i = 1, 2$. Consider the optimization
problem $\mathbf{P} : \inf \{ p_1(\mathbf{x})/q_1(\mathbf{x}) - p_2(\mathbf{x})/q_2(\mathbf{x}) : \mathbf{x} \in \mathbf{K} \}$.

(a) By reducing to the same denominator and using the results of Chap-
ter 10, reformulate $\mathbf{P}$ as a polynomial optimization problem on $\mathbb{R}^n$. What is the
number of variables in the first semidefinite relaxation of the hierarchy (10.3)?

(b) Reformulate $\mathbf{P}$ as a polynomial optimization problem $\mathbf{P}'$ in $\mathbb{R}^{n+2}$ with
additional constraints and variables. Show how to obtain bounds on the two
additional variables so as to make the feasible set of $\mathbf{P}'$ a compact basic semi-
algebraic set of $\mathbb{R}^{n+2}$. What is the number of variables (y_α) in the first semidefinite relaxation of the hierarchy (6.3) associated with $\mathbf{P}'$? Compare with (a) and
when is this number smaller?

Exercise 11.2   Let $\mathbf{K} \subset \mathbb{R}^n$ as in (11.1) be compact and let $p_i \in \mathbb{R}[\mathbf{x}]$, $i = 1, \ldots, 4$.

(a) Consider the optimization problem $\mathbf{P} : \inf_{\mathbf{x}} \{ p_1(\mathbf{x}) \vee p_2(\mathbf{x}) : \mathbf{x} \in \mathbf{K} \}$. Write $\mathbf{P}$ as a polynomial optimization problem in $\mathbb{R}^{n+1}$. Do we need to
introduce an equality constraint to model the semi-algebraic function $f(\mathbf{x}) = p_1(\mathbf{x}) \vee p_2(\mathbf{x})$? Can we simplify?

(b) Consider the optimization problem **P** : $\inf_{\mathbf{x}} \{ p_1(\mathbf{x}) \vee p_2(\mathbf{x}) + p_3(\mathbf{x}) \wedge p_4(\mathbf{x}) : \mathbf{x} \in \mathbf{K} \}$. Write **P** as a polynomial optimization problem in $\mathbb{R}^{n+2}$. Do we need to introduce equality constraints? What about **P** : $\inf_{\mathbf{x}} \{ p_1(\mathbf{x}) \vee p_2(\mathbf{x}) - p_3(\mathbf{x}) \wedge p_4(\mathbf{x}) : \mathbf{x} \in \mathbf{K} \}$?

11.6 Notes and sources

The material of this chapter is from Lasserre and Putinar (2010, 2012) where the case of some transcendental functions is also discussed. A more general framework for Positivstellensätze described in terms of weighted sums of squares of real-valued functions, has been considered in the recent works of Burgdorf et al. (2012) and Marshall and Netzer (2012). In the latter reference the authors extend results of this chapter and Lasserre and Putinar (2010) to a larger class of functions (e.g. some noncontinuous semi-algebraic functions on $\mathbf{K} \subset \mathbb{R}^n$ can be handled).

Again, from an optimization viewpoint, an important issue is the compromise between the generality of a Positivstellensatz and its practical use in an efficient numerical procedure. Indeed Positivstellensätze exist in the much more general analytic setting (and also on *O-minimal* structures); see for example Acquistapace et al. (2000, 2002). But of course, and in contrast to the semi-algebraic setting, there is no analogue of the LP and SDP hierarchies for testing positivity.

12

Polynomial optimization as an eigenvalue problem

In Chapter 6 we have seen how to generate a monotone nondecreasing sequence of lower bounds which converges to the global minimum f^* of the polynomial optimization problem

$$\mathbf{P}: \quad f^* = \inf_{\mathbf{x}} \{ f(\mathbf{x}) : \mathbf{x} \in \mathbf{K} \}, \tag{12.1}$$

where $\mathbf{K} \subseteq \mathbb{R}^n$ is a compact basic semi-algebraic set and $f \in \mathbb{R}[\mathbf{x}]_d$. For this purpose we have used powerful results on the representation of polynomials positive on $\mathbf{K}$ described in Chapter 2, and in particular Putinar's Positivstellensatz (Theorem 2.15). A resulting sequence of lower bounds (f_k^*), $k \in \mathbb{N}$, is obtained by maximizing λ over all scalars λ such that $f - \lambda \in \mathcal{C}_k$, where the convex cones $\mathcal{C}_k$ form a nested and converging sequence of inner approximations of the convex cone $C_d(\mathbf{K})$ of polynomials of degree at most d, nonnegative on $\mathbf{K}$; see Theorem 4.1.

However, recall that from Corollary 4.3 in Chapter 4 one may also define a nested sequence of outer approximations of $C_d(\mathbf{K})$. In this chapter we describe how to use these outer approximations to provide a monotone nonincreasing sequence of *upper bounds* which converges to f^* via solving a hierarchy of semidefinite programs. These semidefinite programs are very specific and in fact reduce to computing the smallest *generalized eigenvalue* of a pair of real symmetric matrices (of increasing size). And by a suitable change of basis, it reduces to computing the smallest eigenvalue of a single real symmetric matrix. Therefore powerful specialized packages can be used in lieu of a general semidefinite solver.

Observe that for nonconvex problems, providing a sequence of upper bounds that converges to the global minimum f^* is in general impossible. (Of course, for some special cases of compact sets $\mathbf{K}$ a brute force approach is possible; one can build up grids $\Omega_k \subset \mathbf{K}$, $k \in \mathbb{N}$, that approximate $\mathbf{K}$ better and better as $k \to \infty$; one then evaluates f at each point of Ω_k and records the minimum value.) For local minimization algorithms which provide a sequence of feasible points (and so upper bounds on f^*) convergence usually takes place to a local minimum only. So again, in the polynomial setting the representation results of Chapter 4 permit overcoming the difficulty when the (not necessarily compact) set $\mathbf{K}$ is such that one is able to compute moments of a measure μ whose support is exactly $\mathbf{K}$.

12.1 A converging sequence of upper bounds

Let $\mathbf{K} \subset \mathbb{R}^n$ be a closed subset (not necessarily semi-algebraic). If $\mathbf{K}$ is compact let μ be a finite Borel measure whose support[1] supp μ is exactly $\mathbf{K}$. If $\mathbf{K}$ is not compact, let φ be an arbitrary finite Borel measure with supp $\varphi = \mathbf{K}$ and let μ be as in (3.7). In both cases, the sequence of moments $\mathbf{y} = (y_\alpha)$, $\alpha \in \mathbb{N}^n$, is well defined, and we assume that y_α is available or can be computed, for every $\alpha \in \mathbb{N}^n$.

Consider the hierarchy of semidefinite programs:

$$\lambda_d = \sup_{\lambda \in \mathbb{R}} \{\lambda : \mathbf{M}_d(f_\lambda \, \mathbf{y}) \succeq 0\}, \tag{12.2}$$

where $f_\lambda \in \mathbb{R}[\mathbf{x}]$ is the polynomial $\mathbf{x} \mapsto f(\mathbf{x}) - \lambda$, and $\mathbf{M}_d(f_\lambda \, \mathbf{y})$ is the localizing matrix associated with f_λ and the sequence $\mathbf{y}$. Notice that (12.2) has only one variable! Obviously, $\lambda_d \geq \lambda_{d+1}$ for all d because if $\mathbf{M}_{d+1}(f_\lambda \, \mathbf{y}) \succeq 0$, then $\mathbf{M}_d(f_\lambda \, \mathbf{y}) \succeq 0$.

Theorem 12.1 *Let $f^* > -\infty$ and consider the hierarchy of semidefinite programs (12.2) indexed by $d \in \mathbb{N}$. Then*

(a) *(12.2) has an optimal solution $\lambda_d \geq f^*$ for every $d \in \mathbb{N}^n$,*
(b) *the sequence (λ_d), $d \in \mathbb{N}$, is monotone nonincreasing and $\lambda_d \downarrow f^*$ as $d \to \infty$.*

Proof (a) Since $f - f^* \geq 0$ on $\mathbf{K}$, by Theorem 3.2, $\lambda := f^*$ is a feasible solution of (12.2) for every d. Hence $\lambda_d \geq f^*$ for every $d \in \mathbb{N}$. Next, let $d \in \mathbb{N}$ be fixed, and let λ be an arbitrary feasible solution of (12.2). From the condition $\mathbf{M}_d(f_\lambda \, \mathbf{y}) \succeq 0$, the diagonal entry $\mathbf{M}_d(f_\lambda \, \mathbf{y})(1, 1)$ must be nonnegative, i.e., $\lambda y_0 \leq \sum_\alpha f_\alpha \, y_\alpha$. As $y_0 > 0$ then $\lambda \leq L_\mathbf{y}(f)/y_0$ and so as we maximize λ, (12.2) must have an optimal solution λ_d.

---

[1] Recall that the support supp μ of a finite Borel measure μ on $\mathbb{R}^n$ is the smaller closed set $\mathbf{K}$ such that $\mu(\mathbb{R}^n \setminus \mathbf{K}) = 0$.

(b) As already noticed, $\lambda_d \leq \lambda_m$ whenever $d \geq m$, because $\mathbf{M}_d(f_\lambda \mathbf{y}) \succeq 0$ implies $\mathbf{M}_m(f_\lambda \mathbf{y}) \succeq 0$. Therefore, the sequence (λ_d), $d \in \mathbb{N}$, is monotone nonincreasing and, being bounded below by f^*, converges to $\lambda^* \geq f^*$. Next, suppose that $\lambda^* > f^*$; fix $k \in \mathbb{N}$, arbitrary. The convergence $\lambda_d \downarrow \lambda^*$ implies $\mathbf{M}_k(f_{\lambda^*} \mathbf{y}) \succeq 0$. As k was arbitrary, we obtain that $\mathbf{M}_d(f_{\lambda^*} \mathbf{y}) \succeq 0$ for every $d \in \mathbb{N}$. But then by Theorem 3.2 or Theorem 3.4, $f - \lambda^* \geq 0$ on $\mathbf{K}$, and so $\lambda^* \leq f^*$, in contradiction with $\lambda^* > f^*$. Therefore $\lambda^* = f^*$.

$\square$

For each $d \in \mathbb{N}$, the semidefinite program (12.2) is very specific as it has only one variable and can be rewritten as

$$\lambda_d = \sup_{\lambda} \{\, \lambda : \lambda \mathbf{M}_d(\mathbf{y}) \preceq \mathbf{M}_d(f\,\mathbf{y}) \,\}, \qquad (12.3)$$

in which one recognizes a *generalized eigenvalue* problem for the pair of real symmetric matrices $\mathbf{M}_d(\mathbf{y})$ and $\mathbf{M}_d(f\,\mathbf{y})$. Therefore (12.2) can be solved via specialized eigenvalue software which is more efficient than a general purpose semidefinite solver. The optimal value λ_d^* provides an upper bound on the optimal value f^* only. We next show that the dual of (12.3) contains some information on global minimizers, at least when d is sufficiently large.

12.1.1 Duality

Recall that $s(d) = \binom{n+d}{n}$ and let $\mathcal{S}_d$ be the space of real symmetric $s(d) \times s(d)$ matrices. The dual of (12.3) is the semidefinite program:

$$\lambda_d^* = \inf_{\mathbf{x} \in \mathcal{S}_d} \{\, \langle \mathbf{X}, \mathbf{M}_d(f\,\mathbf{y}) \rangle : \langle \mathbf{X}, \mathbf{M}_d(\mathbf{y}) \rangle = 1; \ \mathbf{X} \succeq 0 \,\}. \qquad (12.4)$$

Let $\mathbf{X}$ be an arbitrary feasible solution of (12.4). By the spectral decomposition of $\mathbf{X}$, one may write $\mathbf{x} = \sum_\ell \mathbf{q}_\ell \mathbf{q}_\ell^T$ for some vectors $\mathbf{q}_\ell \in \mathbb{R}^{s(d)}$, $\ell = 1, \ldots, s$. Letting $\mathbf{x} \mapsto q_\ell(\mathbf{x}) := \mathbf{v}_d(\mathbf{x})^T \mathbf{q}_\ell$, $\ell = 1, \ldots, s$, one obtains

$$\langle \mathbf{X}, \mathbf{M}_d(f\,\mathbf{y}) \rangle = \int_{\mathbf{K}} \underbrace{\sum_{\ell=1}^{s} q_\ell(\mathbf{x})^2}_{\sigma(\mathbf{x}) \in \Sigma[\mathbf{x}]_d} f(\mathbf{x})\, d\mu(\mathbf{x}), \quad \langle \mathbf{X}, \mathbf{M}_d(\mathbf{y}) \rangle = \int_{\mathbf{K}} \sigma(\mathbf{x})\, d\mu(\mathbf{x}),$$

and so (12.4) has the equivalent formulation

$$\lambda_d^* = \inf_{\sigma} \left\{ \int_{\mathbf{K}} f\sigma\, d\mu : \int_{\mathbf{K}} \sigma\, d\mu = 1; \ \sigma \in \Sigma[\mathbf{x}]_d \right\}, \qquad (12.5)$$

and by weak duality in conic programming, $\lambda_d^* \geq \lambda_d \geq f^*$ for all d. So the dual problem (12.5) is to find an SOS polynomial σ of degree at most $2d$ (normalized to satisfy $\int \sigma d\mu = 1$) that minimizes the integral $\int_{\mathbf{K}} f\sigma d\mu$, and a simple interpretation of (12.5) is as follows.

Let $M(\mathbf{K})$ be the space of Borel *probability* measures on $\mathbf{K}$. In Chapter 5 we have seen that $f^* = \inf_{\psi \in M(\mathbf{K})} \int_{\mathbf{K}} f d\psi$. Next, let $M_d(\mu) \subset M(\mathbf{K})$ be the space of probability measures on $\mathbf{K}$ which have a density $\sigma \in \Sigma[\mathbf{x}]_d$ with respect to μ. Then (12.5) reads $\inf_{\psi \in M_d(\mu)} \int_{\mathbf{K}} f d\psi$, which clearly shows why λ_d^* is an upper bound on f^*. Indeed, instead of searching ψ in $M(\mathbf{K})$ one searches ψ in its subset $M_d(\mu)$. What is not obvious at all is whether the obtained upper bound obtained in (12.5) converges to f^* when the degree of $\sigma \in \Sigma[\mathbf{x}]_d$ is allowed to increase!

Theorem 12.2   *Suppose that $f^* > -\infty$ and $\mathbf{K}$ has nonempty interior.*

(a) *There is no duality gap between (12.2) and (12.5), and (12.5) has an optimal solution $\sigma^* \in \Sigma[\mathbf{x}]_d$ which satisfies $\int_{\mathbf{K}} (f(\mathbf{x}) - \lambda_d)\,\sigma^*(\mathbf{x})\,d\mu(\mathbf{x}) = 0.$*

(b) *If $\mathbf{K}$ is convex and f is convex, let $\mathbf{x}_d^* := \int_{\mathbf{K}} \mathbf{x}\,\sigma^*(\mathbf{x})d\mu(\mathbf{x})$. Then $\mathbf{x}_d^* \in \mathbf{K}$ and $f^* \leq f(\mathbf{x}_d^*) \leq \lambda_d$, so that $f(\mathbf{x}_d^*) \to f^*$ as $d \to \infty$. Moreover, if the set $\{\mathbf{x} \in \mathbf{K} : f(\mathbf{x}) \leq f_0\}$ is compact for some $f_0 > f^*$, then any accumulation point $\mathbf{x}^* \in \mathbf{K}$ of the sequence $(\mathbf{x}_d^*)$, $d \in \mathbb{N}$, is a minimizer of problem (12.1), that is, $f(\mathbf{x}^*) = f^*$.*

Proof   (a) Any scalar $\lambda < f^*$ is a feasible solution of (12.2) and in addition, $\mathbf{M}_d(f_\lambda\,\mathbf{y}) \succ 0$ because since $\mathbf{K}$ has nonempty interior and $f - \lambda > 0$ on $\mathbf{K}$,

$$\langle \mathbf{g}, \mathbf{M}_d(f_\lambda\,\mathbf{y})\,\mathbf{g} \rangle = \int_{\mathbf{K}} (f(\mathbf{x}) - \lambda)g(\mathbf{x})^2 \mu(d\mathbf{x}) > 0, \qquad \forall 0 \neq g \in \mathbb{R}[\mathbf{x}]_d.$$

But this means that Slater's condition[2] holds for (12.2) which in turn implies that there is no duality gap and (12.5) has an optimal solution $\sigma^* \in \Sigma[\mathbf{x}]_d$; see Appendix A. Therefore

$$\int_{\mathbf{K}} (f(\mathbf{x}) - \lambda_d)\,\sigma^*(\mathbf{x})\,d\mu(\mathbf{x}) = \int_{\mathbf{K}} f\,\sigma^*\,d\mu - \lambda_d = 0.$$

(b) Let ν be the Borel probability measure on $\mathbf{K}$ defined by $\nu(B) = \int_B \sigma^* d\mu$, $B \in \mathcal{B}$. As f is convex, by Jensen's inequality,

---

[2] Recall that for an optimization problem $\mathbf{P} : \inf_{\mathbf{x}}\{f_0(\mathbf{x}) : f_j(\mathbf{x}) \geq 0, j = 1, \dots, m\}$, Slater's condition holds if there exists $\mathbf{x}_0$ such that $f_j(\mathbf{x}_0) > 0$ for every $j = 1, \dots, m$. If f_0 and $-f_j$ are convex, $j = 1, \dots, m$, then there is no duality gap between $\mathbf{P}$ and its dual $\mathbf{P}^*$, and $\mathbf{P}^*$ has an optimal solution if the optimal value is finite.

$$\int_K f\sigma^* d\mu = \int_K f\, dv \geq f\left(\int_K \mathbf{x}\, dv\right) = f(\mathbf{x}_d^*).$$

In addition, if $\mathbf{K}$ is convex then $\mathbf{x}_d^* \in \mathbf{K}$ and so $\lambda_d \geq f(\mathbf{x}_d^*) \geq f^*$. Finally, if the set $\mathcal{H} := \{\mathbf{x} \in \mathbf{K} : f(\mathbf{x}) \leq f_0\}$ is compact for some $f_0 > f^*$ and since $\lambda_d \to f^*$, then $f(\mathbf{x}_d^*) \leq f_0$ for d sufficiently large, i.e., $\mathbf{x}_d^* \in \mathcal{H}$ for sufficiently large d. By compactness there is a subsequence $(d_\ell), \ell \in \mathbb{N}$, and a point $\mathbf{x}^* \in \mathcal{H} \subset \mathbf{K}$ such that $\mathbf{x}_{d_\ell}^* \to \mathbf{x}^*$ as $\ell \to \infty$. Continuity of f combined with the convergence $f(\mathbf{x}_d^*) \to f^*$ yields $f(\mathbf{x}_{d_\ell}^*) \to f(\mathbf{x}^*) = f^*$ as $\ell \to \infty$. As the convergent subsequence $(\mathbf{x}_{d_\ell}^*)$ was arbitrary, the proof is complete. $\qquad\square$

Hence solving the hierarchy of semidefinite programs (12.5), $d \in \mathbb{N}$, amounts to approximating the Dirac measure at a global minimizer $\mathbf{x}^* \in \mathbf{K}$ by a probability measure $\sigma\, d\mu$ whose density σ with respect to μ is an SOS polynomial $\sigma \in \Sigma[\mathbf{x}]_d$. Of course the larger d is the better the approximation.

In the case where f is a convex polynomial and $\mathbf{K}$ is a convex set, Theorem 12.2 provides a means of approximating not only the optimal value f^*, but also a global minimizer $\mathbf{x}^* \in \mathbf{K}$.

In the more subtle nonconvex case, one can still obtain some information on global minimizers from an optimal solution $\sigma^* \in \Sigma[\mathbf{x}]_d$ of (12.5). Let $\epsilon > 0$ be fixed, and suppose that d is large enough so that $f^* \leq \lambda_d < f^* + \epsilon$. Then, by Theorem 12.2(a),

$$\int_K (f(\mathbf{x}) - f^*)\,\sigma^*(\mathbf{x})\, d\mu(\mathbf{x}) = \lambda_d - f^* < \epsilon.$$

As $f - f^* \geq 0$ on $\mathbf{K}$, necessarily the measure $dv = \sigma^* d\mu$ gives very small weight to regions of $\mathbf{K}$ where $f(\mathbf{x})$ is significantly larger than f^*. For instance, if $\epsilon = 10^{-2}$ and $\Delta := \{\mathbf{x} \in \mathbf{K} : f(\mathbf{x}) \geq f^* + 1\}$, then $v(\Delta) \leq 10^{-2}$, i.e., the set Δ contributes to less than 1% of the total mass of v. So if μ is uniformly distributed on $\mathbf{K}$ (which is a reasonable choice for μ if one has to compute all moments of μ) then a simple inspection of the values of $\sigma^*(\mathbf{x})$ provides some rough indication on where (in $\mathbf{K}$) $f(\mathbf{x})$ is close to f^*.

The interpretation (12.5) of the dual shows that in general the monotone convergence is only asymptotic and cannot be finite. Indeed if $\mathbf{K}$ has a nonempty interior then the probability measure $dv = \sigma d\mu$ cannot be a Dirac measure at any global minimizer $\mathbf{x}^* \in \mathbf{K}$. An exception is the discrete case, i.e., when $\mathbf{K}$ is a finite number of points, for example in 0/1 programs. Indeed we have the following.

Corollary 12.3 *Let* $\mathbf{K} \subset \mathbb{R}^n$ *be a discrete set* $(\mathbf{x}(k)) \subset \mathbb{R}^n$, $k \in J$, *and let* μ *be the probability measure uniformly distributed in* $\mathbf{K}$, *i.e.,*

$$\mu = \frac{1}{s} \sum_{k=1}^{s} \delta_{\mathbf{x}(k)},$$

where $s = |J|$ *and* $\delta_{\mathbf{x}}$ *denotes the Dirac measure at the point* $\mathbf{x}$. *Then the convergence* $\lambda_d \to f^*$ *is finite, i.e., the optimal value* λ_d *of (12.2) satisfies* $\lambda_d = f^*$ *for some integer* d.

Proof Let $\mathbf{x}^* = \mathbf{x}(j^*)$ (for some $j^* \in J$) be the global minimizer of $\inf\{ f(\mathbf{x}) : \mathbf{x} \in \mathbf{K} \}$. For each $k = 1, \ldots, s$ there exists a polynomial $q_k \in \mathbb{R}[\mathbf{x}]$ such that $q_k(\mathbf{x}(j)) = \delta_{k=j}$ for every $j = 1, \ldots, s$ (where $\delta_{k=j}$ denotes the Kronecker symbol). The polynomials (q_k) are called *interpolation polynomials*. Let $\sigma^* := s q_{j^*}^2 \in \Sigma[\mathbf{x}]$ so that

$$\int_{\mathbf{K}} f(\mathbf{x}) \sigma^*(\mathbf{x}) d\mu(\mathbf{x}) = f(\mathbf{x}(j^*)) = f^* \quad \text{and} \quad \int_{\mathbf{K}} \sigma^* d\mu = 1.$$

Hence as soon as $d \geq \deg q_{j^*}$, $\sigma^* \in \Sigma[\mathbf{x}]_d$ is a feasible solution of (12.5), and so from $f^* = \int f \sigma^* d\mu \geq \lambda_d^* \geq \lambda_d \geq f^*$ we deduce that $\lambda_d^* = \lambda_d = f^*$, the desired result. $\qquad\square$

12.1.2 Discussion

In nonlinear programming, a sequence of upper bounds on the global minimum f^* is usually obtained from feasible points $\mathbf{x} \in \mathbf{K}$, for example, via some (local) minimization algorithm. But for nonconvex problems, providing a sequence of upper bounds that converges to the global minimum f^* is in general impossible unless one computes points on a grid whose mesh size tends to zero (if $\mathbf{K}$ is bounded and one knows how to build up such a mesh).

In the above methodology one provides a monotone nonincreasing sequence of upper bounds converging to f^* for polynomial optimization problems on sets $\mathbf{K}$, not necessarily compact but such that one may compute all moments of some finite Borel measure μ with $\operatorname{supp} \mu = \mathbf{K}$. In fact, if there are only finitely many moments available (say up to order $2d$) then one obtains a finite sequence of upper bounds (λ_k), $k \leq d$.

In contrast to the hierarchy of semidefinite relaxations of Chapter 6 which provide lower bounds converging to f^* when $\mathbf{K}$ is a compact basic semi-algebraic set, the convergence of the upper bounds to f^* is only asymptotic and never finite, except when $\mathbf{K}$ is a discrete set; however it provides a systematic approach to obtaining converging upper bounds. Even if the convergence is expected to be rather slow when λ_d is close to f^*, in a few iterations one may obtain upper bounds which (even if crude) complement the lower bounds obtained in Chapter 6.

Cases of interest There are several interesting cases where the above described methodology can be applied, i.e., cases where $\mathbf{y}$ can be obtained either explicitly in closed form or numerically, in particular, when $\mathbf{K}$ is one of the following.

- A box $\mathbf{B} := \prod_{i=1}^{n}[a_i, b_i] \subset \mathbb{R}^n$, with μ being the normalized Lebesgue measure on $\mathbf{B}$. The sequence $\mathbf{y} = (y_\alpha)$ is trivial to obtain in closed form.
- The discrete set $\{-1, 1\}^n$ with μ being uniformly distributed and normalized. Again the sequence $\mathbf{y} = (y_\alpha)$ is trivial to obtain in closed form. Notice that in this case we obtain a new hierarchy of semidefinite relaxations (with only one variable) for the celebrated MAXCUT problem (and any nonlinear 0/1 program $f^* = \inf\{ f(\mathbf{x}) : \mathbf{x} \in \{0, 1\}^n\}$).
- The unit Euclidean ball $\mathbf{B} := \{ \mathbf{x} : \|\mathbf{x}\|^2 \leq 1 \}$ with μ uniformly distributed, and similarly the unit sphere $\mathbf{S} := \{ \mathbf{x} : \|\mathbf{x}\|^2 = 1 \}$, with μ being the rotation invariant probability measure on $\mathbf{S}$. In both cases the moments $\mathbf{y} = (y_\alpha)$ are obtained easily.
- A simplex $\Delta \subset \mathbb{R}^n$, in which case if one takes μ as the Lebesgue measure then all moments of μ can be computed numerically. In particular, with d fixed, this computation can be done in polynomial time. See for example the recent work of Baldoni et al. (2011).
- The whole space $\mathbb{R}^n$ in which case μ may be chosen to be the product measure $\otimes_{i=1}^{n} \nu_i$ with each ν_i being the normal distribution. Observe that one then obtains a new hierarchy of semidefinite approximations (upper bounds) for unconstrained global optimization. The corresponding monotone sequence of upper bounds converges to f^* no matter if f has a global minimizer or not. This may be an alternative and/or a complement to the convex relaxations of Schweighofer (2006) and Vui and Son (2008) which also work when f^* is not attained and provide a convergent sequence of *lower* bounds.
- The positive orthant $\mathbb{R}_+^n$, in which case μ may be chosen to be the product measure $\otimes_{i=1}^{n} \nu_i$ with each ν_i being the exponential distribution with density $x \mapsto \exp(-x)$. In particular, if $\mathbf{x} \mapsto f_{\mathbf{A}}(\mathbf{x}) := \mathbf{x}^T \mathbf{A} \mathbf{x}$ where $\mathbf{A} \in \mathcal{S}_n$, then one obtains a hierarchy of numerical tests to check whether $\mathbf{A}$ is a *copositive* matrix. Indeed, if λ_d is an optimal solution of (12.3) then $\mathbf{A}$ is copositive if and only if $\lambda_d \geq 0$ for all $d \in \mathbb{N}$. Recall from Chapter 4 that we also obtain a hierarchy of *outer approximations* $(\mathcal{C}_d) \subset \mathcal{S}$ of the cone $\mathcal{C}$ of $n \times n$ copositive

matrices. Indeed, for every d, the set

$$\mathcal{C}_d := \{ \mathbf{A} \in \mathcal{S} : \mathbf{M}_d(f_{\mathbf{A}} \, \mathbf{y}) \succeq 0 \} \tag{12.6}$$

is a convex cone defined only in terms of the coefficients of the matrix $\mathbf{A}$. It is even a spectrahedron since $\mathbf{M}_d(f_{\mathbf{A}} \, \mathbf{y})$ is a linear matrix inequality in the coefficients of $\mathbf{A}$. And by Theorem 4.10, $\mathcal{C} = \bigcap_{d \in \mathbb{N}} \mathcal{C}_d$.

12.1.3 Examples

In this section we provide three simple illustrative examples.

Example 12.4 Consider the global minimization on $\mathbf{K} = \mathbb{R}_+^2$ of the Motzkin-like polynomial $\mathbf{x} \mapsto f(\mathbf{x}) = x_1^2 x_2^2 (x_1^2 + x_2^2 - 1)$ whose global minimum is $f^* = -1/27 \approx -0.037$, attained at $(x_1^*, x_2^*) = (\pm \sqrt{1/3}, \pm \sqrt{1/3})$. Choose for μ the probability measure $\mu(B) := \int_B \mathbf{e}^{-x_1 - x_2} d\mathbf{x}$, $B \in \mathcal{B}(\mathbb{R}_+^2)$, for which the sequence of moments $\mathbf{y} = (y_{ij})$, $i, j \in \mathbb{N}$, is easy to obtain, namely, up to a constant, $y_{ij} = i! j!$ for every $i, j \in \mathbb{N}$. Then the semidefinite relaxations (12.2) yield $\lambda_0 = 92$, $\lambda_1 = 1.5097$, and $\lambda_{14} = -0.0113$, showing (as displayed in Figure 12.1) a significant and rapid decrease in first iterations with a long tail close to f^*. Then after $d = 14$, one encounters some numerical problems and we cannot trust the results anymore.

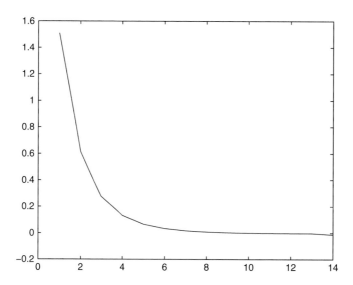

Figure 12.1 Minimizing the Motzkin-like polynomial in $\mathbb{R}_+^2$.

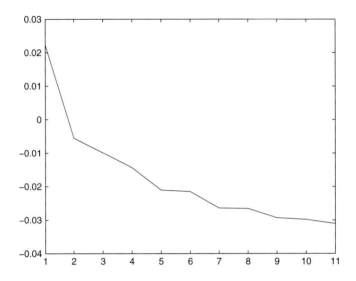

Figure 12.2 Minimizing the Motzkin-like polynomial in $[0, 1]^2$.

If we now minimize the same polynomial f on the box $[0, 1]^2$, one can choose for μ the probability uniformly distributed on $[0, 1]^2$, whose moments $\mathbf{y} = (y_{ij})$, $i, j \in \mathbb{N}$, are also easily obtained by $y_{ij} = (i + 1)^{-1}(j + 1)^{-1}$. Then one obtains $\lambda_0 = 0.222$, $\lambda_1 = -0.055$, and $\lambda_{10} = -0.0311$, showing again a rapid decrease in first iterations with a long tail close to f^*, illustrated in Figure 12.2.

Example 12.5 Still on $\mathbf{K} = \mathbb{R}_+^2$, consider the global minimization of the polynomial $\mathbf{x} \mapsto x_1^2 + (1 - x_1 x_2)^2$ whose global minimum $f^* = 0$ is not attained. Again, choose for μ the probability measure $\mu(B) := \int_B e^{-x_1 - x_2} d\mathbf{x}$, $B \in \mathcal{B}(\mathbb{R}_+^2)$. Then the semidefinite relaxations (12.2) yield $\lambda_0 = 5$, $\lambda_1 = 1.9187$ and $\lambda_{15} = 0.4795$, showing again a significant and rapid decrease in first iterations with a long tail close to f^*, as illustrated in Figure 12.3; numerical problems occur after $d = 15$. However, this kind of problem, where the global minimum f^* is not attained, is notoriously difficult. Even the semidefinite relaxations defined in Vui and Son (2008) (which provide lower bounds on f^*), which are especially devised for such problems, encounter numerical difficulties; see Vui and Son (2008, Example 4.8).

Example 12.6 The following toy example illustrates the duality results of Section 12.1.1. The univariate polynomial $x \mapsto f(x) := 0.375 - 5x + 21x^2 - 32x^3 + 16x^4$ displayed in Figure 12.4 has two global minima at $x_1^* = 0.1939$ and $x_2^* = 0.8062$, with $f^* = -0.0156$. The sequence of upper bounds

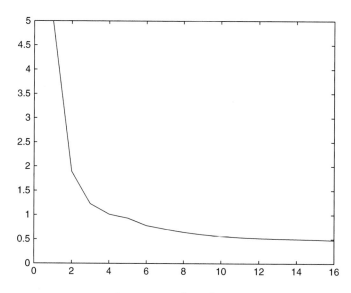

Figure 12.3 Minimizing $x_1^2 + (1 - x_1 x_2)^2$ on $\mathbb{R}_+^2$.

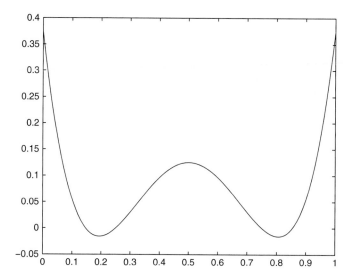

Figure 12.4 $f(x) = 0.375 - 5x + 21x^2 - 32x^3 + 16x^4$ on $[0, 1]$.

$\lambda_d \to f^*$ as $d \to \infty$ is plotted in Figure 12.5, again with a rapid decrease in first iterations. Figure 12.6 shows the plot of the SOS polynomial $x \mapsto \sigma_{10}(x)$, optimal solution of (12.5) with $d = 10$, associated with the probability density $\sigma_{10}(x)dx$ as explained in Section 12.1.1. As expected, two sharp peaks appear at the points $\tilde{x}_i \approx x_i^*, i = 1, 2$.

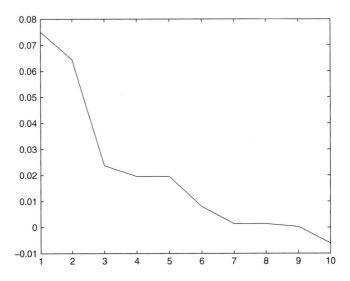

Figure 12.5 Minimizing $0.375 - 5x + 21x^2 - 32x^3 + 16x^4$ on $[0, 1]$.

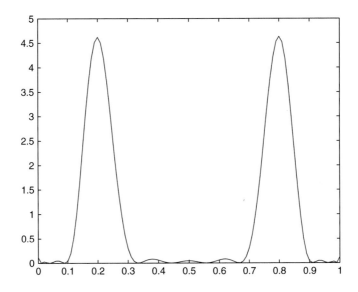

Figure 12.6 The probability density $\sigma_{10}(x)dx$ on $[0, 1]$.

Example 12.7   We finally consider a discrete optimization problem of the celebrated MAXCUT type,

$$f^* = \inf_{\mathbf{x}} \{ \mathbf{x}^T \mathbf{Q} \mathbf{x} : \mathbf{x} \in \{-1, 1\}^n \},$$

where $\mathbf{Q} = (Q_{ij}) \in \mathbb{R}^{n \times n}$ is a real symmetric matrix whose all diagonal elements vanish. The measure μ is uniformly distributed on $\{-1, 1\}^n$ so that its moments are readily available. We first consider the equal weights case, i.e., $Q_{ij} = 1/2$ for all (i, j) with $i \neq j$ in which case $f^* = -\lfloor n/2 \rfloor$. With $n = 11$ the successive values for λ_d, $d \leq 4$, are displayed in Table 12.1 and λ_4 is relatively close to f^*. We have also generated five random instances of MAXCUT with $n = 11$ but $Q_{ij} = 0$ with probability $1/2$; if $Q_{ij} \neq 0$ it is randomly generated using the MATLAB "rand" function. The successive values of λ_d, $d \leq 4$, are displayed in Table 12.2.

The above examples seem to indicate that even though one chooses a measure μ uniformly distributed on $\mathbf{K}$, one obtains a rapid decrease in the first iterations and then a slow convergence close to f^*. Again, if on the one hand the convergence to f^* is likely to be slow, on the other hand one has to solve a hierarchy of *generalized eigenvalue* problems associated with the pair of real symmetric matrices $(\mathbf{M}_d(f\,\mathbf{y}), \mathbf{M}_d(\mathbf{y}))$, for which specialized codes are available (instead of using a solver for semidefinite programs). However, one has to remember that the choice is limited to measures μ with $\operatorname{supp}\mu = \mathbf{K}$ with moments which are available or easy to compute. Hence, the present methodology is currently limited to simple sets $\mathbf{K}$ as described before. Finally, analyzing how the convergence to f^* depends on μ remains to be done.

Table 12.1 *MAXCUT: $n = 11$, $Q(i, j) = 1$ for all $i \neq j$*

d	0	1	2	3	4	f^*
λ_d	0	-1	-2.662	-3.22	-4	-5

Table 12.2 *MAXCUT: $n = 11$, Q random*

d	λ_0	λ_1	λ_2	λ_3	λ_4	f^*
Ex1	0	-1.928	-3.748	-5.22	-6.37	-7.946
Ex2	0	-1.56	-3.103	-4.314	-5.282	-6.863
Ex3	0	-1.910	-3.694	-5.078	-6.161	-8.032
Ex4	0	-2.164	-4.1664	-5.7971	-7.06	-9.198
Ex5	0	-1.825	-3.560	-4.945	-5.924	-7.467

12.2 The associated eigenvalue problem

We have seen that for each $d \in \mathbb{N}$, the semidefinite program (12.2) is very specific as it has only one variable and can be rewritten as

$$\lambda_d = \sup_{\lambda} \{ \lambda : \lambda \, \mathbf{M}_d(\mathbf{y}) \preceq \mathbf{M}_d(f \, \mathbf{y}) \}, \tag{12.7}$$

which in fact reduces to computing the smallest *generalized eigenvalue* associated with the pair of real symmetric matrices $\mathbf{M}_d(\mathbf{y})$ and $\mathbf{M}_d(f \, \mathbf{y})$. Hence powerful and more efficient specialized (eigenvalue) software packages can be used instead of a general purpose semidefinite solver.

In fact one may even simplify the problem and reduce (12.7) to a standard eigenvalue problem, i.e., computing the smallest eigenvalue of a *single* real symmetric matrix. It suffices to rewrite the moment (respectively localizing) matrix $\mathbf{M}_d(\mathbf{y})$ (respectively $\mathbf{M}_d(f \, \mathbf{y})$) in an appropriate basis so that $\mathbf{M}_d(\mathbf{y})$ becomes the identity matrix in the new basis.

Such a basis is precisely provided by the *orthonormal polynomials* up to degree d associated with the Riesz linear functional $L_{\mathbf{y}}$ (or, equivalently, associated with the measure μ). Fortunately such a basis is easy to compute once the moments $\mathbf{y}$ of μ are available.

12.2.1 Orthonormal polynomials

The *graded lexicographic order* (Glex) $\alpha \leq_{gl} \beta$ on $\mathbb{N}^n$, first creates a partial order by $|\alpha| \leq |\beta|$, and then refines this to a total order by breaking ties when $|\alpha| = |\beta|$ as would a dictionary with $x_1 = a$, $x_2 = b$, etc. For instance with $n = 2$, the ordering reads $1, x_1, x_2, x_1^2, x_1 x_2, x_2^2, \ldots$.

Given μ, let $\mathbf{y} = (y_\alpha)$, $\alpha \in \mathbb{N}_{2d}^n$, be the sequence of its moments and assume that $\mathbf{M}_d(\mathbf{y}) \succ 0$ for every d. Recall from Section 2.7.1 the scalar product $\langle \cdot, \cdot \rangle_{\mathbf{y}}$ on $\mathbb{R}[\mathbf{x}]_d$:

$$\langle f, g \rangle_{\mathbf{y}} := \langle \mathbf{f}, \mathbf{M}_d(\mathbf{y}) \, \mathbf{g} \rangle = \int f g \, d\mu, \qquad \forall f, g \in \mathbb{R}[\mathbf{x}]_d,$$

where $\mathbf{f}$ and $\mathbf{g}$ are the vectors of coefficients associated with the polynomials $f, g \in \mathbb{R}[\mathbf{x}]_d$, respectively. With $d \in \mathbb{N}$ fixed, one may also associate a *unique* family of polynomials $(P_\alpha) \subset \mathbb{R}[\mathbf{x}]_d$, $\alpha \in \mathbb{N}_d^n$, orthonormal with respect to μ, as follows:

$$\begin{cases} P_\alpha \in \text{lin.span} \{ \mathbf{x}^\beta : \beta \leq_{gl} \alpha \} \\ \langle P_\alpha, P_\beta \rangle_{\mathbf{y}} = \delta_{\alpha=\beta}, \quad \alpha, \beta \in \mathbb{N}_d^n \\ \langle P_\alpha, \mathbf{x}^\beta \rangle_{\mathbf{y}} = 0 \quad \text{if } \beta <_{gl} \alpha \\ \langle P_\alpha, \mathbf{x}^\alpha \rangle_{\mathbf{y}} > 0, \quad \alpha \in \mathbb{N}_d^n, \end{cases} \tag{12.8}$$

where $\delta_{\alpha=\beta}$ denotes the usual Kronecker symbol. Existence and uniqueness of such a family is guaranteed by the Gram–Schmidt orthonormalization process following the "Glex" order of monomials, and by positivity of the moment matrix $\mathbf{M}_d(\mathbf{y})$; see for example Dunkl and Xu (2001, Theorem 3.1.11, p. 68).

Computation   Computing the family of orthonormal polynomials is relatively easy once the moment matrix $\mathbf{M}_d(\mathbf{y})$ is available. Suppose that one wants to compute p_σ for some $\sigma \in \mathbb{N}^n_d$.

Build up the submatrix $\mathbf{M}^\sigma_d(\mathbf{y})$ obtained from $\mathbf{M}_d(\mathbf{y})$ by keeping only those columns indexed by $\beta \leq_{gl} \sigma$ and with rows indexed by $\alpha <_{gl} \sigma$. Hence $\mathbf{M}^\sigma_d(\mathbf{y})$ has one row less than columns. Complete $\mathbf{M}^\sigma_d(\mathbf{y})$ with an additional last row described by $\mathbf{M}^\sigma_d(\mathbf{y})[\sigma, \beta] = \mathbf{x}^\beta$, $\beta \leq_{gl} \sigma$. Then, up to a normalizing constant, $P_\sigma = \det \mathbf{M}^\sigma_d(\mathbf{y})$. For instance with $n = 2$, $d = 2$ and $\sigma = (11)$, one has

$$
\mathbf{x} \mapsto P_{(11)}(\mathbf{x}) = \rho \cdot \det
\begin{bmatrix}
y_{00} & y_{10} & y_{01} & y_{20} & y_{11} \\
y_{10} & y_{20} & y_{11} & y_{30} & y_{21} \\
y_{01} & y_{11} & y_{02} & y_{21} & y_{12} \\
y_{20} & y_{30} & y_{21} & y_{40} & y_{31} \\
1 & x_1 & x_2 & x_1^2 & x_1 x_2
\end{bmatrix}
\tag{12.9}
$$

where the constant ρ is chosen so that $\int P^2_{(11)}\,d\mu = 1$; see for example Dunkl and Xu (2001) and Helton et al. (2008).

Example 12.8   With $n = 2$ let us compute the constant and affine orthonormal polynomials (P_{00}, P_{10}, P_{01}) associated with a sequence $\mathbf{y} = (y_\alpha)$, $\alpha \in \mathbb{N}^n$, with $y_0 = 1$ and such that $\mathbf{M}_d(\mathbf{y}) \succ 0$ for all d:

$$
P_{00} = 1, \quad \mathbf{x} \mapsto P_{10}(\mathbf{x}) = a \det
\begin{bmatrix}
1 & y_{10} \\
1 & x_1
\end{bmatrix} = a\,(x_1 - y_{10}),
$$

with a chosen to make $L_\mathbf{y}(P^2_{10}) = 1$, i.e., $a = (y_{20} - y_{10}^2)^{-1/2}$,

$$
\mathbf{x} \mapsto P_{01}(\mathbf{x}) = b \det
\begin{bmatrix}
1 & y_{10} & y_{01} \\
y_{10} & y_{20} & y_{11} \\
1 & x_1 & x_2
\end{bmatrix} = b\,\big(x_1(y_{10}y_{01} - y_{11})
$$

$$
+ x_2(y_{20} - y_{10}^2) + y_{10}y_{11} - y_{01}y_{20}\big),
$$

where b is chosen to make $L_\mathbf{y}(P^2_{01}) = 1$.

12.2.2 The eigenvalue problem

So let us take the orthonormal polynomials (P_α), $\alpha \in \mathbb{N}^n_d$, as a basis of $\mathbb{R}[\mathbf{x}]_d$, and let $\overline{\mathbf{M}}_d(\mathbf{y})$ be the moment matrix in that basis, i.e., $\overline{\mathbf{M}}_d(\mathbf{y})$ is the real symmetric matrix with rows and columns index in $\alpha \in \mathbb{N}^n_d$, and with entries

$$\overline{\mathbf{M}}_d(\mathbf{y})[\boldsymbol{\alpha}, \boldsymbol{\beta}] = L_{\mathbf{y}}(P_{\boldsymbol{\alpha}} P_{\boldsymbol{\beta}}) = \int_{\mathbb{R}^n} P_{\boldsymbol{\alpha}} P_{\boldsymbol{\beta}} \, d\mu = \delta_{\boldsymbol{\alpha}=\boldsymbol{\beta}}, \qquad \boldsymbol{\alpha}, \boldsymbol{\beta} \in \mathbb{N}_d^n,$$

and so $\overline{\mathbf{M}}_d(\mathbf{y})$ is indeed the identity matrix $\mathbf{I}$. Similarly, one defines the localizing matrix $\overline{\mathbf{M}}_d(f\,\mathbf{y})$ with respect to f and $\mathbf{y}$, by

$$\overline{\mathbf{M}}_d(f\,\mathbf{y})[\boldsymbol{\alpha}, \boldsymbol{\beta}] = L_{\mathbf{y}}(f\, P_{\boldsymbol{\alpha}} P_{\boldsymbol{\beta}}) = \int_{\mathbb{R}^n} f\, P_{\boldsymbol{\alpha}} P_{\boldsymbol{\beta}} \, d\mu, \qquad \boldsymbol{\alpha}, \boldsymbol{\beta} \in \mathbb{N}_d^n.$$

$$(12.10)$$

And so the entry $(\boldsymbol{\alpha}, \boldsymbol{\beta})$ of $\overline{\mathbf{M}}_d(f\,\mathbf{y})$ can be computed by expanding the polynomial $f\, P_{\boldsymbol{\alpha}} P_{\boldsymbol{\beta}}$ in the usual basis of monomials $(\mathbf{x}^{\boldsymbol{\alpha}})$ (or in the basis $(P_{\boldsymbol{\alpha}})$), and applying the Riesz functional $L_{\mathbf{y}}$.

Therefore solving the semidefinite program (or generalized eigenvalue problem) (12.7) reduces to solving

$$\lambda_d = \sup_{\lambda} \{\, \lambda \,:\, \lambda\, \mathbf{I} \preceq \overline{\mathbf{M}}_d(f\,\mathbf{y}) \,\},$$

that is, λ_d is the smallest eigenvalue of the localizing matrix $\overline{\mathbf{M}}_d(f\,\mathbf{y})$.

Powerful packages can be used to compute the smallest eigenvalue of a real symmetric matrix.

12.3 On copositive programs

Recall from Chapter 4 that a real symmetric matrix $\mathbf{A} \in \mathcal{S}^n$ is copositive if and only if $\mathbf{x}^T \mathbf{A} \mathbf{x} \geq 0$ for all $\mathbf{x} \in \mathbb{R}_+^n$. Let $\mathcal{C}$ be the convex cone of $n \times n$ copositive matrices whose dual $\mathcal{C}^*$ is the convex cone of $n \times n$ completely positive matrices; see Chapter 4.

A copositive program is a conic optimization problem of the form:

$$\mathbf{P}: \qquad \rho = \inf_{\mathbf{X}} \{\, \langle \mathbf{C}, \mathbf{X} \rangle \,:\, \langle \mathbf{A}_i, \mathbf{X} \rangle = b_i, \quad i = 1, \ldots, m; \quad \mathbf{X} \in \mathcal{C} \,\},$$

$$(12.11)$$

and its dual $\mathbf{P}^*$ reads

$$\mathbf{P}^*: \qquad \rho^* = \sup_{\lambda} \left\{\, \mathbf{b}^T \lambda \,:\, \mathbf{C} - \sum_{i=1}^{m} \lambda_i\, \mathbf{A}_i \in \mathcal{C}^* \,\right\}. \qquad (12.12)$$

Even though **P** is a (conic) convex optimization problem, **P** (as well as **P***) is in general NP-hard. Indeed, consider the general quadratic binary problem

$$\begin{aligned}
\inf_{\mathbf{x}} \quad & \mathbf{x}^T \mathbf{Q} \mathbf{x} + 2 \mathbf{c}^T \mathbf{x} \\
\text{s.t.} \quad & \mathbf{a}_i^T \mathbf{x} = b_i, \quad i = 1, \ldots, m \\
& \mathbf{x} \geq 0, \quad x_j \in \{0, 1\}, \quad j \in J,
\end{aligned} \tag{12.13}$$

for some vector $\mathbf{c} \in \mathbb{R}^n$, some real $n \times n$ symmetric matrix $\mathbf{Q}$ and some subset $J \subseteq \{1, \ldots, m\}$. It turns out that (12.13) can be reformulated in the equivalent completely positive program:

$$\begin{aligned}
\inf_{\mathbf{x}} \quad & \langle \mathbf{Q}, \mathbf{X} \rangle + 2 \mathbf{c}^T \mathbf{x} \\
\text{s.t.} \quad & \mathbf{a}_i^T \mathbf{x} = b_i, \ i = 1, \ldots, m \\
& \langle \mathbf{a}_i \mathbf{a}_i^T, \mathbf{X} \rangle = b_i^2, \ i = 1, \ldots, m \\
& x_j = X_{jj}, \quad j \in J, \quad \begin{bmatrix} 1 & \mathbf{x}^T \\ \mathbf{x} & \mathbf{X} \end{bmatrix} \in \mathcal{C}^*,
\end{aligned} \tag{12.14}$$

provided that problem (12.13) satisfies the condition:

$$\left[\mathbf{x} \geq 0 \text{ and } \mathbf{a}_i^T \mathbf{x} = b_i, \quad i = 1, \ldots, m \right] \quad \Rightarrow \quad x_j \leq 1, \quad \forall j \in J.$$

And in fact, the latter condition can be enforced with no loss of generality.

A hierarchy of upper and lower bounds Therefore a typical approach is to approximate ρ (respectively ρ^*) by solving a simpler problem where $\mathcal{C}$ is replaced with an appropriate convex cone which is easier to handle. Precisely, in Chapter 4 we have described tractable outer and inner (respectively inner and outer) approximations $(\mathcal{C}_d)$ and $(\mathcal{K}_d^2)$ (respectively $(\mathcal{C}_d^*)$ and $((\mathcal{K}_d^2)^*)$), $d \in \mathbb{N}$, of $\mathcal{C}$ (respectively $\mathcal{C}^*$). So for instance, consider the hierarchy of convex optimization problems indexed by $d \in \mathbb{N}$:

$$\rho_d = \inf_{\mathbf{X}} \left\{ \langle \mathbf{C}, \mathbf{X} \rangle \ : \ \langle \mathbf{A}_i, \mathbf{X} \rangle = b_i, \quad i = 1, \ldots, m; \ \mathbf{X} \in \mathcal{K}_d^2 \right\},$$

$$\tag{12.15}$$

where $\mathcal{K}_d^2$ is the convex cone defined in (4.19) (with $\mathcal{K}_0 := \mathcal{N} + \mathcal{S}_+$) and where $\mathcal{N}$ (respectively $\mathcal{S}_+$) denotes the cone of real symmetric matrices with all entries nonnegative (respectively the cone of real symmetric positive semidefinite matrices).

As $\mathcal{K}_d^2 \subset \mathcal{C}$ for all $d \in \mathbb{N}$, the sequence (ρ_d), $d \in \mathbb{N}$, is monotone nonincreasing and provides a hierarchy of upper bounds on ρ. Moreover, in view of the definition (4.19) of $\mathcal{K}_d^2$, for each $d \in \mathbb{N}$, (12.15) is a semidefinite program. Observe that in (12.15) one may also use the convex cone $\mathcal{K}_d^1$ instead of $\mathcal{K}_d^2$, in which case one obtains a hierarchy of linear programs instead of semidefinite programs. The dual of (12.15) is the conic program:

$$\rho_d^* = \sup_{\lambda} \left\{ \mathbf{b}^T \lambda \ : \ \mathbf{C} - \sum_{i=1}^m \lambda_i \, \mathbf{A}_i \ \in \ (\mathcal{K}_d^2)^* \right\}. \qquad (12.16)$$

On the other hand, let $\mathbf{y} = (y_\alpha)$, $\alpha \in \mathbb{N}^n$, be as in (4.15), i.e., $\mathbf{y}$ is the moment sequence of the exponential probability measure on $\mathbb{R}_+^n$ (or, alternatively, on the simplex $\Delta = \{ \mathbf{x} \in \mathbb{R}_+^n \ : \ \sum_i x_i \leq 1 \}$). Associated with $\mathbf{S} \in \mathcal{S}^n$, let $f_{\mathbf{S}} \in \mathbb{R}[\mathbf{x}]_2$ be the quadratic form $\mathbf{x} \mapsto f_{\mathbf{S}}(\mathbf{x}) := \mathbf{x}^T \mathbf{S} \mathbf{x}$, and consider the hierarchy of semidefinite programs indexed by $d \in \mathbb{N}$:

$$\psi_d = \inf_{\mathbf{X}} \{ \langle \mathbf{C}, \mathbf{X} \rangle \ : \ \langle \mathbf{A}_i, \mathbf{X} \rangle = b_i, \ i = 1, \ldots, m; \ \mathbf{M}_d(f_{\mathbf{X}} \, \mathbf{y}) \succeq 0 \},$$
$$(12.17)$$

where $\mathbf{M}_d(f_{\mathbf{X}} \, \mathbf{y})$ is the localizing matrix associated with the quadratic form $f_{\mathbf{X}}$ and the sequence $\mathbf{y}$. The constraint $\mathbf{M}_d(f_{\mathbf{X}} \, \mathbf{y}) \succeq 0$ can be stated in the equivalent form $\mathbf{X} \in \mathcal{C}_d$, where $\mathcal{C}_d \supset \mathcal{C}$ is the convex cone defined in (12.6). Therefore, the dual of (12.17) is the conic program:

$$\psi_d^* = \sup_{\lambda} \left\{ \mathbf{b}^T \lambda \ : \ \mathbf{C} - \sum_{i=1}^m \lambda_i \, \mathbf{A}_i \ \in \ \mathcal{C}_d^* \right\}, \qquad (12.18)$$

where $\mathcal{C}_d^*$ is the dual cone of $\mathcal{C}_d$ described in Theorem 4.11. As $\mathbf{M}_d(f_{\mathbf{X}} \, \mathbf{y}) \succeq 0$ is a relaxation of the constraint $\mathbf{X} \in \mathcal{C}$, then necessarily $\psi_d \leq \rho$ for every $d \in \mathbb{N}$, and the sequence (ψ_d), $d \in \mathbb{N}$, which is monotone nondecreasing, provides tighter and tighter lower bounds on ρ.

Theorem 12.9 *Suppose that the optimal value ρ of the copositive program (12.11) is finite, the $\mathbf{A}_i$ are linearly independent, and there exists $\mathbf{X}_0 \in \mathcal{S}^n$ and $\lambda_0 \in \mathbb{R}^m$ such that:*

- $\langle \mathbf{A}_i, \mathbf{X}_0 \rangle = b_i$, $i = 1, \ldots, m$, and $\mathbf{X}_0 \in \text{int} \, (\mathcal{K}_{d_0}^2)$ for some d_0.
- $\mathbf{C} - \sum_{i=1}^{m} (\lambda_0)_i \, \mathbf{A}_i \in \text{int} \, (\mathcal{C}_{d_1}^*)$, for some d_1.

Then both $\mathbf{P}$ and $\mathbf{P}^*$ have an optimal solution. Moreover, consider the hierarchies of semidefinite programs (12.15) and (12.17) with associated respective optimal value ρ_d and ψ_d, $d \in \mathbb{N}$.

(a) *The sequence (ρ_d) (respectively (ψ_d)), $d \in \mathbb{N}$, is monotone nonincreasing (respectively nondecreasing) and converges to the optimal value ρ of the copositive program $\mathbf{P}$, as $d \to \infty$.*

(b) *In addition, for all $d \geq d_1$, (12.17) has an optimal solution $\mathbf{X}_d^* \in \mathcal{C}_d$ and every accumulation point of the sequence $(\mathbf{X}_d^*)$, $d \geq d_1$, is an optimal solution of (12.11). Similarly, for every $d \geq d_0$, (12.16) has an optimal solution $\boldsymbol{\lambda}_d^* \in \mathbb{R}^m$ and every accumulation point of the sequence $(\boldsymbol{\lambda}_d^*)$, $d \geq d_0$, is an optimal solution of (12.12).*

Proof   Observe that Slater's condition holds for both conic programs (12.11) and (12.12) because by Proposition 4.9 and Corollary 4.12, $\mathcal{C} = \overline{\bigcup_{d=0}^{\infty} \mathcal{K}_d^2}$ and $\mathcal{C}^* = \overline{\bigcup_{d=0}^{\infty} \mathcal{C}_d^*}$. Moreover, ρ is finite and so by standard conic duality, $\mathbf{P}$ and $\mathbf{P}^*$ have an optimal solution.

(a) and (b). By the assumption on λ_0, Slater's condition holds for (12.18) for all $d_1 \leq d \in \mathbb{N}$, and so $\psi_d^* = \psi_d$ for all $d \geq d_1$. Moreover, as $\psi_d^* \leq \rho$ for all $d \in \mathbb{N}$ and ρ is finite, (12.17) has an optimal solution $\mathbf{X}_d^* \in \mathcal{C}_d$, for every $d \geq d_1$. Next, we have

$$\left\langle \mathbf{C} - \sum_{i=1}^{m} (\lambda_0)_i \, \mathbf{A}_i \, , \, \mathbf{X}_d^* \right\rangle = \langle \mathbf{C}, \mathbf{X}_d^* \rangle - \sum_{i=1}^{m} (\lambda_0)_i b_i \leq \rho - \sum_{i=1}^{m} (\lambda_0)_i b_i.$$

On the other hand, since $\mathbf{C} - \sum_{i=1}^{m} (\lambda_0)_i \, \mathbf{A}_i \in \text{int} \, (\mathcal{C}_d^*)$ for every $d \geq d_1$, by Lemma[3] I.1.5 in Faraut and Korányi (1994), there exists $\delta > 0$ such that for all $\mathbf{x} \in \mathcal{C}_{d_1}$ (and hence for all $d \geq d_1$ and all $\mathbf{X} \in \mathcal{C}_d$),

$$\left\langle \mathbf{C} - \sum_{i=1}^{m} (\lambda_0)_i \, \mathbf{A}_i \, , \, \mathbf{X} \right\rangle \geq \delta \, \|\mathbf{X}\|.$$

Therefore $\sup_d \|\mathbf{X}_d^*\| \leq \delta^{-1} (\rho - \sum_{i=1}^{m} (\lambda_0)_i b_i)$. Hence there is a subsequence (d_k), $k \in \mathbb{N}$, and a real symmetric matrix $\mathbf{X}^*$ such that $\mathbf{X}_{d_k}^* \to \mathbf{X}^*$ as $k \to \infty$. But then one has

$$b_i = \lim_{k \to \infty} \langle \mathbf{A}_i, \mathbf{X}_{d_k}^* \rangle = \langle \mathbf{A}_i, \mathbf{X}^* \rangle, \qquad \forall i = 1, \ldots, m,$$

---

[3] This lemma states that if C is a convex cone of a finite-dimensional Euclidean space and $U \subset \text{int} \, (C^*)$ is compact, then there exists $\rho > 0$ such that for all $(\mathbf{x}, \mathbf{y}) \in C \times U$, $\langle \mathbf{x}, \mathbf{y} \rangle \geq \rho \|\mathbf{x}\|$.

and $\rho \geq \lim_{k\to\infty} \psi_{d_k} = \lim_{k\to\infty} \langle \mathbf{C}, \mathbf{X}^*_{d_k} \rangle = \langle \mathbf{C}, \mathbf{X}^* \rangle$. Finally, $\mathbf{X}^* \in \mathcal{C}$ since $\mathcal{C} = \bigcap_{d=0}^{\infty} \mathcal{C}_d$ and $\mathbf{X}^*_{d_k} \in \mathcal{C}_{d_k}$ for all k. But this proves that $\mathbf{X}^*$ is an optimal solution of (12.11), and by monotonicity of the sequence (ψ_d), $d \in \mathbb{N}$, $\lim_{d\to\infty} \psi_d = \rho$. As the converging subsequence was arbitrary, every accumulation point $\mathbf{X}^* \in \mathcal{C}$ is an optimal solution of (12.11).

For the sequence (ρ_d), $d \in \mathbb{N}$ we use similar (but dual) arguments. By the assumption on $\mathbf{X}_0$, Slater's condition holds for (12.15) for every $d \geq d_0$, and so $\rho_d = \rho^*_d$ for all $d \geq d_0$. Moreover ρ_d is finite since $\rho_d \geq \rho$. Therefore, for every $d \geq d_0$, the dual (12.16) has an optimal solution $\boldsymbol{\lambda}^*_d$, and in addition, $\boldsymbol{\lambda}^*_d \in \Lambda$ for some compact set $\Lambda \subset \mathbb{R}^m$. Therefore there is a subsequence (d_k), $k \in \mathbb{N}$, and $\boldsymbol{\lambda}^* \in \Lambda$ such that $\boldsymbol{\lambda}^*_{d_k} \to \boldsymbol{\lambda}^*$ as $k\to\infty$. This implies

$$
\mathbf{C} - \sum_{i=1}^{m} \lambda^*_i \mathbf{A}_i = \lim_{k\to\infty} \underbrace{\left(\mathbf{C} - \sum_{i=1}^{m} (\lambda^*_{d_k})_i \mathbf{A}_i \right)}_{\in (\mathcal{K}^2_{d_k})^*} \in \bigcap_{d=0}^{\infty} (\mathcal{K}^2_d)^* = \mathcal{C}^*,
$$

and $\sum_{i=1}^{m} \lambda^*_i b_i = \lim_{k\to\infty} \rho^*_{d_k} = \lim_{k\to\infty} \rho_{d_k} \geq \rho (= \rho^*)$. But this proves that $\boldsymbol{\lambda}^*$ is an optimal solution of (12.12) and so (by monotonicity of the sequence (ρ_d)), $\rho_d \to \rho$ as $d\to\infty$. Finally, as the converging subsequence was arbitrary, every accumulation point $\boldsymbol{\lambda}^* \in \mathbb{R}^m$ is an optimal solution of (12.12).   □

12.4 Exercises

Exercise 12.1   Let $n = 1$, $\mathbf{K} = [0, 1]$ and let μ be the Lebesgue measure on $\mathbf{K}$ with moments $\mathbf{y} = (y_k)$, $k \in \mathbb{N}$. Let $x \mapsto f(x) = x(x - 1)$ so that $f^* = \inf\{ f(x) : x \in \mathbf{K} \} = -1/4$ and $x^* = 1/2 \in \mathbf{K}$ is the global minimizer of f.

(a) Compute a basis of orthonormal polynomials (P_0, P_1, P_2) with respect to μ, of degree up to 2.

(b) Compute the smallest eigenvalue $\lambda_{\min}$ of the localizing matrix $\overline{\mathbf{M}}_2(f, \mathbf{y})$ in (12.10) and its associated eigenvector $\mathbf{u}$. Let $x \mapsto \sigma(x) := (\mathbf{w}_2(x)^T \mathbf{u})^2$ (where $\mathbf{w}_2(x)^T = [P_0(x), P_1(x), P_2(x)]$) and let $m := \int_0^1 x\sigma(x)dx$. Show that $f^* \leq f(m) \leq \lambda_{\min}$; why?

(c) Plot the polynomial $x \mapsto \sigma(x)$ on the box $[0, 1]$; what do you observe? At what point does σ attain its supremum on $[0, 1]$?

(d) Compute the largest eigenvalue $\lambda_{\max}$ of the localizing matrix $\overline{\mathbf{M}}_2(f, \mathbf{y})$ and its associated eigenvector $\mathbf{u}$. Let $x \mapsto \sigma(x) := (\mathbf{w}_2(x)^T \mathbf{u})^2$ and let $s := \int_0^1 x\sigma(x)dx$. Show that $\rho := \sup\{ f(x) : x \in \mathbf{K} \} \geq \lambda_{\max}$. Do we have $\lambda_{\max} \leq f(s) \leq \rho$? why? Where do f and σ attain their supremum on $[0, 1]$?

(e) Repeat steps (a)–(d) now with the localizing matrix $\overline{\mathbf{M}}_4(f\ \mathbf{y})$ in the basis $(P_0, \ldots, P_4)$ of orthonormal polynomials with respect to μ, up to degree 4.

12.5 Notes and sources

12.1 Most of the material is taken from Lasserre (2011). In nonlinear programming, sequences of upper bounds on the global minimum f^* are usually obtained from feasible points $\mathbf{x} \in \mathbf{K}$, for example, via some (local) minimization algorithm. But for nonconvex problems, providing a sequence of upper bounds that converges to the global minimum f^* is in general impossible except for some compact sets where one computes points on a grid whose mesh size tends to zero. In the methodology developed in this chapter, one provides a monotone nonincreasing sequence of upper bounds converging to f^* for polynomial optimization problems on sets $\mathbf{K}$, not necessarily compact but such that one may compute all moments of some finite Borel measure μ with supp $\mu = \mathbf{K}$. In fact, if there are only finitely many (say up to order $2d$) moments available then one obtains a finite sequence of upper bounds.

In contrast to the hierarchy of semidefinite relaxations defined in Chapter 6 which provide lower bounds converging to f^* when $\mathbf{K}$ is a compact basic semialgebraic set, the convergence of the upper bounds to f^* is only asymptotic and never finite, except when $\mathbf{K}$ is a discrete set. However, and even if we expect the convergence to be rather slow when close to f^*, in a few iterations one may obtain upper bounds which (even if crude) complements the lower bounds obtained in Chapter 6 (in the compact case).

12.2. The reader interested in more details on orthogonal polynomials is referred to the excellent book of Dunkl and Xu (2001) and the many references therein. Interestingly, for the optimization problem considered in Chapter 6 one also computes global minimizers via solving a series of eigenvalue problems as described in Section 6.1.2. Indeed in Algorithm 6.9, one computes eigenvalues of "multiplication-by-x_i" matrices, $i = 1, \ldots, n$; the latter are nothing less than localizing matrices with respect to the sequence $\mathbf{y}$ and the polynomial $\mathbf{x} \mapsto x_i$.

12.3. The formulation of quadratic binary programs as the equivalent completely positive program (12.14) is due to Burer (2009), while the hierarchy of semidefinite programs (12.15) to approximate copositive programs is due to Parrilo (2000). If in (12.15) one uses the convex cones $\mathcal{K}_d^1$ instead of $\mathcal{K}_d^2$ one obtains the hierarchy of LP-approximations of copositive programs due to de Klerk and Pasechnik (2002). For more details and results on copositive programming, the interested reader is referred to the nice surveys of Bomze (2012), Burer (2012), and Dürr (2010).

PART III

Specializations and extensions

13

Convexity in polynomial optimization

The moment-SOS approach described in Chapter 6 as well as its extensions in Chapter 11 aim at solving difficult nonconvex (and in general NP-hard) optimization problems. On the other hand, a large class of convex optimization problems can be solved efficiently and sometimes in time polynomial in the input size of the problem. Therefore the moment-SOS approach should have the highly desirable feature of somehow *recognizing* "easy" problems, for example convex problems. That is, when applied to such problems it should show some significant improvement or a particular nice behavior not necessarily valid in the general case. Otherwise

would one trust an optimization method aimed at solving difficult problems which behaves poorly on relatively "easy" problems?

Probably not! The impact of convexity on the moment-SOS approach is one of the issues investigated in this chapter. We will show that indeed the moment-SOS approach recognizes a large class of convex problems. In particular, the first semidefinite relaxation of the hierarchy (6.3) is exact for SOS-convex problems (among which are linear and convex quadratic programs), i.e., finite convergence takes place at the first step in the hierarchy. But we also provide algebraic characterization of convexity for basic semi-algebraic sets and polynomials as well as some of their specific properties.

13.1 Convexity and polynomials

We first introduce some material which shows that in the presence of convexity some of the results presented in Chapter 2 have a nice specialization.

13.1.1 Algebraic certificates of convexity for sets

We first consider the problem of detecting whether some given basic semi-algebraic set **K** as in (5.2) is convex. By detecting, we mean that if **K** is convex then one can obtain a *certificate* (or a proof) of convexity by some algorithm. The geometric characterization of convexity

$$\lambda\, \mathbf{x} + (1 - \lambda)\, \mathbf{y} \in \mathbf{K}, \qquad \forall\, \mathbf{x}, \mathbf{y} \in \mathbf{K}, \ \lambda \in (0, 1) \tag{13.1}$$

is not a certificate because it cannot be checked by an algorithm. In fact, detection of convexity is a difficult problem.

Theorem 13.1 *Let* $\mathbf{K} \subset \mathbb{R}^n$ *be the basic closed semi-algebraic set defined by*

$$\mathbf{K} = \{\, \mathbf{x} \in \mathbb{R}^n : g_j(\mathbf{x}) \geq 0, \quad j = 1, \ldots, m \,\}, \tag{13.2}$$

for some polynomials $(g_j) \subset \mathbb{R}[\mathbf{x}]$. *Then deciding whether* **K** *is convex is NP-hard whenever at least one of the polynomials* g_j *has degree four.*

Even though deciding convexity is NP-hard we next describe how to obtain algebraic certificates of convexity. Given the basic closed semi-algebraic set **K** defined in (13.2), let $\widehat{\mathbf{K}} := \mathbf{K} \times \mathbf{K}$, i.e.,

$$\widehat{\mathbf{K}} = \{\, (\mathbf{x}, \mathbf{y}) : \hat{g}_j(\mathbf{x}, \mathbf{y}) \geq 0, \quad j = 1, \ldots, 2m \,\} \tag{13.3}$$

(a basic closed semi-algebraic set of $\mathbb{R}^{2n}$), where:

$$(\mathbf{x}, \mathbf{y}) \mapsto \hat{g}_j(\mathbf{x}, \mathbf{y}) := g_j(\mathbf{x}), \qquad \text{for } j = 1, \ldots, m$$
$$(\mathbf{x}, \mathbf{y}) \mapsto \hat{g}_j(\mathbf{x}, \mathbf{y}) := g_{j-m}(\mathbf{y}), \quad \text{for } j = m + 1, \ldots, 2m.$$

Let $P(\hat{g}) \subset \mathbb{R}[\mathbf{x}, \mathbf{y}]$ be the preordering associated with the polynomials $\hat{g}_j$ that define $\widehat{\mathbf{K}}$ in (13.3), i.e.,

$$P(\hat{g}) := \left\{ \sum_{J \subseteq \{1, \ldots, 2m\}} \left(\prod_{k \in J} \hat{g}_k \right) \sigma_J : \quad \sigma_J \in \Sigma[\mathbf{x}, \mathbf{y}] \right\}. \tag{13.4}$$

An algebraic certificate of convexity can be obtained by using Stengle's Positivstellensätze of Chapter 2.

Theorem 13.2 *Let* $\mathbf{K} \subset \mathbb{R}^n$ *be the basic closed semi-algebraic set defined in (13.2). Then* **K** *is convex if and only if for every* $j = 1, \ldots, m$ *and all* $(\mathbf{x}, \mathbf{y}) \in \mathbb{R}^n \times \mathbb{R}^n$:

$$\theta_j(\mathbf{x}, \mathbf{y})\, g_j\!\left(\frac{\mathbf{x} + \mathbf{y}}{2} \right) = g_j\!\left(\frac{\mathbf{x} + \mathbf{y}}{2} \right)^{2p_j} + h_j(\mathbf{x}, \mathbf{y}) \tag{13.5}$$

for some polynomials $\theta_j, h_j \in P(\hat{g})$ *and some integer* $p_j \in \mathbb{N}$.

Proof The set **K** is convex if and only if (13.1) holds, or equivalently, if and only if $(\mathbf{x} + \mathbf{y})/2 \in \mathbf{K}$ whenever $\mathbf{x}, \mathbf{y} \in \mathbf{K}$. That is, if and only if for every $j = 1, \ldots, m$,

$$g_j \left(\frac{\mathbf{x} + \mathbf{y}}{2} \right) \geq 0, \qquad \forall (\mathbf{x}, \mathbf{y}) \in \widehat{\mathbf{K}}. \qquad (13.6)$$

But then (13.5) is just an application of Stengle's Nichtnegativstellensatz, i.e., Theorem 2.12(a), applied to (13.6). □

The polynomials $\theta_j, h_j \in P(\hat{g})$, $j = 1, \ldots, m$, obtained in (13.5) indeed provide an obvious algebraic certificate of convexity for **K**. This is because if (13.5) holds then for every $\mathbf{x}, \mathbf{y} \in \mathbf{K}$ one has $\theta_j(\mathbf{x}, \mathbf{y}) \geq 0$ and $h_j(\mathbf{x}, \mathbf{y}) \geq 0$ because $\theta_j, h_j \in P(\hat{g})$; and so $\theta_j(\mathbf{x}, \mathbf{y})g_j((\mathbf{x} + \mathbf{y})/2) \geq 0$. Therefore if $\theta_j(\mathbf{x}, \mathbf{y}) > 0$ then $g_j((\mathbf{x} + \mathbf{y})/2) \geq 0$ whereas if $\theta_j(\mathbf{x}, \mathbf{y}) = 0$ then $g_j((\mathbf{x} + \mathbf{y})/2)^{2p_j} = 0$ which in turn implies $g_j((\mathbf{x} + \mathbf{y})/2) = 0$. Hence for every $j = 1, \ldots, m$, $g_j((\mathbf{x} + \mathbf{y})/2) \geq 0$ for every $\mathbf{x}, \mathbf{y} \in \mathbf{K}$, which implies that **K** is convex.

In principle, the algebraic certificate can be obtained numerically. Indeed, there is a bound on $p_j \in \mathbb{N}$, and the degrees of the SOS weights σ_J in the representation (13.4) of the polynomial certificates $\theta_j, h_j \in P(\hat{g})$ in (13.5); see the discussion just after Theorem 2.12. Therefore, checking whether (13.5) holds reduces to checking whether, for each j, a single semidefinite program has a feasible solution. However, the bound is so large that for pratical implementation one should proceed as follows. Fix an a priori bound M on $p_j \in \mathbb{N}$ and on the degrees of the SOS polynomial weights σ_J that define $h_j, \theta_j \in P(\hat{g})$; then check whether (13.5) holds true by solving the associated semidefinite program. If **K** is convex and the degrees of the certificates are small then by increasing M, eventually one finds a feasible solution. However, in practice such a certificate can be obtained only up to some machine precision, because of numerical inaccuracies inherent to semidefinite programming solvers.

We next provide another algebraic certificate of convexity for the class of basic closed semi-algebraic sets **K** whose defining polynomials (g_j) satisfy the following nondegeneracy assumption on the boundary $\partial\mathbf{K}$.

Assumption 13.3 (Nondegeneracy) *Let $\mathbf{K} \subset \mathbb{R}^n$ be as in (13.2). For every $j = 1, \ldots, m$, $\nabla g_j(\mathbf{x}) \neq 0$ whenever $\mathbf{x} \in \mathbf{K}$ and $g_j(\mathbf{x}) = 0$.*

We first have the following characterization of convexity.

Lemma 13.4 *With **K** as in (13.2), let Slater's condition[1] and Assumption 13.3 hold. Then **K** is convex if and only if for every $j = 1, \ldots, m$,*

---

[1] Recall that Slater's condition holds if there exists $\mathbf{x}_0 \in \mathbf{K}$ such that $g_j(\mathbf{x}_0) > 0$ for all $j = 1, \ldots, m$ (if g_j is affine then $g_j(\mathbf{x}_0) \geq 0$ is tolerated).

$$\langle \nabla g_j(\mathbf{y}), \mathbf{x} - \mathbf{y} \rangle \geq 0, \qquad \forall \mathbf{x}, \mathbf{y} \in \mathbf{K} \ \textit{with} \ g_j(\mathbf{y}) = 0. \qquad (13.7)$$

Proof The *only if* part is obvious. Indeed if $\langle \nabla g_j(\mathbf{y}), \mathbf{x} - \mathbf{y} \rangle < 0$ for some $\mathbf{x} \in \mathbf{K}$ and $\mathbf{y} \in \mathbf{K}$ with $g_j(\mathbf{y}) = 0$, then there is some $\bar{t} > 0$ such that $g_j(\mathbf{y} + t(\mathbf{x} - \mathbf{y})) < 0$ for all $t \in (0, \bar{t})$ and so the point $\mathbf{x}' := t\mathbf{x} + (1 - t)\mathbf{y}$ does not belong to $\mathbf{K}$, which in turn implies that $\mathbf{K}$ is not convex.

For the *if* part, (13.7) implies that at every point of the boundary, there exists a supporting hyperplane for $\mathbf{K}$. As $\mathbf{K}$ is closed with nonempty interior, the result follows from Schneider (1994, Theorem 1.3.3). □

As a consequence we obtain the following algebraic certificate of convexity.

Corollary 13.5 (Algebraic certificate of convexity)   *With $\mathbf{K}$ as in (13.2), let Slater's condition and Assumption 13.3 hold. Then $\mathbf{K}$ is convex if and only if for every $j = 1, \ldots, m$,*

$$h_j(\mathbf{x}, \mathbf{y}) \langle \nabla g_j(\mathbf{y}), \mathbf{x} - \mathbf{y} \rangle = \langle \nabla g_j(\mathbf{y}), \mathbf{x} - \mathbf{y} \rangle^{2\ell} + \theta_j(\mathbf{x}, \mathbf{y}) + \varphi_j(\mathbf{x}, \mathbf{y}) g_j(\mathbf{y})$$
$$(13.8)$$

for some integer $\ell \in \mathbb{N}$, some polynomial $\varphi_j \in \mathbb{R}[\mathbf{x}, \mathbf{y}]$ and some polynomials h_j, θ_j in the preordering of $\mathbb{R}[\mathbf{x}, \mathbf{y}]$ generated by the family of polynomials $(g_k(\mathbf{x}), g_p(\mathbf{y})), k, p \in \{1, \ldots, m\}, p \neq j$.

Proof By Lemma 13.4, $\mathbf{K}$ is convex if and only if for every $j = 1, \ldots, m$, the polynomial $(\mathbf{x}, \mathbf{y}) \mapsto \langle \nabla g_j(\mathbf{y}), \mathbf{x} - \mathbf{y} \rangle$ is nonnegative on the set Ω_j defined by:

$$\Omega_j := \{ (\mathbf{x}, \mathbf{y}) \in \mathbf{K} \times \mathbf{K} : g_j(\mathbf{y}) = 0 \}. \qquad (13.9)$$

Then (13.8) follows from Theorem 2.12(a). □

So if $\mathbf{K}$ is convex then $(\ell, h_j, \theta_j, \varphi_j)$ provides us with the desired certificate of convexity which, in principle, can be obtained numerically as for the algebraic certificate (13.5). However, again in practice such a certificate can be obtained only up to some machine precision, because of numerical inaccuracies inherent to semidefinite programming solvers. See the discussion after Theorem 13.2.

Observe that in Corollary 13.5, $\mathbf{K}$ is not necessarily compact. There is also a sufficient condition that allows the conclusion that $\mathbf{K}$ is convex, using Putinar's representation of strictly positive polynomials. It is particularly interesting for compact sets $\mathbf{K}$ because it is almost necessary. Recall that $g_0 = 1$.

> **Corollary 13.6** (Certificate of convexity)   *With* $\mathbf{K} \subset \mathbb{R}^n$ *as in (13.2) let Slater's condition and Assumption 13.3 hold. Then* $\mathbf{K}$ *is convex if for every* $j = 1, \ldots, m$,
>
> $$\langle \nabla g_j(\mathbf{y}), \mathbf{x} - \mathbf{y} \rangle = \sum_{k=0}^{m} \sigma_{jk}\, g_k(\mathbf{x}) + \sum_{k=0, k \neq j}^{m} \psi_{jk}\, g_k(\mathbf{y}) + \psi_j\, g_j(\mathbf{y}),$$
>
> $$(13.10)$$
>
> *for some SOS* (σ_{jk}), $(\psi_{jk})_{k \neq j} \subset \Sigma[\mathbf{x}, \mathbf{y}]$, *and some* $\psi_j \in \mathbb{R}[\mathbf{x}, \mathbf{y}]$.

Proof   Obviously (13.10) implies (13.7) which in turn by Lemma 13.4 implies that $\mathbf{K}$ is convex.   □

By fixing an a priori bound $2d_j$ on the polynomials $(\sigma_{jk}g_k, \psi_{jk}g_k, \psi_j g_j)$, checking whether (13.10) holds reduces to solving a semidefinite program which is simpler than that for checking whether (13.8) holds. For instance, and with the notation of Chapter 6, for every $j = 1, \ldots, m$, it suffices to solve the semidefinite program:

$$\rho_j^* = \sup_{\lambda} \Big\{ \lambda : \quad \langle \nabla g_j(\mathbf{y}), \mathbf{x} - \mathbf{y} \rangle - \lambda = \sum_{k=0}^{m} \sigma_{jk}\, g_k(\mathbf{x}) + \sum_{k=0, k \neq j}^{m} \psi_{jk}\, g_k(\mathbf{y})$$
$$+ \psi_j\, g_j(\mathbf{y}); \quad \deg \sigma_{jk}\, g_k,\ \deg g_k\, \psi_{jk},\ \deg \psi_j \leq 2d_j \Big\},$$

which is the dual of the semidefinite program:

$$\rho_j := \inf_{\mathbf{z}} \ L_{\mathbf{z}}(\langle \nabla g_j(\mathbf{y}), \mathbf{x} - \mathbf{y} \rangle)$$
$$\text{s.t. } \mathbf{M}_{d_j}(\mathbf{z}) \succeq 0$$
$$\mathbf{M}_{d_j - r_k}(g_k(\mathbf{x})\, \mathbf{z}) \succeq 0, \quad k = 1, \ldots, m$$
$$\mathbf{M}_{d_j - r_k}(g_k(\mathbf{y})\, \mathbf{z}) \succeq 0, \quad k = 1, \ldots, m,\ k \neq j \qquad (13.11)$$
$$\mathbf{M}_{d_j - r_j}(g_j(\mathbf{y})\, \mathbf{z}) = 0$$
$$y_0 = 1,$$

where $r_k = \lceil (\deg g_k)/2 \rceil$, $k = 1, \ldots, m$, and $\mathbf{z} = (z_{\alpha\beta})$, $(\alpha, \beta) \in \mathbb{N}_{2d_j}^{2n}$. Then by weak duality $\rho_j^* \leq \rho_j$, and $\rho_j \leq 0$.

If $\rho_j = 0$ for every $j = 1, \ldots, m$, then necessarily $\langle \nabla g_j(\mathbf{y}), \mathbf{x} - \mathbf{y} \rangle \geq 0$ on Ω_j. Therefore if Slater's condition and Assumption 13.3 hold then $\mathbf{K}$ is convex. However, again because of the numerical inaccuracies inherent to the SDP solvers, one would only get $\rho_j \approx 0$, and so this certificate of convexity is valid only up to machine precision.

Again in Corollary 13.6 the set $\mathbf{K}$ is not necessarily compact. However, if $\mathbf{K}$ is compact then (13.10) is almost necessary. Indeed let $\mathbf{K}$ in (13.2) be convex and compact, let Slater's condition and Assumptions 2.14 and 13.3 hold. Then for every $j = 1, \ldots, m$, the polynomial $h_j(\mathbf{x}, \mathbf{y}) := \langle \nabla g_j(\mathbf{y}), \mathbf{x} - \mathbf{y} \rangle$ is nonnegative on Ω_j in (13.9) and vanishes on Ω_j. Next, Assumption 2.14 also holds for Ω_j, for all $j = 1, \ldots, m$. So if the ideal $\langle g_j \rangle$ is real radical then by Theorem 7.5, (13.10) holds whenever linear independence, strict complementarity, and second-order sufficiency conditions hold at every minimizer of h_j on Ω_j (which by Theorem 7.6 happens generically).

Example 13.7   Consider the following simple illustrative example in $\mathbb{R}^2$:

$$\mathbf{K} := \{ \mathbf{x} \in \mathbb{R}^2 : x_1 x_2 - 1/4 \geq 0; \ 0.5 - (x_1 - 0.5)^2 - (x_2 - 0.5)^2 \geq 0 \}. \tag{13.12}$$

Obviously $\mathbf{K}$ is convex but its defining polynomial $\mathbf{x} \mapsto g_1(\mathbf{x}) := x_1 x_2 - 1/4$ is not concave.

With $d_1 = 3$, solving (13.11) using GloptiPoly3[2] yields the optimal value $\rho_1 \approx -4.58 \times 10^{-11}$ which, in view of the machine precision for the SDP solvers used in GloptiPoly, could be considered to be zero, but of course with no guarantee. For $j = 2$ there is no test to perform because $-g_2$, being quadratic and convex, yields

$$\langle \nabla g_2(\mathbf{y}), \mathbf{x} - \mathbf{y} \rangle = g_2(\mathbf{x}) - g_2(\mathbf{y}) + \underbrace{(\mathbf{x} - \mathbf{y})'(-\nabla^2 g_2(\mathbf{y}))(\mathbf{x} - \mathbf{y})}_{\text{SOS}} \tag{13.13}$$

which is in the form (13.10) with $d_2 = 1$, and so $\rho_2 = 0$.

13.1.2 Notions of convexity for polynomials

If $C \subseteq \mathbb{R}^n$ is a nonempty convex set, a function $f : C \to \mathbb{R}$ is convex on C if

$$f(\lambda \mathbf{x} + (1 - \lambda)\mathbf{y}) \leq \lambda f(\mathbf{x}) + (1 - \lambda)f(\mathbf{y}), \qquad \forall \lambda \in (0, 1), \ \mathbf{x}, \mathbf{y} \in C.$$

Similarly, f is strictly convex on C if the above inequality is strict for every $\mathbf{x}, \mathbf{y} \in C$, $\mathbf{x} \neq \mathbf{y}$, and all $\lambda \in (0, 1)$.

If $C \subseteq \mathbb{R}^n$ is an open convex set and f is twice differentiable on C, then f is convex on C if and only if its Hessian[3] $\nabla^2 f$ is positive semidefinite on C (denoted $\nabla^2 f \succeq 0$ on C). Finally, if $\nabla^2 f$ is positive definite on C (denoted $\nabla^2 f \succ 0$ on C) then f is strictly convex on C.

A polynomial of odd degree $d \geq 3$ cannot be convex and the following result states that, like the convexity of basic closed semi-algebraic sets, deciding the convexity of a polynomial is a difficult problem.

---

[2] GloptiPoly3, a software package dedicated to solving the generalized problem of moments, is described in Appendix B.

[3] Recall that the Hessian $\nabla^2 f(\mathbf{x})$ is the $n \times n$ symmetric matrix whose entry (i, j) is $\partial^2 f / \partial x_i \partial x_j$ evaluated at $\mathbf{x}$.

Theorem 13.8 *It is NP-hard to check the convexity of polynomials of any fixed even degree $d \geq 4$.*

However, it is still possible to provide a means to check convexity and provide an algebraic certificate.

Theorem 13.9 *A polynomial $f \in \mathbb{R}[\mathbf{x}]$ is convex if and only if there exist nontrivial SOS polynomials $p, q \in \Sigma[\mathbf{x}, \mathbf{y}]$ such that*

$$p(\mathbf{x}, \mathbf{y})\, (f(\mathbf{x}) + f(\mathbf{y}) - 2\, f((\mathbf{x}+\mathbf{y})/2)) = q(\mathbf{x}, \mathbf{y}), \qquad \forall \mathbf{x}, \mathbf{y} \in \mathbb{R}^n. \tag{13.14}$$

Proof f is convex if and only if

$$f\left(\frac{\mathbf{x}+\mathbf{y}}{2}\right) \leq \frac{1}{2} f(\mathbf{x}) + \frac{1}{2} f(\mathbf{y}), \qquad \forall \mathbf{x}, \mathbf{y} \in \mathbb{R}^n,$$

or, equivalently, if and only if the polynomial

$$(\mathbf{x}, \mathbf{y}) \mapsto h(\mathbf{x}, \mathbf{y}) := f(\mathbf{x}) + f(\mathbf{y}) - 2\, f((\mathbf{x}+\mathbf{y})/2),$$

is nonnegative. But then h is a sum of squares of rational functions, and so, clearing up denominators, $p\,h = q$ for some nontrivial SOS polynomials $p, q \in \Sigma[\mathbf{x}, \mathbf{y}]$. Conversely, if $p\,h = q$ for some nontrivial SOS polynomials $p, q \in \Sigma[\mathbf{x}, \mathbf{y}]$, then h is nonnegative (and so f is convex). Indeed, suppose that $h(\mathbf{x}_0, \mathbf{y}_0) < 0$ for some $(\mathbf{x}_0, \mathbf{y}_0)$ (hence $h < 0$ in some open neighborhood of $(\mathbf{x}_0, \mathbf{y}_0)$). Then necessarily $p(\mathbf{x}, \mathbf{y}) = 0$ in that neighborhood which in turn implies $p(\mathbf{x}, \mathbf{y}) = 0$ for all $(\mathbf{x}, \mathbf{y})$, and so $q = 0$; this contradicts that p and q are nontrivial. $\qquad\square$

In view of the discussion in Chapter 2 on nonnegative polynomials, checking whether (13.14) has a nontrivial SOS solution $(p, q) \in \Sigma[\mathbf{x}, \mathbf{y}]^2$ reduces to solving a single semidefinite program. But of course, the size of the resulting semidefinite program is too large for practical computation. On the other hand, one may solve (13.14) with an a priori fixed and "small" degree bound on p and q. If a solution (p, q) is found then f is convex, otherwise try with a larger degree bound, etc.

Quasiconvex and pseudoconvex polynomials The two (convexity related) weaker notions of quasiconvexity and pseudoconvexity are also useful in optimization. Recall that a function $f : \mathbb{R}^n \to \mathbb{R}$ is *quasiconvex* if its level set

$\{\mathbf{x} : f(\mathbf{x}) \leq s\}$ is convex for every $s \in \mathbb{R}$, and a differentiable function $f : \mathbb{R}^n \to \mathbb{R}$ is *pseudoconvex* if the implication

$$\nabla f(\mathbf{x})^T (\mathbf{y} - \mathbf{x}) \geq 0 \quad \Longrightarrow \quad f(\mathbf{y}) \geq f(\mathbf{x}),$$

holds for all $\mathbf{x}, \mathbf{y} \in \mathbb{R}^n$. Moreover one has the implication

$$\text{convexity} \quad \Rightarrow \quad \text{pseudoconvexity} \quad \Rightarrow \quad \text{quasiconvexity},$$

but none of the converse implications holds in general. However, the three notions coincide

- for quadratic polynomials (and then can be checked in polynomial time),
- for homogeneous polynomials of even degree.

For even degree larger than two, checking pseudo-convexity or quasiconvexity is difficult.

Theorem 13.10 *It is NP-hard to decide pseudoconvexity or quasiconvexity of polynomials of even degree $d \geq 4$.*

On the other hand, pseudoconvex or quasiconvex polynomials of odd degree have a nice characterization that can be checked in polynomial time.

Theorem 13.11 (a) *An odd degree polynomial f is pseudoconvex or quasiconvex if and only if it can be written*

$$f(\mathbf{x}) = h(\mathbf{y}^T \mathbf{x})$$

for some nonzero vector $\mathbf{y} \in \mathbb{R}^n$ and some monotonic univariate polynomial h. In addition, if f is pseudoconvex then the derivative h' has no real root.

(b) *For any fixed odd degree d, quasiconvexity or pseudoconvexity of degree d polynomials can be checked in polynomial time.*

Example 13.12 The polynomial

$$\mathbf{x} \mapsto f(\mathbf{x}) := 8x_1^3 - 12x_1^2 x_2 + 6x_1 x_2^2 - x_2^3,$$

is not convex because its Hessian

$$\mathbf{x} \mapsto \nabla^2 f(\mathbf{x}) = \begin{bmatrix} 48x_1 - 24x_2 & 6x_1 - 24x_2 \\ 6x_1 - 24x_1 & 12x_1 - 6x_2 \end{bmatrix},$$

is not positive semidefinite. On the other hand, f is quasiconvex as one may check that $f(\mathbf{x}) = (2x_1 - x_2)^3$.

Finally it is worth noticing that quasiconvex homogeneous polynomials of even degree $d \geq 2$ are nonnegative.

SOS-convex polynomials We now detail the very useful notion of SOS-convexity first introduced by Helton and Nie (2010, 2012).

Definition 13.13 A polynomial $f \in \mathbb{R}[\mathbf{x}]_{2d}$ is said to be SOS-convex if its Hessian $\nabla^2 f$ is an SOS matrix-polynomial, that is, $\nabla^2 f = \mathbf{LL}^T$ for some real matrix-polynomial $\mathbf{L} \in \mathbb{R}[\mathbf{x}]^{n \times s}$ (for some $s \in \mathbb{N}$).

Example 13.14 A polynomial $f \in \mathbb{R}[\mathbf{x}]$ is said to be *separable* if $f = \sum_{i=1}^{n} f_i$ for some univariate polynomials $f_i \in \mathbb{R}[x_i]$. Therefore every convex separable polynomial f is SOS-convex because its Hessian $\nabla^2 f(\mathbf{x})$ is a positive semidefinite diagonal matrix, with all diagonal entries $(f_i^{''}(x_i))$, $i = 1, \ldots, n$, nonnegative for all $\mathbf{x} \in \mathbb{R}^n$. Hence, being univariate and nonnegative, $f_i^{''}$ is SOS for every $i = 1, \ldots, n$, from which one easily concludes that f is SOS-convex.

Of course every SOS-convex polynomial is convex, and we also have the following useful equivalences for SOS-convexity.

Theorem 13.15 *The following four propositions are equivalent.*

(a) *The polynomial g is SOS-convex.*
(b) *The Hessian $\nabla^2 g(\mathbf{x})$ is SOS.*
(c) *The polynomial $(\mathbf{x}, \mathbf{y}) \mapsto g(\mathbf{x})/2 + g(\mathbf{y})/2 - g((\mathbf{x} + \mathbf{y})/2)$ is SOS.*
(d) *The polynomial $(\mathbf{x}, \mathbf{y}) \mapsto g(\mathbf{x}) - g(\mathbf{y}) - \nabla g(\mathbf{y})^T (\mathbf{x} - \mathbf{y})$ is SOS.*

In particular, an important feature of SOS-convexity is that it can be checked numerically by solving a semidefinite program. For instance, by Theorem 13.15(c) this reduces to checking whether the polynomial $g(\mathbf{x})/2 + g(\mathbf{y})/2 - g((\mathbf{x}+\mathbf{y})/2)$ is SOS, a semidefinite optimization (feasibility) problem; see (2.2).

Also, the SOS-convex polynomials have the following interesting property.

Lemma 13.16 *If a symmetric matrix-polynomial $\mathbf{P} \in \mathbb{R}[\mathbf{x}]^{r \times r}$ is SOS then the double integral*

$$(\mathbf{x}, \mathbf{u}) \mapsto \quad \mathbf{F}(\mathbf{x}, \mathbf{u}) := \int_0^1 \int_0^t \mathbf{P}(\mathbf{u} + s(\mathbf{x} - \mathbf{u}))\, ds\, dt \qquad (13.15)$$

is also a symmetric SOS matrix-polynomial $\mathbf{F} \in \mathbb{R}[\mathbf{x}, \mathbf{u}]^{r \times r}$.

Proof Writing $\mathbf{P} = (p_{ij})_{1 \le i,j \le n}$ with $p_{ij} \in \mathbb{R}[\mathbf{x}]$ for every $1 \le i, j \le n$, let $d := \max_{i,j} \deg p_{ij}$. With $\mathbf{x}, \mathbf{u}$ fixed, introduce the univariate matrix-polynomial $s \mapsto \mathbf{Q}(s) := \mathbf{P}(\mathbf{u} + s(\mathbf{x} - \mathbf{u}))$ so that $\mathbf{Q} = (q_{ij})$ where $q_{ij} \in \mathbb{R}[s]$ has degree at most d for every $1 \le i, j \le n$. Observe that

$$\int_0^1 \int_0^t \mathbf{P}(\mathbf{u} + s(\mathbf{x} - \mathbf{u})) \, ds \, dt = \int_0^1 \int_0^t \mathbf{Q}(s) \, ds \, dt = \int_\Delta \mathbf{Q}(s) \, d\mu(s, t),$$

where μ is the measure uniformly distributed on the set $\Delta := \{ (s, t) : 0 \le t \le 1; \ 0 \le s \le t \}$ with $\mu(\Delta) = 1/2$. By an extension of Tchakaloff's theorem Tchakaloff (1957), there exists a measure φ finitely supported on, say m, points $\{(s_k, t_k)\}_{k=1}^m \subset \Delta$ whose moments up to degree d match exactly those of μ. That is, there exist positive weights $h_k, k = 1, \ldots, m$, such that

$$\int_\Delta f \, d\mu = \sum_{k=1}^m h_k \, f(s_k, t_k), \qquad \forall f \in \mathbb{R}[s, t], \ \deg f \le d.$$

So let $\mathbf{x}, \mathbf{u} \in \mathbb{R}^n$ be fixed. If $\mathbf{P} = \mathbf{L}\mathbf{L}^T$ for some $\mathbf{L} \in \mathbb{R}[\mathbf{x}]^{r \times q}$, one obtains:

$$\int_0^1 \int_0^t \mathbf{P}(\mathbf{u} + s(\mathbf{x} - \mathbf{u})) \, ds \, dt = \sum_{k=1}^m h_k \, \mathbf{Q}(s_k)$$

$$= \sum_{k=1}^m h_k \, \mathbf{L}(\mathbf{u} + s_k(\mathbf{x} - \mathbf{u}) \, \mathbf{L}^T(\mathbf{u} + s_k(\mathbf{x} - \mathbf{u})$$

$$= \mathbf{A}\mathbf{A}^T$$

for some $\mathbf{A} \in \mathbb{R}[\mathbf{x}, \mathbf{u}]^{r \times q}$. $\qquad \square$

And as a consequence we have the following lemma.

Lemma 13.17 *For a polynomial $f \in \mathbb{R}[\mathbf{x}]$ and every $\mathbf{x}, \mathbf{u} \in \mathbb{R}^n$:*

$$f(\mathbf{x}) = f(\mathbf{u}) + \nabla f(\mathbf{u})^T (\mathbf{x} - \mathbf{u})$$

$$+ (\mathbf{x} - \mathbf{u})^T \underbrace{\int_0^1 \int_0^t \nabla^2 f(\mathbf{u} + s(\mathbf{x} - \mathbf{u})) \, ds \, dt}_{F(\mathbf{x}, \mathbf{u})} (\mathbf{x} - \mathbf{u}).$$

If f is SOS-convex and $f(\mathbf{u}) = 0, \nabla f(\mathbf{u}) = 0$, then f is an SOS polynomial.

Let $\mathbf{K}$ be the basic semi-algebraic set in (13.2), $Q(g)$ be the quadratic module defined in (2.9), and let $Q_c(g) \subset Q(g)$ be the set:

$$Q_c(g) := \left\{ \sigma_0 + \sum_{j=1}^{m} \lambda_j \, g_j : \quad \boldsymbol{\lambda} \in \mathbb{R}_+^m; \ \sigma_0 \in \Sigma[\mathbf{x}], \ \sigma_0 \text{ convex} \right\}.$$

$$(13.16)$$

The set $Q_c(g)$ is a specialization of the quadratic module $Q(g)$ in the convex case, in that the weights associated with the g_j are nonnegative scalars, i.e., SOS polynomials of degree zero, and the SOS polynomial σ_0 is convex. In particular, every $f \in Q_c(g)$ is nonnegative on $\mathbf{K}$. Let $C(\mathbf{K}) \subset \mathbb{R}[\mathbf{x}]$ be the convex cone of convex polynomials nonnegative on $\mathbf{K}$. Recall that $\|f\|_1$ denotes the ℓ_1-norm of the vector of coefficients, i.e., $\|f\|_1 = \sum_\alpha |f_\alpha|$.

Theorem 13.18   *With $\mathbf{K}$ as in (13.2), let Slater's condition hold and let g_j be concave for every $j = 1, \ldots, m$.*

Then with $Q_c(g)$ as in (13.16), the set $Q_c(g) \cap C(\mathbf{K})$ is dense in $C(\mathbf{K})$ for the ℓ_1-norm $\| \cdot \|_1$. In particular, if $\mathbf{K} = \mathbb{R}^n$ (so that $C(\mathbb{R}^n) =: C$ is now the set of nonnegative convex polynomials), then $\Sigma[\mathbf{x}] \cap C$ is dense in C.

Theorem 13.18 states that if f is convex and nonnegative on $\mathbf{K}$ (including the case $\mathbf{K} \equiv \mathbb{R}^n$) then one may approximate f by a sequence $(f_{\epsilon r}) \subset Q_c(g) \cap C(\mathbf{K})$ with $\|f - f_{\epsilon r}\|_1 \to 0$ as $\epsilon \to 0$ (and $r \to \infty$). For instance, with $r_0 := \lfloor (\deg f)/2 \rfloor + 1$, the polynomial $f_{\epsilon r}$ may be defined as $\mathbf{x} \mapsto f_{\epsilon r}(\mathbf{x}) := f + \epsilon(\theta_{r_0}(\mathbf{x}) + \theta_r(\mathbf{x}))$ with

$$\mathbf{x} \mapsto \quad \theta_r(\mathbf{x}) := 1 + \sum_{i=1}^{n} \sum_{k=1}^{r} \frac{x_i^{2k}}{(2k)!}.$$

(For more details see for example Lasserre (2005, 2006c) and Lasserre and Netzer (2007).)

Observe that Theorem 13.18 provides f with a *certificate* of nonnegativity on $\mathbf{K}$. Indeed, let $\mathbf{x} \in \mathbf{K}$ be fixed arbitrary. Then as $f_{\epsilon r} \in Q_c(g)$ one has $f_{\epsilon r}(\mathbf{x}) \geq 0$. Letting $\epsilon \downarrow 0$ yields $0 \leq \lim_{\epsilon \to 0} f_{\epsilon r}(\mathbf{x}) = f(\mathbf{x})$. And as $\mathbf{x} \in \mathbf{K}$ was arbitrary, $f \geq 0$ on $\mathbf{K}$.

For the class of SOS-convex polynomials, we have the following more precise result.

Theorem 13.19   *With $\mathbf{K}$ as in (13.2) let Slater's condition hold. Let $f \in \mathbb{R}[\mathbf{x}]$ be such that $f^* := \inf_{\mathbf{x}}\{ f(\mathbf{x}) : \mathbf{x} \in \mathbf{K} \} = f(\mathbf{x}^*)$ for some $\mathbf{x}^* \in \mathbf{K}$. If f and $(-g_j)_{j=1}^m$ are SOS-convex then $f - f^* \in Q_c(g)$.*

Proof   Recall from Section 7.1 that since Slater's condition holds, there exists a vector of Lagrange KKT multipliers $\boldsymbol{\lambda} \in \mathbb{R}_+^m$ such that the pair $(\mathbf{x}^*, \boldsymbol{\lambda}) \in$

$\mathbf{K} \times \mathbb{R}^m_+$ satisfies the KKT optimality conditions (7.3). And so the Lagrangian polynomial

$$\mathbf{x} \mapsto \quad L(\mathbf{x}, \lambda) := f(\mathbf{x}) - f^* - \sum_{j=1}^m \lambda_j g_j(\mathbf{x}) \qquad (13.17)$$

is nonnegative and satisfies $L(\mathbf{x}^*, \lambda) = 0$ as well as $\nabla_\mathbf{x} L(\mathbf{x}^*, \lambda) = 0$. Moreover, as f and $(-g_j)$ are SOS-convex then so is L because $\nabla^2_{\mathbf{xx}} L = \nabla^2 f - \sum_j \lambda_j \nabla^2 g_j$ with $\lambda \geq 0$. Therefore, by Lemma 13.17, $L \in \Sigma[\mathbf{x}]$, i.e.,

$$f - f^* - \sum_{j=1}^m \lambda_j g_j = \sigma_0,$$

for some SOS polynomial $\sigma_0 \in \Sigma[\mathbf{x}]$. As L is convex then so is σ_0 and therefore $f - f^* \in Q_c(g)$. □

Hence the class of SOS-convex polynomials is very interesting because a nice representation result is available. Notice that the above representation holds with $f \geq 0$ on $\mathbf{K}$ and $\mathbf{K}$ is not required to be compact. Another interesting case is when the Hessian of the Lagrangian L defined in (13.17) is positive definite at every minimizer $\mathbf{x}^* \in \mathbf{K}$.

Theorem 13.20   *With $\mathbf{K}$ as in (13.2), let Assumption 2.14 and Slater's condition hold. Let f and $-g_j$ be convex, $j = 1, \ldots, m$, and assume that there are finitely many minimizers of f on $\mathbf{K}$. If the Hessian of the Lagrangian defined in (13.17) is positive definite (in particular if $\nabla^2 f(\mathbf{x}^*) \succ 0$) at every minimizer $\mathbf{x}^* \in \mathbf{K}$ then $f - f^* \in Q(g)$, i.e.,*

$$f - f^* = \sigma_0 + \sum_{j=}^m \sigma_j g_j, \qquad (13.18)$$

for some SOS polynomials $(\sigma_j)_{j=0}^m \subset \Sigma[\mathbf{x}]$.

Proof   Under Assumption 2.14 the set $\mathbf{K}$ is compact and therefore $f^* = f(\mathbf{x}^*)$ for some $\mathbf{x}^* \in \mathbf{K}$. Moreover, as Slater's condition holds and $f, -g_j$ are convex, every minimizer $\mathbf{x}^*$ satisfies the KKT optimality conditions for some vector $\lambda^* \in \mathbb{R}^m_+$, and $(\mathbf{x}^*, \lambda^*)$ is a saddle point of L defined in (13.17), i.e.,

$$L(\mathbf{x}^*, \lambda) \leq L(\mathbf{x}^*, \lambda^*) \leq L(\mathbf{x}, \lambda^*),$$

for all $\mathbf{x} \in \mathbb{R}^n, \lambda \in \mathbb{R}^m_+$. As $\nabla^2_{\mathbf{xx}} L(\mathbf{x}^*, \lambda^*) \succ 0$ at every minimizer $\mathbf{x}^* \in \mathbf{K}$, the result follows from Theorem 3.2 in de Klerk and Laurent (2011). □

So for the class of compact basic semi-algebraic sets $\mathbf{K}$ defined by concave polynomials, one obtains a refinement of Putinar's Positivstellensatz, Theorem 2.15, for the class of convex polynomials for which the Hessian of the associated Lagrangian is positive definite at every minimizer $\mathbf{x}^* \in \mathbf{K}$, and that are only nonnegative on $\mathbf{K}$.

13.1.3 An extension of Jensen's inequality

Recall that if μ is a probability measure on $\mathbb{R}^n$ with $E_\mu(\mathbf{x}) < \infty$, Jensen's inequality states that if $f \in L_1(\mu)$ and f is convex, then

$$E_\mu(f) = \int_{\mathbb{R}^n} f \, d\mu \geq f(E_\mu(\mathbf{x})),$$

a very useful property in many applications. We next provide an extension of Jensen's inequality to a larger class of linear functionals when one restricts its application to the class of SOS-convex polynomials (as opposed to all convex functions). Recall the notation $\mathbb{N}_{2d}^n = \{ \alpha \in \mathbb{N}^n : |\alpha| \leq 2d \}$.

Theorem 13.21 *Let $f \in \mathbb{R}[\mathbf{x}]_{2d}$ be SOS-convex and let $\mathbf{y} = (y_\alpha)$, $\alpha \in \mathbb{N}_{2d}^n$, satisfy $y_0 = 1$ and $\mathbf{M}_d(\mathbf{y}) \succeq 0$. Then*

$$L_\mathbf{y}(f) \geq f(L_\mathbf{y}(\mathbf{x})), \tag{13.19}$$

where $L_\mathbf{y}(\mathbf{x}) = (L_\mathbf{y}(x_1), \ldots, L_\mathbf{y}(x_n))$.

Proof Let $\mathbf{z} \in \mathbb{R}^n$ be fixed, arbitrary, and consider the polynomial $\mathbf{x} \mapsto f(\mathbf{x}) - f(\mathbf{z})$. Then by Lemma 13.17,

$$f(\mathbf{x}) - f(\mathbf{z}) = \langle \nabla f(\mathbf{z}), \mathbf{x} - \mathbf{z} \rangle + \langle (\mathbf{x} - \mathbf{z}), \mathbf{F}(\mathbf{x})(\mathbf{x} - \mathbf{z}) \rangle, \tag{13.20}$$

with $\mathbf{F} : \mathbb{R}^n \to \mathbb{R}[\mathbf{x}]^{n \times n}$ being the matrix-polynomial

$$\mathbf{x} \mapsto \mathbf{F}(\mathbf{x}) := \int_0^1 \int_0^t \nabla^2 f(\mathbf{z} + s(\mathbf{x} - \mathbf{z})) \, ds \, dt.$$

As f is SOS-convex, by Lemma 13.16, $\mathbf{F}$ is an SOS matrix-polynomial and so the polynomial $\mathbf{x} \mapsto g(\mathbf{x}) := \langle (\mathbf{x} - \mathbf{z}), \mathbf{F}(\mathbf{x})(\mathbf{x} - \mathbf{z}) \rangle$ is SOS, i.e., $g \in \Sigma[\mathbf{x}]$. Then applying $L_\mathbf{y}$ to the polynomial $\mathbf{x} \mapsto f(\mathbf{x}) - f(\mathbf{z})$ and using (13.20) yields (recall that $y_0 = 1$)

$$L_\mathbf{y}(f) - f(\mathbf{z}) = \langle \nabla f(\mathbf{z}), L_\mathbf{y}(\mathbf{x}) - \mathbf{z} \rangle + L_\mathbf{y}(g)$$
$$\geq \langle \nabla f(\mathbf{z}), L_\mathbf{y}(\mathbf{x}) - \mathbf{z} \rangle$$

because $L_\mathbf{y}(g) \geq 0$ (since g is SOS). As $\mathbf{z} \in \mathbb{R}^n$ was arbitrary, taking $\mathbf{z} := L_\mathbf{y}(\mathbf{x})$ ($= (L_\mathbf{y}(x_1), \dots, L_\mathbf{y}(x_n))$) yields the desired result.   □

Hence (13.19) is Jensen's inequality extended to linear functionals $L_\mathbf{y}$: $\mathbb{R}[\mathbf{x}]_{2d} \to \mathbb{R}$ in the dual cone of $\Sigma[\mathbf{x}]_d$, that is, vectors $\mathbf{y} = (y_\alpha)$ such that $\mathbf{M}_d(\mathbf{y}) \succeq 0$ and $y_0 = L_\mathbf{y}(1) = 1$; hence $\mathbf{y}$ is *not* necessarily the (truncated) moment sequence of some probability measure μ. As a consequence we also get the following result.

Corollary 13.22   *Let f be a convex univariate polynomial, $g \in \mathbb{R}[\mathbf{x}]$ (and so $f \circ g \in \mathbb{R}[\mathbf{x}]$). Let $d := \lceil (\deg f \circ g)/2 \rceil$, and let $\mathbf{y} = (y_\alpha)$, $\alpha \in \mathbb{N}^n_{2d}$, be such that $y_0 = 1$ and $\mathbf{M}_d(\mathbf{y}) \succeq 0$. Then*

$$L_\mathbf{y}(f \circ g) \geq f(L_\mathbf{y}(g)). \tag{13.21}$$

13.2 Semidefinite representation of convex sets

Convex sets that can be represented by linear matrix inequalities (LMIs) are very important because optimizing a linear functional on such sets can be done efficiently by semidefinite programming. Therefore an important issue is the characterization of convex sets that have a *semidefinite representation* (SDr), called SDr sets.

Definition 13.23   A convex set $\Omega \subset \mathbb{R}^n$ is SDr if there exist integers m, p and real $p \times p$ symmetric matrices $(\mathbf{A}_i)_{i=0}^n$, $(\mathbf{B}_j)_{j=1}^m$ such that:

$$\Omega = \left\{ \mathbf{x} \in \mathbb{R}^n : \exists \mathbf{y} \in \mathbb{R}^m \text{ s.t. } \mathbf{A}_0 + \sum_{i=1}^n \mathbf{A}_i \, x_i + \sum_{j=1}^m \mathbf{B}_j \, y_j \succeq 0 \right\}.$$
$$\tag{13.22}$$

In other words, Ω is the linear projection on $\mathbb{R}^n$ of the convex set

$$\Omega' := \left\{ (\mathbf{x}, \mathbf{y}) \in \mathbb{R}^n \times \mathbb{R}^m : \mathbf{A}_0 + \sum_{i=1}^n \mathbf{A}_i \, x_i + \sum_{j=1}^m \mathbf{B}_j \, y_j \succeq 0 \right\}$$

in the *lifted* space $\mathbb{R}^{n+m}$, and Ω' is a *semidefinite* representation (SDr) of Ω.

The set Ω' is also called a *lifted* SDr or a lifted LMI because one sometimes needs additional variables $\mathbf{y} \in \mathbb{R}^m$ to obtain a description of Ω via appropriate LMIs. Here are some examples of SDr sets.

- The intersection of half-spaces, i.e., a polyhedron $\{\mathbf{x} \in \mathbb{R}^n : \mathbf{A}\mathbf{x} \leq \mathbf{b}\}$, is a trivial example of convex sets whose SDr is readily available without lifting. Indeed $\mathbf{A}\mathbf{x} \leq \mathbf{b}$ is an LMI with diagonal matrices $\mathbf{A}_i$ in (13.22).
- The intersection of ellipsoids

$$\Omega := \{\mathbf{x} \in \mathbb{R}^n : \mathbf{x}^T \mathbf{Q}_j \mathbf{x} + \mathbf{b}^T \mathbf{x} + c_j \geq 0, \ j = 1, \ldots, m\}$$

(where $-\mathbf{Q}_j \succeq 0$ for all $j = 1, \ldots, m$) is an SDr set with lifted LMI representation in $\mathbb{R}^{(n+1)(n+2)/2-1}$:

$$\Omega' = \left\{ (\mathbf{x}, \mathbf{Y}) : \begin{bmatrix} 1 & \mathbf{x}' \\ \mathbf{x} & \mathbf{Y} \end{bmatrix} \succeq 0, \ \mathrm{trace}(\mathbf{Q}_j \mathbf{Y}) + \mathbf{b}^T \mathbf{x} + c_j \geq 0, \quad j = 1, \ldots, m \right\}.$$

- The epigraph of a univariate convex polynomial, convex sets of $\mathbb{R}^2$ described from genus-zero plane curves are SDr sets and *hyperbolic* cones obtained from three-variable hyperbolic homogeneous polynomials are SDr sets.

Of course in view of Definition 13.23, if $\Omega \subset \mathbb{R}^n$ is SDr then necessarily it is semi-algebraic (as the projection to $\mathbb{R}^n$ of $\Omega' \subset \mathbb{R}^{n+m}$ which is itself basic semi-algebraic).

A first important result states that one can proceed with elementary pieces.

Theorem 13.24 *Let $T_1, \ldots, T_m$ be bounded semi-algebraic sets. If the convex hull of each T_k is SDr then the convex hull of $\cup_{k=1}^m T_k$ is also SDr.*

The Putinar bounded degree representation We next consider semidefinite representations for the convex hull co(**K**) of compact basic semi-algebraic sets $\mathbf{K} \subset \mathbb{R}^n$ as defined in (13.2) for some polynomials $(g_j) \subset \mathbb{R}[\mathbf{x}]$. We show that co(**K**) has an SDr if the defining polynomials $(g_j) \subset \mathbb{R}[\mathbf{x}]$ of **K** satisfy what we call the *Putinar's bounded degree representation* (P-BDR) of *affine* polynomials.

Recall the truncated version $Q_d(g) \subset \mathbb{R}[\mathbf{x}]$ of the quadratic module $Q(g)$ generated by the g_j, that is, with $d \in \mathbb{N}$ and $g_0 = 1$,

$$Q_d(g) = \left\{ \sum_{j=0}^m \sigma_j g_j : \sigma_j \in \Sigma[\mathbf{x}], \ \deg \sigma_j g_j \leq 2d, \ \forall j = 0, \ldots, m \right\}.$$

(13.23)

For an affine polynomial $f \in \mathbb{R}[\mathbf{x}]_1$ write $\mathbf{x} \mapsto f(\mathbf{x}) = f_0 + \sum_i f_i x_i$ with $\mathbf{f} = (f_0, \mathbf{f}_1) \in \mathbb{R}^{n+1}$.

Definition 13.25 (P-BDR property)   Given a compact set $\mathbf{K} \subset \mathbb{R}^n$ as defined in (13.2), the *Putinar bounded degree representation* (P-BDR) of affine polynomials holds for $\mathbf{K}$ if there exists $d \in \mathbb{N}$ such that

$$[\, f \text{ affine and positive on } \mathbf{K} \,] \quad \Rightarrow \quad f \in Q_d(g), \qquad (13.24)$$

except perhaps for a set of vectors $\mathbf{f}_1 \in \mathbb{R}^n$ with Lebesgue measure zero. Call d its order.

Example 13.26   Let $n = 2$ and let $\mathbf{K} \subset \mathbb{R}^2$ be the TV screen $\{ \mathbf{x} \in \mathbb{R}^2 : 1 - x_1^4 - x_2^4 \geq 0 \}$. It is well known that $\mathbf{K}$ does not have a semidefinite representation with no lifting. On the other hand the P-BDR property holds for $\mathbf{K}$. Indeed let $f^* = \inf\{ f(\mathbf{x}) : \mathbf{x} \in \mathbf{K} \}$ where $f \in \mathbb{R}[\mathbf{x}]_1$ is arbitrary (written $\mathbf{x} \mapsto \mathbf{f}^T \mathbf{x}$ for some $\mathbf{f} \in \mathbb{R}^n$). As $\mathbf{K}$ is compact and Slater's condition holds, there exists $(\mathbf{x}^*, \lambda^*) \in \mathbb{R}^2 \times \mathbb{R}_+$ such that the KKT optimality conditions hold. That is, $f^* = f(\mathbf{x}^*)$, $\mathbf{f} = \lambda^* \nabla g(\mathbf{x}^*)$ (with $\mathbf{x} \mapsto g(\mathbf{x}) = 1 - x_1^4 - x_2^4$). But then the convex polynomial $\mathbf{x} \mapsto L(\mathbf{x}) := f(\mathbf{x}) - f^* - \lambda^* g(\mathbf{x})$ satisfies $\nabla L(\mathbf{x}^*) = 0$ and $L(\mathbf{x}^*) = 0$. Therefore, $L(\mathbf{x}) \geq 0$ for all $\mathbf{x} \in \mathbb{R}^2$ and as L is a quartic bivariate polynomial, it must be an SOS. And so the P-BDR property holds for $\mathbf{K}$ (with order 2).

The importance of the P-BDR property comes from the following result. Recall that the convex hull of a set $\mathbf{K} \subset \mathbb{R}^n$ is denoted by $\mathrm{co}(\mathbf{K})$. Let $v_j := \lceil (\deg g_j)/2 \rceil$, $j = 1, \ldots, m$.

Theorem 13.27   *Let* $\mathbf{K} \subset \mathbb{R}^n$ *be as in (13.2) and assume that the P-BDR property with order d holds for* $\mathbf{K}$. *Then* $\mathrm{co}(\mathbf{K})$ *is SDr and more precisely:*

$$\mathrm{co}(\mathbf{K}) = \Big\{ \mathbf{z} \in \mathbb{R}^n : \exists\, \mathbf{y} \in \mathbb{R}^{s(2d)} \quad \text{such that}$$
$$\mathbf{M}_d(\mathbf{y}) \succeq 0, \ \mathbf{M}_{d-v_j}(g_j\, \mathbf{y}) \succeq 0, \ j = 1, \ldots, m$$
$$L_\mathbf{y}(1) = 1; \ L_\mathbf{y}(x_i) = z_i, \ i = 1, \ldots, n \Big\}. \quad (13.25)$$

Hence when the P-BDR property holds one obtains an SDr of $\mathrm{co}(\mathbf{K})$ in explicit form and, moreover, its description is only in terms of the polynomials g_j that define the set $\mathbf{K}$. Therefore an important issue is to determine under which conditions on the g_j that define $\mathbf{K}$, the P-BDR property holds.

But notice already that for general basic semi-algebraic sets $\mathbf{K}$ one cannot expect the P-BDR property to hold (if it ever holds) for *nice* values of the order d. Otherwise one could minimize almost all affine polynomials on **K** *efficiently.* Indeed with $f \in \mathbb{R}[\mathbf{x}]_1$, let f^* be the minimum value of f on **K**. Then for every $0 < k \in \mathbb{N}$, $f - f^* + 1/k$ is affine and positive on **K**, and so if $f - f^* + 1/k \in Q_d(g)$ for all k then the optimal value ρ_d of the dth semidefinite relaxation (6.3) (and of its dual) equals f^*.

So from a practical point of view, the most interesting case is essentially the convex case, that is, when **K** is convex, and even more, when the defining polynomials g_j in (13.2) are concave, in which case one may hope that the P-BDR property holds for nice values of d. And we get the following important result.

Definition 13.28   A function $f : \mathbb{R}^n \rightarrow \mathbb{R}$ is strictly quasiconcave at **x** if

$$-\mathbf{u}^T \nabla^2 f(\mathbf{x}) \mathbf{u} > 0, \quad \forall 0 \neq \mathbf{u} \in \nabla f(\mathbf{x})^\perp,$$

where $\nabla f(\mathbf{x})^\perp = \{\mathbf{u} \in \mathbb{R}^n : \nabla f(\mathbf{x})^T \mathbf{u} = 0\}$. If $\nabla f(\mathbf{x}) = 0$ it requires $\nabla^2 f(\mathbf{x}) \succ 0$. And f is strictly quasiconcave on $\mathbf{K} \subset \mathbb{R}^n$ if it is strictly quasiconcave at every point $\mathbf{x} \in \mathbf{K}$.

Theorem 13.29   *Let* **K** *in (13.2) be convex and compact, with nonempty interior, and let Assumption 2.14 hold.*

(a) *Suppose that for every* $j = 1, \ldots, m$, *either* $-g_j$ *is SOS-convex or* g_j *is strictly quasiconcave on* **K**. *Then* **K** *is SDr. More precisely, there exists* $d \in \mathbb{N}$ *such that:*

$$\mathbf{K} = \Big\{ \mathbf{z} \in \mathbb{R}^n : \exists \mathbf{y} \in \mathbb{R}^{s(2d)} \quad such\ that$$
$$\mathbf{M}_d(\mathbf{y}) \succeq 0, \ \mathbf{M}_{d-v_j}(g_j\,\mathbf{y}) \succeq 0, \ j = 1, \ldots, m$$
$$L_\mathbf{y}(1) = 1; \quad L_\mathbf{y}(x_i) = z_i, \ i = 1, \ldots, n \Big\}. \quad (13.26)$$

(b) *Suppose that* $-g_j$ *is SOS-convex for every* $j = 1, \ldots, m$. *Then* **K** *is SDr and with* $d \geq \max_j v_j$,

$$\mathbf{K} = \Big\{ \mathbf{z} \in \mathbb{R}^n : \exists \mathbf{y} \in \mathbb{R}^{s(2d)} \quad such\ that$$
$$\mathbf{M}_d(\mathbf{y}) \succeq 0, \ L_\mathbf{y}(g_j) \succeq 0, \ j = 1, \ldots, m \quad (13.27)$$
$$L_\mathbf{y}(1) = 1; \quad L_\mathbf{y}(x_i) = z_i, \ i = 1, \ldots, n \Big\}.$$

Once again Theorem 13.29(b) illustrates a nice property of the class of SOS-convex polynomials. Indeed, when the $-g_j$ are SOS-convex then not only is the set **K** SDr but also its representation (13.27) is "minimal" as it contains only one semidefinite constraint (or LMI) (namely $\mathbf{M}_d(\mathbf{y}) \succeq 0$) which, in addition, does *not* depend on the data g_j! The latter appear in the linear constraints

$L_\mathbf{y}(g_j) \geq 0$ which is in fact the smallest semidefinite constraint $\mathbf{M}_k(g_j\,\mathbf{y}) \succeq 0$ with $k = 0$.

If in Theorem 13.29(a) one replaces strict quasiconcavity of g_j on $\mathbf{K}$ with the weaker condition of strict quasiconcavity of g_j at every point $\mathbf{x}$ on the boundary $\partial\mathbf{K}$ where $g_j(\mathbf{x}) = 0$, then $\mathbf{K}$ is also SDr but (13.26) is not necessarily an SDr of $\mathbf{K}$.

Example 13.30   Recall the TV screen of Example 13.26 in $\mathbb{R}^2$ which is SDr (with order 2). Moreover $\mathbf{x} \mapsto x_1^4 + x_2^4 - 1$ is SOS-convex. Hence (13.27) reads:

$$
\mathbf{K} = \left\{ \mathbf{z} \in \mathbb{R}^2 : \exists\, \mathbf{y} \text{ s.t. } 1 - y_{40} - y_{04} \geq 0, \begin{bmatrix} 1 & z_1 & z_2 & y_{20} & y_{11} & y_{02} \\ z_1 & y_{20} & y_{11} & y_{30} & y_{21} & y_{12} \\ z_2 & y_{11} & y_{02} & y_{21} & y_{12} & y_{03} \\ y_{20} & y_{30} & y_{21} & y_{40} & y_{31} & y_{22} \\ y_{11} & y_{21} & y_{12} & y_{31} & y_{22} & y_{13} \\ y_{02} & y_{12} & y_{03} & y_{22} & y_{13} & y_{04} \end{bmatrix} \succeq 0 \right\}.
$$

Of course the lifted semidefinite representation (13.26) (or (13.27)) is not unique and other SDR's with fewer variables $\mathbf{y}$ may exist.

13.3 Convex polynomial programs

We now address the issue raised by the question in the introduction of this chapter. We show that for problem (5.1) the moment-SOS approach based on the semidefinite relaxations (6.3) *recognizes* a large class of convex "easy" problems. That is, when applied to such easy problems (5.1), finite convergence takes place, i.e., the optimal value f^* is obtained by solving a particular semidefinite program of the hierarchy. Recall from the discussion in Chapter 9 that this is not the case in general for the moment approach based on the LP-relaxations (9.2); see Proposition 9.5.

We first observe that convex polynomial programs have an optimal solution as soon as the optimal value is finite. Indeed, we have the following result.

Theorem 13.31   *Consider the convex polynomial optimization problem* $f^* = \inf_\mathbf{x}\{ f(\mathbf{x}) : \mathbf{x} \in \mathbf{K} \}$ *with* $\mathbf{K}$ *as in (13.2), and where* f *and* $-g_j$ *are convex,* $j = 1, \ldots, m$. *If* f^* *is finite then* $f^* = f(\mathbf{x}^*)$ *for some* $\mathbf{x}^* \in \mathbf{K}$.

This is in contrast to convex functions where Theorem 13.31 is not true in general. For instance let $\mathbf{K} = [1, +\infty) \subset \mathbb{R}$ and $x \mapsto f(x) = 1/x$. Then $f^* = 0$ but f^* is not attained on $\mathbf{K}$.

We next consider the case where the convex problem (5.1) is defined by SOS-convex polynomials and next when f and the $-g_j$ are convex and a strict convexity condition holds for the Lagrangian in a neighborhood of an optimal solution.

13.3.1 The SOS-convex case

With $f \in \mathbb{R}[\mathbf{x}]$ and $2d \geq \max[\deg f, \max_j \deg g_j]$, consider the semidefinite program:

$$\rho_d = \inf_{\mathbf{y}} \{ L_{\mathbf{y}}(f) : \mathbf{M}_d(\mathbf{y}) \succeq 0; \; L_{\mathbf{y}}(g_j \, \mathbf{y}) \geq 0, \; j = 1, \ldots, m; \; y_0 = 1 \}, \tag{13.28}$$

and its dual

$$\rho_d^* = \sup_{\lambda, \gamma, \sigma_0} \left\{ \gamma : \; f - \gamma = \sigma_0 + \sum_{j=1}^{m} \lambda_j \, g_j \, ; \right. \\ \left. \gamma \in \mathbb{R}, \; \lambda \in \mathbb{R}_+^m, \; \sigma_0 \in \Sigma[\mathbf{x}]_d \right\}. \tag{13.29}$$

Theorem 13.32   *Let $\mathbf{K}$ be as in (13.2) and let Slater's condition hold. Let $f \in \mathbb{R}[\mathbf{x}]$ be such that $f^* := \inf_{\mathbf{x}} \{ f(\mathbf{x}) : \mathbf{x} \in \mathbf{K} \} > -\infty$. Assume that f and $-g_j$ are SOS-convex, $j = 1, \ldots, m$.*

Then $f^* = \rho_d = \rho_d^*$. Moreover, if $\mathbf{y}$ is an optimal solution of (13.28) then $\mathbf{x}^* := (L_{\mathbf{y}}(x_1), \ldots, L_{\mathbf{y}}(x_n)) \in \mathbf{K}$ is a global minimizer of f on $\mathbf{K}$.

Proof   By Theorem 13.31 $f^* = f(\mathbf{z})$ for some minimizer $\mathbf{z} \in \mathbf{K}$. Next, recall the definition of $Q_c(g)$ in (13.16). By Theorem 13.19, $f - f^* \in Q_c(g)$, i.e., $f - f^* = \sigma_0 + \sum_j \lambda_j g_j$ for some $\lambda \in \mathbb{R}_+^m$ and some $\sigma_0 \in \Sigma[\mathbf{x}]$. Therefore, (f^*, λ, σ_0) is a feasible solution of (13.29), which yields $\rho_d^* \geq f^*$, and which combined with $\rho_d^* \leq \rho_d \leq f^*$ yields $\rho_d^* = \rho_d = f^*$. Obviously (13.28) has an optimal solution (e.g. the vector of moments $\mathbf{y}$ of the Dirac measure $\delta_{\mathbf{z}}$ at any minimizer $\mathbf{z} \in \mathbf{K}$) and so, let $\mathbf{y}$ be an optimal solution. By Theorem 13.21, $f^* = \rho_d = L_{\mathbf{y}}(f) \geq f(L_{\mathbf{y}}(\mathbf{x}))$, and similarly, $0 \geq L_{\mathbf{y}}(-g_j) \geq -g_j(L_{\mathbf{y}}(\mathbf{x}))$, $j = 1, \ldots, m$, which shows that $\mathbf{x}^* := L_{\mathbf{y}}(\mathbf{x}) (= L_{\mathbf{y}}(x_1), \ldots, L_{\mathbf{y}}(x_n)) \in \mathbf{K}$ is a global minimizer. $\square$

Therefore when the polynomials f and $(-g_j)$ are SOS-convex, the first semidefinite program in the hierarchy of semidefinite relaxations (6.3) is exact as it is either identical to (13.28) or more constrained, hence with optimal value $\rho_d = f^*$. In other words, the moment-SOS approach recognizes SOS-convexity.

Moreover, observe that (13.28) has only one semidefinite constraint $(\mathbf{M}_d(\mathbf{y}) \succeq 0)$ which, in addition, does not depend on the data f or g_j of the problem (it just states that the moment matrix is positive semidefinite). The data of the problem only appear in the linear constraints $L_{\mathbf{y}}(g_j) \geq 0$ and the linear criterion $L_{\mathbf{y}}(f)$ to minimize.

Recognizing linear and convex quadratic programs Every affine or convex quadratic polynomial is SOS-convex and so by Theorem 13.32, the first semidefinite relaxation of the hierarchy (6.3) is exact for linear programs and convex quadratic programs.

When f and the g_j are affine polynomials, i.e., $f(\mathbf{x}) = \mathbf{f}^T\mathbf{x}$ and $g_j(\mathbf{x}) = g_{j0} + \mathbf{g}_j^T\mathbf{x}$ for some vectors $\mathbf{f}, \mathbf{g}_j \in \mathbb{R}^n$ and some scalars g_{j0}, $j = 1, \ldots, m$, then (5.1) is the linear programming (LP) problem:

$$\mathbf{P}: \quad \inf\{\mathbf{f}^T\mathbf{x} : \mathbf{g}_j^T\mathbf{x} \geq -g_{j0}, \quad j = 1, \ldots, m\}. \tag{13.30}$$

In this case the first semidefinite relaxation of the hierarchy (6.3) reads:

$$\inf_{\mathbf{y},\mathbf{Y}} \left\{ \mathbf{f}^T\mathbf{y} : \begin{bmatrix} 1 & \mathbf{y}^T \\ \mathbf{y} & \mathbf{Y} \end{bmatrix} \succeq 0; \quad \mathbf{g}_j^T\mathbf{y} \geq -g_{j0}, \quad j = 1, \ldots, m \right\},$$

which is equivalent to the linear program $\mathbf{P}$ because the real symmetric matrix $\mathbf{Y}$ plays no role. Given $\mathbf{y} \in \mathbb{R}^n$ arbitrary, one may always find $\mathbf{Y}$ such that the matrix $\begin{bmatrix} 1 & \mathbf{y}^T \\ \mathbf{y} & \mathbf{Y} \end{bmatrix}$ is positive semidefinite. Equivalently, the dual of (6.3) reads:

$$\sup \left\{ \gamma : \mathbf{f}^T\mathbf{x} - \sum_{j=1}^m \lambda_j(g_{j0} + \mathbf{g}_j^T\mathbf{x}) - \gamma = \sigma(\mathbf{x}), \ \forall \mathbf{x} \in \mathbb{R}^n; \lambda \geq 0; \ \sigma \in \Sigma[\mathbf{x}]_1 \right\}.$$

And the only way a linear polynomial can be SOS is when it is constant and nonnegative. Therefore, the above semidefinite program is just

$$\sup_{\gamma,\lambda} \left\{ \gamma : \mathbf{f}^T = \sum_{j=1}^m \lambda_j\mathbf{g}_j^T; \quad \gamma + \sum_{j=1}^m \lambda_j g_{j0} \leq 0; \ \lambda \geq 0 \right\},$$

which is the same as

$$\sup_{\lambda} \left\{ -\sum_{j=1}^{m} \lambda_j \, g_{j0} : \mathbf{f}^T = \sum_{j=1}^{m} \lambda_j \, \mathbf{g}_j^T; \ \lambda \geq 0 \right\},$$

i.e., the dual of the LP (13.30).

When f and the $-g_j$ are convex quadratic polynomials, $f(\mathbf{x}) = \mathbf{f}^T \mathbf{x} + \mathbf{x}^T \mathbf{Q}_0 \mathbf{x}$ and $g_j(\mathbf{x}) = g_{j0} + \mathbf{g}_j^T \mathbf{x} - \mathbf{x}^T \mathbf{Q}_j \mathbf{x}$ for some vectors $\mathbf{f}, \mathbf{g}_j \in \mathbb{R}^n$, some scalars g_{j0}, $j = 1, \ldots, m$, and some real symmetric positive semidefinite matrices $\mathbf{Q}_j$, $j = 0, \ldots, m$, then (5.1) is a convex quadratic programming (QP) problem. In this case the first semidefinite relaxation of the hierarchy (6.3) reads:

$$\begin{aligned} \inf_{\mathbf{y}, \mathbf{Y}} \quad & \mathbf{f}^T \mathbf{y} + \langle \mathbf{Q}_0, \mathbf{Y} \rangle \\ \text{s.t.} \quad & \mathbf{g}_j^T \mathbf{y} - \langle \mathbf{Q}_j, \mathbf{Y} \rangle \geq -g_{j0}, \quad j = 1, \ldots, m \\ & \begin{bmatrix} 1 & \mathbf{y}^T \\ \mathbf{y} & \mathbf{Y} \end{bmatrix} \succeq 0, \end{aligned}$$

whose dual is

$$\sup_{\lambda \geq 0, \gamma, \sigma} \left\{ \gamma : f(\mathbf{x}) - \gamma - \sum_{j=1}^{m} \lambda_j \, g_j(\mathbf{x}) = \sigma(\mathbf{x}), \quad \forall \mathbf{x} \in \mathbb{R}^n; \quad \sigma \in \Sigma[\mathbf{x}]_1 \right\}.$$

Recall that $f - \sum_j \lambda_j g_j$ is convex quadratic for every $\lambda \geq 0$, and every non-negative quadratic polynomial is SOS. Therefore solving the above semidefinite program is the same as solving

$$\sup_{\lambda \geq 0, \gamma} \left\{ \gamma : f(\mathbf{x}) - \sum_{j=1}^{m} \lambda_j \, g_j(\mathbf{x}) \geq \gamma, \quad \forall \mathbf{x} \in \mathbb{R}^n \right\},$$

or, equivalently,

$$\sup_{\lambda \geq 0} \left\{ \inf_{\mathbf{x}} \left\{ f(\mathbf{x}) - \sum_{j=1}^{m} \lambda_j \, g_j(\mathbf{x}) : \mathbf{x} \in \mathbb{R}^n \right\} \right\},$$

which is also the dual $\mathbf{P}^*$ of (5.1). And so if Slater's condition holds for (5.1) then there is no duality gap.

13.3.2 The strictly convex case

If f or some of the $-g_j$ is not SOS-convex but f and $-g_j$ are convex, $j = 1, \ldots, m$, and the Hessian $\nabla^2 f$ is positive definite at every minimizer $\mathbf{x}^* \in \mathbf{K}$, then finite convergence of the semidefinite relaxations (6.3) still takes place.

Theorem 13.33 *Let f be convex and $\mathbf{K}$ be as in (13.2) with $-g_j$ convex, $j = 1, \ldots, m$. Let Assumption 2.14 and Slater's condition hold, and $\nabla^2 f(\mathbf{x}^*) \succ 0$ at every global minimizer $\mathbf{x}^* \in \mathbf{K}$ (assumed to be finitely many).*

  Then the hierarchy of semidefinite relaxations (6.3) and its dual (6.4) have finite convergence. That is, $f^* = \rho_d^* = \rho_d$ for some index d. In addition, both (6.3) and (6.4) have an optimal solution.

Proof By Theorem 13.20 (and with $g_0 = 1$), $f - f^* = \sum_{j=0}^m \sigma_j g_j$ for some SOS polynomials $(\sigma_j) \subset \Sigma[\mathbf{x}]$. If d is the smallest integer such that $2d := \max_j [\deg \sigma_j g_j]$ then $(f^*, (\sigma_j))$ is an optimal solution of (6.4) because $\rho_d^* \leq f^*$ for all d. But then as $\rho_d^* \leq \rho_d \leq f^*$, we also have $\rho_d = f^*$, which in turn implies that (6.3) has an optimal solution (e.g. the vector $\mathbf{y} = (y_\alpha)$ of moments of the Dirac measure $\delta_{\mathbf{z}}$ at any global minimizer $\mathbf{z} \in \mathbf{K}$). $\qquad\square$

When compared to Theorem 13.32 for the SOS-convex case, in the strictly convex case the simplified semidefinite relaxation (13.28) is not guaranteed to be exact. However, finite convergence still occurs for the standard semidefinite relaxations (6.3). Hence the hierarchy of semidefinite relaxations exhibits particularly nice behavior for convex problems, a highly desirable property since convex optimization problems are easier to solve.

13.4 Exercises

Exercise 13.1 In Example 13.30 exhibit a lifted semidefinite representation of $\mathbf{K}$ with fewer variables $\mathbf{y}$.

Exercise 13.2 Let $n = 2$ and $\mathbf{K} = \{\mathbf{x} \in \mathbb{R}^2 : g_j(\mathbf{x}) \geq 0, j = 1, \ldots, m\}$ where each polynomial g_j is concave of degree at most 4, $j = 1, \ldots, m$. Assume that $\mathbf{K}$ is compact and Slater's condition holds.

  (a) Show that the P-BDR property holds for $\mathbf{K}$. (*Hint:* See Example 13.26.)
  (b) Write an explicit semidefinite representation of $\mathbf{K}$.

Exercise 13.3 Let $f \in \mathbb{R}[x]$ be a univariate polynomial and let $I \subseteq \mathbb{R}$ be the interval $[0, 1]$, or $[0, \infty)$, or $(-\infty, \infty)$. Show that checking whether f is convex on I reduces to solving a single semidefinite program.

Exercise 13.4 Let $f \in \mathbb{R}[\mathbf{x}]$ be a bivariate polynomial of degree 4. Show that checking whether f is convex on $\mathbb{R}^2$ reduces to solving two semidefinite programs. (*Hint:* Use the result that nonnegative bivariate quartic polynomials are SOS.)

Exercise 13.5 Let $f \in \mathbb{R}[\mathbf{x}]$ be separable, i.e., $f = \sum_{i=1}^{n} f_i$ where each $f_i \in \mathbb{R}[x_i]$, $i = 1, \ldots, n$, is a univariate polynomial. Show that f is convex on $\mathbb{R}^n$ if and only if each f_i is convex on $\mathbb{R}$, $i = 1, \ldots, n$. Let $n = 3$. Show how to check whether f is convex on $[0, 1] \times [0, \infty) \times (-\infty, \infty)$.

13.5 Notes and sources

13.1 Results in this section are taken from Ahmadi et al. (2013) and Ahmadi and Parrilo (2010), as well as Lasserre (2009b) and Lasserre (2010a). In particular, the complexity results in Theorem 13.10 and Theorem 13.11(b) as well as the nice characterization of pseudoconvex and quasiconvex polynomials of odd degree in Theorem 13.11(a) are from Ahmadi et al. (2013), whereas Theorem 13.15 is from Ahmadi and Parrilo (2010). Lemmas 13.16 and 13.17 are from Helton and Nie (2010, 2012) who first introduced the important class of SOS-convex polynomials.

13.2 In view if the power of semidefinite programming, the important issue of semidefinite representation of convex sets was raised in Ben-Tal and Nemirovski (2001). Theorem 13.27 is from Lasserre (2009a) while Theorem 13.24 and 13.29 are from Helton and Nie (2012) and are based on previous results from Helton and Nie (2009, 2010). The main message in Helton and Nie (2012) is that a compact convex set with nonempty interior has an SDr if every irreducible piece of its boundary is "nonsingular" and has "positive curvature." In particular, the authors have developed a *calculus* for the semidefinite representation of the convex hull conv $\left(\bigcup_{k=1}^{m} W_k \right)$ from semidefinite representations of the convex sets W_k; see Helton and Nie (2012, Theorem 4.15). These weak conditions which are nearly necessary for existence of an SDr led to the *Helton–Nie* conjecture that every compact and convex basic semi-algebraic set has an SDr. Under slightly stronger conditions the semidefinite representations obtained in Theorems 13.27 and 13.29 are *explicit* in terms of the polynomials that define the set. Finally, Netzer et al. (2010) have proved that if $\mathbf{K}$ has the semidefinite representation (13.26) then necessarily all faces of $\mathbf{K}$ must be exposed, a nice necessary geometric condition; see also Gouveia and Netzer (2011) for more details. Finally, Theorem 13.20 is from de Klerk and Laurent (2011), a nice refinement of Theorem 3.4 in Lasserre (2009b).

Finally, in the particular case where $\mathbf{K}$ is a real algebraic variety, i.e., the real zero set of finitely many algebraic equations $g_j(\mathbf{x}) = 0$, $j = 1, \ldots, m$, Gouveia et al. (2010) have considered a hierarchy of *Theta bodies* to approximate $\mathrm{conv}(V_{\mathbb{R}}(I))$ (*equal to* $\mathrm{conv}(\mathbf{K})$) for the ideal $I \subset \mathbb{R}[\mathbf{x}]$ generated by the g_j. Namely, the kth Theta body $\mathrm{TH}_k(I)$ is the set

$$\mathrm{TH}_k = \{\, \mathbf{x} \in \mathbb{R}^n : \ell(\mathbf{x}) \geq 0, \quad \forall \ell \in \mathbb{R}[\mathbf{x}]_1, \ \ell \text{ is } k\text{-SOS mod } I \,\},$$

where ℓ is k-SOS mod I if $\ell = \sigma \bmod I$ for some SOS polynomial $\sigma \in \Sigma[\mathbf{x}]_k$. Then

$$T H_1 \supseteq T H_2 \supseteq \cdots \supseteq T H_k \supseteq \cdots \supset \mathbf{K}.$$

The ideal I is said to be $(1 - k)$-SOS if every $\ell \in \mathbb{R}[\mathbf{x}]_1$ which is nonnegative on $\mathrm{conv}(V_{\mathbb{R}}(I))$ is also k-SOS mod I. The name Theta body comes from a paper by Lovász (2003) who asked which ideals of $\mathbb{R}[\mathbf{x}]$ are $(1, 1)$-SOS. Under a certain technical condition, the Theta body hierarchy is a modified version of the semidefinite relaxations defined in (6.3)–(6.4). As an application they could define specific semidefinite relaxations for some combinatorial problems in graphs; see Gouveia et al. (2012) and the survey in Gouveia and Thomas (2012).

13.3 Belousov (1977) was one of the first to investigate the properties of convex polynomials. Theorem 13.31 is from Belousov and Klatte (2002). Theorem 13.32 is from Lasserre (2008) whereas Theorem 13.33 which is from de Klerk and Laurent (2011) is a refinement of Theorem 3.4 in Lasserre (2009b). Some properties of convex polynomials with applications to polynomial programming were also investigated in Andronov et al. (1982) and Belousov and Klatte (2002).

14

Parametric optimization

14.1 Introduction

Consider the *parametric* optimization problem:

$$\mathbf{P_y}: \quad J(\mathbf{y}) = \inf_{\mathbf{x}} \{ f(\mathbf{x}, \mathbf{y}) : \mathbf{x} \in \mathbf{K_y} \} \tag{14.1}$$

where $\mathbf{y}$ is a *parameter* vector which lives in some parameter set $\mathbf{Y} \subset \mathbb{R}^p$, $f \in \mathbb{R}[\mathbf{x}, \mathbf{y}]$, and for every $\mathbf{y} \in \mathbf{Y}$, $\mathbf{K_y} \subset \mathbb{R}^n$ is the basic semi-algebraic set defined by:

$$\mathbf{K_y} := \{ \mathbf{x} \in \mathbb{R}^n : g_j(\mathbf{x}, \mathbf{y}) \geq 0, \ j = 1, \dots, m \},$$

for some polynomials $(g_j) \subset \mathbb{R}[\mathbf{x}, \mathbf{y}]$. Parametric programming is concerned with:

- the *optimal value function* $\mathbf{y} \mapsto J(\mathbf{y})$,
- the *global minimizer set-valued mapping* $\mathbf{y} \mapsto \mathbf{X}^*(\mathbf{y})$, and/or
- the *Lagrange-multiplier set-valued mapping* $\mathbf{y} \mapsto \mathbf{\Lambda}^*(\mathbf{y})$,

as functions of the parameter vector $\mathbf{y}$. But computing the mappings $\mathbf{y} \to J(\mathbf{y})$ and $\mathbf{y} \mapsto \mathbf{X}^*(\mathbf{y}), \mathbf{\Lambda}^*(\mathbf{y})$ is a very challenging infinite-dimensional problem and therefore, in general, a less ambitious and more realistic goal is to obtain information on $J(\mathbf{y})$ and $\mathbf{X}^*(\mathbf{y}), \mathbf{\Lambda}^*(\mathbf{y})$, only locally around a nominal value $\mathbf{y}_0$ of the parameter.

However, in the polynomial context we will see that much more is possible. In particular, one can approximate the mappings $J(\mathbf{y})$ and $\mathbf{X}^*(\mathbf{y})$ by using what we call the "joint+marginal" approach which consists of solving a hierarchy of semidefinite relaxations, very much in the spirit of those developed in Chapter 6 for the moment-SOS approach in nonparametric polynomial optimization. We will then see how to use results from parametric programming for the following problems.

- For the *robust* polynomial optimization problem:

$$J^* = \sup_{\mathbf{y} \in \Omega} \inf_{\mathbf{x}} \{ f(\mathbf{x}, \mathbf{y}) : \mathbf{x} \in \mathbf{K_y} \}, \qquad (14.2)$$

which may be considered as a game against nature. Indeed, after $\mathbf{y} \in \Omega$ (player-1's action) is chosen, player-2 (nature) picks an action $\mathbf{x}^*(\mathbf{y}) \in \mathbf{K_y}$ whose value $J(\mathbf{y}) = f(\mathbf{x}^*(\mathbf{y}), \mathbf{y})$ minimizes $f(\mathbf{x}, \mathbf{y})$ on $\mathbf{K_y}$ (the worst possible outcome for $\mathbf{y}$). However, notice that if (14.2) is a robust optimization problem for player-1, then it is a parametric optimization problem for player-2.
- For the nonparametric optimization problem $\mathbf{P} : \inf_{\mathbf{x}} \{ f(\mathbf{x}) : \mathbf{x} \in \mathbf{K} \}$, when $\mathbf{K} \subset \mathbb{R}^n$ is convex or when $\mathbf{P}$ is a discrete optimization problem for which feasibility is easy to detect (for example the MAXCUT, k-cluster and 0/1 knapsack problems).

14.2 Parametric optimization

Let $\mathbf{Y} \subset \mathbb{R}^p$ be a compact set, called the *parameter* set, and let

$$\mathbf{K} := \{ (\mathbf{x}, \mathbf{y}) \in \mathbb{R}^{n+p} : \mathbf{y} \in \mathbf{Y}; \ h_j(\mathbf{x}, \mathbf{y}) \geq 0, \quad j = 1, \ldots, m \}, \quad (14.3)$$

for some polynomials $(h_j) \subset \mathbb{R}[\mathbf{x}, \mathbf{y}]$. For each $\mathbf{y} \in \mathbf{Y}$, fixed, consider the following optimization problem:

$$\mathbf{P_y} : \quad J(\mathbf{y}) := \inf_{\mathbf{x}} \{ f(\mathbf{x}, \mathbf{y} : (\mathbf{x}, \mathbf{y}) \in \mathbf{K} \}. \qquad (14.4)$$

The interpretation is as follows: $\mathbf{Y}$ is a set of parameters and for each value $\mathbf{y} \in \mathbf{Y}$ of the parameter, one wishes to compute an optimal *decision* vector $\mathbf{x}^*(\mathbf{y})$ that solves problem (14.4). As already mentioned, parametric optimization is concerned with:

- the *optimal value function* $\mathbf{y} \mapsto J(\mathbf{y})$, as well as
- the set-valued *minimizer mapping* $\mathbf{y} \mapsto \mathbf{X}^*(\mathbf{y})$ (respectively *dual multiplier mapping* $\mathbf{y} \mapsto \Lambda^*(\mathbf{y})$) where for each $\mathbf{y} \in \mathbf{Y}$, $\mathbf{X}^*(\mathbf{y}) \subset \mathbb{R}^n$ (respectively $\Lambda^*(\mathbf{y}) \subset \mathbb{R}^m$) is the set of global minimizers (respectively dual optimal solutions) of (14.4).

Of course, this is a difficult infinite-dimensional problem as one tries to approximate *functions* and so, in general, a less ambitious goal is to obtain information locally around some nominal value $\mathbf{y}_0$ of the parameter. However, as one will see below, in the polynomial context, i.e., when $f \in \mathbb{R}[\mathbf{x}, \mathbf{y}]$ and $\mathbf{K}, \mathbf{Y}$ are basic semi-algebraic sets, much more is possible.

14.2.1 The "joint+marginal" approach

Let φ be a Borel probability measure on $\mathbf{Y}$, with a positive density with respect to the Lebesgue measure on the smallest affine variety that contains $\mathbf{Y}$. For instance, choose for φ the probability measure uniformly distributed on $\mathbf{Y}$. Of course, one may also treat the case of a *discrete* set of parameters $\mathbf{Y}$ (finite or countable) by taking for φ a discrete probability measure on $\mathbf{Y}$ with strictly positive weight at each point of the support. Sometimes, for example in the context of optimization with data uncertainty, φ is already specified. We will use φ (or more precisely, its moments) to obtain information on the distribution of optimal solutions $\mathbf{x}^*(\mathbf{y})$ of $\mathbf{P_y}$, viewed as random vectors. In the rest of this section we assume that for every $\mathbf{y} \in \mathbf{Y}$, the set

$$\mathbf{K_y} := \{\, \mathbf{x} \in \mathbb{R}^n : h_j(\mathbf{x}, \mathbf{y}) \geq 0, \quad j = 1, \ldots, m \,\} \qquad (14.5)$$

is not empty.

A related infinite-dimensional linear program Let $\mathcal{M}(\mathbf{K})_+$ be the set of finite Borel measures on $\mathbf{K}$, and consider the following infinite-dimensional linear program

$$\mathbf{P}: \qquad \rho := \inf_{\mu \in \mathcal{M}(\mathbf{K})_+} \left\{ \int_{\mathbf{K}} f \, d\mu \, : \, \pi\mu = \varphi \right\} \qquad (14.6)$$

where $\pi\mu$ denotes the marginal of μ on $\mathbb{R}^p$, that is, $\pi\mu$ is a probability measure on $\mathbb{R}^p$ defined by $\pi\mu(B) = \mu(\mathbb{R}^n \times B)$ for all $B \in \mathcal{B}(\mathbb{R}^p)$. Notice that $\mu(\mathbf{K}) = 1$ for any feasible solution μ of $\mathbf{P}$. Indeed, as φ is a probability measure and $\pi\mu = \varphi$ one has $1 = \varphi(\mathbf{Y}) = \mu(\mathbb{R}^n \times \mathbb{R}^p) = \mu(\mathbf{K})$.

Theorem 14.1   *Let both $\mathbf{Y} \subset \mathbb{R}^p$ and $\mathbf{K}$ in (14.3) be compact and assume that for every $\mathbf{y} \in \mathbf{Y}$, the set $\mathbf{K_y} \subset \mathbb{R}^n$ in (14.5) is nonempty. For problem $\mathbf{P}$ in (14.6), let $\mathbf{X}^*(\mathbf{y}) := \{\, \mathbf{x} \in \mathbb{R}^n : f(\mathbf{x}, \mathbf{y}) = J(\mathbf{y})\,\}$, $\mathbf{y} \in \mathbf{Y}$.*

(a) $\rho = \displaystyle\int_{\mathbf{Y}} J(\mathbf{y}) \, d\varphi(\mathbf{y})$ *and $\mathbf{P}$ has an optimal solution.*

(b) *For every optimal solution μ^* of $\mathbf{P}$, and for φ-almost all $\mathbf{y} \in \mathbf{Y}$, there is a probability measure $\psi^*(d\mathbf{x}\,|\,\mathbf{y})$ on $\mathbb{R}^n$, concentrated on $\mathbf{X}^*(\mathbf{y})$, such that*

$$\mu^*(C \times B) = \int_B \psi^*(C\,|\,\mathbf{y}) \, d\varphi(\mathbf{y}), \qquad \forall B \in \mathcal{B}(\mathbf{Y}), \; C \in \mathcal{B}(\mathbb{R}^n).$$

$$(14.7)$$

(c) *Assume that for φ-almost all $\mathbf{y} \in \mathbf{Y}$, the set of minimizers $\mathbf{X}^*(\mathbf{y})$ is the singleton $\{\mathbf{x}^*(\mathbf{y})\}$ for some $\mathbf{x}^*(\mathbf{y}) \in \mathbf{K_y}$. Then there is a measurable mapping $g : \mathbf{Y} \to \mathbf{K_y}$ such that $g(\mathbf{y}) = \mathbf{x}^*(\mathbf{y})$ for φ-almost all $\mathbf{y} \in \mathbf{Y}$ and*

$$\rho = \int_{\mathbf{Y}} f(g(\mathbf{y}), \mathbf{y}) \, d\varphi(\mathbf{y}) = \int_{\mathbf{Y}} f(\mathbf{x}^*(\mathbf{y}), \mathbf{y}) \, d\varphi(\mathbf{y}) = \int_{\mathbf{Y}} J(\mathbf{y}) \, d\varphi(\mathbf{y}).$$
$$(14.8)$$

For every $\boldsymbol{\alpha} \in \mathbb{N}^n$ and $\boldsymbol{\beta} \in \mathbb{N}^p$:

$$\int_{\mathbf{K}} \mathbf{x}^{\alpha} \mathbf{y}^{\beta} \, d\mu^*(\mathbf{x}, \mathbf{y}) = \int_{\mathbf{Y}} \mathbf{y}^{\beta} \, g(\mathbf{y})^{\alpha} \, d\varphi(\mathbf{y}). \qquad (14.9)$$

As one may see, any optimal solution μ^* of $\mathbf{P}$ encodes *all* information on the global minimizers $\mathbf{x}^*(\mathbf{y})$ of $\mathbf{P_y}$. For instance, let B be a given Borel set of $\mathbb{R}^n$. Then from Theorem 14.1,

$$\text{Prob}\,(\mathbf{x}^*(\mathbf{y}) \in B) = \mu^*(B \times \mathbb{R}^p) = \int_{\mathbf{Y}} \psi^*(B \mid \mathbf{y}) \, d\varphi(\mathbf{y}),$$

with ψ^* as in Theorem 14.1(b).

Now the reader can understand where the name *"joint+marginal"* is coming from. Indeed, one considers a related linear programming problem $\mathbf{P}$ over the space of Borel probability measures μ on $\mathbb{R}^n \times \mathbb{R}^p$ (i.e., *joint* probability distributions on $\mathbf{x}$ *and* $\mathbf{y}$) whose *marginal* on $\mathbf{Y}$ is fixed and equal to the given probability measure φ. Incidentally, when φ is the Lebesgue measure, such measures μ are called *Young measures* and have been introduced to characterize nonclassical solutions of problems in the calculus of variations; see Section 14.6.

Duality Consider the following infinite-dimensional linear program $\mathbf{P}^*$:

$$\rho^* := \sup_{p \in \mathbb{R}[\mathbf{y}]} \left\{ \int_{\mathbf{Y}} p \, d\varphi \; : \; f(\mathbf{x}, \mathbf{y}) - p(\mathbf{y}) \geq 0, \quad \forall (\mathbf{x}, \mathbf{y}) \in \mathbf{K} \right\}. \quad (14.10)$$

Observe that any feasible solution $p \in \mathbb{R}[\mathbf{y}]$ of (14.10) is dominated by J, i.e., $J(\mathbf{y}) \geq p(\mathbf{y})$ for all $\mathbf{y} \in \mathbf{Y}$. Then $\mathbf{P}^*$ is a *dual* of $\mathbf{P}$ and the next result states that there is no duality gap between $\mathbf{P}$ and $\mathbf{P}^*$.

Lemma 14.2 *Let both $\mathbf{Y} \subset \mathbb{R}^p$ and $\mathbf{K}$ in (14.3) be compact with $\mathbf{K_y} \neq \emptyset$ for all $\mathbf{y} \in \mathbf{Y}$ and let $\mathbf{P}$ and $\mathbf{P}^*$ be as in (14.6) and (14.10), respectively. Then there is no duality gap, i.e., $\rho = \rho^*$.*

Moreover, any maximizing sequence of $\mathbf{P}^*$ allows approximation of the optimal value function $\mathbf{y} \mapsto J(\mathbf{y})$ in the following precise sense.

Recall that $L_1(\varphi)$ is the space of φ-integrable functions on $\mathbf{Y}$ (a Banach space when equipped with the norm $\| f \| := \int_{\mathbf{Y}} |f| d\varphi$).

Definition 14.3   A sequence of measurable functions (h_n) on a measure space $(\mathbf{Y}, \mathcal{B}(\mathbf{Y}), \varphi)$ converges to h, φ-almost uniformly, if for every $\epsilon > 0$ there is a set $A \in \mathcal{B}(\mathbf{Y})$ such that $\varphi(A) < \epsilon$ and $h_n \to h$ uniformly on $\mathbf{Y} \setminus A$.

Corollary 14.4   *Let both $\mathbf{Y} \subset \mathbb{R}^p$ and $\mathbf{K}$ in (14.3) be compact and assume that for every $\mathbf{y} \in \mathbf{Y}$, the set $\mathbf{K}_\mathbf{y}$ in (14.5) is nonempty. Let $\mathbf{P}^*$ be as in (14.10). If $(p_d)_{d \in \mathbb{N}} \subset \mathbb{R}[\mathbf{y}]$ is a maximizing sequence of (14.10) then $p_d \leq J$ on $\mathbf{Y}$ for every d and*

$$\int_{\mathbf{Y}} |J(\mathbf{y}) - p_d(\mathbf{y})| \, d\varphi(\mathbf{y}) \to 0 \quad as \ d \to \infty, \qquad (14.11)$$

i.e., $p_d \to J(\cdot)$ for the $L_1(\varphi)$-norm.

   Moreover, define the functions $(\tilde{p}_d)$, $d \in \mathbb{N}$, as follows:

$$\tilde{p}_0 := p_0, \quad \mathbf{y} \mapsto \tilde{p}_d(\mathbf{y}) := \max[\tilde{p}_{d-1}(\mathbf{y}), p_d(\mathbf{y})], \quad d = 1, 2, \ldots.$$

Then $\tilde{p}_d \to J(\cdot)$ φ-almost uniformly on $\mathbf{Y}$.

Theorem 14.1 and Corollary 14.4 are even valid in the more general context where f is continuous and $\mathbf{K}, \mathbf{Y}$ are compact sets. So, theoretically, any optimal solution of $\mathbf{P}$ and its dual $\mathbf{P}^*$ provide all the information required in parametric optimization. It is worth noticing that in view of the (strong) L_1-norm convergence $\| J - p_d \| \to 0$ in (14.11) and the almost-uniform convergence $\tilde{p}_d \to J$, both p_d and $\tilde{p}_d$ provide rather strong approximations of J, which is one of the main goals of parametric optimization.

However, in full generality $\mathbf{P}$ and $\mathbf{P}^*$ are intractable numerically. But as we will see next, when the data are polynomials and basic semi-algebraic sets, we can obtain strong approximations of the optimal value function $\mathbf{y} \mapsto J(\mathbf{y})$ and of the global minimizer mapping $\mathbf{y} \mapsto \mathbf{x}^*(\mathbf{y})$ as well (when the latter is well defined).

14.2.2 A hierarchy of semidefinite relaxations

With $\mathbf{K} \subset \mathbb{R}^n \times \mathbb{R}^p$ as in (14.3), let $\mathbf{Y} \subset \mathbb{R}^p$ be the compact semi-algebraic set:

$$\mathbf{Y} := \{\mathbf{y} \in \mathbb{R}^p : h_k(\mathbf{y}) \geq 0, \quad k = m+1, \ldots, t\} \qquad (14.12)$$

for some polynomials $(h_k)_{k=m+1}^t \in \mathbb{R}[\mathbf{y}]$. For instance if $\mathbf{Y}$ is the box $\mathbf{B} = [-a, a]^p$ for some $a \in \mathbb{R}$, then $t = m+p$ and $h_{m+i}(\mathbf{y}) = a^2 - y_i^2, i = 1, \ldots, p$.

Semidefinite relaxations Let $v_k := \lceil (\deg h_k)/2 \rceil$ for every $k = 1, \ldots, t$. Next, let $\gamma = (\gamma_\beta)$ with

$$\gamma_\beta = \int_Y \mathbf{y}^\beta \, d\varphi(\mathbf{y}), \qquad \forall \beta \in \mathbb{N}^p, \qquad (14.13)$$

be the moments of a probability measure φ on $\mathbf{Y}$, absolutely continuous with respect to the Lebesgue measure, and let $d_0 := \max[\lceil (\deg f)/2 \rceil, \max_k v_k]$. For $d \geq d_0$, consider the following hierarchy of semidefinite relaxations:

$$
\begin{aligned}
\rho_d = \quad &\inf_{\mathbf{z}} \quad L_{\mathbf{z}}(f) \\
&\text{s.t.} \quad \mathbf{M}_d(\mathbf{z}) \succeq 0 \\
&\qquad \mathbf{M}_{d-v_j}(h_j \, \mathbf{z}) \succeq 0, \quad j = 1, \ldots, t \\
&\qquad L_{\mathbf{z}}(\mathbf{y}^\beta) = \gamma_\beta, \quad \forall \beta \in \mathbb{N}^p_{2d}.
\end{aligned}
\qquad (14.14)
$$

Theorem 14.5 *Let $\mathbf{K}, \mathbf{Y}$ be as in (14.3) and (14.12) respectively, and let $(h_k)_{k=1}^t$ satisfy Assumption 2.14. Assume that for every $\mathbf{y} \in \mathbf{Y}$ the set $\mathbf{K_y}$ in (14.5) is nonempty and consider the semidefinite relaxations (14.14).*

(a) The sequence (ρ_d), $d \in \mathbb{N}$, is monotone nondecreasing and $\rho_d \uparrow \rho$ as $d \to \infty$.

(b) Let $\mathbf{z}^d$ be a nearly optimal solution of (14.14), for example, such that $L_{\mathbf{z}^d}(f) \leq \rho_d + 1/d$, and assume the same condition as in Theorem 14.1(c). Then with $g : \mathbf{Y} \to \mathbf{K_y}$ being the measurable mapping in Theorem 14.1(c):

$$\lim_{d \to \infty} z^d_{\alpha\beta} = \int_Y \mathbf{y}^\beta \, g(\mathbf{y})^\alpha \, d\varphi(\mathbf{y}), \qquad \forall \alpha \in \mathbb{N}^n, \; \beta \in \mathbb{N}^p. \qquad (14.15)$$

In particular, for every $k = 1, \ldots, n$,

$$\lim_{i \to \infty} z^d_{e(k)\beta} = \int_Y \mathbf{y}^\beta \underbrace{g_k(\mathbf{y}) \, d\varphi(\mathbf{y})}_{d\phi_k(\mathbf{y})} = \int_Y \mathbf{y}^\beta \, d\phi_k(\mathbf{y}), \qquad \forall \beta \in \mathbb{N}^p,$$

$$(14.16)$$

where $e(k)_j = \delta_{j=k}$, $j = 1, \ldots, n$ (with $\delta.$ being the Kronecker symbol).

Among other things, Theorem 14.5 states that if one lets $d \to \infty$ then one can approximate, as closely as desired, all moments of the (possibly signed) measures $d\phi_k(\mathbf{y}) = g_k(\mathbf{y}) d\mathbf{y}$ on $\mathbf{Y}$, $k = 1, \ldots, n$ (recall that $\mathbf{y} \mapsto g(\mathbf{y})$ is the global minimizer mapping of **P**). As we will see below, this precious moment information can be used to approximate the function g_k.

The dual of semidefinite relaxations Consider the dual of the semidefinite program (14.14) which reads:

$$
\rho_d^* = \sup_{p_d, (\sigma_j)} \int_{\mathbf{Y}} p_d \, d\varphi \;\; = \sum_{\beta \in \mathbb{N}_{2d}^p} p_{d\beta} \, \gamma_\beta
$$

$$
\text{s.t. } f - p_d = \sigma_0 + \sum_{j=1}^{t} \sigma_j \, h_j \tag{14.17}
$$

$$
p_d \in \mathbb{R}[\mathbf{y}]; \;\; \sigma_j \subset \Sigma[\mathbf{x}, \mathbf{y}], \quad j = 1, \ldots, t
$$

$$
\deg p_d \leq 2d, \; \deg \sigma_j h_j \leq 2d, \quad j = 1, \ldots, t.
$$

Observe that (14.17) is a strengthening of (14.10) as one is restricted to polynomials $p \in \mathbb{R}[\mathbf{y}]$ of degree at most $2d$ and the nonnegativity of $f - p$ in (14.10) is replaced in (14.17) with the stronger weighted SOS Putinar representation of Chapter 2. Therefore $\rho_d^* \leq \rho^*$ for every $d \in \mathbb{N}$.

Theorem 14.6 *Let* **K**, **Y** *be as in (14.3) and (14.12) respectively, and let* $(h_k)_{k=1}^{t}$ *satisfy Assumption 2.14. Assume that for every* $\mathbf{y} \in \mathbf{Y}$ *the set* $\mathbf{K_y}$ *in (14.5) is nonempty, and consider the semidefinite programs (14.17).*

(a) The sequence (ρ_d^*), $d \in \mathbb{N}$, *is monotone nondecreasing and* $\rho_d^* \uparrow \rho$ *as* $d \to \infty$.

(b) Let $(p_d, (\sigma_j^d))$ *be a nearly optimal solution of (14.17), for example such that* $\int_{\mathbf{Y}} p_d d\varphi \geq \rho_d^* - 1/d$. *Then* $p_d(\cdot) \leq J(\cdot)$ *and*

$$
\lim_{d \to \infty} \int_{\mathbf{Y}} |J(\mathbf{y}) - p_d(\mathbf{y})| \, d\varphi(\mathbf{y}) = 0, \tag{14.18}
$$

i.e., $\| p_d - J \|_1 \to 0$ *for the* $L_1(\varphi)$-*norm. Moreover, if one defines*

$$
\tilde{p}_0 := p_0, \quad \mathbf{y} \mapsto \tilde{p}_d(\mathbf{y}) := \max [\, \tilde{p}_{d-1}(\mathbf{y}), p_d(\mathbf{y}) \,], \quad d = 1, 2, \ldots,
$$

then $\tilde{p}_d \to J(\cdot)$ φ-*almost uniformly on* **Y**.

So if the semidefinite relaxation (14.14) provides useful information on the global minimizer mapping $\mathbf{y} \mapsto g(\mathbf{y})$, its dual (14.17) provides lower polynomial (respectively piecewise polynomial) approximations $(p_d(\mathbf{y}))$ (respectively $(\tilde{p}_d)$), $d \in \mathbb{N}$, of the optimal value function $J(\mathbf{y})$ of **P**, with strong convergence properties as d increases.

Remark 14.7 Let $\mathbf{Y}$ be the box $[-a, a]^p$ and let the compact set $\mathbf{K}$ in (14.3) have nonempty interior. Let $M > 0$ be sufficiently large that $M - x_i^2 \geq 0$, $i = 1, \ldots, n$, whenever $(\mathbf{x}, \mathbf{y}) \subset \mathbf{K}$. If one includes the redundant quadratic constraints $M - x_i^2 \geq 0, i = 1, \ldots, n$, in the definition of $\mathbf{K}$, then the semidefinite relaxation (14.14) and its dual (14.17) have an optimal solution $\mathbf{z}^*$ and $p_d^* \in \mathbb{R}[\mathbf{y}]$, respectively.

This is because Slater's condition holds for (14.14) and the set of feasible solutions of (14.14) is bounded. Indeed, from $M - x_i^2 \geq 0$ and $a^2 - y_j^2 \geq 0$, for all $i = 1, \ldots, n$, $j = 1, \ldots, p$, and the constraints $\mathbf{M}_{d-v_j}(h_j\,\mathbf{z}) \succeq 0$, one has

$$L_{\mathbf{z}}(y_\ell^{2k}) \leq a^{2k}, \quad \forall \ell = 1, \ldots, p; \qquad L_{\mathbf{z}}(x_i^{2k}) \leq M^k, \quad \forall i = 1, \ldots, n,$$

for all $k \leq d$. So by Lasserre and Netzer (2007, Lemma 4.3, p. 111) this implies $\sup\{\,|z_{\alpha\beta}| \,:\, (\boldsymbol{\alpha}, \boldsymbol{\beta}) \in \mathbb{N}_{2d}^{n+p}\} < \infty$. Therefore, the feasible set is compact since it is closed and bounded, which in turn implies that (14.14) has an optimal solution $\mathbf{z}^*$ if it has a feasible solution. But as $\mathbf{K}$ has a nonempty interior it contains an open set $O \subset \mathbb{R}^n \times \mathbb{R}^m$. Let $O_{\mathbf{y}} \subset \mathbf{Y}$ be the projection of O onto $\mathbf{Y}$, so that its Lebesgue volume is positive. Let μ be the finite Borel measure on $\mathbf{K}$ defined by

$$\mu(A \times B) := \int_A \phi(B \mid \mathbf{y})\,d\mathbf{y}, \qquad A \in \mathcal{B}(\mathbb{R}^p), \; B \in \mathcal{B}(\mathbb{R}^n),$$

where for every $\mathbf{y} \in O_{\mathbf{y}}$, $\phi(d\mathbf{x} \mid \mathbf{y})$ is the probability measure on $\mathbb{R}^n$, supported on $\mathbf{K}_{\mathbf{y}}$, and defined by:

$$\phi(B \mid \mathbf{y}) = \mathrm{vol}(\mathbf{K}_{\mathbf{y}} \cap B)/\mathrm{vol}(\mathbf{K}_{\mathbf{y}}), \qquad \forall B \in \mathcal{B}(\mathbb{R}^n)$$

(and where again vol$(\cdot)$ denotes the Lebesgue volume in $\mathbb{R}^n$). On $\mathbf{Y} \setminus O_{\mathbf{y}}$, the probability $\phi(d\mathbf{x} \mid \mathbf{y})$ is an arbitrary probability measure on $\mathbf{K}_{\mathbf{y}}$. The vector $\mathbf{z}$ of moments of μ is a feasible solution of (14.14). In addition, $\mathbf{M}_d(\mathbf{z}) \succ 0$ and $\mathbf{M}_{d-v_j}(h_j\,\mathbf{z}) \succ 0$, $j = 1, \ldots, t$, i.e., Slater's condition holds and so (14.17) has an optimal solution.

Consequences

Theorem 14.5 has some interesting consequences.

Functionals of optimal solutions If φ is a probability measure one may wish to evaluate the φ-mean or variance of optimal solutions $\mathbf{x}^*(\mathbf{y})$ of $\mathbf{P}_{\mathbf{y}}$, or more generally some polynomial functional of $\mathbf{x}^*(\mathbf{y})$.

Corollary 14.8 *Under the assumptions of Theorem 14.5, let $h \in \mathbb{R}[\mathbf{x}]$ be the polynomial $\mathbf{x} \mapsto h(\mathbf{x}) := \sum_{\alpha \in \mathbb{N}^n} h_\alpha \mathbf{x}^\alpha$, and let $\mathbf{z}^d$ be a nearly optimal solution of the semidefinite relaxations (14.14). Then*

$$\lim_{d \to \infty} \sum_{\alpha \in \mathbb{N}^n} h_\alpha z^d_{\alpha 0} = \int_Y h(\mathbf{x}^*(\mathbf{y})) \, d\varphi(\mathbf{y}) = E_\varphi[h(\mathbf{x}^*(\mathbf{y}))],$$

where E_φ stands for the expectation operator associated with the probability distribution φ.

Persistency for Boolean variables Suppose that for some $I \subseteq \{1, \ldots, n\}$, the variables (x_i), $i \in I$, are Boolean, that is, the definition of $\mathbf{K}$ in (14.3) includes the quadratic equality constraints $x_i^2 - x_i = 0$, for every $i \in I$. An interesting issue is to determine whether in any optimal solution $\mathbf{x}^*(\mathbf{y})$ of $\mathbf{P_y}$, and for some index $i \in I$, one has $x_i^*(\mathbf{y}) = 1$ (or $x_i^*(\mathbf{y}) = 0$) for φ-almost all values of the parameter $\mathbf{y} \in \mathbf{Y}$. The probability that $x_i^*(\mathbf{y})$ is 1 is called the *persistency* of the Boolean variable $x_i^*(\mathbf{y})$.

Corollary 14.9 *Under the assumptions of Theorem 14.5, let $\mathbf{z}^d$ be a nearly optimal solution of the semidefinite relaxations (14.14). Then for $k \in I$ fixed:*

(a) $x_k^*(\mathbf{y}) = 1$ *(respectively equal to 0) in any optimal solution and for φ-almost all $\mathbf{y} \in \mathbf{Y}$, only if $\lim_{d \to \infty} z^d_{e(k)0} = 1$ (respectively equal to 0).*

Assume that for φ-almost all $\mathbf{y} \in \mathbf{Y}$, $J(\mathbf{y})$ is attained at a unique optimal solution $\mathbf{x}^*(\mathbf{y}) \in \mathbf{X}^*_\mathbf{y}$. Then $\mathrm{Prob}\,(x_k^*(\mathbf{y}) = 1) = \lim_{i \to \infty} z^d_{e(k)0}$, and so:

(b) $x_k^*(\mathbf{y}) = 1$ *(respectively $= 0$) for φ-almost all $\mathbf{y} \in \mathbf{Y}$, if and only if $\lim_{d \to \infty} z^d_{e(k)0} = 1$ (respectively $= 0$).*

The difference between (a) and (b) is due to the possible nonuniqueness of an optimal solution $\mathbf{x}^*(\mathbf{y})$ all $\mathbf{y} \in A$ with $\varphi(A) > 0$, in which case one may have $0 < \lim_{d \to \infty} z^d_{e(k)0} < 1$.

14.2.3 Approximating the global minimizer mapping

By Corollary 14.8 one may approximate any polynomial functional of the optimal solutions. But one may also wish to approximate (in some sense) the function $\mathbf{y} \mapsto x_k^*(\mathbf{y})$, that is, the "curve" described by the kth coordinate $x_k^*(\mathbf{y})$ of the optimal solution $\mathbf{x}^*(\mathbf{y})$ when $\mathbf{y}$ varies in $\mathbf{Y}$.

So let $g : \mathbf{Y} \to \mathbb{R}^n$ be the measurable mapping in Theorem 14.1(c) and suppose that one knows some lower bound vector $\mathbf{a} = (a_k) \in \mathbb{R}^n$, where:

$$a_k \leq \inf \{ x_k : (\mathbf{x}, \mathbf{y}) \in \mathbf{K} \}, \quad k = 1, \ldots, n.$$

Then for every $k = 1, \ldots, n$, the measurable function $\hat{g}_k : \mathbf{Y} \to \mathbb{R}^n$ defined by

$$\mathbf{y} \mapsto \quad \hat{g}_k(\mathbf{y}) := g_k(\mathbf{y}) - a_k, \qquad \mathbf{y} \in \mathbf{Y}, \tag{14.19}$$

is nonnegative and φ-integrable. Hence for every $k = 1, \ldots, n$, one may consider $d\lambda := \hat{g}_k d\varphi$ as a Borel measure on $\mathbf{Y}$ with unknown density $\hat{g}_k$ with respect to φ, but with known moments $\mathbf{u} = (u_\beta)$. Indeed, using (14.16),

$$u_\beta := \int_{\mathbf{Y}} \mathbf{y}^\beta \, d\lambda(\mathbf{y}) = -a_k \int_{\mathbf{Y}} \mathbf{y}^\beta \, d\varphi(\mathbf{y}) + \int_{\mathbf{Y}} \mathbf{y}^\beta \, g_k(\mathbf{y}) \, d\varphi(\mathbf{y})$$

$$= -a_k \gamma_\beta + z_{e(k)\beta}, \qquad \forall \beta \in \mathbb{N}^p, \tag{14.20}$$

where for every $k = 1, \ldots, n$, $z_{e(k)\beta} = \lim_{d \to \infty} z^d_{e(k)\beta}$ for all $\beta \in \mathbb{N}^n$, with $\mathbf{z}^d$ being an optimal (or nearly optimal) solution of the semidefinite relaxation (14.14).

Hence we are now faced with a density approximation problem, that is, given the sequence of moments $u_\beta = \int_{\mathbf{Y}} \mathbf{y}^\beta \hat{g}_k(\mathbf{y}) d\varphi$, $\beta \in \mathbb{N}^p$, of the unknown nonnegative measurable function $\hat{g}_k$ on $\mathbf{Y}$, "approximate" $\hat{g}_k$.

One possibility is to use the *maximum entropy* approach where given finitely many moments $\mathbf{u} = (\mathbf{u}_\beta)$ of $\hat{g}_k$ on $\mathbf{Y}$, one obtains an approximation of $\hat{g}_k$ as an optimal solution of a convex optimization problem. Another possibility is to compute the best approximation of g_k for the L_2-norm, i.e., one computes the polynomial g that minimizes $\int_{\mathbf{Y}} (g - \hat{g}_k)^2 d\mathbf{x}$ (assuming that $g_k \in L_2(\varphi)$).

The maximum entropy approach For simplicity of exposition suppose that $\mathbf{Y} = [0, 1]$ and $2d + 1$ moments $\mathbf{u} = (\mathbf{u}_j)_{j=0}^{2d}$ are available.

It turns out that an estimate h that maximizes the Boltzmann–Shannon entropy $E(h) := \int_0^1 h \log h \, dx$ is of the form $h_d^*(y) = \exp \sum_{j=0}^{2d} \lambda_j^* y^j$, where $\boldsymbol{\lambda}^* \in \mathbb{R}^{2d+1}$ is an optimal solution of the convex optimization problem

$$\sup_{\boldsymbol{\lambda} \in \mathbb{R}^{2d+1}} \left\{ \boldsymbol{\lambda}'\mathbf{u} - \int_0^1 \exp \left(\sum_{k=0}^{2d} \lambda_k x^k \right) dx \right\}. \tag{14.21}$$

Example 14.10 In this example, $\mathbf{Y} = [0, 1]$, $(\mathbf{x}, y) \mapsto f(\mathbf{x}, y) := yx_1 + (1 - y)x_2$, and $\mathbf{K} := \{ (\mathbf{x}, y) : yx_1^2 + x_2^2 - y \le 0; \ x_1^2 + yx^2 - y \le 0 \}$. That is, for each $y \in \mathbf{Y}$ the set $\mathbf{K}_y$ is the intersection of two ellipsoids. It is easy to check that $1 + x_i^*(y) \ge 0$ for all $y \in \mathbf{Y}$, $i := 1, 2$.

With $d = 2$ (i.e., four moments), the maximum entropy estimate $h_4^*(y)$ for $1 + x_1^*(y)$ reads $y \mapsto h_4^*(y) = \exp \sum_{j=0}^4 \lambda_j^* y^j$, and is obtained with the optimal solution $\boldsymbol{\lambda}^* = (-0.2894, 1.7192, -19.8381, 36.8285, -18.4828)$

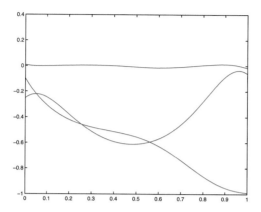

Figure 14.1 Example 14.10: $x_1^*(y)$, $x_2^*(y)$ and $h_1(\mathbf{x}^*(y), y)$ on $[0, 1]$.

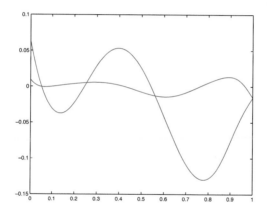

Figure 14.2 Example 14.10: $h_1(\mathbf{x}^*(y), y)$ and $h_2(\mathbf{x}^*(y), y)$ on $[0, 1]$.

of (14.21). Similarly, the maximum entropy estimate $h_4^*(y)$ for $1 + x_2^*(y)$ is obtained with

$$\lambda^* = (-0.1018, -3.0928, 4.4068, 1.7096, -7.5782).$$

Figure 14.1 displays the curves of $x_1^*(y)$ and $x_2^*(y)$, as well as the constraint $h_1(\mathbf{x}^*(y), y)$. Observe that $h_1(\mathbf{x}^*(y), y) \approx 0$ on $[0, 1]$ which means that for φ-almost all $y \in [0, 1]$, at an optimal solution $\mathbf{x}^*(y)$, the constraint $h_1 \leq 0$ is saturated. Figure 14.2 displays the curves of $h_1(\mathbf{x}^*(y), y)$ and $h_2(\mathbf{x}^*(y), y)$.

The L_2-norm approach We now show how to approximate the function $g_k :$ $\mathbf{Y} \rightarrow \mathbb{R}$ by a polynomial of degree d, based on knowledge of moments up to order d,

$$z^d_{e(k),\beta} \approx u_\beta := \int_Y \mathbf{y}^\beta \underbrace{g_k(\mathbf{y})\,d\varphi(\mathbf{y})}_{d\phi_k(\mathbf{y})}, \qquad \forall \beta \in \mathbb{N}^n_d,$$

defined in Theorem 14.5(b). Recall that for every $k = 1,\ldots,n$, $u_\beta = \lim_{d\to\infty} z^d_{e(k)\beta}$ for all $\beta \in \mathbb{N}^n$, with $\mathbf{z}^d$ being an optimal (or nearly optimal) solution of the semidefinite relaxation (14.14). Denote by $\mathbf{u}^d$ the vector of such moments up to order d.

Assume that $g_k \in L_2(\varphi)$ and let $p_d \in \mathbb{R}[\mathbf{x}]_d$ with coefficient vector $\mathbf{p}_d \in \mathbb{R}^{s(d)}$. Observe that

$$\int_Y (p_d(\mathbf{y}) - g_k(\mathbf{y}))^2 d\varphi(\mathbf{y}) = \int_Y p_d^2\,d\varphi \;-\; 2\underbrace{\int_Y p_d\,g_k\,d\varphi}_{\langle \mathbf{p}_d, \mathbf{u}^d \rangle} + \underbrace{\int_Y g_k^2 d\varphi}_{\text{cte}}$$

$$= \langle \mathbf{p}_d, \mathbf{M}_d(\varphi)\,\mathbf{p}_d \rangle - 2\,\langle \mathbf{p}_d, \mathbf{u}^d \rangle + \text{cte},$$

where $\mathbf{M}_d(\varphi)$ denotes the moment matrix or order d associated with the measure φ, i.e.,

$$\mathbf{M}_d(\varphi)(\alpha, \beta) = \int_Y \mathbf{y}^{\alpha+\beta} d\varphi(\mathbf{y}), \qquad \forall \alpha, \beta \in \mathbb{N}^n_d.$$

Therefore the coefficient vector of the polynomial p^*_d which minimizes the L_2-norm

$$\|p - g_k\|_2 = \left(\int_Y (p(\mathbf{y}) - g_k(\mathbf{y}))^2 d\varphi(\mathbf{y}) \right)^{1/2}$$

of $p - g_k$ over all polynomials $p \in \mathbb{R}[\mathbf{x}]_d$, satisfies $\mathbf{M}_d(\varphi)\,\mathbf{p}^*_d = \mathbf{u}^d$. And so $\mathbf{p}^*_d = \mathbf{M}_d(\varphi)^{-1}\,\mathbf{u}^d$ because $\mathbf{M}_d(\varphi)$ is nonsingular. In other words, we have just proved the following.

Lemma 14.11 *Under the assumptions of Theorem 14.5, let $\mathbf{y} \mapsto g_k(\mathbf{y})$ be the k-coordinate of the global minimizer mapping (assumed to be in $L_2(\varphi)$). Let $\mathbf{M}_d(\varphi)$ be the (order d) moment matrix of the measure φ on $\mathbf{Y}$ and let $\mathbf{u}^d$ be the vector of generalized moments up to order d, defined by:*

$$u^d_\beta = \int_Y \mathbf{y}^\beta\,g_k(\mathbf{y})\,d\varphi(\mathbf{y}), \qquad \beta \in \mathbb{N}^p_d.$$

Let $p^*_d \in \mathbb{R}[\mathbf{x}]_d$ with vector of coefficients $\mathbf{p}^*_d = \mathbf{M}_d(\varphi)^{-1}\,\mathbf{u}^d$. Then

$$\int_Y (p^*_d - g_k)^2 d\varphi = \inf_{p\in\mathbb{R}[\mathbf{x}]_d} \int_Y (p - g_k)^2 d\varphi.$$

Alternatively, in the basis of orthonormal polynomials associated with φ, the vector of coefficients of p^*_d is exactly $\mathbf{u}^d$.

The last statement is important because, in general, inverting $\mathbf{M}_d(\varphi)$ is an ill-conditioned problem. On the other hand, if one defines the moment matrix of φ in the basis of *orthonormal polynomials* associated with φ, it becomes the identity matrix, and so no inversion is needed. Moreover, computing the orthonormal polynomials of degree up to d associated with φ reduces to computing determinants of some submatrices of $\mathbf{M}_d(\varphi)$; see Section 12.2.1.

14.3 On robust polynomial optimization

Let $\mathbf{K} \subset \mathbb{R}^{n+p}$ be the basic semi-algebraic set defined in (14.3) and let $\mathbf{K_y}$ be as in (14.5). In this section we are interested in the min-max optimization problem:

$$\mathbf{PR}: \quad J^* = \sup_{\mathbf{y} \in \Omega} \inf_{\mathbf{x}} \{ f(\mathbf{x}, \mathbf{y}) : \mathbf{x} \in \mathbf{K_y} \}, \qquad (14.22)$$

where $\Omega \subset \mathbb{R}^p$ is a compact basic semi-algebraic set.

Problem **PR** can be viewed as a two-player zero sum *game* problem where Ω is player-1's action set and $\mathbf{K_y}$ is player-2's action set, once player-1 has chosen action $\mathbf{y} \in \Omega$. So from the point of view of player-1, **PR** is a *robust optimization* problem since an optimal action $\mathbf{y}^* \in \Omega$ has to be made *before* player-2 chooses an action $\mathbf{x}^*(\mathbf{y})$ that minimizes his objective function $\mathbf{x} \mapsto f(\mathbf{x}, \mathbf{y}^*)$. Hence player-1 has to choose an action which is the best possible against the worst action by player-2.

As already mentioned, in the general context of (14.22) and without further assumptions on the data Ω, $\mathbf{K_y}$, f, and the description of the uncertainty set, **PR** is a difficult and challenging problem for which there is no general methodology. On the other hand, from the point of view of player-2, **PR** is a parametric optimization problem because player-2 waits for the decision $\mathbf{y}$ of player-1 and then subsequently chooses the best possible action $\mathbf{x}^*(\mathbf{y})$ that minimizes $f(\mathbf{x}, \mathbf{y})$. Thus observe that if player-1 knows the optimal value function $\mathbf{y} \mapsto J(\mathbf{y}) = \inf_{\mathbf{x}} \{ f(\mathbf{x}, \mathbf{y}) : \mathbf{x} \in \mathbf{K_y} \}$, then **PR** reduces to the optimization problem

$$J^* = \sup_{\mathbf{y}} \{ J(\mathbf{y}) : \mathbf{y} \in \Omega \}.$$

If in general there is no closed-form expression for J, we have just seen in Section 14.2 that in some polynomial context one may approximate (in a strong sense) the optimal value function $J(\cdot)$ by polynomials or piecewise polynomials; see Theorem 14.6.

So it makes sense to replace **PR** with the optimization problem $\sup_{\mathbf{y}} \{ p_d(\mathbf{y}) : \mathbf{y} \in \Omega \}$, where p_d is some polynomial approximation of the optimal value function $\mathbf{y} \mapsto J(\mathbf{y})$. Inspired by results of Section 14.2, to obtain such an approximation p_d we need a set $\mathbf{Y} \subset \mathbb{R}^p$ that contains Ω and is simple enough so that one may compute easily all moments of a probability measure φ on $\mathbf{Y}$, absolutely continuous with respect to the Lebesgue measure on the smallest affine variety that contains Ω. In general Ω is a complicated set and so the choice $\mathbf{Y} = \Omega$ is not appropriate (except if Ω is a basic semi-algebraic set of small dimension, in which case the moments of φ can be approximated as described in Henrion et al. (2009a)).

So we assume that $\Omega \subseteq \mathbf{Y}$ where $\mathbf{Y}$ is a box or an ellipsoid, and φ is the probability with uniform distribution on $\mathbf{Y}$, so that all its moments $\boldsymbol{\gamma} = (\gamma_\beta)$, $\beta \in \mathbb{N}^p$, in (14.13) can be computed.

Proposition 14.12 *Under the assumptions of Theorem 14.5, the optimal value function $\mathbf{y} \mapsto J(\mathbf{y})$ is lower semi-continuous (l.s.c.) on $\mathbf{Y}$.*

Proof Let $(\mathbf{y}_\ell) \subset \mathbf{Y}$ be a sequence with $\mathbf{y}_\ell \to \mathbf{y} \in \mathbf{Y}$ as $\ell \to \infty$ and such that

$$\liminf_{\mathbf{z} \to \mathbf{y}} J(\mathbf{z}) = \lim_{\ell \to \infty} J(\mathbf{y}_\ell).$$

Next let $\mathbf{x}^*(\mathbf{y}_\ell) \in \mathbf{X}^*(\mathbf{y}_\ell)$ be an arbitrary minimizer in $\mathbf{K}_{\mathbf{y}_\ell}$. By compactness of $\mathbf{K}$, there is a subsequence denoted (ℓ_i) and a point $(\mathbf{a}, \mathbf{y}) \in \mathbf{K}$ such that $(\mathbf{x}^*(\mathbf{y}_{\ell_i}), \mathbf{y}_{\ell_i}) \to (\mathbf{a}, \mathbf{y})$. Consequently,

$$\liminf_{\mathbf{z} \to \mathbf{y}} J(\mathbf{z}) = \lim_{i \to \infty} J(\mathbf{y}_{\ell_i}) = \lim_{i \to \infty} f(\mathbf{x}^*(\mathbf{y}_{\ell_i}), \mathbf{y}_{\ell_i})$$
$$= f(\mathbf{a}, \mathbf{y}) \geq \inf_{\mathbf{x}} \{ f(\mathbf{x}, \mathbf{y}) : (\mathbf{x}, \mathbf{y}) \in \mathbf{K} \} = J(\mathbf{y}),$$

which proves that J is l.s.c. $\qquad\qquad\qquad\qquad\qquad\qquad\qquad\qquad\square$

Theorem 14.13 *Let $\Omega \subseteq \mathbf{Y}$ and under the assumptions of Theorem 14.5, let $p_d \in \mathbb{R}[\mathbf{y}]$ be the polynomial defined in Theorem 14.6(b). Consider the optimization problem:*

$$\mathbf{PR}_d : \quad \sup_{\mathbf{y}} \{ p_d(\mathbf{y}) : \mathbf{y} \in \Omega \}, \qquad\qquad (14.23)$$

with optimal value J_d^*. Let $\mathbf{y}_d^* \in \Omega$ be some global maximizer of $\mathbf{PR}_d$ and for every d, let $\hat{J}_d^* := \sup_{\ell=1,\dots,d} J_\ell^* = p_{t(d)}(\mathbf{y}_{t(d)}^*)$ for some $t(d) \in \{1, \dots, d\}$.

(a) $\hat{J}_d^* \to J^*$ as $d \to \infty$.

(b) *If J is continuous, so that $J^* = J(\mathbf{y}^*)$ for some $\mathbf{y}^* \in \Omega$, then any accumulation point $\bar{\mathbf{y}}$ of the sequence $(\mathbf{y}_{t(d)}^*) \subset \Omega$ is a global maximizer of **PR**. In particular, if $\mathbf{y}^*$ is the unique maximizer of **PR** then $\mathbf{y}_{t(d)}^* \to \mathbf{y}^*$ as $d \to \infty$.*

Proof   (a) Observe that p_d being continuous on $\mathbf{Y}$ (hence on Ω), **PR**$_d$ has a global minimizer $\mathbf{y}_d^*$, for every d. Next, recall that from Theorem 14.6(b), $p_d \overset{L_1(\varphi)}{\to} J$ (i.e., convergence in the $L_1(\varphi)$-norm). Hence, by Ash (1972, Theorem 2.5.1), $p_d \overset{\varphi}{\to} J$ (i.e., convergence in probability) and in turn, by Ash (1972, Theorem 2.5.3), there exists a subsequence (d_ℓ) such that $p_{d_\ell} \to J$, φ-almost uniformly.

Next, by Proposition 14.12, the optimal value mapping J is l.s.c. on $\mathbf{Y}$ (hence on $\Omega \subseteq \mathbf{Y}$). With $\epsilon > 0$ fixed, arbitrary, let $\mathbf{B}(\epsilon) := \{\mathbf{y} \in \Omega : J(\mathbf{y}) > J^* - \epsilon\}$ and let $\kappa := \varphi(\mathbf{B}(\epsilon))$. As J is l.s.c., $B(\epsilon)$ is nonempty, open, and therefore $\kappa > 0$. As $p_{d_\ell} \to J$, φ-almost uniformly on $\mathbf{Y}$, there exists a Borel set $A_\kappa \in \mathcal{B}(\mathbf{Y})$ such that $\varphi(A_\kappa) < \kappa$ and $p_{d_\ell} \to J$, uniformly on $\mathbf{Y} \setminus A_\kappa$. Hence, as $\Delta := (\mathbf{Y} \setminus A_\kappa) \cap \mathbf{B}(\epsilon) \neq \emptyset$, one has

$$\lim_{\ell \to \infty} p_{d_\ell}(\mathbf{y}) = J(\mathbf{y}) \geq J^* - \epsilon, \qquad \forall \mathbf{y} \in \Delta,$$

and so, as $J_d^* \geq p_d(\mathbf{y})$ on Δ, one obtains $\lim_{\ell \to \infty} J_{d_\ell}^* \geq J^* - \epsilon$. As $\epsilon > 0$ was arbitrary, one finally gets $\lim_{\ell \to \infty} J_{d_\ell}^* = J^*$. On the other hand, by monotonicity of the sequence $(\hat{J}_d^*)$,

$$J^* \geq \lim_{i \to \infty} \hat{J}_d^* = \lim_{\ell \to \infty} \hat{J}_{d_\ell}^* \geq \lim_{\ell \to \infty} J_{d_\ell}^* = J^*,$$

and so the conclusion of (a) follows.

(b) Let $\mathbf{y}_d^* \in \Omega$ be a maximizer of **PR**$_d$. As Ω is compact, there exists $\bar{\mathbf{y}} \in \Omega$ and a subsequence d_ℓ such that $\mathbf{y}_{t(d_\ell)}^* \to \bar{\mathbf{y}}$ as $\ell \to \infty$. In addition, from $p_d(\mathbf{y}) \leq J(\mathbf{y})$ for every d and every $\mathbf{y} \in \mathbf{Y}$,

$$J^* \geq J(\mathbf{y}_{t(d_\ell)}^*) \geq p_{t(d_\ell)}(\mathbf{y}_{t(d_\ell)}^*) = \hat{J}_{d_\ell}^*.$$

So using (a) and letting $\ell \to \infty$ yields the desired result $J^* = J(\bar{\mathbf{y}})$. Since the converging subsequence (d_ℓ) was arbitrary, the desired result follows. The last statement is straightforward. $\square$

Continuity of J is restrictive and is obtained if, for example, the set-valued mapping $\mathbf{y} \mapsto \mathbf{X}^*(\mathbf{y})$ is l.s.c., that is, $(\mathbf{X}^*)^{-1}(O)$ is open whenever $O \subset \mathbb{R}^n$

is open; see for example Proposition D 6(b) in Hernández-Lerma and Lasserre (1996, p. 183).

Hence Theorem 14.13 states that one may approximate as closely as desired the optimal value J^* of the robust optimization problem **PR** by minimizing the polynomial p_d (a nearly optimal solution of the semidefinite program (14.17)) provided that d is sufficiently large.

14.4 A "joint+marginal" algorithm in optimization

Consider the (nonparametric) polynomial program

$$\mathbf{P}: \quad f^* := \inf_{\mathbf{x}} \{ f(\mathbf{x}) : \mathbf{x} \in \mathbf{K} \} \tag{14.24}$$

where $f \in \mathbb{R}[\mathbf{x}]$ is a polynomial, $\mathbf{K} \subset \mathbb{R}^n$ is a basic semi-algebraic set

$$\mathbf{K} := \{ \mathbf{x} \in \mathbb{R}^n : h_j(\mathbf{x}) \geq 0, \quad j = 1, \ldots, m \}, \tag{14.25}$$

for some polynomials $(h_j) \subset \mathbb{R}[\mathbf{x}]$, and f^* is the *global* minimum of **P**. One way to approximate f^* is to solve a hierarchy of either semidefinite relaxations or LP-relaxations as proposed in Chapter 6 and Chapter 9 respectively.

Although practice with the semidefinite relaxations seems to reveal that convergence is fast, the matrix size in the dth semidefinite relaxation of the hierarchy increases as $O(n^d)$. Hence, for large size (and sometimes even medium size) problems where sparsity and/or symmetry cannot be exploited, only a few relaxations of the hierarchy can be implemented (typically the first, second or third relaxation). In that case, one only obtains a lower bound on the optimal value f^*, and no feasible solution.

So an important issue is: *how can we use the result of the ith semidefinite relaxation to find an approximate solution of the original problem?*

For some well-known special cases in 0/1 optimization, for example the celebrated MAXCUT problem, one may generate a feasible solution, with guaranteed performance, from a randomized rounding procedure that uses an optimal solution of the first semidefinite relaxation (i.e., with $d = 1$); see Goemans and Williamson (1995). But in general there is no such procedure.

In this section we provide a relatively simple algorithm for polynomial programs with a convex feasible set **K** and for 0/1 programs for which feasibility is easy to detect, for example the MAXCUT, k-CLUSTER or 0/1 knapsack problems. The algorithm builds upon the "joint+marginal" approach (in short (J+M)) developed in Section 14.2 for *parametric* polynomial optimization.

The "joint+marginal" strategy Let the compact interval $\mathbf{Y}_1 := [\underline{x}_1, \overline{x}_1] \subset \mathbb{R}$ be the projection of **K** on the x_1-coordinate axis, and consider the parametric optimization problem:

$$\mathbf{P}(y): \quad J^1(y) = \inf_{\mathbf{x}} \{ f(\mathbf{x}) : \mathbf{x} \in \mathbf{K}; \ x_1 = y \}, \tag{14.26}$$

or, equivalently $J^1(y) = \inf\{ f(\mathbf{x}) : \mathbf{x} \in \mathbf{K}_y \}$. Observe that (14.26) is a particular case of the parametric problem $\mathbf{P_y}$ in (14.4) with $\mathbf{Y} = [\underline{x}_1, \overline{x}_1]$, and $\mathbf{K}_y \neq \emptyset$ whenever $y \in \mathbf{Y}_1$. By definition, $f^* = \inf_x \{ J^1(x) : x \in \mathbf{Y}_1 \}$.

In the context of the (nonparametric) polynomial optimization (14.24), the approach of Section 14.2 can be used as follows in what we call the (J+M)-algorithm.

Algorithm 14.14 (J+M) algorithm. Let $d \in \mathbb{N}$ be fixed.

1. Treat x_1 as a parameter in the compact interval $\mathbf{Y}_1 = [\underline{x}_1, \overline{x}_1]$ with associated probability distribution φ_1, uniformly distributed on $\mathbf{Y}_1$. (For 0/1 programs $\mathbf{Y}_1 = \{0, 1\}$.)
2. Solve the dth semidefinite relaxation of the (J+M) hierarchy (14.14) applied to the parametric optimization problem (14.26) with $n - 1$ variables $(x_2, \ldots, x_n)$ and parameter x_1. The dual (14.17) provides a univariate polynomial $x_1 \mapsto p_d^1(x_1)$ that converges to J^1 in the $L_1(\varphi_1)$-norm, as d increases. (The map $v \mapsto J^1(v)$ denotes the optimal value function of $\mathbf{P}(v)$, i.e., the optimal value of $\mathbf{P}$ given that the variable x_1 is fixed at the value v.) Compute a global minimizer $\tilde{x}_1 \in \mathbf{Y}_1$ of the univariate polynomial p_d^1 on $\mathbf{Y}_1$ (e.g. this can be done by solving a single semidefinite program). Ideally, when d is large enough, $\tilde{x}_1$ should be close to the first coordinate x_1^* of a global minimizer $\mathbf{x}^* = (x_1^*, \ldots, x_n^*)$ of $\mathbf{P}$.
3. Consider now the new optimization problem $\mathbf{P}(\tilde{x}_1)$ in the $n - 1$ variables $(x_2, \ldots, x_n)$, obtained from $\mathbf{P}$ by fixing the variable x_1 at the value $\tilde{x}_1$. The feasible set of $\mathbf{P}(\tilde{x}_1)$ is now the convex set $\mathbf{K}_1 := \mathbf{K} \cap \{ \mathbf{x} : x_1 = \tilde{x}_1 \}$. For 0/1 programs where feasibility is easy to detect, one checks whether $\mathbf{P}(\tilde{x}_1)$ has a feasible solution and if not, then one sets $\tilde{x}_1 := 1 - \tilde{x}_1$. Let $\mathbf{Y}_2$ be the projection of $\mathbf{K}_1$ on the x_2-coordinate axis (or $\mathbf{Y}_2 = \{0, 1\}$ for 0/1 programs). Then go back to step 2 with now $x_2 \in \mathbf{Y}_2$ as parameter and $(x_3, \ldots, x_n)$ as variables, to obtain a point $\tilde{x}_2 \in \mathbf{Y}_2$, etc.

When $\mathbf{K}$ is convex one ends up with a feasible point $\tilde{\mathbf{x}} \in \mathbf{K}$. Then one can use $\tilde{\mathbf{x}}$ as initial point in a local optimization procedure to find a local minimum $\hat{\mathbf{x}} \in \mathbf{K}$. The rationale behind the (J+M) algorithm is that if k is large enough and $\mathbf{P}$ has a unique global minimizer $\mathbf{x}^* \in \mathbf{K}$, then $\tilde{\mathbf{x}}$ as well as $\hat{\mathbf{x}}$ should be close to $\mathbf{x}^*$. In practice d is the largest value for which one is able to solve the semidefinite program (14.14). The computational cost before the local optimization procedure is less than solving n times the dth semidefinite relaxation in the

(J+M) hierarchy, which is itself the same as the dth semidefinite relaxation (6.3) with only an additional $2d$ linear constraints.

Numerical experiments

Continuous optimization problems The (J+M) algorithm has been tested on nonconvex test problems taken from Floudas et al. (1999, §2) in which **K** is a convex polytope and f is a nonconvex quadratic polynomial $\mathbf{x} \mapsto \mathbf{x}'\mathbf{Q}\mathbf{x} + \mathbf{b}'\mathbf{x}$ for some real symmetric matrix **Q** and vector **b**. The relaxation order d is chosen to be the smallest one, i.e., $2d$ is the smallest integer that is larger than the maximum of the degree in the data of the problem. We have used the GloptiPoly software described in Appendix B. In Table 14.1 n (respectively m) stands for the number of variables (respectively constraints), and the value displayed in the "(J+M)-algorithm" column is obtained in running a local min-imization algorithm of the MATLAB toolbox with the output $\tilde{\mathbf{x}}$ of the (J+M) algorithm as initial guess. When running GloptiPoly, and as recommended for numerical stability and precision, the problem data were rescaled to obtain a polytope $\mathbf{K} \subset [-1, 1]^n$.

For 0/1 programs Consider the following three NP-hard 0/1 problems **P** for which detecting feasibility is easy:

$$\text{MAXCUT}: \quad \sup_{\mathbf{x}} \left\{ \sum_{1 \le i \le j \le n} (1 - Q_{ij}x_i x_j) : \mathbf{x} \in \{-1, 1\}^n \right\}, \quad (14.27)$$

$$t\text{-cluster}: \quad \sup_{\mathbf{x}} \left\{ \mathbf{x}'\mathbf{Q}\mathbf{x} : \mathbf{x} \in \{0, 1\}^n; \sum_{\ell=1}^{n} x_\ell = t \right\}, \quad (14.28)$$

Table 14.1 *(J+M) algorithm for convex set* **K**

Problem	n	m	f^*	d	(J+M)-algorithm	relative error
2.2	5	11	−17	2	−17.00	0%
2.3	6	8	−361.5	1	−361.50	0%
2.6	10	21	−268.01	1	−267.00	0.3%
2.9	10	21	0	1	0.00	0%
2.8C1	20	30	−394.75	1	−385.30	2.4%
2.8C2	20	30	−884.75	1	−871.52	1.5%
2.8C3	20	30	−8695	1	−8681.7	0.15%
2.8C4	20	30	−754.75	1	−754.08	0.09%
2.8C5	20	30	−4150.41	1	−3678.2	11%

Table 14.2 *Relative error for the MAXCUT problem*

n	20	30	40	50
$(\rho_1 - \mathbf{P}_1)/|\rho_1|$	3.23%	3.28%	3.13%	2.92%
$(\rho_1 - GW)/|\rho_1|$	2.58%	2.60%	2.84%	2.60%

$$\text{0/1-knapsack}: \quad \sup_{\mathbf{x}} \left\{ \mathbf{c}'\mathbf{x} : \mathbf{x} \in \{0, 1\}^n; \sum_{\ell=1}^{n} a_\ell x_\ell \leq b \right\}, \quad (14.29)$$

for some real symmetric matrix $\mathbf{Q} = (Q_{ij}) \in \mathbb{R}^{n \times n}$, vector $\mathbf{c} \in \mathbb{R}^n$, $\mathbf{a} = (a_\ell) \in \mathbb{N}^n$ and integers $t, b \in \mathbb{N}$. In the sample of randomly generated problems, the entry Q_{ij} of the real symmetric matrix $\mathbf{Q}$ is set to zero with probability $1/2$ and when different from zero, Q_{ij} is randomly (and independently) generated according to the uniform probability distribution on the interval $[0, 10]$. The probability φ is uniform and again $d = 1$, i.e., in Algorithm 14.14 one considers the first semidefinite relaxation (14.14).

Let ρ_1 denote the optimal value of the first semidefinite relaxation of the hierarchy (6.3). So for MAXCUT, ρ_1 is the optimal value of Shor's relaxation with Goemans and Williamson's famous 0.878 performance guarantee. Let $(\rho_1 - GW)/|\rho_1|$ be the relative error associated with the reward GW of the solution $\mathbf{x} \in \{1, 1\}^n$ obtained by the randomized rounding algorithm of Goemans and Williamson (1995). Similarly, let $(\rho_1 - \mathbf{P}_1)/|\rho_1|$ be the relative error associated with the reward $\mathbf{P}_1$ of the solution $\mathbf{x} \in \{-1, 1\}^n$ (or $\mathbf{x} \in \{0, 1\}^n$) generated by Algorithm 14.14

- For the *MAXCUT* problem, we have run the (J+M) algorithm for problems on random graphs with $n = 20, 30, 40$ and 50 variables. For each value of n, we generated 50 problems. In Table 14.2 we show the average relative error $(\rho_1 - \mathbf{P}_1)/|\rho_1|$, which, as one may see, is comparable with the relative error $(\rho_1 - GW)/|\rho_1|$ obtained for the GW solution.
- For the *t-cluster problem* one can add the n constraints $x_j(t - \sum_\ell x_\ell) = 0$, $j = 1, \ldots, n$, in the definition (14.28) of $\mathbf{P}$ because they are redundant. We have tested the (J+M) algorithm on problems randomly generated as for MAXCUT, and with $t = n/2 = 10$. As for MAXCUT, let ρ_1 denote the optimal value of the first semidefinite relaxation of the hierarchy (6.3) applied to problem (14.28). The average relative error $(\rho_1 - \mathbf{P}_1)/|\rho_1|$ was:
  - 5.7% on four randomly generated problems with $n = 60$ variables,
  - 4.5% and 5.6% on two randomly generated problems with $n = 70$ variables,
  - 5.7% on a problem with $n = 80$ variables.

Table 14.3 *Relative error for the 0/1-knapsack problem*

n	50	60
$(\rho_1 - \mathbf{P}_1)/|\rho_1|$	2.1%	0.62%

- For the *0/1 knapsack problem* one can add the n redundant constraints x_j $(b - \sum_\ell a_\ell x_\ell) \geq 0$, and $(1 - x_j)(b - \sum_\ell a_\ell x_\ell) \geq 0$, $j = 1, \ldots, n$, in the definition (14.29) of $\mathbf{P}$. We have tested the (J+M) algorithm on a sample of 16 problems with $n = 50$ variables and three problems with $n = 60$ variables where $b = \sum_\ell a_\ell/2$, and the integers a_ℓ are generated uniformly in $[10, 100]$. The vector $\mathbf{c}$ is generated by $c_\ell = s\epsilon + a_\ell$ with $s = 0.1$ and ϵ is a random variable uniformly distributed in $[0, 1]$. From the results reported in Table 14.3 one can see that very good relative errors are obtained.

14.5 Exercises

Exercise 14.1 Let $\mathbf{K} \subset \mathbb{R}^n$ and $\mathbf{Y} \subset \mathbb{R}^p$ be nonempty compact basic semi-algebraic sets. Show that the function $\mathbf{y} \mapsto J(\mathbf{y}) := \inf_{\mathbf{x}} \{ f(\mathbf{x}, \mathbf{y}) : \mathbf{x} \in \mathbf{K} \}$ is continuous on $\mathbf{Y}$. What may we conclude for the robust optimization problem $J^* = \sup_{\mathbf{y} \in \mathbf{Y}} \{ \inf_{\mathbf{x}} \{ f(\mathbf{x}, \mathbf{y}) : \mathbf{x} \in \mathbf{K} \}\} = \sup_{\mathbf{y}} \{ J(\mathbf{y}) : \mathbf{y} \in \mathbf{Y} \}$? Describe a numerical scheme to approximate J^* as closely as desired.

Exercise 14.2 Let $f, g_j \in \mathbb{R}[\mathbf{x}]$, $j = 1, \ldots, m$. With $\mathbf{Y} := [-1, 1]^m$ consider the *value* function $\theta : \mathbb{R}^m \to \mathbb{R} \cup \{-\infty, +\infty\}$ defined by $\theta(\mathbf{y}) = +\infty$ if $\mathbf{y} \notin \mathbf{Y}$ and

$$\mathbf{y} \mapsto \quad \theta(\mathbf{y}) := \inf_{\mathbf{x}} \{ f(\mathbf{x}) : g_j(\mathbf{x}) \geq y_j, \quad j = 1, \ldots, m \}$$

otherwise.

(a) Assume that θ is finite on $\mathbf{Y}$. Show that in $\mathbf{Y}$ one may approximate θ by a sequence of polynomials $(p_d) \subset \mathbb{R}[\mathbf{y}]$, $d \in \mathbb{N}$, such that $p_d \leq \theta$ on $\mathbf{Y}$ and $p_d \to \theta$ as $d \to \infty$, for the $L_1(\mathbf{Y})$-norm and almost-uniformly as well.

(b) The Legendre–Fenchel conjugate of θ is the function

$$\boldsymbol{\lambda} \mapsto \theta^*(\boldsymbol{\lambda}) = \sup_{\mathbf{y}} \{ \langle \boldsymbol{\lambda}, \mathbf{y} \rangle - \theta(\mathbf{y}) \}, \quad \boldsymbol{\lambda} \in \mathbb{R}^m.$$

Consider the "dual" problem $(\theta^*)^*(\mathbf{y}) := \sup_{\boldsymbol{\lambda}} \{ \langle \boldsymbol{\lambda}, \mathbf{y} \rangle - \theta^*(\boldsymbol{\lambda}) \}$, $\mathbf{y} \in \mathbf{Y}$. Show that $(\theta^*)^*$ provides a convex underestimator of θ on $\mathbf{Y}$.

(c) Let $(p_d) \subset \mathbb{R}[\mathbf{y}]$ be as in (a) and define $\tilde{p}_d := p_d(\mathbf{y})$ if $\mathbf{y} \in \mathbf{Y}$ and $\tilde{p}_d(\mathbf{y}) = +\infty$ otherwise. Compare the Legendre–Fenchel conjugates θ^* and $\tilde{p}_d^*$ and compare $(\theta^*)^*$ and $(\tilde{p}^*)^*$ as well.

(d) Devise a numerical scheme that provides a polynomial approximation $q_d \in \mathbb{R}[\boldsymbol{\lambda}]$ of $\tilde{p}_d^*$ on a box $\mathbf{B} \subset \mathbb{R}^m$.

14.6 Notes and sources

14.2 There is a vast and rich literature on the topic of parametric optimization and for a detailed treatment, the interested reader is referred for example to Bonnans and Shapiro (2000) and the many references therein. Most of the material in this section is from Lasserre (2010b). This section has some links with *Young measures*. If $\boldsymbol{\Omega} \subset \mathbb{R}^p$ is a Borel set, a measure on $\boldsymbol{\Omega} \times \mathbb{R}^n$ whose marginal on $\boldsymbol{\Omega}$ is the Lebesgue measure λ on $\boldsymbol{\Omega}$ is called a Young measure. For a measurable function $\mathbf{u} : \boldsymbol{\Omega} \to \mathbb{R}^n$, the Young measure ν associated with $\mathbf{u}$ is the unique Young measure supported on the graph of $\mathbf{u}$. Equivalently, ν is the image of the Lebesgue measure on $\boldsymbol{\Omega}$ by the mapping $\mathbf{x} \mapsto (\mathbf{x}, \mathbf{u}(\mathbf{x}))$. And for every $\psi : \boldsymbol{\Omega} \times \mathbb{R}^n \to \overline{\mathbb{R}}$, measurable and nonnegative, or ν-integrable,

$$\int_{\boldsymbol{\Omega} \times \mathbb{R}^n} \psi \, d\nu = \int_{\boldsymbol{\Omega}} \psi(\mathbf{x}, \mathbf{u}(\mathbf{x})) \, \lambda(d\mathbf{x}).$$

Such measures have been introduced to characterize nonclassical solutions of some problems in the calculus of variations, and later have also been used to analyze the limit of minimizing sequences of functions for some optimal control problems, controlled partial differential equations, shape optimization problems, etc. For more details on this fascinating topic the interested reader is referred for example to Pedregal (1999) and the more recent Castaing et al. (2004). So the measure μ in the infinite-dimensional linear programming problem (14.6) is a Young measure, and the measure μ^* in Theorem 14.1 is the Young measure associated with the global minimizer mapping $\mathbf{y} \mapsto \mathbf{x}^*(\mathbf{y})$ (when $\mathbf{x}^*(\mathbf{y})$ is unique for almost all $\mathbf{y} \in \mathbf{Y}$).

14.3 In view of its many important applications, in recent years "robust optimization" has become a central topic in optimization and operations research. In the general context (14.2), robust optimization is a difficult and challenging problem for which there is still no general methodology. We have shown that the methodology developed in Section 14.3 can solve (at least in principle) robust optimization problems in a relatively general framework with polynomial data. But one has to keep in mind that again in view of the present limitations of the semidefinite programming solvers, this methodology is limited to problems of modest size. However, as shown in Ben-Tal et al. (2000, 2006) and Bertsimas and Sim (2004, 2006), when the description of the uncertainty set takes some special form, one may obtain *computationally tractable* robust counterparts of some "easy" convex problems (like linear, second-order cone and semidefinite programs) which in general provide upper bounds on J^*, and even the exact value J^* in some special cases.

14.4 This section is from Lasserre and Thanh (2012) where a variant of the algorithm was also tested on continuous problems with nonconvex feasible set **K**.

15

Convex underestimators of polynomials

15.1 Introduction

We have already mentioned that the moment-SOS approach developed in Chapter 6 is limited to problems **P** of modest size unless some sparsity or symmetry in the problem data can be exploited, for example, as described in Chapter 8. In this chapter we consider polynomial optimization problems for which the hierarchy of semidefinite relaxations (6.3) cannot be implemented (for size reasons) and show how the moment-SOS approach can still be used to help solve such problems.

Consider the general polynomial optimization problem **P**:

$$
\begin{aligned}
f^* = \inf_{\mathbf{x}} \quad & f(\mathbf{x}) \\
\text{s.t} \quad & g_j(\mathbf{x}) \geq 0, \quad j = 1, \ldots, m \\
& \mathbf{x} \in [\mathbf{x}^L, \mathbf{x}^U] \subset \mathbb{R}^n,
\end{aligned}
\tag{15.1}
$$

where f, g_i are polynomials and $\mathbf{x}^l, \mathbf{x}^U \in \mathbb{R}^n$ define the box $\mathbf{B} = [\mathbf{x}^L, \mathbf{x}^U] \subset \mathbb{R}^n$.

When the feasible set is convex but the objective function f is not, an appealing idea is to replace f with some convex underestimator $\tilde{f}$ on **B** and solve **P'** which is the same as **P** except for the new objective function $\tilde{f}$ in lieu of f. As the resulting optimization problem **P'** is convex, it can be solved via efficient techniques of convex optimization. Its solution provides a lower bound and one may even be satisfied with the associated global minimizer.

Similarly, a nonconvex feasible set $\mathbf{K} \subset \mathbb{R}^n$ of the form $\{\mathbf{x} \in \mathbb{R}^n : g_j(\mathbf{x}) \geq 0, \ j \in J\}$, can be replaced with the convex superset $\mathbf{K} \subset \tilde{\mathbf{K}} := \{\mathbf{x} \in \mathbb{R}^n : \tilde{g}_j(\mathbf{x}) \geq 0, \ j \in J\}$ where each $-\tilde{g}_j$ is a convex underestimator of $-g_j$. For instance in 0/1 programs, if the initial feasible set $\mathbf{K} \cap \{0, 1\}^n$ is replaced with the new convex feasible set $\tilde{\mathbf{K}} \cap [0, 1]^n$, lower bounds on the optimal value can also be computed efficiently by minimizing a convex underestimator of f on $\tilde{\mathbf{K}} \cap [0, 1]^n$. This is what is typically done in the search tree of the Branch

243

and Bound scheme associated with the global optimization algorithm α**BB** for large scale non linear programs.[1] Of course, the overall efficiency of such an algorithm depends heavily on the quality of lower bounds computed at nodes of the branch and bound search tree, and so, ultimately, on the quality of the underestimators that are used. Therefore developing tight convex underestimators for nonconvex polynomials on some given feasible region is of crucial importance.

In many (if not most) large scale optimization problems, each function g_j of the constraint $g_j \geq 0$ involves a few variables only and, similarly, the objective function f is a sum $\sum_k f_k$ where each f_k also involves a few variables only. Hence if $\tilde{f}_k$ is a convex underestimator of f_k, then $\sum_k \tilde{f}_k$ is a convex underestimator of f. Therefore the computational effort in computing convex underestimators of f and the g_j is not too demanding as one may treat functions with a few variables only. As we will next see, this permits us to use the moment-SOS approach to find convex polynomial underestimators with good properties.

15.2 Convex polynomial underestimators

For simplicity (and possibly after a change of variable and scaling) we may and will assume that the box $\mathbf{B} = [\mathbf{x}^L, \mathbf{x}^U]$ in (15.1) is the standard unit box $[0, 1]^n$ described as the compact basic semi-algebraic set $\mathbf{B} = \{\mathbf{x} \in \mathbb{R}^n : x_i (1 - x_i) \geq 0, \ i = 1, \ldots, n \}$.

Let $\mathbf{K} \in \mathbb{R}^n$ be the closure of some open bounded set, and let $\mathbf{U} := \{\mathbf{x} \in \mathbb{R}^n : \|\mathbf{x}\|^2 \leq 1\}$ (the Euclidean unit ball). Recall that $f \in \mathbb{R}[\mathbf{x}]_d$ is convex on $\mathbf{K}$ if and only if its Hessian $\nabla^2 f$ is positive semidefinite on $\mathbf{K}$. Equivalently, f is convex if and only if $\mathbf{T}f \geq 0$ on $\mathbf{K} \times \mathbf{U}$, where $\mathbf{T} : \mathbb{R}[\mathbf{x}] \to \mathbb{R}[\mathbf{x}, \mathbf{y}]$ is the mapping:

$$h \mapsto \mathbf{T}h (\mathbf{x}, \mathbf{y}) := \mathbf{y}^T \nabla^2 h(\mathbf{x}) \, \mathbf{y}, \quad \forall h \in \mathbb{R}[\mathbf{x}]. \tag{15.2}$$

The best degree-d convex underestimator So let $f \in \mathbb{R}[\mathbf{x}]$ be given and consider the associated optimization problem $\mathbf{P}_d$:

$$\rho_d = \inf_{h \in \mathbb{R}[\mathbf{x}]_d} \left\{ \int_{\mathbf{B}} (f - h) \, d\mathbf{x} : \ f - h \geq 0 \text{ on } \mathbf{B}; \quad h \text{ convex on } \mathbf{B} \right\},$$

or equivalently

$$\rho_d = \inf_{h \in \mathbb{R}[\mathbf{x}]_d} \left\{ \int_{\mathbf{B}} (f - h) \, d\mathbf{x} : \ f - h \geq 0 \text{ on } \mathbf{B}; \quad \mathbf{T}h \geq 0 \text{ on } \mathbf{S} \right\}, \tag{15.3}$$

---

[1] The α**BB** algorithm is based on a global optimization method for general constrained nonconvex problems, and is detailed for example in Androulakis et al. (1995).

where **T** is defined in (15.2) and $\mathbf{S} = \mathbf{B} \times \mathbf{U}$.

That is, one searches for a polynomial h of degree at most d which is (a) bounded above by f on **B**, (b) convex on **B** and minimizes $\int_{\mathbf{B}} (f - h)\, d\mathbf{x}$. Any such h is a polynomial convex underestimator of f with the nice property that it minimizes the L_1-norm $\|f - h\|_1$ on $\mathbb{R}[\mathbf{x}]_d$ because as $f \geq h$ on **B**,

$$\int_{\mathbf{B}} (f - h)\, d\mathbf{x} = \int_{\mathbf{B}} |f - h|\, d\mathbf{x} =: \|f - h\|_1.$$

Hence an optimal solution of (15.3) should provide a *tight* convex underestimator of f on **B** because the criterion $\int_{\mathbf{B}} |f - h|\, d\mathbf{x}$ is a good measure of *tightness*. In particular, if f is a polynomial with $\deg f \leq d$ and f is convex on **B** then $\rho_d = 0$ and $f_d^* := f$ is an optimal solution of (15.3).

Lemma 15.1   *The optimization problem (15.3) has an optimal solution* $f_d^* \in \mathbb{R}[\mathbf{x}]_d$.

Proof   For simplicity of exposition we will assume that $\deg f \leq d$. Observe that for every feasible solution $h \in \mathbb{R}[\mathbf{x}]_d$, $h \leq f$ on **B** and so

$$\int_{\mathbf{B}} (f - h)\, d\mathbf{x} = \int_{\mathbf{B}} |f - h|\, d\mathbf{x} = \|f - h\|_1,$$

where $\|\cdot\|_1$ denotes the norm on $L_1([0, 1]^n)$, and which also defines a norm on $\mathbb{R}[\mathbf{x}]_d$ (or, equivalently, on $\mathbb{R}^{s(d)}$). Indeed, if $f, g \in \mathbb{R}[\mathbf{x}]$ and $\|f - g\|_1 = 0$ then $f = g$, almost everywhere on **B**, and so on all of **B** because both are polynomials and **B** has nonempty interior. So if $(h_k) \subset \mathbb{R}[\mathbf{x}]_d$, $k \in \mathbb{N}$, is a minimizing sequence then $(h_k) \subset \Delta_a := \{h : \|f - h\|_1 \leq a\}$ (where $a := \int_{\mathbf{B}} (f - h_0)\, d\mathbf{x})$, and $\int_{\mathbf{B}} (f - h_k)\, d\mathbf{x} \to \rho_d$ as $k \to \infty$. Notice that $\Delta_a \subset \mathbb{R}[\mathbf{x}]_d$ is a ball and a compact set. Therefore there is a subsequence (k_i), $i \in \mathbb{N}$, and an element $f_d^* \in \Delta_a$ such that $h_{k_i} \to f_d^*$ as $i \to \infty$. Therefore, $h_{k_i}(\mathbf{x}) \to f_d^*(\mathbf{x})$ for every $\mathbf{x} \in \mathbf{B}$. Next, since **B** is bounded we also have $\sup_i \|h_{k_i}\|_\infty < M$ for some M, and as a consequence of the Bounded Convergence Theorem

$$\rho_d = \lim_{i \to \infty} \int_{\mathbf{B}} (f - h_{k_i})\, d\mathbf{x} \to \int_{\mathbf{B}} (f - f_d^*)\, d\mathbf{x}.$$

It remains to prove that f_d^* is a feasible solution of (14.4). So, let $\mathbf{x} \in \mathbf{B}$ be fixed, arbitrary. Then since $f - h_k \geq 0$ on **B**, the pointwise convergence $h_{k_i} \to f_d^*$ yields $f(\mathbf{x}) - f_d^*(\mathbf{x}) \geq 0$. Hence $f - f_d^* \geq 0$ on **B**. Similarly, let $(\mathbf{x}, \mathbf{y}) \in \mathbf{S}$ be fixed, arbitrary. Again, from $\mathbf{T}h_k(\mathbf{x}, \mathbf{y}) \geq 0$, the convergence $h_{k_i} \to f_d^*$, and the definition of **T** in (15.2), it immediately follows that $\mathbf{T}f_d^*(\mathbf{x}, \mathbf{y}) \geq 0$. Therefore, $\mathbf{T}f_d^* \geq 0$ on **S**, and so f_d^* is feasible for (14.4). $\square$

Semidefinite approximations As usual, let g_0 be the constant polynomial equal to 1, and let $g_k(\mathbf{x}, \mathbf{y}) := x_k(1-x_k), k = 1, \dots, n, g_{n+1}(\mathbf{x}, \mathbf{y}) := 1-\|\mathbf{y}\|^2$. Let $Q_\mathbf{B} \subset \mathbb{R}[\mathbf{x}]$ be the quadratic module associated with the g_j that define the box $\mathbf{B} = [0, 1]^n$, i.e.,

$$Q_\mathbf{B} := \left\{ \sum_{j=0}^{n} \sigma_j\, g_j : \sigma_j \in \Sigma[\mathbf{x}],\ j = 1, \dots, n \right\}.$$

Similarly, let $\mathbf{S} := \mathbf{B} \times \mathbf{U} = \{ (\mathbf{x}, \mathbf{y}) \in \mathbb{R}^n \times \mathbb{R}^n : g_j(\mathbf{x}, \mathbf{y}) \geq 0,\ j = 1, \dots, n + 1 \}$, and let $Q_\mathbf{S}$ be the quadratic module associated with $\mathbf{S}$, i.e.,

$$Q_\mathbf{S} := \left\{ \sum_{j=0}^{n+1} \theta_j\, g_j : \theta_j \in \Sigma[\mathbf{x}, \mathbf{y}] \right\}.$$

To approximate problem (15.3), a natural idea already used several times in this book, is to replace:

- the nonnegativity constraint $f - h \geq 0$ on $\mathbf{B}$ with $f - h \in Q_\mathbf{B}$, and
- the constraint "h convex on $\mathbf{B}$" with $\mathbf{T}h \in Q_\mathbf{S}$.

So we introduce the following approximation $\mathbf{P}_{dk}$ of $\mathbf{P}_d$, defined by:

$$\sup_{h \in \mathbb{R}[\mathbf{x}]_d, \sigma_j, \theta_\ell} \int_\mathbf{B} h\, d\mathbf{x}$$

$$\text{s.t.} \qquad f = h + \sum_{j=0}^{n} \sigma_j\, g_j$$

$$\mathbf{T}h = \sum_{j=0}^{n+1} \theta_j\, g_j$$

$$\sigma_0 \in \Sigma[\mathbf{x}]_k,\ \sigma_j \in \Sigma[\mathbf{x}]_{k-1},\quad j = 1, \dots, n$$

$$\theta_0 \in \Sigma[\mathbf{x}, \mathbf{y}]_k,\ \theta_j \in \Sigma[\mathbf{x}, \mathbf{y}]_{k-1},\quad j = 1, \dots, n + 1,$$

$$\tag{15.4}$$

with $k \geq \max[\lceil d/2 \rceil, \lceil (\deg f)/2 \rceil]$ and with optimal value denoted by ρ_{dk}. Observe that P_{dk} is a semidefinite program.

> **Lemma 15.2** *For every k such that $2k \geq \deg f$, the semidefinite program (15.4) has an optimal solution $f_{dk} \in \mathbb{R}[\mathbf{x}]_d$.*

Proof From the definition (15.2) of $\mathbf{T}$, define the operator $\mathbf{T}^* : \mathbb{R}[\mathbf{x}, \mathbf{y}]_{2k}^* \to \mathbb{R}[\mathbf{x}]_{2k}^*$ by:

$$\langle g, \mathbf{T}^* \mathbf{z} \rangle = \langle \mathbf{T} g, \mathbf{z} \rangle, \qquad \forall g \in \mathbb{R}[\mathbf{x}]_{2k}, \ \mathbf{z} \in \mathbb{R}[\mathbf{x}, \mathbf{y}]_{2k}^*.$$

The dual of (15.4) is a semidefinite program whose feasible set is described by:

$$
\begin{aligned}
\mathbf{M}_k(\mathbf{u}), \ \mathbf{M}_{k-1}(g_j \, \mathbf{u}) &\succeq 0, \quad j = 1, \dots, n \\
\mathbf{M}_k(\mathbf{z}), \ \mathbf{M}_{k-1}(g_j \, \mathbf{z}) &\succeq 0, \quad j = 1, \dots, n+1 \\
-(\mathbf{T}^*\mathbf{z})_\alpha + u_\alpha &= \gamma_\alpha, \quad \forall \alpha \in \mathbb{N}_d^n,
\end{aligned}
\tag{15.5}
$$

where

- $\mathbf{u} \in \mathbb{R}[\mathbf{x}]_{2k}^*$, $\mathbf{z} \in \mathbb{R}[\mathbf{x}, \mathbf{y}]_{2k}^*$ and $\gamma_\alpha = \int_{\mathbf{B}} \mathbf{x}^\alpha d\mathbf{x}$ for every $\alpha \in \mathbb{N}^n$, and
- $\mathbf{M}_k(\mathbf{u})$ (respectively $\mathbf{M}_{k-1}(g_j \, \mathbf{u})$) is the moment (respectively localizing) matrix associated with the sequence $\mathbf{u}$ and the polynomial g_j, already encountered in Chapter 6. And similarly with $\mathbf{M}_k(\mathbf{z})$ and $\mathbf{M}_{k-1}(g_j \, \mathbf{z})$.

It suffices to show that the set (15.5) has a strictly feasible solution $(\mathbf{u}, \mathbf{z})$, in which case (a) there is no duality gap between (15.4) and its dual, and (b) the dual has an optimal solution if the optimal value is finite. So with $\epsilon > 0$ fixed, let $\mathbf{z} \in \mathbb{R}[\mathbf{x}, \mathbf{y}]_{2k}^*$ be the moment sequence associated with the Borel measure $\epsilon \cdot \lambda \otimes \nu$ on $\mathbf{B} \times \mathbf{U}$, where λ and ν are Borel probability measures uniformly supported on $\mathbf{B}$ and $\mathbf{U}$ respectively. Hence $\mathbf{M}_k(\mathbf{z}), \mathbf{M}_{k-1}(g_j \, \mathbf{z}) \succ 0$, $j = 1, \dots, n+1$. Observe that

$$-(\mathbf{T}^*\mathbf{z})_\alpha = \langle -\mathbf{x}^\alpha, \mathbf{T}^* \mathbf{z} \rangle = \langle -\mathbf{T}\mathbf{x}^\alpha, \mathbf{z} \rangle = -\epsilon\, \theta_\alpha,$$

with

$$\theta_\alpha := \sum_{i,j=1}^{n} \int_{\mathbf{B}} \frac{\partial^2 \mathbf{x}^\alpha}{\partial x_i \partial x_j} d\lambda(\mathbf{x}) \int_{\mathbf{U}} y_i y_j d\nu(\mathbf{y}), \qquad \forall \alpha \in \mathbb{N}_{2k}^n.$$

In particular, $(\mathbf{T}^*\mathbf{z})_\alpha = 0$ whenever $|\alpha| < 2$. Next, let $\mathbf{u} \in \mathbb{R}[\mathbf{x}]_{2k}^*$ be such that

$$
u_\alpha = \begin{cases} \gamma_\alpha + \epsilon\, \theta_\alpha & \forall \alpha \in \mathbb{N}_d^n, \\ \gamma_\alpha & \forall \alpha \in \mathbb{N}_{2k}^n, \ |\alpha| > d. \end{cases}
\tag{15.6}
$$

So the linear constraints $-(\mathbf{T}^*\mathbf{z})_\alpha + u_\alpha = \gamma_\alpha$ of (15.5) are all satisfied. Moreover, from (15.6), the moment matrix $\mathbf{M}_k(\mathbf{u})$ reads $\mathbf{M}_k(\gamma) + \epsilon \Delta_k$ for some matrix Δ_k, and the localizing matrix $\mathbf{M}_{k-1}(g_j \, \mathbf{u})$ reads $\mathbf{M}_{k-1}(g_j \, \mathbf{u}) = \mathbf{M}_{k-1}(g_j \, \gamma) + \epsilon \Theta_{jk}$ for some appropriate matrix Θ_{jk}, $j = 1, \dots, n$. Since $\mathbf{M}_k(\gamma) \succ 0$ and $\mathbf{M}_{k-1}(g_j \, \gamma) \succ 0$ we also have $\mathbf{M}_k(\mathbf{u}) \succ 0$ and $\mathbf{M}_{k-1}(g_j \, \mathbf{u}) \succ 0$ provided that $\epsilon > 0$ is sufficiently small. Hence we have found a strictly feasible solution $(\mathbf{u}, \mathbf{z})$ for the set (15.5), i.e., Slater's condition holds for the dual of (15.4).

Next, (15.4) has a feasible solution. Indeed there exists $M > 0$ such that

$$f(\mathbf{x}) + M = \sigma_0(\mathbf{x}) + \sum_{j=1}^{n} \sigma_i(\mathbf{x}) x_j (1 - x_j), \qquad \forall \mathbf{x} \in \mathbb{R}^n,$$

for some SOS polynomials $\sigma_0 \in \Sigma[\mathbf{x}]_k$ and $\sigma_j \in \Sigma[\mathbf{x}]_{k-1}$, $j = 1, \ldots, n$ (where $k = \lceil (\deg f)/2 \rceil$); see Exercise 6.2. Therefore the constant polynomial $\mathbf{x} \mapsto h(\mathbf{x}) := -M$ is a feasible solution of (15.4) (as $\mathbf{T}h = 0$). Finally the optimal value of (15.4) is finite. Indeed, every feasible solution h of (15.4) is bounded above by f on $\mathbf{B}$. Therefore the objective value $\int_{\mathbf{B}} h d\mathbf{x}$ is bounded above by $\int_{\mathbf{B}} f d\mathbf{x}$ which is finite. Hence (15.4) has an optimal solution because Slater's condition holds for its dual. $\qquad \square$

Theorem 15.3   *Let ρ_d be the optimal value of (15.3) and consider the hierarchy of semidefinite relaxations (15.4) with associated sequence of optimal values (ρ_{dk}), $k \in \mathbb{N}$. Then:*

(a) *$\int_{\mathbf{B}} f d\mathbf{x} - \rho_{dk} \downarrow \rho_d$ as $k \to \infty$, so that $\| f - f_{dk} \|_1 \downarrow \rho_d$ if $f_{dk} \in \mathbb{R}[\mathbf{x}]_d$ is any optimal solution of (15.4);*

(b) *moreover, any accumulation point $\varphi^* \in \mathbb{R}[\mathbf{x}]_d$ of the sequence $(f_{dk}) \subset \mathbb{R}[\mathbf{x}]_d$, is an optimal solution of (15.3), and $f_{dk_i} \to \varphi^*$ pointwise for some subsequence (k_i), $i \in \mathbb{N}$.*

Proof (a) Let $f_d^* \in \mathbb{R}[\mathbf{x}]_d$ be an optimal solution of (15.3), which by Lemma 15.1, is guaranteed to exist. As f_d^* is convex on $\mathbf{B}$, $\nabla^2 f_d^* \succeq 0$ on $\mathbf{B}$. Let $\epsilon > 0$ be fixed and such that $\epsilon \|\mathbf{x}\|^2 < 1$ on $\mathbf{B}$. Let $g_\epsilon := f_d^* - \epsilon + \epsilon^2 \|\mathbf{x}\|^2$, so that $\nabla^2 g_\epsilon \succeq \epsilon^2 \mathbf{I}$ on $\mathbf{B}$. Hence, by the matrix version of Putinar's Theorem (see Theorem 2.19), there exist SOS matrix-polynomials $\mathbf{F}_j$ such that

$$\nabla^2 g_\epsilon(\mathbf{x}) = \mathbf{F}_0(\mathbf{x}) + \sum_{j=1}^{n} \mathbf{F}_j(\mathbf{x}) g_j(\mathbf{x}).$$

(Recall that an SOS matrix-polynomial $\mathbf{F} \in \mathbb{R}[\mathbf{x}]^{q \times q}$ is a matrix-polynomial of the form $\mathbf{x} \mapsto \mathbf{L}(\mathbf{x})\mathbf{L}(\mathbf{x})^T$ for some matrix-polynomial $\mathbf{L} \in \mathbb{R}[\mathbf{x}]^{q \times p}$ for some $p \in \mathbb{N}$.)

And so, for every $j = 0, \ldots, n$, the polynomial $(\mathbf{x}, \mathbf{y}) \mapsto \theta_j^\epsilon(\mathbf{x}, \mathbf{y}) := \mathbf{y}'\mathbf{F}_j\mathbf{y}$ is SOS, and

$$\mathbf{T}g_\epsilon = \sum_{j=0}^{n} \theta_j^\epsilon(\mathbf{x}, \mathbf{y}) g_j(\mathbf{x}) + \theta_{n+1}^\epsilon(\mathbf{x}, \mathbf{y}) (1 - \|\mathbf{y}\|^2),$$

with $\theta_{n+1}^\epsilon = 0$. Moreover, observe that $f - g_\epsilon = f - f_d^* + \epsilon(1 - \epsilon\|\mathbf{x}\|^2)$ is strictly positive on $\mathbf{B}$. Hence by Putinar's Theorem, $f - g_\epsilon = \sum_{j=0}^n \sigma_j^\epsilon g_j$, for some SOS polynomials $\sigma_j \in \mathbb{R}[\mathbf{x}]$, $j = 0, 1, \ldots, n$.

Let $2t \geq \max\{[\max_k \deg \sigma_k + 2, \max_j[\deg \mathbf{F}_j + 4]\}$. Then the polynomial g_ϵ is a feasible solution of (15.4) whenever $k \geq t$. Its value satisfies

$$\int_{\mathbf{B}} f_d^* \, d\mathbf{x} \geq \rho_{dt} \geq \int_{\mathbf{B}} g_\epsilon d\mathbf{x} = \int_{\mathbf{B}} (f_d^* - \epsilon + \epsilon^2\|\mathbf{x}\|^2) \, d\mathbf{x} \geq \int_{\mathbf{B}} f_d^* d\mathbf{x} - \epsilon \, \mathrm{vol}\,(\mathbf{B}),$$

and so $\rho_{dt} \geq \int_{\mathbf{B}} f_d^* \, d\mathbf{x} - \epsilon \, \mathrm{vol}\,(\mathbf{B})$. As $\epsilon > 0$ was arbitrary and the sequence (ρ_{dk}) is monotone nondecreasing, the first result (a) follows.

(b) Next, any optimal solution $f_{dk} \in \mathbb{R}[\mathbf{x}]_d$ of (14.14) satisfies $\|f - f_{dk}\|_1 \leq \int_{\mathbf{B}} f \, d\mathbf{x} - \rho_{d1} =: a$ and so belongs to the ball $\Delta_a := \{h : \|f - h\|_1 \leq a\}$. Let $\varphi^* \in \Delta_a$ be an arbitrary accumulation point of the sequence (f_{dk}) for some subsequence (k_i), $i \in \mathbb{N}$. Proceeding as in the proof of Lemma 15.1, $f_{dk_i} \to \varphi^*$ pointwise, $f - \varphi^* \geq 0$ and $\nabla^2 \varphi^* \succeq 0$ on $\mathbf{B}$. Moreover, by the Bounded Convergence Theorem:

$$\rho_d = \lim_{i\to\infty} \rho_{dk_i} = \lim_{i\to\infty} \int_{\mathbf{B}} (f - f_{dk_i}) \, d\mathbf{x} = \int_{\mathbf{B}} (f - \varphi^*) \, d\mathbf{x},$$

which proves that φ^* is an optimal solution of (15.3). $\qquad\square$

Theorem 15.3 states that the optimal value ρ_{dk} of the semidefinite approximation (15.4) can be as close as desired to that of problem (15.3), and accumulation points of solutions of (15.4) are also optimal solutions of (15.3). The price to pay is the size of the semidefinite program (15.4) which increases with k. In practice one keeps k fixed at a small value. Some computational experiments presented below indicate that even with k fixed at its smallest possible value, the polynomial underestimator f_{dk} provides better lower bounds than the popular αBB underestimator.

15.3 Comparison with the αBB convex underestimator

We compare the convex underestimators $(f_{dk}) \subset \mathbb{R}[\mathbf{x}]_d$ obtained with the moment-SOS approach in Lemma 15.2, with the popular and so-called αBB underestimator.

The αBB convex underestimator To obtain a convex underestimator of a nonconvex polynomial, the αBB method is based on a decomposition of f into a sum of nonconvex terms of special type (e.g. linear, bilinear, trilinear, fractional, fractional trilinear and quadrilinear) and nonconvex terms of arbitrary type. The terms of special type are replaced with their convex envelopes which are already known.

For an arbitrary type f, the underestimator $\mathcal{L}$ is obtained by adding a separable quadratic polynomial nonpositive on $\mathbf{B}$, i.e.,

$$\mathbf{x} \mapsto \quad \mathcal{L}(\mathbf{x}) = f(\mathbf{x}) + \sum_{i=1}^{n} \alpha_i \, (x_i - x_i^L)(x_i - x_i^U), \qquad (15.7)$$

where the real positive coefficients α_i are determined so as to make the polynomial underestimator $\mathcal{L}$ convex. Among the most efficient methods is the so-called scaled Gershgorin method where $\boldsymbol{\alpha} = (\alpha_i) \in \mathbb{R}^n$ is determined by

$$\alpha_i = \max\left\{0, -\frac{1}{2}\left(\underline{f}_{ii} - \sum_{j \neq i} \max\{|\underline{f}_{ii}|, |\overline{f}_{ij}|\}\right)\frac{d_j}{d_i}\right\}, \quad i = 1, \ldots, n, \quad (15.8)$$

where $\underline{f}_{ii}$ and $\overline{f}_{ij}$ are the lower and upper bounds of $\partial^2 f / \partial x_i \partial x_j$ on $\mathbf{B}$ and $d_i, i = 1, 2, \ldots, n$ are some chosen positive parameters. The choice $d_i = u_i^U - u_i^L$ reflects the fact that the underestimator is more sensitive to variables with a wide range than to variables with a small range.

Comparison with the moment-SOS approach Given an arbitrary polynomial $f \in \mathbb{R}[\mathbf{x}]$ and $d \in \mathbb{N}$, one searches for an *ideal* polynomial $f_d^* \in \mathbb{R}[\mathbf{x}]_d$ convex on $\mathbf{B}$, which is an optimal solution of $\mathbf{P}_d$, i.e., f_d^* solves (15.3) (see Lemma 15.1). In practice, one obtains a convex underestimator $f_{dk} \in \mathbb{R}[\mathbf{x}]_d$ by solving the semidefinite relaxation (15.4) of $\mathbf{P}_d$ for a small value of k.

We next compare f_{dk} with the αBB underestimator $\mathcal{L}$ in (15.7), with $x_i^L = 0$ and $x_i^U = 1, i = 1, \ldots, n$ (possibly after scaling).

Lemma 15.4 *With f being a nonconvex polynomial, let $f_{dk} \in \mathbb{R}[\mathbf{x}]_d$ be an optimal solution of (15.4) and let $\mathcal{L}$ be as in (15.7). If $\nabla^2 \mathcal{L}(\mathbf{x}) \succ 0$ for all $\mathbf{x} \in \mathbf{B}$ then*

$$\|f - f_{dk}\|_1 \leq \|f - \mathcal{L}\|_1, \qquad (15.9)$$

whenever k is sufficiently large. That is, the convex underestimator $f_{dk} \in \mathbb{R}[\mathbf{x}]_d$ is tighter than $\mathcal{L}$ when evaluated for the L_1-norm $\int_{\mathbf{B}} |f - g| \, d\mathbf{x}$.

Proof Observe that

$$f(\mathbf{x}) - \mathcal{L}(\mathbf{x}) = \sum_{i=1}^{n} \underbrace{\alpha_i}_{\sigma_i \in \Sigma[\mathbf{x}]_0} x_i \, (1 - x_i),$$

that is, $f - \mathcal{L}$ is a very specific element of the quadratic module $Q_{\mathbf{B}}$, where the SOS weights σ_j are the constant polynomials $\alpha_j, j = 1, \ldots, n$.

Moreover, if $\mathbf{T}\mathcal{L}\succ 0$ on $\mathbf{B}$ then by Theorem 2.19, $\nabla^2\mathcal{L}(\mathbf{x})=\sum_{j=0}^{n}\mathbf{F}_j(\mathbf{x})g_j(\mathbf{x})$, for some SOS polynomial matrices $\mathbf{x}\mapsto\mathbf{F}_j(\mathbf{x})$ (i.e., of the form $\mathbf{L}_j(\mathbf{x})\mathbf{L}_j(\mathbf{x})^T$ for some matrix-polynomials $\mathbf{L}_j$) and so

$$\mathbf{T}\mathcal{L}(\mathbf{x},\mathbf{y}) \;=\; \mathbf{y}'\nabla^2\mathcal{L}(\mathbf{x})\mathbf{y} \;=\; \sum_{j=0}^{n}\underbrace{(\mathbf{L}_j(\mathbf{x})^T\mathbf{y})^2}_{\theta_j\in\Sigma[\mathbf{x},\mathbf{y}]}\,g_j(\mathbf{x}).$$

Hence $\mathbf{T}\mathcal{L}\in Q_\mathbf{S}$ and $\mathcal{L}$ is a feasible solution of (15.4) as soon as $2k\geq\max_j\deg\mathbf{F}_j+4$. Therefore, at least for k sufficiently large,

$$\int_\mathbf{B} f_{dk}\,d\mathbf{x} \;\geq\; \int_\mathbf{B}\mathcal{L}\,d\mathbf{x},$$

and so as $f\geq f_{dk}$ and $f\geq\mathcal{L}$ on $\mathbf{B}$, (15.9) holds.   □

Some computational experiments Among possible choices of d, consider the two natural choices, $d=\deg f$ or $d=2$, and $k\geq\max[\lceil d/2\rceil,\lceil(\deg f)/2\rceil]$. With the former choice one searches for the best convex underestimator of same degree as f, while with the latter choice one searches for the best quadratic underestimator of f.

Recall that the main motivation for computing underestimators is to compute "good" lower bounds on a box $\mathbf{B}$ for nonconvex problems and, for instance, to use these lower bounds in a branch and bound algorithm. Therefore, to compare the moment and αBB underestimators, we have chosen nonconvex optimization problems in the literature, and replaced the original nonconvex objective function by its moment-SOS underestimator f_{dk} and its αBB underestimator $\mathcal{L}$, respectively. We then compare:

- the minimum f^*_{mom} (respectively $f^*_{\alpha\mathbf{BB}}$) obtained by minimizing f_{dk} (respectively $\mathcal{L}$) on the box $\mathbf{B}^2$;
- the respective values of the "tightness score," i.e., the L_1-norm $\int_\mathbf{B}|f-f_{dk}|\,d\mathbf{x}$ and $\int_\mathbf{B}|f-\mathcal{L}|\,d\mathbf{x}$; in view of (15.7), the latter is easy to compute.

Figure 15.1 displays an illustrative example with the Six-Hump Camelback function $\mathbf{x}\mapsto f(\mathbf{x})=4x_1^2-2.1x_1^4+\frac{1}{3}x_1^6+x_1x_2-4x_2^2+4x_2^4$ in the box $\mathbf{B}=[0,1].^2$ The global minimum is $f^*=-1$ to be compared with $f^*_{mom}=-1.36$, and $f^*_{\alpha\mathbf{BB}}=-3.33$.

- **Choice 1:** $d=2$ **(quadratic underestimator)** Given $f\in\mathbb{R}[\mathbf{x}]$, one searches for a convex quadratic polynomial $f_d\in\mathbb{R}[\mathbf{x}]_2$, i.e., of the form

---

^{2} All computations were made by running the GloptiPoly software described in Appendix B. The αBB underestimator was computed via the scaled Gershgorin method.

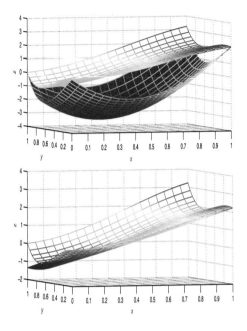

Figure 15.1 Comparing f and $\mathcal{L}$ and then f and f_{dk}.

$\mathbf{x} \mapsto f_d(\mathbf{x}) = \mathbf{x}^T \mathbf{A} \mathbf{x} + \mathbf{a}^T \mathbf{x} + b$ for some real positive semidefinite symmetric matrix $\mathbf{A} \in \mathbb{R}^{n \times n}$, vector $\mathbf{a} \in \mathbb{R}^n$ and scalar b. Let $\mathbf{M}_\lambda$ be the moment matrix of order 1 of the (normalized) Lebesgue measure λ on $\mathbf{B}$, i.e.,

$$\mathbf{M}_\lambda = \begin{bmatrix} 1 & \boldsymbol{\gamma}^T \\ \boldsymbol{\gamma} & \boldsymbol{\Lambda} \end{bmatrix}$$

with $\gamma_i = \int_{\mathbf{B}} x_i \, d\lambda$ for all $i = 1, \ldots, n$, and $\Lambda_{ij} = \int_{\mathbf{B}} x_i x_j \, d\lambda$ for all $1 \leq i, j \leq n$. The semidefinite relaxation $\mathbf{P}_{dk}$ in (15.4) reads:

$$\sup_{b, \mathbf{a}, \mathbf{A}} b + \mathbf{a}^T \boldsymbol{\gamma} + \langle \mathbf{A}, \boldsymbol{\Lambda} \rangle$$

$$\text{s.t.} \ f(\mathbf{x}) = b + \mathbf{a}^T \mathbf{x} + \mathbf{x}^T \mathbf{A} \mathbf{x} + \sum_{j=0}^n \sigma_j(\mathbf{x}) \, g_j(\mathbf{x}), \qquad \forall \mathbf{x} \qquad (15.10)$$

$$\mathbf{A} \succeq 0; \ \sigma_0 \in \Sigma[\mathbf{x}]_k, \ \sigma_j \in \Sigma[\mathbf{x}]_{k-1}, \ j \geq 1.$$

Table 15.1 displays results for some examples with $d = 2$. The test functions f are taken from Floudas (2000) and Gounaris and Floudas (2008). On a box $\mathbf{B}$ that contains the feasible set, we compute the convex $\alpha\mathbf{BB}$ underestimator $\mathcal{L}$ and the (only degree 2) moment-SOS underestimator f_{dk} of the initial objective function f via solving (15.10) with the smallest value of $k = \lceil (\deg f)/2 \rceil$.

Table 15.1 *Comparing f^*_{mom} and $f^*_{\alpha \mathbf{BB}}$, $d = 2$*

Problem	n	deg f	d	$[\mathbf{x}^L, \mathbf{x}^U]$	$f^*_{\alpha \mathbf{BB}}$	f^*_{mom}	f^*
Test2	4	3	2	[0,1]	−1.54	−1.22	−1
Test3	5	4	2	[−1,1]	−15	−13.95	−6
Test4	6	6	2	[−1,1]	−60.15	−10.06	−3
Test5	3	6	2	[−2,2]	−411.2	−12.66	−1
Test10	4	4	2	[0,1]	−197.54	−0.9698	0
Test11	4	4	2	[0,1]	−33.02	−0.623	0
Test14(1)	3	4	2	[−5,2]	−2409	−300	−300
Test14(2)	4	4	2	[−5,2]	−3212	−400	−400
Test14(3)	5	4	2	[−5,2]	−4015	−500	−500
Fl.2.2	5	2	2	[0,1]	−18.9	−18.9	−17
Fl.2.3	6	2	2	[0,1]	−5270.9	−361.50	−361
Fl.2.4	13	2	2	[0,3]	−592	−195	−195
Fl.2.6	10	2	2	[0,1]	−269.83	−269.45	−268.01
Fl.2.8C1	20	2	2	[0,3]	−560	−560	−394.75
Fl.2.8C2	20	2	2	[0,10]	−1050	−1050	−884
Fl.2.8C3	20	2	2	[0,30]	−13600	−12000	−8695
Fl.2.8C4	20	2	2	[0,30]	−920	−920	−754.75
Fl.2.8C5	20	2	2	[0,30]	−16645	−10010	−4150.41

Table 15.2 *Comparing f^*_{mom} and $f^*_{\alpha \mathbf{BB}}$, $d = $ deg f*

Problem	n	deg f	d	$[\mathbf{x}^L, \mathbf{x}^U]$	$f^*_{\alpha \mathbf{BB}}$	f^*_{mom}	f^*
Test2	4	3	3	[0,1]	−1.54	−1.22	−1
Test3	5	4	4	[−1,1]	−15	−11.95	−6
Test4	6	6	2	[−1,1]	−60.15	−10.06	−3
Test5	3	6	6	[−2,2]	−411.2	−12.07	−1
Test10	4	4	4	[0,1]	−197.54	−0.778	0
Test11	4	4	4	[0,1]	−33.02	0	0
Test14(1)	3	4	4	[−5,2]	−2409	−300	−300
Test14(2)	4	4	4	[−5,2]	−3212	−400	−400
Test14(3)	5	4	4	[−5,2]	−4015	−500	−500

We then compute their respective minima $f^*_{\alpha \mathbf{BB}}$ and f^*_{mom} on **B**. All examples were run on an Intel(R) Core(TM) i5 2.53 GHz processor with 4Gb of Ram. In a typical example with degree 4 and 5 variables, the CPU time was 0.68 s to find the underestimator f_{dk} with $d = 2$.

• **Choice 2:** $d = $ deg f Table 15.2 displays results for some examples taken from Table 15.1 but now using the moment-SOS underestimator f_d with choice $d = $ deg f. Again k in (14.14) is set to its smallest possible value $\lceil d/2 \rceil$. As one may see and as expected, the lower bound f^*_{mom} is better, and in several

Table 15.3 *Comparing $V_{\mathbf{B}}^{-1}\int_{\mathbf{B}}|f-\mathcal{L}|\,d\lambda$ and $V_{\mathbf{B}}^{-1}\int_{\mathbf{B}}|f-f_d|\,d\lambda$*

Problem	$V_{\mathbf{B}}^{-1}\int_{\mathbf{B}}|f-\mathcal{L}|\,d\mathbf{x}$	$V_{\mathbf{B}}^{-1}\int_{\mathbf{B}}|f-f_d|\,d\mathbf{x}$
Test2	1	0.625
Test3	11.67	3.33
Test4	60	7.29
Test5	99.00	23.20
Test10	133.33	57.00
Test11	46.33	1
Test14(1)	1641.4×10^3	149.2711
Test14(2)	2186.6	199.08
Test14(3)	2731	248.71
Fl.2.2	41.66	41.66
Fl.2.3	67500	833.33
Fl.2.4	0.005	0.0005
Fl.2.6	8.33	5.83
Fl.2.8C1	3.4989×10^{14}	3.4989×10^{14}
Fl.2.8C2	12200	12200
Fl.2.8C3	6.9979×10^{-5}	6.9979×10^{-5}
Fl.2.8C4	3.4989×10^{-6}	3.4989×10^{-6}
Fl.2.8C5	9.6077×10^{-15}	9.6077×10^{-15}

examples f^*_{mom} is very close to the global minimum f^*. However, depending
on the degree of f, the computing time is now larger than with $d = 2$; recall
that the size of the semidefinite program (15.4) is parametrized by k, chosen
here to be equal to its smallest possible value $\lceil d/2 \rceil$. For a typical example with
degree 4 and 5 variables, the CPU time was 1.68 s to find the underestimator
f_{dk} with $d = 4$ and $k = 2$.

Finally, Table 15.3 displays the respective values of the "tightness" score
$\int_{\mathbf{B}}|f-\mathcal{L}|\,d\mathbf{x}$ and $\int_{\mathbf{B}}|f-f_{d1}|\,d\mathbf{x}$ with $d = 2$, normalized with respect
to the volume $V_{\mathbf{B}}$ of the box $\mathbf{B}$. Again, the tightness score of the moment
underestimator f_d with $d = 2$ is significantly better than that of the α**BB**
underestimator $\mathcal{L}$.

In view of the above experimental results (even though they are limited) it
seems that the practical choice $d = 2$ combined with the smallest possible
value $k := \lceil(\deg f)/2\rceil$ in (14.14), is enough to obtain a good convex polynomial underestimator.

15.4 Exercises

Exercise 15.1 Let $n = 1$, $\mathbf{B} = [0, 1]$ and let $p, q \in \mathbb{R}[x]$ be such that
$q > 0$ on $\mathbf{B}$. Show that computing the best degree d convex polynomial

underestimator h of the rational function $f = p/q$ on **B** reduces to solving a single semidefinite program. Run an example with $x \mapsto p(x) = x$ and $x \mapsto q(x) = 1 + x^2$. (*Hint:* The constraint $f - h \geq 0$ on **B** is identical to $p - qh \geq 0$ on **B** and convexity reduces to $h'' \geq 0$ on **B**. Then use positivity certificates for univariate polynomials.)

Exercise 15.2 Let $n = 1$, **B** $:= [0, 1]$ and let $p_1, p_2 \in \mathbb{R}[x]$. Show that computing the best degree d convex polynomial underestimator h of the function $f = p_1 \wedge p_2$ on **B** reduces to solving a single semidefinite program. Run an example with $x \mapsto p_1(x) = x\,(x - 1/3)\,(x - 1)$ and $x \mapsto p_2(x) = x\,(x - 1/3)\,(3/4 - x)$. (*Hint:* The constraint $f - h \geq 0$ on **B** is identical to $p_1 - h \geq 0$ and $p_2 - h \geq 0$ on **B** and convexity reduces to $h'' \geq 0$ on **B**. Then use positivity certificates for univariate polynomials.)

Exercise 15.3 Let $n = 1$, **B** $:= [0, 1]$ and let $p_1, p_2 \in \mathbb{R}[x]$. Provide a hierarchy of semidefinite programs to approximate the best degree d convex polynomial underestimator h of the function $f = p_1 \vee p_2$ on **B**. Run an example with $x \mapsto p_1(x) = x\,(x - 1/3)\,(x - 1)$ and $x \mapsto p_2(x) = x\,(x - 1/3)\,(3/4 - x)$. (*Hint:* Model the function f on **B** as $(x, z) \mapsto (z + p_1(x) + p_2(x))/2$ on the set

$$\mathbf{K} = \{\, (x, z) : x\,(1 - x) \geq 0;\ z^2 - (p_1(x) - p_2(x))^2 = 0;\ z \geq 0 \,\} \subset \mathbb{R}^2.$$

The constraint $f - h \geq 0$ on **B** is identical to $(z + p_1(x) + p_2(x))/2 - h \geq 0$ on **K**.)

Exercise 15.4 Let $n = 2$ and let $f \in \mathbb{R}[\mathbf{x}]_4$. Show that computing the best degree 4 convex underestimator of f on $\mathbb{R}^2$ reduces to solving a single semidefinite program. Describe this semidefinite program. (*Hint:* Use the fact that nonnegative quartic bivariate polynomials are SOS.)

15.5 Notes and sources

Most of the material of this chapter is from Lasserre and Thanh (2011, 2013). The deterministic global optimization algorithm α**BB** (which uses α**BB** convex underestimators) is used for large scale nonlinear optimization, and particularly for MINLP (Mixed Integer Nonlinear Problems). Using a branch and bound scheme, lower bounds computed at nodes of the search tree are obtained by solving a *convex* problem where the objective function f as well as each $-g_j$ of a nonconvex constraint $g_j(\mathbf{x}) \geq 0$, is replaced with some convex underestimator $\tilde{f}$ and $\tilde{g}_j$ on a box **B** $\subset \mathbb{R}^n$. Hence a lower bound can be obtained efficiently by minimizing $\tilde{f}$ on $\tilde{\mathbf{K}}$ (or $\tilde{\mathbf{K}} \cap [0, 1]^n$ in 0/1 programs), a convex problem that can be solved efficiently via techniques from convex optimization; see for example Floudas and Pardalos (2001, pp. 1–12), Floudas (2000), Androulakis et al. (1995). Of course, the overall efficiency of the α**BB** algorithm depends heavily on the quality of the lower bounds computed in the branch and bound tree search, and so, ultimately, on the quality of the underestimators of f that are used. There are also several results available in the literature for computing convex envelopes of specific simple functions in *explicit* form, on a box $\mathbf{B} \subset \mathbb{R}^n$. See for instance Floudas (2000) for convex envelopes of bilinear, trilinear, multilinear monomials, Tawarmalani and Sahinidis (2002) for fractional terms, and Cafieri et al. (2010) for quadrilinear terms. Finally see Guzman et al. (2014) for a recent comparison of several techniques (including the techniques presented in this chapter) to build up convex underestimators used in a branch and bound procedure for global optimization.

16

Inverse polynomial optimization

16.1 Introduction

Again let $\mathbf{P}$ be the polynomial optimization problem $f^* = \inf\{\, f(\mathbf{x}) : \mathbf{x} \in \mathbf{K}\,\}$, whose feasible set is the basic semi-algebraic set:

$$\mathbf{K} := \{\, \mathbf{x} \in \mathbb{R}^n : g_j(\mathbf{x}) \geq 0, \ j = 1, \ldots, m \,\}, \qquad (16.1)$$

for some polynomials $f, (g_j) \subset \mathbb{R}[\mathbf{x}]$. As already mentioned, $\mathbf{P}$ is in general NP-hard and one goal of this book is precisely to describe methods to obtain (or at least approximate) f^* and whenever possible a global minimizer $\mathbf{x}^* \in \mathbf{K}$.

However, in many cases one is often satisfied with a local minimum only (for instance because the methods described in Chapter 6 are computationally too expensive and cannot be implemented for the problem on hand). On the other hand a local minimum can be obtained by running some local minimization algorithm choosen among those available in the literature. Typically in such algorithms, at a current iterate (i.e., some feasible solution $\mathbf{y} \in \mathbf{K}$) one checks whether some optimality conditions (e.g. the Karush–Kuhn–Tucker (KKT) optimality conditions of Chapter 7) are satisfied within some ϵ-tolerance. However, those conditions are valid for any local minimum, and in fact, even for any stationary point of the Lagrangian. Moreover, in some practical situations the criterion f to minimize is subject to modeling errors or is questionable. In such a situation the practical meaning of a local (or global) minimum f^* (and local (or global) minimizer) also becomes questionable. It could well be that the current solution $\mathbf{y}$ is in fact a global minimizer of an optimization problem $\mathbf{P}'$ with the same feasible set as $\mathbf{P}$ but with a different criterion $\tilde{f}$. Therefore if $\tilde{f}$ is "close enough" to f one might not be willing to spend an enormous amount of computing time and effort to find the global (or even local) minimum f^* because one might be already satisfied with the current iterate $\mathbf{y}$ as a global minimizer of $\mathbf{P}'$.

Inverse optimization is precisely concerned with the above issue of determining a criterion $\tilde{f}$ as close to f as possible, and for which a given current solution $\mathbf{y} \in \mathbf{K}$ is an optimal solution of $\mathbf{P}'$ with this new criterion $\tilde{f}$. Hence the inverse optimization problem associated with $\mathbf{P}$ could be stated informally as:

given $\mathbf{y} \in \mathbf{K}$ and a space of functions F, find $\tilde{f} \in F$ such that

$$\tilde{f} = \mathrm{argmin}_{g \in F} \left\{ \|f - g\| : g(\mathbf{y}) = \inf_{\mathbf{x}} \{ g(\mathbf{x}) : \mathbf{x} \in \mathbf{K} \} \right\},$$

for some norm $\| \cdot \|$ on F. If $\|f - \tilde{f}\|$ is small then one might be satisfied with the solution $\mathbf{y} \in \mathbf{K}$. In fact, with $\epsilon > 0$ fixed, a test of the form $\|f - \tilde{f}\| < \epsilon$ could be used as a stopping criterion in any (local) minimization algorithm to solve $\mathbf{P}$.

As the reader may immediately guess, in inverse optimization the main difficulty lies in having a tractable characterization of global optimality for a given current point $\mathbf{y} \in \mathbf{K}$ and some candidate criterion $g \in F$. This is why most previous works have addressed linear programs or combinatorial optimization problems for which some characterization of global optimality is available and can (sometimes) be effectively used for practical computation. This perhaps also explains why inverse (nonlinear) optimization has not attracted much attention in the past, a pity since inverse optimality could provide an alternative stopping criterion at a feasible solution $\mathbf{y} \in \mathbf{K}$ obtained by a (local) optimization algorithm.

In this chapter we provide a systematic numerical scheme for computing an inverse optimal solution associated with the polynomial program $\mathbf{P}$ and a given feasible solution $\mathbf{y} \in \mathbf{K}$. It consists of solving a semidefinite program whose size can be adapted to the problem on hand, and so is tractable (at least for moderate size problems and possibly for larger size problems if sparsity is taken into account). Moreover, if one uses the ℓ_1-norm then the inverse optimal objective function exhibits a simple *canonical* (and sparse) form.

16.2 Computing an inverse optimal solution

For $f \in \mathbb{R}[\mathbf{x}]_d$, let

$$\|f\|_k = \begin{cases} \sum_{\alpha \in \mathbb{N}_d^n} |f_\alpha| & \text{if } k = 1, \\ \sum_{\alpha \in \mathbb{N}_d^n} f_\alpha^2 & \text{if } k = 2, \\ \sup\{ |f_\alpha| : \alpha \in \mathbb{N}_d^n \} & \text{if } k = \infty. \end{cases}$$

Observe that $\| \cdot \|_k$ is a norm on $\mathbb{R}^{s(d)}$ which is the vector space for the coefficient vector $\mathbf{f} = (f_\alpha)$, $\alpha \in \mathbb{N}^n_d$, of the polynomial $f \in \mathbb{R}[\mathbf{x}]_d$. For notational convenience (and by a slight abuse of notation) we have written $\| f \|_k$ instead of $\| \mathbf{f} \|_k$.

16.2.1 The ideal inverse optimization problem

Let $\mathbf{P}$ be the global optimization problem $f^* = \inf_{\mathbf{x}}\{ f(\mathbf{x}) : \mathbf{x} \in \mathbf{K} \}$ with $\mathbf{K} \subset \mathbb{R}^n$ as in (16.1), and $f \in \mathbb{R}[\mathbf{x}]_{d_0}$ where $d_0 := \deg f$. As in Chapter 4, let $C_d(\mathbf{K}) \subset \mathbb{R}[\mathbf{x}]_d$ be the convex cone of polynomials of degree at most d, nonnegative on $\mathbf{K}$. Recall that g_0 is the constant polynomial $g_0(\mathbf{x}) = 1$ for all $\mathbf{x}$ and for every $j = 0, \ldots, m$, let $v_j := \lceil (\deg g_j)/2 \rceil$.

Next, let $\mathbf{y} \in \mathbf{K}$ and $k \in \{1, 2, \infty\}$ be fixed, and consider the following optimization problem:

$$\mathcal{P}: \quad \rho^k = \inf_{h \in \mathbb{R}[\mathbf{x}]_{d_0}} \{ \| h - f \|_k : \mathbf{x} \mapsto h(\mathbf{x}) - h(\mathbf{y}) \in C_{d_0}(\mathbf{K}) \}. \quad (16.2)$$

Theorem 16.1 *Problem (16.2) has an optimal solution $\tilde{f}^* \in \mathbb{R}[\mathbf{x}]_{d_0}$. In addition, $\rho^k = 0$ if and only if $\mathbf{y}$ is an optimal solution of $\mathbf{P}$.*

Proof Obviously the constant polynomial $\mathbf{x} \mapsto h_0(\mathbf{x}) := 1$ is a feasible solution with associated value $\delta := \| h_0 - f \|_k$. Moreover the optimal value of (16.2) is bounded below by 0. Consider a minimizing sequence $(h_j) \subset \mathbb{R}[\mathbf{x}]_{d_0}$, $j \in \mathbb{N}$, hence such that $\| h_j - f \|_k \to \rho^k$ as $j \to \infty$. As we have $\| h_j - f \|_k \leq \delta$ for every j, the sequence (h_j) belongs to the ℓ_k-ball $\mathbf{B}_k(f) := \{ h \in \mathbb{R}[\mathbf{x}]_{d_0} : \| h - f \|_k \leq \delta \}$, a compact set (where $\mathbb{R}[\mathbf{x}]_{d_0}$ is identified with $\mathbb{R}^{s(d_0)}$). Therefore, there is an element $\tilde{f}^* \in \mathbf{B}_k(f)$ and a subsequence (j_t), $t \in \mathbb{N}$, such that $h_{j_t} \to \tilde{f}^*$ as $t \to \infty$. Let $\mathbf{x} \in \mathbf{K}$ be fixed arbitrary. Obviously $(0 \leq) h_{j_t}(\mathbf{x}) - h_{j_t}(\mathbf{y}) \to \tilde{f}^*(\mathbf{x}) - \tilde{f}^*(\mathbf{y})$ as $t \to \infty$, which implies $\tilde{f}^*(\mathbf{x}) - \tilde{f}^*(\mathbf{y}) \geq 0$, and so as $\mathbf{x} \in \mathbf{K}$ was arbitrary, $\tilde{f}^* - \tilde{f}^*(\mathbf{y}) \geq 0$ on $\mathbf{K}$, i.e., $\tilde{f}^* - \tilde{f}^*(\mathbf{y}) \in C_{d_0}(\mathbf{K})$. Finally, we also obtain the desired result

$$\rho^k = \lim_{j \to \infty} \| h_j - f \|_k = \lim_{t \to \infty} \| h_{j_t} - f \|_k = \| \tilde{f}^* - f \|_k.$$

Next, if $\mathbf{y}$ is an optimal solution of $\mathbf{P}$ then $\tilde{f}^* := f$ is an optimal solution of $\mathcal{P}$ with value $\rho^k = 0$. Conversely, if $\rho^k = 0$ then $\tilde{f}^* = f$, and so by feasibility of $\tilde{f}^*$ *(equal to f)* for (16.2), $f(\mathbf{x}) \geq f(\mathbf{y})$ for all $\mathbf{x} \in \mathbf{K}$, which shows that $\mathbf{y}$ is an optimal solution of $\mathbf{P}$. $\square$

Theorem 16.1 states that the ideal inverse optimization problem is well-defined. However, even though $C_{d_0}(\mathbf{K})$ is a finite-dimensional convex cone, it has no simple and tractable characterization to be used for practical computation. Therefore one needs an alternative and more tractable version of problem $\mathcal{P}$ and, fortunately, in the polynomial context we have seen in Chapter 2 and Chapter 4 that one may define tractable inner approximations for $C_{d_0}(\mathbf{K})$ thanks to the powerful Putinar Positivstellensatz of Theorem 2.15.

16.2.2 A practical inverse problem

We first start with the following definition.

Definition 16.2   If $\mathbf{y} \in \mathbf{K}$ and $h \in \mathbb{R}[\mathbf{x}]$ satisfy

$$h(\mathbf{x}) - h(\mathbf{y}) = \sum_{j=0}^{m} \sigma_j(\mathbf{x}) \, g_j(\mathbf{x}), \quad \forall \mathbf{x} \in \mathbb{R}^n, \tag{16.3}$$

for some SOS weights $(\sigma_j) \subset \Sigma[\mathbf{x}]_d$, then one says that (16.3) is a "Putinar certificate of global optimality on $\mathbf{K}$ with degree bound d" for $\mathbf{y}$ and h.

Indeed from (16.3) one easily deduces that $h(\mathbf{x}) \geq h(\mathbf{y})$ for all $\mathbf{x} \in \mathbf{K}$ and so $\mathbf{y}$ is a global minimizer of h on $\mathbf{K}$. Moreover, the SOS weights σ_j provide a *certificate with degree bound d* of such an assertion.

A rationale for using the global optimality certificate (16.3) is provided by Theorem 7.5 and Theorem 7.6 which assert that this global optimality certificate is generic in polynomial optimization.

With no loss of generality and up to a change of coordinates $\mathbf{x}' = \mathbf{x} - \mathbf{y}$, we may and will assume that $\mathbf{y} = 0 \in \mathbf{K}$. With $d \in \mathbb{N}$ fixed, consider the following optimization problem $\mathbf{P}_d$.

$$\rho_d^k = \inf_{h, \sigma_j \in \mathbb{R}[\mathbf{x}]} \quad \|f - h\|_k$$

$$\text{s.t.} \quad h(\mathbf{x}) - h(0) = \sum_{j=0}^{m} \sigma_j(\mathbf{x}) \, g_j(\mathbf{x}), \quad \forall \mathbf{x} \in \mathbb{R}^n$$

$$h \in \mathbb{R}[\mathbf{x}]_{d_0}; \ \sigma_j \in \Sigma[\mathbf{x}]_{d-v_j}, \quad j = 0, 1, \ldots, m, \tag{16.4}$$

where $d_0 = \deg f$, and $v_j = \lceil (\deg g_j)/2 \rceil$, $j = 0, \ldots, m$ (recall that $g_0 = 1$).

Observe that the constraint of (16.4) is nothing less than a Putinar certificate of global optimality on $\mathbf{K}$ with degree bound d, for h and $\mathbf{y} = 0 \in \mathbf{K}$. The parameter $d \in \mathbb{N}$ impacts (16.4) by the degree bounds $d - u_j$ allowed for the SOS weights $(\sigma_j) \subset \Sigma[\mathbf{x}]$ in the Putinar certificate of global optimality on $\mathbf{K}$ for h and $0 \in \mathbf{K}$; and so the higher d the lower (and the better) ρ_d^k.

So $\mathbf{P}_d$ is a strengthening of $\mathcal{P}$ in that one has replaced the (hard) constraint $h - h(0) \in C_{d_0}(\mathbf{K})$ with the stronger Putinar certificate of global optimality with degree bound d. And so $\rho^k \leq \rho_d^k$ for all $d \in \mathbb{N}$.

Therefore an optimal solution $\tilde{f}_d$ of (16.4) is a polynomial of $\mathbb{R}[\mathbf{x}]_{d_0}$ which is the closest to f for the ℓ_k-norm, and is such that $\mathbf{y} = 0 \in \mathbf{K}$ and $\tilde{f}_d$ have a Putinar certificate of global optimality with degree bound d.

Remark 16.3 Observe that in any feasible solution h of $\mathbf{P}_d$, the coefficient h_0 plays no role in the constraint, and since we minimize $\|h - f\|_k$ then it is always optimal to set $h_0 = f_0$. That is, $h(0) = h_0 = f_0 = f(0)$.

Moreover, for each $d \in \mathbb{N}$, $\mathbf{P}_d$ is a tractable optimization problem with nice properties because it is a semidefinite program. Indeed, with $\mathbf{v}_d(\mathbf{x}) = (\mathbf{x}^\alpha)$, $\alpha \in \mathbb{N}_d^n$, recall the matrices $(\mathbf{C}_\alpha^j)$:

$$g_j(\mathbf{x})\, \mathbf{v}_{d-v_j}(\mathbf{x})\, \mathbf{v}_{d-v_j}(\mathbf{x})^T = \sum_{\alpha \in \mathbb{N}_{2d}^n} \mathbf{C}_\alpha^j\, \mathbf{x}^\alpha, \qquad j = 0, 1, \ldots, m.$$

Then for instance, if $k = 1$ one may rewrite $\mathbf{P}_d$ as:

$$
\begin{aligned}
\rho_d^1 := \inf_{h, \lambda, \mathbf{Z}_j} \quad & \sum_{\alpha \in \mathbb{N}_{d_0}^n \setminus \{0\}} \lambda_\alpha \\
\text{s.t.} \quad & \lambda_\alpha + h_\alpha \geq f_\alpha, \quad \forall \alpha \in \mathbb{N}_{d_0}^n \setminus \{0\} \\
& \lambda_\alpha - h_\alpha \geq -f_\alpha, \quad \forall \alpha \in \mathbb{N}_{d_0}^n \setminus \{0\} \\
& \langle \mathbf{Z}_0, \mathbf{B}_\alpha \rangle + \sum_{j=1}^m \langle \mathbf{Z}_j, \mathbf{C}_\alpha^j \rangle = \begin{cases} h_\alpha & \text{if } 0 < |\alpha| \leq d_0 \\ 0 & \text{if } \alpha = 0 \text{ or } |\alpha| > d_0 \end{cases} \\
& \mathbf{Z}_j \succeq 0, \quad j = 0, 1, \ldots, m.
\end{aligned}
$$

(16.5)

Similarly, if $k = \infty$ then one may rewrite $\mathbf{P}_d$ as:

$$
\begin{aligned}
\rho_d^\infty := \inf_{h, \lambda, \mathbf{Z}_j} \quad & \lambda \\
\text{s.t.} \quad & \lambda + h_\alpha \geq f_\alpha, \quad \forall \alpha \in \mathbb{N}_{d_0}^n \setminus \{0\} \\
& \lambda - h_\alpha \geq -f_\alpha, \quad \forall \alpha \in \mathbb{N}_{d_0}^n \setminus \{0\} \\
& \langle \mathbf{Z}_0, \mathbf{B}_\alpha \rangle + \sum_{j=1}^m \langle \mathbf{Z}_j, \mathbf{C}_\alpha^j \rangle = \begin{cases} h_\alpha & \text{if } 0 < |\alpha| \leq d_0 \\ 0 & \text{if } \alpha = 0 \text{ or } |\alpha| > d_0 \end{cases} \\
& \mathbf{Z}_j \succeq 0, \quad j = 0, 1, \ldots, m,
\end{aligned}
$$

(16.6)

and finally, if $k = 2$ then one may rewrite $\mathbf{P}_d$ as:

$$\rho_d^2 := \inf_{h, \lambda, \mathbf{Z}_j} \sum_{\alpha \in \mathbb{N}_{d_0}^n \setminus \{0\}} \lambda_\alpha$$

$$\text{s.t.} \quad \begin{bmatrix} \lambda_\alpha & h_\alpha - f_\alpha \\ h_\alpha - f_\alpha & 1 \end{bmatrix} \succeq 0, \quad \forall \alpha \in \mathbb{N}_{d_0}^n \setminus \{0\}$$

$$\langle \mathbf{Z}_0, \mathbf{B}_\alpha \rangle + \sum_{j=1}^{m} \langle \mathbf{Z}_j, \mathbf{C}_\alpha^j \rangle = \begin{cases} h_\alpha & \text{if } 0 < |\alpha| \leq d_0 \\ 0 & \text{if } \alpha = 0 \text{ or } |\alpha| > d_0 \end{cases}$$

$$\mathbf{Z}_j \succeq 0, \quad j = 0, 1, \ldots, m.$$

$$(16.7)$$

16.2.3 Duality

The semidefinite program dual of (16.5) reads

$$\begin{cases} \sup_{\mathbf{u}, \mathbf{v} \geq 0, \mathbf{z}} \sum_{\alpha \in \mathbb{N}_{d_0}^n \setminus \{0\}} f_\alpha (u_\alpha - v_\alpha) \; (= L_\mathbf{z}(f(0) - f)) \\ \text{s.t.} \quad u_\alpha + v_\alpha \leq 1, \quad \forall \alpha \in \mathbb{N}_{d_0}^n \setminus \{0\} \\ \quad u_\alpha - v_\alpha + z_\alpha = 0, \quad \forall \alpha \in \mathbb{N}_{d_0}^n \setminus \{0\} \\ \quad \mathbf{M}_d(\mathbf{z}), \mathbf{M}_{d-v_j}(g_j \, \mathbf{z}) \succeq 0, \quad j = 1, \ldots, m, \end{cases} \qquad (16.8)$$

while the semidefinite program dual of (16.6) reads

$$\begin{cases} \sup_{\mathbf{u}, \mathbf{v} \geq 0, \mathbf{z}} \sum_{\alpha \in \mathbb{N}_{d_0}^n \setminus \{0\}} f_\alpha (u_\alpha - v_\alpha) \; (= L_\mathbf{z}(f(0) - f)) \\ \text{s.t.} \quad \sum_{\alpha \in \mathbb{N}_{d_0}^n \setminus \{0\}} u_\alpha + v_\alpha \leq 1 \\ \quad u_\alpha - v_\alpha + z_\alpha = 0, \quad \forall \alpha \in \mathbb{N}_{d_0}^n \setminus \{0\} \\ \quad \mathbf{M}_d(\mathbf{z}), \mathbf{M}_{d-v_j}(g_j \, \mathbf{z}) \succeq 0, \quad j = 1, \ldots, m, \end{cases} \qquad (16.9)$$

and the semidefinite program dual of (16.7) reads

$$\begin{cases} \sup_{\mathbf{z}, \Delta_\alpha} \sum_{\alpha \in \mathbb{N}_{d_0}^n \setminus \{0\}} \left\langle \Delta_\alpha, \begin{bmatrix} 0 & f_\alpha \\ f_\alpha & -1 \end{bmatrix} \right\rangle \\[2mm] \text{s.t.} \quad \left\langle \Delta_\alpha, \begin{bmatrix} 1 & 0 \\ 0 & 0 \end{bmatrix} \right\rangle \leq 1, \quad \forall \alpha \in \mathbb{N}_{d_0}^n \setminus \{0\} \\[3mm] \quad \left\langle \Delta_\alpha, \begin{bmatrix} 0 & 1 \\ 1 & 0 \end{bmatrix} \right\rangle + z_\alpha = 0, \quad \forall \alpha \in \mathbb{N}_{d_0}^n \setminus \{0\} \\[2mm] \quad \mathbf{M}_d(\mathbf{z}), \mathbf{M}_{d-v_j}(g_j \, \mathbf{z}) \succeq 0, \quad j = 1, \ldots, m \\ \quad \Delta_\alpha \succeq 0, \quad \forall \alpha \in \mathbb{N}_{d_0}^n \setminus \{0\}. \end{cases} \qquad (16.10)$$

In fact in (16.10) one may replace the criterion with the equivalent criterion

$$\sup_{\mathbf{z}} \left\{ L_{\mathbf{z}}(f(0) - f)) - \frac{1}{4} \sum_{\alpha \in \mathbb{N}^n_{d_0} \setminus \{0\}} z^2_\alpha \right\}.$$

Lemma 16.4   *Assume that $\mathbf{K} \subset \mathbb{R}^n$ has nonempty interior. Then there is no duality gap between the semidefinite programs (16.5) and (16.8), (16.6) and (16.9), and (16.7) and (16.10). Moreover, all semidefinite programs (16.5), (16.6) and (16.7) have an optimal solution $\tilde{f}_d \in \mathbb{R}[\mathbf{x}]_{d_0}$.*

Proof   The proof is detailed for the case $k = 1$ and omitted for the cases $k = 2$ and $k = \infty$ because it is very similar. Observe that $\rho^1_d \geq 0$ and the constant polynomial $h(\mathbf{x}) = 0$, for all $\mathbf{x} \in \mathbb{R}^n$, is a feasible solution of (16.4) (hence of (16.5)). Therefore ρ^1_d being finite, it suffices to prove that Slater's condition holds for the dual (16.8) and then the conclusion of Lemma 16.4 follows from a standard result of convex optimization. So let μ be the finite Borel measure defined by

$$\mu(B) := \int_{B \cap \mathbf{K}} e^{-\|\mathbf{x}\|^2} d\mathbf{x}, \qquad \forall B \in \mathcal{B}$$

(where $\mathcal{B}$ is the usual Borel σ-field), and let $\mathbf{z} = (z_\alpha), \alpha \in \mathbb{N}^n_{2d}$, with

$$z_\alpha := \kappa \int_{\mathbf{K}} \mathbf{x}^\alpha d\mu(\mathbf{x}), \qquad \alpha \in \mathbb{N}^n_{2d},$$

for some $\kappa > 0$ sufficiently small to ensure that

$$\kappa \left| \int \mathbf{x}^\alpha d\mu(\mathbf{x}) \right| < 1, \qquad \forall \alpha \in \mathbb{N}^n_{2d} \setminus \{0\}. \tag{16.11}$$

Define $u_\alpha = \max[0, -z_\alpha]$ and $v_\alpha = \max[0, z_\alpha], \alpha \in \mathbb{N}^n_{d_0} \setminus \{0\}$, so that $u_\alpha + v_\alpha < 1, \alpha \in \mathbb{N}^n_{2d} \setminus \{0\}$. Hence $(\mathbf{u}, \mathbf{v}, \mathbf{z})$ is a feasible solution of (16.8). In addition, $\mathbf{M}_d(\mathbf{z}) \succ 0$ and $\mathbf{M}_{d-v_j}(g_j\,\mathbf{z}) \succ 0, \ j = 1, \ldots, m$, because $\mathbf{K}$ has nonempty interior, and so Slater's condition holds for (16.8), the desired result.   □

Theorem 16.5   *Assume that $\mathbf{K}$ in (16.1) has nonempty interior. Let $\mathbf{x}^* \in \mathbf{K}$ be a global minimizer of $\mathbf{P}$ with optimal value f^*, and let $\tilde{f}_d \in \mathbb{R}[\mathbf{x}]_{d_0}$ be an optimal solution of $\mathbf{P}_d$ in (16.4) with optimal value ρ^k_d.*

> (a) $0 \in \mathbf{K}$ is a global minimizer of the problem $\tilde{f}_d^* = \inf_{\mathbf{x}} \{ \tilde{f}_d(\mathbf{x}) : \mathbf{x} \in \mathbf{K} \}$. In particular, if $\rho_d^k = 0$ then $\tilde{f}_d = f$ and 0 is a global minimizer of **P**.
>
> (b) If $k = 1$ then $f^* \leq f(0) \leq f^* + \rho_d^1 \sup_{\alpha \in \mathbb{N}_{d_0}^n} |(\mathbf{x}^*)^\alpha|$. In particular,
>
> if $\mathbf{K} \subseteq [-1, 1]^n$ then $f^* \leq f(0) \leq f^* + \rho_d^1$.
>
> (c) If $k = \infty$ then $f^* \leq f(0) \leq f^* + \rho_d^\infty \sum_{\alpha \in \mathbb{N}_{d_0}^n} |(\mathbf{x}^*)^\alpha|$. In particular,
>
> if $\mathbf{K} \subseteq [-1, 1]^n$ then $f^* \leq f(0) \leq f^* + s(d_0) \rho_d^\infty$.

Proof (a) Existence of $\tilde{f}_d$ is guaranteed by Lemma 16.4. From the constraints of (16.4) we have $\tilde{f}_d(\mathbf{x}) - \tilde{f}_d(0) = \sum_{j=0}^m \sigma_j(\mathbf{x}) g_j(\mathbf{x})$ which implies that $\tilde{f}_d(\mathbf{x}) \geq \tilde{f}_d(0)$ for all $\mathbf{x} \in \mathbf{K}$, and so $\mathbf{y} = 0$ is a global minimizer of the optimization problem $\mathbf{P'} : \inf_{\mathbf{x}} \{ \tilde{f}_d(\mathbf{x}) : \mathbf{x} \in \mathbf{K} \}$.

(b) Let $\mathbf{x}^* \in \mathbf{K}$ be a global minimizer of **P**. Observe that with $k = 1$,

$$f^* = f(\mathbf{x}^*) = \underbrace{f(\mathbf{x}^*) - \tilde{f}_d(\mathbf{x}^*)}_{} + \underbrace{\tilde{f}_d(\mathbf{x}^*) - \tilde{f}_d(0)}_{\geq 0} + \tilde{f}_d(0)$$

$$\geq \tilde{f}_d(0) - |\tilde{f}_d(\mathbf{x}^*) - f(\mathbf{x}^*)|$$

$$\geq \tilde{f}_d(0) - \|\tilde{f}_d - f\|_1 \times \sup_{\alpha \in \mathbb{N}_{d_0}^n} |(\mathbf{x}^*)^\alpha|$$

$$\geq f(0) - \rho_d^1 \sup_{\alpha \in \mathbb{N}_{d_0}^n} |(\mathbf{x}^*)^\alpha| \tag{16.12}$$

since $\tilde{f}_d(0) = f(0)$; see Remark 16.3.

(c) The proof is similar to that of (b) using the result that with $k = \infty$,

$$|\tilde{f}_d(\mathbf{x}) - f(\mathbf{x})| \leq \underbrace{\left(\sup_{\alpha \in \mathbb{N}_{d_0}^n} |\tilde{f}_{d\alpha} - f_\alpha| \right)}_{= \rho_d^\infty} \times \sum_{\alpha \in \mathbb{N}_{d_0}^n} |\mathbf{x}^\alpha|.$$

$\square$

So Theorem 16.5 not only states that $\mathbf{y} = 0$ is the global optimum of the optimization problem $\inf \{ \tilde{f}_d(\mathbf{x}) : \mathbf{x} \in \mathbf{K} \}$, but it also states that the optimal value ρ_d^k measures how far is $f(0)$ from the optimal value f^* of the initial problem **P**. Moreover, observe that Theorem 16.5 merely requires existence of a minimizer and nonemptyness of **K**. In particular, **K** may not be compact.

Finally, it is worth noticing that the description of $\mathbf{K}$ is important. Different representations of $\mathbf{K}$ may yield different results in (16.4) for the same $d \in \mathbb{N}$.

Example 16.6   Let $n = 2$ and consider the optimization problem $\mathbf{P} : f^* = \inf_{\mathbf{x}} \{ f(\mathbf{x}) : \mathbf{x} \in \mathbf{K} \}$ with $\mathbf{x} \mapsto f(\mathbf{x}) = x_1 + x_2$, and

$$\mathbf{K} = \{ \mathbf{x} \in \mathbb{R}^2 : x_1 x_2 \geq 1; \; 1/2 \leq \mathbf{x} \leq 2 \}.$$

The polynomial f is convex and the set $\mathbf{K}$ is convex as well, but the polynomials that define $\mathbf{K}$ are not all concave. That is, $\mathbf{P}$ is a convex optimization problem, but not a convex programming problem. The point $\mathbf{y} = (1, 1) \in \mathbf{K}$ is a global minimizer and the KKT optimality conditions at $\mathbf{y}$ are satisfied, i.e., $\lambda = (1, 0, 0, 0, 0) \in \mathbb{R}_+^5$,

$$\nabla f(\mathbf{x}) - \lambda_1 \nabla g_1(\mathbf{x}) = 0 \quad \text{with } \mathbf{x} = (1, 1) \text{ and } \lambda_1 = 1.$$

Next, if we now consider the inverse optimization problem with $d \in \mathbb{N}$, do we find that $\mathbf{y}$ is a global minimizer? That is, do we find that $\rho_d^1 = 0$ for some d? With $d = 1$ one searches for a polynomial $\tilde{f}_d$ of degree at most $d_0 = 1$, and such that

$$\tilde{f}_d(\mathbf{x}) - \tilde{f}_d(1, 1) = \sigma_0(\mathbf{x}) + \sigma_1(\mathbf{x})(x_1 x_2 - 1) + \sum_{i=1}^{2} \psi_i(\mathbf{x})(2 - x_i) + \phi_i(\mathbf{x})(x_i - 1/2),$$

for some SOS polynomials $\sigma_1, \psi_i, \phi_i \in \Sigma[\mathbf{x}]_0$ and some SOS polynomial $\sigma_0 \in \Sigma[\mathbf{x}]_1$. But then necessarily $\sigma_1 = 0$ and ψ_i, ϕ_i are constant, which in turn implies that σ_0 is a constant polynomial. A straightforward calculation shows that $\tilde{f}_1(\mathbf{x}) = 0$ for all $\mathbf{x}$, and so $\rho_1^1 = 2$. And indeed this is confirmed when solving (16.8) with $d = 1$ (with the GloptiPoly software described in Appendix B). Solving (16.8) again, now with $d = 2$, yields $\rho_2^1 = 2$ (no improvement) but with $d = 3$ we obtain the desired result $\rho_3^1 = 0$.

Example 16.7   Again consider Example 16.6 but now with $\mathbf{y} = (1.1, 1/1.1) \in \mathbf{K}$ which is not a global optimum of f on $\mathbf{K}$ any more, and with $\mathbf{K}$ now defined by:

$$\mathbf{K} = \{ \mathbf{x} \in \mathbb{R}^2 : x_1 x_2 \geq 1; \; (x_i - 1/2)(2 - x_i) \geq 0, \; i = 1, 2 \}.$$

By solving (16.8) with $d = 1$ we still find $\rho_1^1 = 2$ (i.e., $\tilde{f}_1 = 0$), and with $d = 2$ we find $\rho_2 \approx 0.1734$ and $\tilde{f}_2(\mathbf{x}) \approx 0.8266 \, x_1 + x_2$. And indeed by solving (using GloptiPoly) the new optimization problem with criterion $\tilde{f}_2$ we find the global minimizer $(1.1, 0.9091) \approx \mathbf{y}$. With $d = 3$ we obtain the same value $\rho_3^1 = 0.1734$, suggesting that $\tilde{f}_2$ is already an optimal solution of the ideal inverse problem.

Example 16.8 Consider now the disconnected set $\mathbf{K} := \{ \mathbf{x} : x_1 x_2 \geq 1; x_1^2 + x_2^2 \leq 3 \}$ and the nonconvex criterion $\mathbf{x} \mapsto f(\mathbf{x}) := -x_1 - x_2^2$ for which $\mathbf{x}^* = (\ 0.618, -1/0.618) \in \partial \mathbf{K}$ is the unique global minimizer. Let $\mathbf{y} := (-0.63, -1/0.63) \in \partial \mathbf{K}$ for which the constraint $x_1 x_2 \geq 1$ is active. At steps $d = 2$ and $d = 3$ one finds that $\tilde{f}_d = 0$ and $\rho_d^1 = \| f \|_1$. That is, $\mathbf{y}$ is a global minimizer of the trivial criterion $\tilde{f}(\mathbf{x}) = 0$ for all $\mathbf{x}$, and cannot be a global minimizer of some nontrivial polynomial criterion.

Now let $\mathbf{y} = (-0.63, -\sqrt{3 - 0.63^2})$ so that the constraint $x_1^2 + y_2^2 \leq 3$ is active. With $d = 1$ one obtains $\rho_1^1 = \| f \|_1$ and $\tilde{f}_1 = 0$. With $d = 2$ one obtains $\tilde{f}_2 = 1.26\,x_1 - x_2^2$. With $d = 3$ one obtains the same result, suggesting that $\tilde{f}_2$ is already an optimal solution of the ideal inverse optimization problem.

16.2.4 Structural constraints

It may happen that the initial criterion $f \in \mathbb{R}[\mathbf{x}]$ has some structure that one wishes to keep in the inverse problem. For instance, in MAXCUT problems on $\mathbf{K} = \{-1, 1\}^n$, f is a quadratic form $\mathbf{x} \mapsto \mathbf{x}'\mathbf{A}\mathbf{x}$ for some real symmetric matrix $\mathbf{A}$ associated with a graph (V, E), where $\mathbf{A}_{ij} \neq 0$ if and only if $(i, j) \in E$. Therefore in the inverse optimization problem one may wish that h in (16.4) is also a quadratic form associated with the same graph (V, E), so that $h(\mathbf{x}) = \mathbf{x}'\tilde{\mathbf{A}}\mathbf{x}$ with $\tilde{\mathbf{A}}_{ij} = 0$ for all $(i, j) \notin E$.

So if $\Delta_f \subset \mathbb{N}^n_{d_0}$ denotes the subset of (structural) multi-indices for which f and h should have same coefficient, then in (16.4) one includes the additional constraint $h_\alpha = f_\alpha$ for all $\alpha \in \Delta_f$. Notice that $0 \in \Delta_f$ because $h_0 = f_0$; see Remark 16.3. For instance, if $k = 1$ then (16.5) reads:

$$
\rho_d^1 := \inf_{h, \lambda, \mathbf{Z}_j} \sum_{\alpha \in \mathbb{N}^n_{d_0} \setminus \Delta_f} \lambda_\alpha
$$

$$
\text{s.t.} \quad \lambda_\alpha + h_\alpha \geq f_\alpha, \quad \forall \alpha \in \mathbb{N}^n_{d_0} \setminus \Delta_f
$$

$$
\lambda_\alpha - h_\alpha \geq -f_\alpha, \quad \forall \alpha \in \mathbb{N}^n_{d_0} \setminus \Delta_f
$$

$$
\langle \mathbf{Z}_0, \mathbf{B}_\alpha \rangle + \sum_{j=1}^m \langle \mathbf{Z}_j, \mathbf{C}_\alpha^j \rangle = \begin{cases} f_\alpha & \text{if } 0 < \alpha \in \Delta_f \\ h_\alpha & \text{if } \alpha \in \mathbb{N}^n_{d_0} \setminus \Delta_f \\ 0 & \text{if } \alpha = 0 \text{ or } |\alpha| > d_0 \end{cases}
$$

$$
\mathbf{Z}_j \succeq 0, \quad j = 0, 1, \dots, m,
$$

and its dual has the equivalent form,

$$
\sup_{\mathbf{z}} \quad L_{\mathbf{z}}(f(0) - f)
$$

$$
\text{s.t.} \quad \mathbf{M}_d(\mathbf{z}), \mathbf{M}_{d - v_j}(g_j \, \mathbf{z}) \succeq 0, \quad j = 1, \dots, m,
$$

$$
|z_\alpha| \leq 1, \quad \forall \alpha \in \mathbb{N}^n_{d_0} \setminus \Delta_f.
$$

Example 16.9 Consider the MAXCUT problem $\inf\{\mathbf{x}'\mathbf{A}\mathbf{x} : \mathbf{x}_i^2 = 1, i = 1,\ldots,n\}$ where $\mathbf{A} = \mathbf{A}' \in \mathbb{R}^{n\times n}$ and $\mathbf{A}_{ij} = 1/2$ for all $i \neq j$. For n odd, an optimal solution is $\mathbf{y} = (y_j)$ with $y_j = 1$, $j = 1,\ldots\lceil n/2\rceil$, and $y_j = -1$ otherwise, and with value $-\lfloor n/2\rfloor$. However, the dual (6.4) of the first semidefinite relaxation (6.3), i.e.,

$$\sup\left\{\lambda : \mathbf{x}'\mathbf{A}\mathbf{x} - \lambda = \sigma + \sum_{j=1}^n \gamma_i(x_i^2 - 1); \ \sigma \in \Sigma[\mathbf{x}]_1; \ \lambda, \gamma_i \in \mathbb{R}\right\}$$

provides the lower bound $-n/2$ (with famous Goemans–Williamson ratio guarantee). So $\mathbf{y}$ cannot be obtained from the first semidefinite relaxation even though it is an optimal solution. With $d = 1$, the inverse optimization problem reads: Find a quadratic form $\mathbf{x} \mapsto \mathbf{x}'\tilde{\mathbf{A}}\mathbf{x}$ such that $\mathbf{x}'\tilde{\mathbf{A}}\mathbf{x} - \mathbf{y}'\tilde{\mathbf{A}}\mathbf{y} = \sigma + \sum_{j=1}^n \gamma_i(x_i^2 - 1)$, for some $\sigma \in \Sigma[\mathbf{x}]_1$, $\lambda \in \mathbb{R}$, $\boldsymbol{y} \in \mathbb{R}^n$, and which minimizes the ℓ_1-norm $\|\mathbf{A} - \tilde{\mathbf{A}}\|_1$. This is an inverse optimization problem with structural constraints just described above since we search for a specific quadratic form and not an arbitrary quadratic polynomial $\tilde{f}_2$. Hence, solving (16.5) for $n = 5$ with $\mathbf{y}$ as above, we find that

$$\tilde{\mathbf{A}} = \frac{1}{2}\begin{bmatrix} 0 & 2/3 & 2/3 & 1 & 1 \\ 2/3 & 0 & 2/3 & 1 & 1 \\ 2/3 & 2/3 & 0 & 1 & 1 \\ 1 & 1 & 1 & 0 & 1 \\ 1 & 1 & 1 & 1 & 0 \end{bmatrix},$$

that is, only the entries $(i, j) \in \{(1, 2), (1, 3), (2, 3)\}$ are modified from $1/2$ to $1/3$. Therefore $\mathbf{y}$ is also a global minimizer for MAXCUT with criterion

$$\frac{2}{3}(x_1x_2 + x_1x_3 + x_2x_3) + (x_1x_4 + x_1x_5 + x_2x_4 + x_2x_5 + x_3x_4 + x_3x_5 + x_4x_5),$$

and has a Putinar certificate of optimality with degree bound $d = 1$. For the ℓ_1-norm, the quadratic form $\mathbf{x}^T\tilde{\mathbf{A}}\mathbf{x}$ is the closest to $\mathbf{A}$ among those for which $\mathbf{y}$ has a Putinar certificate of optimality with degree bound $d = 1$.

16.3 A canonical form for the ℓ_1-norm

When $\mathbf{K}$ is compact and if one uses the ℓ_1-norm then an optimal solution $\tilde{f}_d \in \mathbb{R}[\mathbf{x}]_{d_0}$ in Theorem 16.5 (with $k = 1$) takes a particularly simple *canonical* form.

As $\mathbf{K}$ is compact we may and will assume (possibly after some scaling) that $\mathbf{K} \subseteq [-1, 1]^n$ and so in the definition of (16.1) we may and will add the n redundant quadratic constraints $g_{m+i}(\mathbf{x}) \geq 0$, $i = 1, \ldots, n$, with $\mathbf{x} \mapsto g_{m+i}(\mathbf{x}) = 1 - x_i^2$ for every i, that is,

$$\mathbf{K} = \{\mathbf{x} \in \mathbb{R}^n : g_j(\mathbf{x}) \geq 0, \quad j = 1, \ldots, m+n\}, \tag{16.13}$$

and recall the truncated quadratic module associated with the g_j:

$$Q_d(g) = \left\{ \sum_{j=0}^{n+m} \sigma_j g_j : \quad \sigma_j \in \Sigma[\mathbf{x}]_{d-v_j}, \ j = 0, \ldots, m+n \right\}$$

(where $v_j = \lceil (\deg g_j)/2 \rceil$) which is obviously Archimedean.

Theorem 16.10 *Assume that $\mathbf{K}$ in (16.13) has a nonempty interior and let $\tilde{f}_d \in \mathbb{R}[\mathbf{x}]_{d_0}$ be an optimal solution of $\mathbf{P}_d$ in (16.4) (with $m+n$ instead of m) with optimal value ρ_d^1 for the ℓ_1-norm.*

(a) *$\tilde{f}_d$ is of the form*

$$\tilde{f}_d(\mathbf{x}) = \begin{cases} f(\mathbf{x}) + \mathbf{b}^T \mathbf{x} & \text{if } d_0 = 1 \\ f(\mathbf{x}) + \mathbf{b}^T \mathbf{x} + \sum_{i=1}^{n} \lambda_i^* x_i^2 & \text{if } d_0 > 1, \end{cases} \tag{16.14}$$

for some vector $\mathbf{b} \in \mathbb{R}^n$ and some nonnegative vector $\boldsymbol{\lambda}^* \in \mathbb{R}^n$, optimal solution of the semidefinite program:

$$\rho_d^1 := \inf_{\boldsymbol{\lambda}, \mathbf{b}} \quad \|\mathbf{b}\|_1 + \sum_{i=1}^{n} \lambda_i$$

$$\text{s.t.} \quad f - f(0) + \mathbf{b}^T \mathbf{x} + \sum_{i=1}^{n} \lambda_i x_i^2 \in Q_d(g), \quad \boldsymbol{\lambda} \geq 0. \tag{16.15}$$

(b) *The vector $\mathbf{b}$ is of the form $-\nabla f(0) + \sum_{j \in J(0)} \theta_j \nabla g_j(0)$ for some nonnegative scalars (θ_j) (where $j \in J(0)$ if and only if $g_j(0) = 0$).*

Proof (a) Notice that the dual (16.8) of (16.5) is equivalent to:

$$\sup_{\mathbf{z}} \quad L_{\mathbf{z}}(f(0) - f))$$

$$\text{s.t.} \quad \mathbf{M}_d(\mathbf{z}), \ \mathbf{M}_{d-v_j}(g_j \, \mathbf{z}) \succeq 0, \quad j = 1, \ldots, m+n, \tag{16.16}$$

$$|z_\alpha| \leq 1, \quad \forall \alpha \in \mathbb{N}_{d_0}^n \setminus \{0\}.$$

By Proposition 2.38, $\mathbf{M}_d(\mathbf{z}) \succeq 0 \Rightarrow |z_\alpha| \leq \max[L_{\mathbf{z}}(1), \max_i L_{\mathbf{z}}(x_i^{2d})]$ for all $\alpha \in \mathbb{N}_{2d}^n$. In fact similar arguments yield that for every $\alpha \in \mathbb{N}_{2d}^n$ with $|\alpha| > 1$,

$$\mathbf{M}_d(\mathbf{z}) \succeq 0 \Rightarrow \quad |z_\alpha| \leq \max\{\max_i [L_{\mathbf{z}}(x_i^2), L_{\mathbf{z}}(x_i^{2d})]\}.$$

Moreover the constraint $\mathbf{M}_{d-1}(g_{m+i} \, \mathbf{z}) \succeq 0$ implies $\mathbf{M}_{d-1}(g_{m+i} \, \mathbf{z})(\ell, \ell) \geq 0$ for all ℓ, and so, in particular, one obtains $L_{\mathbf{z}}(x_i^{2k-2}) \geq L_{\mathbf{z}}(x_i^{2k})$ for all

$k = 1, \ldots, d$ and all $i = 1, \ldots, n$. Hence $|z_\alpha| \leq \max_{i=1,\ldots,n} L_z(x_i^2)$ for every $\alpha \in \mathbb{N}_{2d}^n$ with $|\alpha| > 1$. Therefore in (16.16) one may replace the constraint $|z_\alpha| \leq 1$ for all $\alpha \in \mathbb{N}_{d_0}^n \setminus \{0\}$ with the $2n$ inequality constraints $\pm L_z(x_i) \leq 1$, $i = 1, \ldots, n$, if $d_0 = 1$ and the $3n$ inequality constraints:

$$\pm L_z(x_i) \leq 1, \ L_z(x_i^2) \leq 1, \ i = 1, \ldots, n \qquad (16.17)$$

if $d_0 > 1$. Consequently, (16.16) is the same as the semidefinite program:

$$\begin{aligned}
\sup_{z} \quad & L_z(f(0) - f)) \\
\text{s.t.} \quad & \mathbf{M}_d(\mathbf{z}), \ \mathbf{M}_{d-v_j}(g_j \, \mathbf{z}) \succeq 0, \quad j = 1, \ldots, m+n \qquad (16.18) \\
& \pm L_z(x_i) \leq 1, \ L_z(x_i^2) \leq 1, \quad i = 1, \ldots, n.
\end{aligned}$$

Let $\mathbf{b}^1 = (b_i^1)$ (respectively $\mathbf{b}^2 = (b_i^2)$) be the nonnegative vector of dual variables associated with the constraints $L_z(x_i) \leq 1$ (respectively $-L_z(x_i) \leq 1$), $i = 1, \ldots, n$. Similarly, let λ_i be the dual variable associated with the constraint $L_z(x_i^2) \leq 1$. Then the dual of (16.18) is the semidefinite program:

$$\begin{aligned}
\sup_{\mathbf{b}^1,\mathbf{b}^2,\lambda} \quad & \sum_{i=1}^{n} \left((b_i^1 + b_i^2) + \lambda_i \right) \\
\text{s.t.} \quad & f - f(0) + (\mathbf{b}^1 - \mathbf{b}^2)^T \mathbf{x} + \sum_{i=1}^{n} \lambda_i \, x_i^2 \in Q_d(g) \\
& \mathbf{b}^1, \mathbf{b}^2, \lambda \geq 0
\end{aligned}$$

which is equivalent to (16.15).

(b) Let $\mathbf{b}, \boldsymbol{\lambda}^*$ be an optimal solution of (16.15), so that

$$f - f(0) + \mathbf{b}^T \mathbf{x} + \sum_{i=1}^{n} \lambda_i^* \, x_i^2 = \sigma_0 + \sum_{j=1}^{m+n} \sigma_j \, g_j,$$

for some SOS polynomials σ_j. Evaluating at $\mathbf{x} = 0$ yields

$$\sigma_0(0) = 0, \quad \underbrace{\sigma_j(0) \, g_j(0)}_{\theta_j \geq 0} = 0, \quad j = 1, \ldots, m+n.$$

Differentiating and evaluating at $\mathbf{x} = 0$ and using that σ_j is SOS and $\sigma_j(0)g_j(0) = 0$, $j = 1, \ldots, n+m$, yields:

$$\nabla f(0) + \mathbf{b} = \sum_{j=1}^{n+m} \sigma_j(0) \, \nabla g_j(0) = \sum_{j \in J(0)} \theta_j \, \nabla g_j(0),$$

which is the desired result.   □

From the proof of Theorem 16.10, this special form of $\tilde{f}_d$ is specific to the ℓ_1-norm, which yields the constraint $|z_\alpha| \leq 1$, $\alpha \in \mathbb{N}^n_{d_0} \setminus \{0\}$ in the dual (16.8) and allows its simplification (16.17) thanks to a property of the moment matrix described in Proposition 2.38. Observe that the canonical form (16.14) of $\tilde{f}_d$ is *sparse* since $\tilde{f}_d$ differs from f in at most $2n$ entries only (recall that f has $\binom{n+d_0}{n}$ entries). This is another example of sparsity properties of optimal solutions of ℓ_1-norm minimization problems, already observed in other contexts (e.g. in some compressed sensing applications). For problems with structural constraints (as described in Section 16.2.4), and for which d_0 is even and Δ_f does not contain the monomials $\alpha \in \mathbb{N}^n_{d_0}$ such that $\mathbf{x}^\alpha = x_i^2$, or $\mathbf{x}^\alpha = x_i^{d_0}$, $i = 1, \ldots, n$, $\tilde{f}_d$ still has the special form described in Theorem 16.10, but with $b_k = 0$ if $\alpha = e_k \in \Delta_f$ (where all entries of e_k vanish except the one at position k). A similar remark applies for 0/1 programs with structural constraints.

16.4 Exercises

Exercise 16.1 Let $n = 1$, $\mathbf{K} = [0, 1]$ and $x \mapsto f(x) := x(x - 1)$. Let $y = 0 \in \mathbf{K}$. Show that solving the ideal inverse optimization problem (16.2) with the ℓ_1-norm reduces to solving a single semidefinite program whose optimal solution $\tilde{f}^* \in \mathbb{R}[x]_2$ is the polynomial $x \mapsto \tilde{f}^*(x) = f(x) + x = x^2$. Show that this solution agrees with Theorem 16.10. (*Hint:* Use a positivity certificate on $[0, 1]$ for the univariate polynomial $x \mapsto h(x) - h(0)$.)

Exercise 16.2 Let $n = 1$ and $f \in \mathbb{R}[x]_d$ be such that $f^* = \inf_x f(x) > -\infty$. Show that solving the ideal inverse optimization problem (16.2) with $y = 0$, $\mathbf{K} = \mathbb{R}$ and with the ℓ_1-norm, reduces to solving a single semidefinite program. Show that there exists an optimal solution $\tilde{f}^* \in \mathbb{R}[x]_d$ of the form $x \mapsto \tilde{f}^*(x) = f(x) - f'(0)x + \lambda x^2$ for some $\lambda > 0$. (*Hint:* Use the result that every nonnegative univariate polynomial is SOS.)

Exercise 16.3 Let $n = 2$ and $f \in \mathbb{R}[\mathbf{x}]_4$ be such that $f^* = \inf_\mathbf{x} f(\mathbf{x}) > -\infty$. Show that solving the ideal inverse optimization problem (16.2) with $y = 0$, $\mathbf{K} = \mathbb{R}^2$ and with the ℓ_1-norm, reduces to solving a single semidefinite program. Show that there exists an optimal solution $\tilde{f}^* \in \mathbb{R}[\mathbf{x}]_4$ of the form $\mathbf{x} \mapsto \tilde{f}^*(\mathbf{x}) = f(\mathbf{x}) - \nabla f(0)^T \mathbf{x} + \lambda_1 x_1^2 + \lambda_2 x_2^2$ for some $0 \leq \lambda \in \mathbb{R}^2$. (*Hint:* Use the result that every nonnegative bivariate quartic polynomial is SOS.)

16.5 Notes and sources

Most of the material of this chapter is from Lasserre (2013c). Pioneering work in inverse optimization dates back to the early 1970s with several contributions in control where given a dynamical system and a control feedback law, the inverse optimal control problem consists of constructing a cost functional which is minimized by this control law; see for example the works of Rugh (1971), Buelens and Hellinckx (1974), as well as Park and Lee (1975) and Moylan and Anderson (1973). Similarly for optimal stabilization, whereas it was known that every value function of an optimal stabilization problem is also a Lyapunov function for the closed-loop system, Freeman and Kokotovic (1996) showed the converse, that is, every Lyapunov function for every stable closed-loop system is also a *value function* for a meaningful optimal stabilization problem. In static optimization, pioneering works in this direction date back to Burton and Toint (1992) for shortest path problems, and Zhang and Liu (1996, 1999), Huang and Liu (1999), and Ahuja and Orlin (2001) for linear programs in the form inf $\{ \mathbf{c}'\mathbf{x} : \mathbf{Ax} \geq \mathbf{b}; \mathbf{r} \leq \mathbf{x} \leq \mathbf{s} \}$ (and with the ℓ_1-norm). For the latter, the inverse problem is again a linear program of the same form. Similar results also hold for inverse linear programs with the ℓ_∞-norm as shown in Ahuja and Orlin (2001), while Zhang et al. (1995) provide a column generation method for the inverse shortest path problem. In Heuberger (2004) the interested reader will find a nice survey on inverse optimization for linear programming and combinatorial optimization problems. For integer programming, Schaefer (2004) characterizes the feasible set of cost vectors $c \in \mathbb{R}^n$ that are candidates for inverse optimality. It is the projection on $\mathbb{R}^n$ of a (lifted) convex polytope obtained from the super-additive dual of integer programs. Unfortunately and as expected, the dimension of the lifted polyhedron (before projection) is exponential in the input size of the problem. Finally, for linear programs Ahmed and Guan (2005) have considered the variant called the *inverse optimal value* problem in which one is interested in finding a linear criterion $c \in C \subset \mathbb{R}^n$ for which the optimal value is the closest to a desired specified value. Perhaps surprisingly, they proved that such a problem is NP-hard.

17

Approximation of sets defined with quantifiers

17.1 Introduction

In many applications of interest one has to handle a class of sets $\mathbf{R} \subset \mathbb{R}^n$ whose definition contains a set $\mathbf{K} \subset \mathbb{R}^{n+p}$ and some quantifier $\exists$ or $\forall$. In other words, $\mathbf{R}$ is described with the help of additional variables $\mathbf{y} \in \mathbb{R}^p$ and constraints linking $\mathbf{x} \in \mathbb{R}^n$ and $\mathbf{y} \in \mathbb{R}^p$. To obtain an exact *explicit* description of $\mathbf{R} \subset \mathbb{R}^n$ solely in terms of the $\mathbf{x}$-variables is a difficult and challenging problem in general. If $\mathbf{K} \subset \mathbb{R}^{n+p}$ is semi-algebraic then it is certainly possible and this is part of *Elimination Theory* in Algebraic Geometry. Most of the associated algorithms are based on Gröbner basis methods (with exact arithmetic) and in practice quantifier elimination is very costly and is limited to small size problems or particular cases.

The goal of this chapter is less ambitious as we only want to obtain *approximations* of $\mathbf{R}$ with some convergence properties. We provide a nested sequence of outer (respectively inner) approximations $\mathbf{R}^k$, $k \in \mathbb{N}$, such that the Lebesgue volume of $\mathbf{R}^k$ converges to that of $\mathbf{R}$ as $k \to \infty$. Moreover the outer (respectively inner) approximations $\mathbf{R}^k$ are of the form $\{\, \mathbf{x} \in \mathbb{R}^n \; : \; p_k(\mathbf{x}) \leq 0 \,\}$ for some polynomial p_k, $k \in \mathbb{N}$, of increasing degree. Therefore approximations (possibly crude if k is small) can be obtained for cases where exact elimination is out of reach.

Problem statement

Consider two sets of variables $\mathbf{x} \in \mathbb{R}^n$ and $\mathbf{y} \in \mathbb{R}^m$ coupled with a constraint $(\mathbf{x}, \mathbf{y}) \in \mathbf{K}$, where $\mathbf{K}$ is the basic semi-algebraic set:

$$\mathbf{K} := \{\, (\mathbf{x}, \mathbf{y}) \in \mathbb{R}^n \times \mathbb{R}^m \; : \; \mathbf{x} \in \mathbf{B}; \; g_j(\mathbf{x}, \mathbf{y}) \geq 0, \; j = 1, \ldots, s \,\} \quad (17.1)$$

for some polynomials g_j, $j = 1, \ldots, s$, and $\mathbf{B} \subset \mathbb{R}^n$ is a simple set (e.g. some box or ellipsoid). With $f : \mathbf{K} \to \mathbb{R}$ a given semi-algebraic function[1] on $\mathbf{K}$, and

$$\mathbf{K_x} := \{ \mathbf{y} \in \mathbb{R}^m : (\mathbf{x}, \mathbf{y}) \in \mathbf{K} \}, \quad \mathbf{x} \in \mathbf{B}, \tag{17.2}$$

consider the following two sets:

$$\mathbf{R}_f = \{ \mathbf{x} \in \mathbf{B} : f(\mathbf{x}, \mathbf{y}) \leq 0 \text{ for all } \mathbf{y} \in \mathbf{K_x} \}, \tag{17.3}$$

$$\mathbf{D}_f = \{ \mathbf{x} \in \mathbf{B} : f(\mathbf{x}, \mathbf{y}) \geq 0 \text{ for some } \mathbf{y} \in \mathbf{K_x} \}. \tag{17.4}$$

Both sets $\mathbf{R}_f$ and $\mathbf{D}_f$ which include a *quantifier* in their definition, are semi-algebraic and are interpreted as *robust* sets of variables $\mathbf{x}$ with respect to the other set of variables $\mathbf{y}$, and to some performance criterion f.

Indeed, in the first case (17.3) one may think of "$\mathbf{x}$" as *decision* variables which should be *robust* with respect to some *noise* (or perturbation) $\mathbf{y}$ in the sense that no matter what the admissible level of noise $\mathbf{y} \in \mathbf{K_x}$ is, the constraint $f(\mathbf{x}, \mathbf{y}) \leq 0$ is satisfied whenever $\mathbf{x} \in \mathbf{R}_f$. For instance, such sets $\mathbf{R}_f$ are fundamental in robust control and robust optimization on a set $\mathbf{\Omega} \subset \mathbf{B}$ (in which case one is interested in $\mathbf{R}_f \cap \mathbf{\Omega}$).

On the other hand, in the second case (17.4) the vector $\mathbf{x}$ should be interpreted as *design* variables (or parameters), and the set $\mathbf{K_x}$ defines a set of admissible decisions $\mathbf{y} \in \mathbf{K_x}$ within the framework of design $\mathbf{x}$. And so $\mathbf{D}_f$ is the set of *robust* design parameters $\mathbf{x}$, in the sense that for every value of the design parameter $\mathbf{x} \in \mathbf{D}_f$, there is at least one admissible decision $\mathbf{y} \in \mathbf{K_x}$ with performance level $f(\mathbf{x}, \mathbf{y}) \geq 0$. Notice that $\mathbf{D}_f \supseteq \overline{\mathbf{B} \setminus \mathbf{R}_f}$, and in a sense robust optimization is dual to design optimization.

The semi-algebraic function f as well as the set $\mathbf{K}$ can be fairly complicated and therefore in general an exact description of both sets $\mathbf{R}_f$ and $\mathbf{D}_f$ is also fairly complicated. Needless to say, robust optimization problems with constraints of the form $\mathbf{x} \in \mathbf{R}_f$ are very difficult to solve. As already mentioned, in principle when $\mathbf{K}$ is a basic semi-algebraic set, quantifier elimination is possible via algebraic techniques. However, in practice quantifier elimination is very costly as it requires symbolic computation with exact arithmetic.

On the other hand, design optimization problems with a constraint of the form $\mathbf{x} \in \mathbf{D}_f$ (or $\mathbf{x} \in \mathbf{D}_f \cap \mathbf{\Omega}$ for some $\mathbf{\Omega} \subset \mathbf{B}$) can be formulated directly in the lifted space of variables $(\mathbf{x}, \mathbf{y}) \in \mathbb{R}^n \times \mathbb{R}^m$ (i.e., by adding the constraints

---

[1] Recall that with $\mathbf{S} \subset \mathbb{R}^n$ being semi-algebraic, $f : \mathbf{S} \to \mathbb{R}$ is a semi-algebraic function if its graph $\{(\mathbf{x}, f(\mathbf{x})) : \mathbf{x} \in \mathbf{S}\}$ is a semi-algebraic set; see Chapter 11.

$f(\mathbf{x}, \mathbf{y}) \geq 0$, $(\mathbf{x}, \mathbf{y}) \in \mathbf{K}$) and so with no approximation. But sometimes one may be interested in finding a description of the set $\mathbf{D}_f$ solely in terms of the $\mathbf{x}$-variables because its "shape" is hidden in the lifted $(\mathbf{x}, \mathbf{y})$-description, or because optimizing over $\mathbf{K} \cap \{(\mathbf{x}, \mathbf{y}) : f(\mathbf{x}, \mathbf{y}) \geq 0\}$ may not be practical. However, if the projection of a basic semi-algebraic set (for example $\mathbf{D}_f$) is semi-algebraic, it is not necessarily *basic* semi-algebraic and could be a complicated union of several basic semi-algebraic sets (hence not very useful in practice).

So a less ambitious but more practical goal is to obtain *tractable* approximations of such sets $\mathbf{R}_f$ (or $\mathbf{D}_f$) which can be used for various purposes, optimization being only one among many potential applications.

We provide a hierarchy $(\mathbf{R}_f^k)$ (respectively $(\mathbf{D}_f^k)$), $k \in \mathbb{N}$, of *inner* approximations for $\mathbf{R}_f$ (respectively *outer* approximations for $\mathbf{D}_f$) with three essential characteristic features.

1. Each set $\mathbf{R}_f^k \subset \mathbb{R}^n$ (respectively $\mathbf{D}_f^k$), $k \in \mathbb{N}$, has a very simple description in terms of the sublevel set $\{\mathbf{x} : p_k(\mathbf{x}) \leq 0\}$ (respectively $\{\mathbf{x} : p_k(\mathbf{x}) \geq 0\}$) associated with a *single* polynomial p_k.
2. Both hierarchies $(\mathbf{R}_f^k)$ and $(\mathbf{D}_f^k)$, $k \in \mathbb{N}$, converge in a strong sense since under some conditions, vol $(\mathbf{R}_f \setminus \mathbf{R}_f^k) \to 0$ (respectively vol $(\mathbf{D}_f^k \setminus \mathbf{D}_f) \to 0$) as $k \to \infty$ (and where "vol$(\cdot)$" denotes the Lebesgue volume). In other words, for k sufficiently large, the inner approximations $\mathbf{R}_f^k$ (respectively outer approximations $\mathbf{D}_f^k$) coincide with $\mathbf{R}_f$ (respectively $\mathbf{D}_f$) up to a set of very small Lebesgue measure.
3. Computing the vector of coefficients of the above polynomial p_k reduces to solving a semidefinite program whose size is parametrized by k.

Hence for instance, the constraint $p_k(\mathbf{x}) \leq 0$ (respectively $p_k(\mathbf{x}) \geq 0$) can be used in any robust (respectively design) polynomial optimization problem on $\mathbf{\Omega} \subset \mathbf{B}$, as a substitute for $\mathbf{x} \in \mathbf{R}_f \cap \mathbf{\Omega}$ (respectively $\mathbf{x} \in \mathbf{D}_f \cap \mathbf{\Omega}$), thereby eliminating the variables $\mathbf{y}$. If $\mathbf{\Omega}$ is a basic semi-algebraic set one then obtains a standard polynomial optimization problem $\mathbf{P}$ for which one may apply the hierarchy of semidefinite relaxations defined in Chapter 6. The sets $\mathbf{R}_f^k$ (or $\mathbf{R}_f^k \cap \mathbf{\Omega}$) can also be used in other applications to provide a certificate for robustness, as membership in $\mathbf{R}_f^k$ is easy to check and the approximation is from inside.

17.2 Inner and outer approximations

Let $\mathbf{K}$ be the basic semi-algebraic set defined in (17.1) for some polynomials $g_j \subset \mathbb{R}[\mathbf{x}, \mathbf{y}]$, $j = 1, \ldots, s$, and with a simple set (e.g. box or ellipsoid)

$\mathbf{B} \subset \mathbb{R}^n$. Denote by $L_1(\mathbf{B})$ the Lebesgue space of measurable functions h : $\mathbf{B} \rightarrow \mathbb{R}$ that are integrable with respect to the Lebesgue measure on $\mathbf{B}$, i.e., such that $\int_{\mathbf{B}} |h| d\mathbf{x} < \infty$.

Given $f \in \mathbb{R}[\mathbf{x}, \mathbf{y}]$, consider the mapping $\overline{J_f} : \mathbf{B} \rightarrow \mathbb{R} \cup \{-\infty\}$ defined by:

$$\mathbf{x} \mapsto \overline{J_f}(\mathbf{x}) := \sup_{\mathbf{y}} \{ f(\mathbf{x}, \mathbf{y}) : \mathbf{y} \in \mathbf{K_x} \}, \qquad \mathbf{x} \in \mathbf{B}.$$

Therefore the set $\mathbf{R}_f$ in (17.3) reads $\{ \mathbf{x} \in \mathbf{B} : \overline{J_f}(\mathbf{x}) \leq 0 \}$ whereas $\mathbf{D}_f$ in (17.4) reads $\{ \mathbf{x} \in \mathbf{B} : \overline{J_f}(\mathbf{x}) \geq 0 \}$. The function $\overline{J_f}$ is upper semi-continuous.

We first recall the following result from Chapter 14, adapted to the maximization case.

Theorem 17.1 (Section 14.2) *Let $\mathbf{K} \subset \mathbb{R}^n \times \mathbb{R}^m$ in (17.1) be compact. If $\mathbf{K_x} \neq \emptyset$ for every $\mathbf{x} \in \mathbf{B}$, there exists a sequence of polynomials $(p_k) \subset \mathbb{R}[\mathbf{x}]$, $k \in \mathbb{N}$, such that $p_k(\mathbf{x}) \geq f(\mathbf{x}, \mathbf{y})$ for all $\mathbf{y} \in \mathbf{K_x}$, $\mathbf{x} \in \mathbf{B}$, and such that*

$$\lim_{k \to \infty} \int_{\mathbf{B}} | p_k(\mathbf{x}) - \overline{J_f}(\mathbf{x}) | \, d\mathbf{x} = 0 \quad \textit{[Convergence in } L_1(\mathbf{B})\textit{]}. \qquad (17.5)$$

17.2.1 Inner approximations of $\mathbf{R}_f$

Let $\mathbf{R}_f$ be as in (17.3) and let $\mathbf{B} \subset \mathbb{R}^n$ be as in Theorem 17.1, assumed to have nonempty interior.

Theorem 17.2 *Let $\mathbf{K} \subset \mathbb{R}^n \times \mathbb{R}^m$ in (17.1) be compact and $\mathbf{K_x} \neq \emptyset$ for every $\mathbf{x} \in \mathbf{B}$. Assume that $\{ \mathbf{x} \in \mathbf{B} : \overline{J_f}(\mathbf{x}) = 0 \}$ has Lebesgue measure zero and let $\mathbf{R}_f^k := \{ \mathbf{x} \in \mathbf{B} : p_k(\mathbf{x}) \leq 0 \}$, with $p_k \in \mathbb{R}[\mathbf{x}]$ as in Theorem 17.1.*

Then $\mathbf{R}_f^k \subset \mathbf{R}_f$ for every k, and

$$\text{vol} \, (\mathbf{R}_f \setminus \mathbf{R}_f^k) \to 0 \quad \textit{as } k \to \infty. \qquad (17.6)$$

Proof By Theorem 17.1, $p_k \to \overline{J_f}$ in $L_1(\mathbf{B})$. Therefore by Ash (1972, Theorem 2.5.1), p_k converges to $\overline{J_f}$ in measure, that is, for every $\epsilon > 0$,

$$\lim_{k \to \infty} \text{vol} \, (\{ \mathbf{x} : | p_k(\mathbf{x}) - \overline{J_f}(\mathbf{x}) | \geq \epsilon \}) = 0. \qquad (17.7)$$

Next, as $\overline{J_f}$ is upper semi-continuous on $\mathbf{B}$, the set $\{ \mathbf{x} : \overline{J_f}(\mathbf{x}) < 0 \}$ is open and as $\text{vol} \, (\{ \mathbf{x} \in \mathbf{B} : \overline{J_f}(\mathbf{x}) = 0 \}) = 0$,

$$\text{vol}\,(\mathbf{R}_f) = \text{vol}\,(\{\,\mathbf{x} \in \mathbf{B} : \overline{J_f}(\mathbf{x}) < 0\,\})$$

$$= \text{vol}\,\left(\bigcup_{\ell=1}^{\infty} \{\,\mathbf{x} \in \mathbf{B} : \overline{J_f}(\mathbf{x}) \le -1/\ell\,\}\right)$$

$$= \lim_{\ell \to \infty} \text{vol}\,(\{\,\mathbf{x} \in \mathbf{B} : \overline{J_f}(\mathbf{x}) \le -1/\ell\,\}) = \lim_{\ell \to \infty} \text{vol}\,(\mathbf{R}_f(\ell)),$$

$$(17.8)$$

where $\mathbf{R}_f(\ell) := \{\,\mathbf{x} \in \mathbf{B} : \overline{J_f}(\mathbf{x}) \le -1/\ell\}$. Next, $\mathbf{R}_f(\ell) \subseteq \mathbf{R}_f$ for every $\ell \ge 1$, and

$$\text{vol}\,(\mathbf{R}_f(\ell)) = \text{vol}\,(\mathbf{R}_f(\ell) \cap \{\,\mathbf{x} : p_k(\mathbf{x}) > 0\}) + \text{vol}\,(\mathbf{R}_f(\ell) \cap \{\mathbf{x} : p_k(\mathbf{x}) \le 0\})\,.$$

Next, by (17.7), $\text{vol}\,(\mathbf{R}_f(\ell) \cap \{\,\mathbf{x} : p_k(\mathbf{x}) > 0\}) \to 0$ as $k \to \infty$. Therefore,

$$\text{vol}\,(\mathbf{R}_f(\ell)) = \lim_{k \to \infty}\, \text{vol}\,\big(\mathbf{R}_f(\ell) \cap \underbrace{\{\,\mathbf{x} \in \mathbf{B} : p_k(\mathbf{x}) \le 0\,\}}_{=\mathbf{R}_f^k}\big)$$

$$\le \lim_{k \to \infty}\, \text{vol}\,(\mathbf{R}_f^k) \le \text{vol}\,(\mathbf{R}_f).$$

As $\mathbf{R}_f^k \subset \mathbf{R}_f$ for all k, letting $\ell \to \infty$ and using (17.8) yields the desired result. $\square$

Hence for robust polynomial optimization problems where one wishes to optimize over the set $\mathbf{R}_f \cap \Omega$ (for some $\Omega \subset \mathbf{B}$), one may reinforce the complicated (and untraceable) constraint $\mathbf{x} \in \mathbf{R}_f \cap \Omega$ by instead considering the inner approximation $\{\,\mathbf{x} \in \Omega : p_k(\mathbf{x}) \le 0\,\}$ obtained with the two much simpler constraints $\mathbf{x} \in \Omega$ and $p_k(\mathbf{x}) \le 0$. By Theorem 17.2, if k is sufficiently large then the resulting conservatism introduced by this approximation is negligible.

17.2.2 Outer approximations of $\mathbf{D}_f$

With $\mathbf{B}$ and $\mathbf{D}_f$ as in (17.4) assumed to have nonempty interior, the next result is the analogue for $\mathbf{D}_f$ of Theorem 17.2 for $\mathbf{R}_f$.

Corollary 17.3 *Let $\mathbf{K} \subset \mathbb{R}^n \times \mathbb{R}^m$ in (17.1) be compact and $\mathbf{K}_\mathbf{x} \ne \emptyset$ for every $\mathbf{x} \in \mathbf{B}$. Assume that $\{\,\mathbf{x} \in \mathbf{B} : \overline{J_f}(\mathbf{x}) = 0\}$ has Lebesgue measure zero and let $\mathbf{D}_f^k := \{\,\mathbf{x} \in \mathbf{B} : p_k(\mathbf{x}) \ge 0\}$, with $p_k \in \mathbb{R}[\mathbf{x}]$ as in Theorem 17.1.*

Then $\mathbf{D}_f^k \supset \mathbf{D}_f$ for every k, and

$$\text{vol}\,(\mathbf{D}_f^k \setminus \mathbf{D}_f) \to 0 \quad as\ k \to \infty. \tag{17.9}$$

Proof The proof uses same arguments as in the proof of Theorem 17.2. Indeed, $\mathbf{D}_f = \mathbf{B} \setminus \Delta_f$ with

$$\Delta_f := \{ \mathbf{x} \in \mathbf{B} : f(\mathbf{x}, \mathbf{y}) < 0 \text{ for all } \mathbf{y} \in \mathbf{K}_\mathbf{x} \}$$
$$= \{ \mathbf{x} \in \mathbf{B} : \sup_{\mathbf{y}} \{ f(\mathbf{x}, \mathbf{y}) : \mathbf{y} \in \mathbf{K}_\mathbf{x} \} < 0 \} \quad \text{[as } \mathbf{K} \text{ is compact]}$$
$$= \{ \mathbf{x} \in \mathbf{B} : \overline{J_f}(\mathbf{x}) < 0 \},$$

and since $\{ \mathbf{x} \in \mathbf{B} : \overline{J_f}(\mathbf{x}) = 0 \}$ has Lebesgue measure zero,

$$\mathrm{vol}\,(\Delta_f) = \mathrm{vol}\left(\{ \mathbf{x} \in \mathbf{B} : \overline{J_f}(\mathbf{x}) \leq 0 \}\right).$$

Hence by Theorem 17.2, $\lim_{k \to \infty} \mathrm{vol}\,(\{ \mathbf{x} \in \mathbf{B} : p_k(\mathbf{x}) \leq 0 \}) = \mathrm{vol}\,(\Delta_f)$, which in turn implies the desired result

$$\lim_{k \to \infty} \mathrm{vol}\,(\{ \mathbf{x} \in \mathbf{B} : p_k(\mathbf{x}) \geq 0 \}) = \lim_{k \to \infty} \mathrm{vol}\,(\{ \mathbf{x} \in \mathbf{B} : p_k(\mathbf{x}) > 0 \})$$
$$= \mathrm{vol}\,(\mathbf{B} \setminus \Delta_f) = \mathrm{vol}\,(\mathbf{D}_f)$$

because $\mathrm{vol}\,(\{ \mathbf{x} \in \mathbf{B} : p_k(\mathbf{x}) = 0 \}) = 0$ for every k. □

Corollary 17.3 states that the set $\mathbf{D}_f$ can be approximated arbitrarily well from outside by sublevel sets of polynomials. In particular, if $\Omega \subset \mathbf{B}$ and one wishes to work with $\mathbf{D}_f \cap \Omega$ and not its lifted representation $\{ f(\mathbf{x}, \mathbf{y}) \geq 0; \mathbf{x} \in \Omega; (\mathbf{x}, \mathbf{y}) \in \mathbf{K} \}$, one may instead use the outer approximation $\{ \mathbf{x} \in \Omega : p_k(\mathbf{x}) \geq 0 \}$. The resulting laxism becomes negligible as k increases.

17.2.3 Practical computation

In this section we show how to compute a sequence of polynomials $(p_k) \subset \mathbb{R}[\mathbf{x}]$, $k \in \mathbb{N}$, as defined in Theorem 17.2 and Corollary 17.3.

Let us proceed as in Chapter 14. With $\mathbf{K} \subset \mathbb{R}^n \times \mathbb{R}^m$ as in (17.1) and compact, we assume that we know some $M > 0$ such that $M - \|\mathbf{y}\|^2 \geq 0$ whenever $(\mathbf{x}, \mathbf{y}) \in \mathbf{K}$. Next, and possibly after a rescaling of the g_j, we may and will set $M = 1$ and $\mathbf{B} = [-1, 1]^n$. Define $\boldsymbol{\gamma} = (\gamma_\alpha)$ with:

$$\gamma_\alpha := \mathrm{vol}\,(\mathbf{B})^{-1} \int_\mathbf{B} \mathbf{x}^\alpha \, d\mathbf{x}, \qquad \alpha \in \mathbb{N}^n.$$

As already mentioned, one may consider any set $\mathbf{B}$ for which all moments of the Lebesgue measure (or any Borel measure with support exactly equal to $\mathbf{B}$) are easy to compute (for instance an ellipsoid).

For convenience, letting $g_{s+1}(\mathbf{y}) := 1 - \|\mathbf{y}\|^2$, and $x_i \mapsto \theta_i(\mathbf{x}) := 1 - x_i^2$, $i = 1, \ldots, n$, we redefine $\mathbf{K} \subset \mathbb{R}^n \times \mathbb{R}^m$ to be the basic semi-algebraic set:

$$\mathbf{K} = \{ (\mathbf{x}, \mathbf{y}) : g_j(\mathbf{x}, \mathbf{y}) \geq 0, \ j = 1, \ldots, s+1; \ \theta_i(\mathbf{x}) \geq 0, \ i = 1, \ldots, n \}.$$
$$\tag{17.10}$$

Let $g_0 = 1$ and $v_j := \lceil \deg(g_j)/2 \rceil$, $j = 0, \ldots, s + 1$, and for fixed $k \geq \max_j [v_j]$, consider the hierarchy of semidefinite programs indexed by $d \in \mathbb{N}$:

$$\rho_d = \sup_{\mathbf{z}} \quad L_{\mathbf{z}}(f)$$

$$\text{s.t.} \quad \mathbf{M}_{d-v_j}(g_j \, \mathbf{z}) \succeq 0, \quad j = 0, \ldots, s + 1 \qquad (17.11)$$
$$\mathbf{M}_{d-1}(\theta_i \, \mathbf{z}) \succeq 0, \quad i = 1, \ldots, n$$
$$L_{\mathbf{z}}(\mathbf{x}^{\alpha}) = \gamma_{\alpha}, \quad \alpha \in \mathbb{N}^n_{2d},$$

and the hierarchy of their duals:

$$\rho_d^* = \inf_{p, \sigma_j, \psi_i} \quad \int_{\mathbf{B}} p(\mathbf{x}) \, d\mathbf{x} \quad \left(= \sum_{\alpha \in \mathbb{N}^m_d} p_{\alpha} \gamma_{\alpha} \right)$$

$$\text{s.t.} \quad p(\mathbf{x}) - f(\mathbf{x}, \mathbf{y}) = \sum_{j=0}^{s+1} \sigma_j(\mathbf{x}, \mathbf{y}) \, g_j(\mathbf{x}, \mathbf{y}) + \sum_{i=1}^{n} \psi_i(\mathbf{x}, \mathbf{y}) \, \theta_i(\mathbf{x})$$

$$p \in \mathbb{R}[\mathbf{x}]_{2d}; \quad \sigma_j \in \Sigma_{d-v_j}[\mathbf{x}, \mathbf{y}], \quad j = 0, \ldots, s + 1$$
$$\psi_i \in \Sigma_{d-1}[\mathbf{x}, \mathbf{y}], \quad i = 1, \ldots, n.$$

$$(17.12)$$

In the present context, (17.11) and (17.12) are the respective analogues of (14.14) and (14.17). Therefore we have the following result.

Proposition 17.4 *Let $\mathbf{K}$ be as in (17.10) with nonempty interior and assume that $\mathbf{K}_{\mathbf{x}} \neq \emptyset$ for every $\mathbf{x} \in \mathbf{B}$. Then:*

(a) *there is no duality gap between (17.11) and its dual (17.12);*
(b) *moreover, (17.11) (respectively (17.12)) has an optimal solution $\mathbf{z}^* = (z^*_{\alpha\beta})$, $(\alpha, \beta) \in \mathbb{N}^{n+m}_{2d}$ (respectively $p^*_d \in \mathbb{R}[\mathbf{x}]_{2d}$), and*

$$\lim_{d \to \infty} \int_{\mathbf{B}} |p_d^*(\mathbf{x}) - \overline{J_f}(\mathbf{x})| \, d\mathbf{x} = 0 \quad [\text{Convergence in } L_1(\mathbf{B})].$$

The proof is an immediate consequence of Theorem 14.5, Theorem 14.15 and Remark 14.7.

Remark 17.5 In fact, in Theorem 17.2 one may impose that the sequence $(p_k) \subset \mathbb{R}[\mathbf{x}]$, $k \in \mathbb{N}$, be monotone, i.e., such that $\overline{J_f} \leq p_k \leq p_{k-1}$ on $\mathbf{B}$, for all $k \geq 2$, and similarly for Corollary 17.3. For practical computation of such a monotone sequence, in the semidefinite program (17.11) it suffices to include the additional constraint (or positivity certificate)

$$p_{d-1}^*(\mathbf{x}) - p(\mathbf{x}) = \sum_{i=0}^{n} \phi_i(\mathbf{x}) \, \theta_i(\mathbf{x}), \quad \phi_0 \in \Sigma[\mathbf{x}]_d, \quad \phi_i \in \Sigma[\mathbf{x}]_{d-1}, \quad i \geq 1,$$

where $\theta_0 = 1$ and $p^*_{d-1} \in \mathbb{R}[\mathbf{x}]_{d-1}$ is the optimal solution computed at the previous step $d - 1$. In this case the inner approximations $(\mathbf{R}^d_f)$, $d \in \mathbb{N}$, form a nested sequence since $\mathbf{R}^d_f \subseteq \mathbf{R}^{d+1}_f$ for all d. Similarly, the outer approximations $(\mathbf{D}^d_f)$, $d \in \mathbb{N}$, also form a nested sequence since $\mathbf{D}^{d+1}_f \subseteq \mathbf{D}^d_f$ for all d.

17.3 Extensions

Semi-algebraic functions Suppose for instance that given $q_1, q_2 \in \mathbb{R}[\mathbf{x}, \mathbf{y}]$, one wants to characterize the set

$$\{\mathbf{x} \in \mathbf{B} : \min[q_1(\mathbf{x}, \mathbf{y}), q_2(\mathbf{x}, \mathbf{y})] \leq 0 \quad \text{for all } \mathbf{y} \in \mathbf{K_x}\},$$

where $\mathbf{K_x}$ has been defined in (17.2), i.e., the set $\mathbf{R}_f$ associated with the semi-algebraic function $(\mathbf{x}, \mathbf{y}) \mapsto f(\mathbf{x}, \mathbf{y}) = \min[q_1(\mathbf{x}, \mathbf{y}), q_2(\mathbf{x}, \mathbf{y})]$. If f is the semi-algebraic function $\max[q_1(\mathbf{x}, \mathbf{y}), q_2(\mathbf{x}, \mathbf{y})]$, characterizing $\mathbf{R}_f$ reduces to the polynomial case as $\mathbf{R}_f = \mathbf{R}_{q_1} \cap \mathbf{R}_{q_2}$. But for $f = \min[q_1, q_2]$ this characterization is not so easy, and in fact is significantly more complicated. However, even though f is no longer a polynomial, we shall next see that the above methodology also works for semi-algebraic functions, a much larger class than the class of polynomials. Of course there is no free lunch, and the resulting computational burden increases because as we have seen in Chapter 11, one needs additional lifting variables to represent the semi-algebraic function f.

With $\mathbf{S} \subset \mathbb{R}^n$ being semi-algebraic, recall that a function $f : \mathbf{S} \to \mathbb{R}$ is semi-algebraic if its graph $\Psi_f = \{(\mathbf{x}, f(\mathbf{x})) : \mathbf{x} \in \mathbf{S}\}$ is a semi-algebraic set. In fact, the graph of every semi-algebraic function is the projection of some *basic* semi-algebraic set in a lifted space; see Remark 11.7 in Chapter 11.

So with $\mathbf{K} \subset \mathbb{R}^n \times \mathbb{R}^m$ as in (17.1), let $f : \mathbf{K} \to \mathbb{R}$ be a semi-algebraic function whose graph $\Psi_f = \{(\mathbf{x}, \mathbf{y}, f(\mathbf{x}, \mathbf{y})) : (\mathbf{x}, \mathbf{y}) \in \mathbf{K}\}$ is the projection $\{(\mathbf{x}, \mathbf{y}, v) \in \mathbb{R}^n \times \mathbb{R}^m \times \mathbb{R}\}$ of a basic semi-algebraic set $\widehat{\mathbf{K}} \subset \mathbb{R}^n \times \mathbb{R}^m \times \mathbb{R}^r$, i.e.:

$$v = f(\mathbf{x}, \mathbf{y}) \text{ and } (\mathbf{x}, \mathbf{y}) \in \mathbf{K} \iff \exists \mathbf{w} \text{ such that } (\mathbf{x}, \mathbf{y}, v, \mathbf{w}) \in \widehat{\mathbf{K}}.$$

Letting $\widehat{f} : \widehat{\mathbf{K}} \to \mathbb{R}$ be such that $\widehat{f}(\mathbf{x}, \mathbf{y}, v, \mathbf{w}) := v$, we have

$$\mathbf{R}_f = \{\mathbf{x} \in \mathbf{B} : f(\mathbf{x}, \mathbf{y}) \leq 0 \text{ for all } \mathbf{y} \text{ such that } (\mathbf{x}, \mathbf{y}) \in \mathbf{K}\}$$

$$= \{\mathbf{x} \in \mathbf{B} : \widehat{f}(\mathbf{x}, \mathbf{y}, v, \mathbf{w})) \leq 0 \text{ for all } (\mathbf{y}, v, \mathbf{w}) \text{ such that } (\mathbf{x}, \mathbf{y}, v, \mathbf{w})) \in \widehat{\mathbf{K}}\}.$$

Hence this is just a special case of what has been considered in Section 17.2 and therefore converging inner approximations of $\mathbf{R}_f$ can be obtained as in Theorem 17.2 and Theorem 17.4.

Example 17.6 For instance suppose that $f : \mathbb{R}^n \times \mathbb{R}^m \to \mathbb{R}$ is the semi-algebraic function $(\mathbf{x}, \mathbf{y}) \mapsto f(\mathbf{x}, \mathbf{y}) := \min[q_1(\mathbf{x}, \mathbf{y}), q_2(\mathbf{x}, \mathbf{y})]$. Then using $a \wedge b = \frac{1}{2}(a + b - |a - b|)$ and $|a - b| = \theta \geq 0$ with $\theta^2 = (a - b)^2$,

$$\widehat{\mathbf{K}} = \{ (\mathbf{x}, \mathbf{y}, v, w) : (\mathbf{x}, \mathbf{y}) \in \mathbf{K}; \; w^2 = (q_1(\mathbf{x}, \mathbf{y}) - q_2(\mathbf{x}, \mathbf{y}))^2; \quad w \geq 0;$$
$$2v = q_1(\mathbf{x}, \mathbf{y}) + q_2(\mathbf{x}, \mathbf{y}) - w \},$$

and

$$\Psi_f = \{ ((\mathbf{x}, \mathbf{y}), f(\mathbf{x}, \mathbf{y})) : (\mathbf{x}, \mathbf{y}) \in \mathbf{K} \} = \{ (\mathbf{x}, \mathbf{y}, v) : (\mathbf{x}, \mathbf{y}, v, w) \in \widehat{\mathbf{K}} \}.$$

Convex inner approximations It is worth mentioning that one may also enforce convexity of inner approximations of $\mathbf{R}_f$. But of course there is some additional computational cost and in general the convergence (17.6) in Theorem 17.2 is lost.

To enforce convexity of the level set $\mathbf{R}_f^d = \{ \mathbf{x} \in \mathbf{B} : p_d^*(\mathbf{x}) \leq 0 \}$ where p_d^* is as in Proposition 17.4, it suffices to require that $p_d^* \in \mathbb{R}[\mathbf{x}]$ is convex on $\mathbf{B}$, i.e., it suffices to add the constraint

$$\langle \mathbf{u}, \nabla^2 p_d^*(\mathbf{x}) \, \mathbf{u} \rangle \geq 0, \qquad \forall (\mathbf{x}, \mathbf{u}) \in \mathbf{B} \times \mathbf{U},$$

where $\mathbf{U} := \{ \mathbf{u} \in \mathbb{R}^n : \|\mathbf{u}\|^2 \leq 1 \}$. The latter constraint can in turn be enforced by the Putinar positivity certificate

$$\langle \mathbf{u}, \nabla^2 p_d^*(\mathbf{x}) \, \mathbf{u} \rangle = \sum_{i=0}^{n} \omega_i(\mathbf{x}, \mathbf{u}) \, \theta_i(\mathbf{x}) + \omega_{n+1}(\mathbf{x}, \mathbf{u}) \, \theta_{n+1}(\mathbf{x}, \mathbf{u}), \quad (17.13)$$

for some SOS polynomials $(\omega_i) \subset \Sigma[\mathbf{x}, \mathbf{u}]$ (and where $\theta_{n+1}(\mathbf{x}, \mathbf{u}) = 1 - \|\mathbf{u}\|^2$).

Then this additional constraint (17.13) can be included in the semidefinite program (17.12) with $\omega_0 \in \Sigma[\mathbf{x}, \mathbf{u}]_d$, and $\omega_i \in \Sigma[\mathbf{x}, \mathbf{u}]_{d-1}, i = 1, \ldots, n + 1$. However, now $\mathbf{z} = (z_{\alpha, \gamma, \beta})$, $(\alpha, \beta, \gamma) \in \mathbb{N}^{2n+m}$, and solving the resulting semidefinite program is more demanding.

Polynomial matrix inequalities Let $\mathbf{A}_\alpha \in \mathcal{S}_m$, $\alpha \in \mathbb{N}_d^n$, be real symmetric matrices and let $\mathbf{B} \subset \mathbb{R}^n$ be a given box. Consider the set

$$\mathbf{S} := \{ \mathbf{x} \in \mathbf{B} : \mathbf{A}(\mathbf{x}) \succeq 0 \}, \qquad (17.14)$$

where $\mathbf{A} \in \mathbb{R}[\mathbf{x}]^{m \times m}$ is the matrix-polynomial

$$\mathbf{x} \mapsto \mathbf{A}(\mathbf{x}) := \sum_{\alpha \in \mathbb{N}_d^n} \mathbf{x}^\alpha \, \mathbf{A}_\alpha.$$

If $\mathbf{A}(\mathbf{x})$ is linear in $\mathbf{x}$ then $\mathbf{S}$ is convex and (17.14) is an LMI description of $\mathbf{S}$ which is very nice as it can be used efficiently in semidefinite programming.

In the general nonconvex case the description (17.14) of $\mathbf{S}$ is called a Polynomial Matrix Inequality (PMI) and cannot be used as efficiently as in the convex case. On the other hand, $\mathbf{S}$ is a basic semi-algebraic set with an alternative description in terms of the box constraint $\mathbf{x} \in \mathbf{B}$ and m additional polynomial inequality constraints (including the constraint $\det(\mathbf{A}(\mathbf{x})) \geq 0$). However, this latter description may not be very appropriate because the degree of polynomials involved in that description is potentially as large as d^m (if each entry of $\mathbf{A}(\mathbf{x})$ has degree d) which precludes its use for practical computation (e.g. for optimization purposes). For instance, the first relaxation in the hierarchy of semidefinite relaxations defined in Chapter 6 may not even be implementable.

Therefore in the general case and when d^m is not small, one may be interested in a description of $\mathbf{S}$ simpler than the PMI (17.14) so that it can be used more efficiently.

So let $\mathbf{Y} := \{\mathbf{y} \in \mathbb{R}^m : \|\mathbf{y}\|^2 = 1\}$ denote the unit sphere of $\mathbb{R}^m$. Then with $(\mathbf{x}, \mathbf{y}) \mapsto f(\mathbf{x}, \mathbf{y}) := -\mathbf{y}^T \mathbf{A}(\mathbf{x})\,\mathbf{y}$, the set $\mathbf{S}$ has the alternative and equivalent description

$$\mathbf{S} = \{\mathbf{x} \in \mathbf{B} : f(\mathbf{x}, \mathbf{y}) \leq 0, \quad \forall \mathbf{y} \in \mathbf{Y}\} =: \mathbf{R}_f, \tag{17.15}$$

which involves the quantifier "$\forall$." In this case one has $\mathbf{K_x} = \mathbf{Y}$. Therefore, even though $\mathbf{K}$ now has empty interior, the machinery developed in Section 17.2 can be applied to define a hierarchy of inner approximations $\mathbf{R}_f^k \subset \mathbf{S}$ as in Theorem 17.2, where for each k, $\mathbf{R}_f^k = \{\mathbf{x} \in \mathbf{B} : p_k(\mathbf{x}) \leq 0\}$ for some polynomial p_k of degree k. If $\mathbf{A}(\mathbf{x})$ is a nontrivial mapping observe that with

$$\mathbf{x} \mapsto \overline{J_f}(\mathbf{x}) := \sup_{\mathbf{y}}\{f(\mathbf{x}, \mathbf{y}) : \mathbf{y} \in \mathbf{Y}\}, \qquad \mathbf{x} \in \mathbf{B},$$

the set $\{\mathbf{x} : \overline{J_f}(\mathbf{x}) = 0\}$ has Lebesgue measure zero because $\overline{J_f}(\mathbf{x})$ is the largest eigenvalue of $-\mathbf{A}(\mathbf{x})$.

Hence by Theorem 17.2

$$\mathrm{vol}\left(\mathbf{R}_f^k\right) \to \mathrm{vol}\,(\mathbf{S}), \qquad \text{as } k \to \infty.$$

Notice that computing p_k requires the introduction of the m additional variables $\mathbf{y}$ but the degree of f is not larger than $d + 2$ if d is the maximum degree of the entries of $\mathbf{A}(\mathbf{x})$.

Importantly for computational purposes, structure sparsity can be exploited to reduce the computational burden. Write the polynomial

$$(\mathbf{x}, \mathbf{y}) \mapsto f(\mathbf{x}, \mathbf{y}) = -\langle \mathbf{y}, \mathbf{A}(\mathbf{x}) \mathbf{y} \rangle = \sum_{\alpha \in \mathbb{N}^n} h_\alpha(\mathbf{y}) \mathbf{x}^\alpha, \quad (\mathbf{x}, \mathbf{y}) \in \mathbb{R}^n \times \mathbb{R}^m,$$

for finitely many quadratic polynomials $\{h_\alpha \in \mathbb{R}[\mathbf{y}]_2 : \alpha \in \mathbb{N}^n\}$. Suppose that the polynomial $\theta \in \mathbb{R}[\mathbf{x}]$:

$$\mathbf{x} \mapsto \theta(\mathbf{x}) := \sum_{\alpha \in \mathbb{N}^n} h_\alpha(\mathbf{y}) \mathbf{x}^\alpha, \quad \mathbf{x} \in \mathbb{R}^n,$$

has some structured sparsity as described in Chapter 8. That is, $\{1, \ldots, n\} = \cup_{\ell=1}^{s} I_\ell$ (with possible overlaps) and $\theta(\mathbf{x}) = \sum_{\ell=1}^{s} \theta_\ell(\mathbf{x}_\ell)$ where $\mathbf{x}_\ell = \{x_i : i \in I_\ell\}$. Then $(\mathbf{x}, \mathbf{y}) \mapsto f(\mathbf{x}, \mathbf{y})$ inherits the same structured sparsity but now with

$$\{1, \ldots, n, n+1, \ldots, n+m\} = \cup_{\ell=1}^{s} I'_\ell,$$

where $I'_\ell = I_\ell \cup \{n+1, \ldots n+m\}$, $\ell = 1, \ldots, s$. And so in particular, to compute p_k one may use the sparse version of the hierarchy of semidefinite relaxations described in Section 8.1 which allows problems with a significantly large number of variables to be handled.

Example 17.7 With $n = 2$, let $\mathbf{B} \subset \mathbb{R}^2$ be the unit disk $\{\mathbf{x} : \|\mathbf{x}\|^2 \leq 1\}$, and

$$\mathbf{A}(\mathbf{x}) := \begin{bmatrix} 1 - 16x_1 x_2 & x_1 \\ x_1 & 1 - x_1^2 - x_2^2 \end{bmatrix}, \quad \mathbf{S} := \{\mathbf{x} \in \mathbf{B} : \mathbf{A}(\mathbf{x}) \succeq 0\}.$$

Figure 17.1 displays $\mathbf{S}$ and the degree-2 $\mathbf{R}_f^1$ and degree-4 $\mathbf{R}_f^2$ inner approximations of $\mathbf{S}$, whereas Figure 17.2 displays the $\mathbf{R}_f^3$ and $\mathbf{R}_f^4$ inner approximations of $\mathbf{S}$. One may see that with $k = 4$, $\mathbf{R}_f^4$ is already a quite good approximation of $\mathbf{S}$.

Next, with $\mathbf{K}$ as in (17.1) consider now sets of the form

$$\mathbf{R}_f := \{\mathbf{x} \in \mathbf{B} : \mathbf{F}(\mathbf{x}, \mathbf{y}) \preceq 0 \text{ for all } \mathbf{y} \text{ in } \mathbf{K_x}\},$$

where $\mathbf{F} \in \mathbb{R}[\mathbf{x}, \mathbf{y}]^{m \times m}$ is a polynomial matrix in the $\mathbf{x}$ and $\mathbf{y}$ variables. Then letting $\mathbf{Z} := \{\mathbf{z} \in \mathbb{R}^m : \|\mathbf{z}\| = 1\}$, $\widehat{\mathbf{K}} := \mathbf{K} \times \mathbf{Z}$, and $f : \widehat{\mathbf{K}} \to \mathbb{R}$, defined by

$$(\mathbf{x}, \mathbf{y}, \mathbf{z}) \mapsto f(\mathbf{x}, \mathbf{y}, \mathbf{z}) := \langle \mathbf{z}, \mathbf{F}(\mathbf{x}, \mathbf{y}) \mathbf{z} \rangle, \quad (\mathbf{x}, \mathbf{y}, \mathbf{z}) \in \widehat{\mathbf{K}},$$

the set $\mathbf{R}_f$ has the equivalent description:

$$\mathbf{R}_f := \{\mathbf{x} \in \mathbf{B} : f(\mathbf{x}, \mathbf{y}, \mathbf{z}) \leq 0 \text{ for all } (\mathbf{y}, \mathbf{z}) \text{ in } \widehat{\mathbf{K}}_\mathbf{x}\},$$

and the methodology of Section 17.2 again applies even though $\mathbf{K}$ has empty interior.

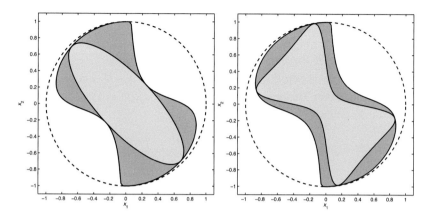

Figure 17.1 Example 17.7: $\mathbf{R}_f^1$ (left) and $\mathbf{R}_f^2$ (right) inner approximations (light gray) of $\mathbf{S}$ (dark gray) embedded in the unit disk $\mathbf{B}$ (dashed line).

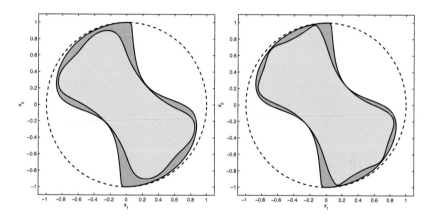

Figure 17.2 Example 17.7: $\mathbf{R}_f^3$ (left) and $\mathbf{R}_f^4$ (right) inner approximations (light gray) of $\mathbf{S}$ (dark gray) embedded in the unit disk $\mathbf{B}$ (dashed line).

Several functions We now consider sets of the form

$$\mathbf{R_F} := \{\, \mathbf{x} \in \mathbf{B} : \ (\mathbf{x}, \mathbf{y}) \in \mathbf{F} \text{ for all } \mathbf{y} \text{ such that } (\mathbf{x}, \mathbf{y}) \in \mathbf{K} \,\}$$

where $\mathbf{F} \subset \mathbb{R}^n \times \mathbb{R}^m$ is a basic semi-algebraic set defined by

$$\mathbf{F} := \{\, (\mathbf{x}, \mathbf{y}) \in \mathbb{R}^n \times \mathbb{R}^m : \ \mathbf{x} \in \mathbf{B}; \ f_\ell(\mathbf{x}, \mathbf{y}) \leq 0, \quad \forall \ell = 1, \ldots, q \,\},$$

for some polynomials $(f_\ell) \subset \mathbb{R}[\mathbf{x}, \mathbf{y}]$, $\ell = 1, \ldots, q$. In other words:

$$\mathbf{R_F} = \{\, \mathbf{x} \in \mathbf{B} : \ \mathbf{K_x} \subseteq \mathbf{F_x} \,\},$$

where $\mathbf{F_x} = \{\mathbf{y} : (\mathbf{x}, \mathbf{y}) \in \mathbf{F}\}$, $\mathbf{x} \in \mathbf{B}$. This is a particular case of the previous section with the semi-algebraic function $\mathbf{x} \mapsto f(\mathbf{x}) := \max[f_1(\mathbf{x}), \ldots, f_q(\mathbf{x})]$, i.e., $\mathbf{R_F} = \mathbf{R}_f$, but in this case a simpler approach is possible.

Let $\mathbf{x} \mapsto \overline{J_{f_\ell}}(\mathbf{x}) := \sup_\mathbf{y}\{f_\ell(\mathbf{x}, \mathbf{y}) : \mathbf{y} \in \mathbf{K_x}\}$, for all $\mathbf{x} \in \mathbf{B}$. Let $p_{k\ell} \in \mathbb{R}[\mathbf{x}]$ be the polynomial in Theorem 17.1 associated with $\overline{J_{f_\ell}}$, $\ell = 1, \ldots, q$, and let the set $\mathbf{R}_\mathbf{F}^k$ be defined by

$$\mathbf{R}_F^k := \{\mathbf{x} \in \mathbf{B} : p_{k\ell}(\mathbf{x}) \leq 0,\ \ell = 1, \ldots, q\} = \bigcap_{\ell=1}^{q} \mathbf{R}_{f_\ell}^k,$$

where for each $\ell = 1, \ldots, q$, the set $\mathbf{R}_{f_\ell}^k$ has been defined in (17.3).

The sets $(\mathbf{R}_\mathbf{F}^k) \subset \mathbf{R}_F$, $k \in \mathbb{N}$, provide a sequence of inner approximations of $\mathbf{R_F}$ with the nice property that

$$\lim_{k \to \infty} \mathrm{vol}\left(\mathbf{R}_\mathbf{F}^k\right) = \mathrm{vol}\left(\mathbf{R_F}\right),$$

whenever the set $\{\mathbf{x} \in \mathbf{B} : \sup_\mathbf{y}\{\max_\ell f_\ell(\mathbf{x}, \mathbf{y}) : \mathbf{y} \in \mathbf{K_x}\} = 0\}$ has Lebesgue measure zero.

Observe that the *containment* problem of deciding whether $\mathbf{K} \subseteq \mathbf{F}$ reduces to checking whether $\mathbf{R}_F = \mathbf{B}$. In particular this happens if $\mathbf{R}_\mathbf{F}^k = \mathbf{B}$ for some k, i.e., if $p_{k\ell}(\mathbf{x}) \leq 0$ for all $\mathbf{x} \in \mathbf{B}$ and all $\ell = 1, \ldots, q$.

17.4 Exercises

Exercise 17.1 With $n = 1$, $\mathbf{B} = \mathbb{R}$ and $p_i \in \mathbb{R}[x]$, $i = 1, 2, 3$, let

$$\mathbf{A}(x) := \begin{bmatrix} p_1(x) & p_2(x) \\ p_2(x) & p_3(x) \end{bmatrix}, \quad \mathbf{S} := \{x \in \mathbb{R} : \mathbf{A}(x) \succeq 0\}.$$

Show that $\mathbf{S} = \mathbb{R}$ if and only if some semidefinite program has a feasible solution.

Exercise 17.2 With $n = 2$, $\mathbf{B} = \mathbb{R}^2$ and $p_i \in \mathbb{R}[\mathbf{x}]_2$, $i = 1, 2, 3$, let $\mathbf{A}(\mathbf{x})$ and $\mathbf{S} \subset \mathbb{R}^2$ be as defined in Exercise 17.1. Show that $\mathbf{S} = \mathbb{R}^2$ if and only if some semidefinite program has a feasible solution. (*Hint:* Use the fact that nonnegative bivariate quartic polynomials are SOS.)

Exercise 17.3 Show that in Corollary 17.3 the assumption that the set $\{\mathbf{x} \in \mathbf{B} : \overline{J_f}(\mathbf{x}) = 0\}$ has Lebesgue measure zero may be discarded. (*Hint:* Use the result that $\{\mathbf{x} : \overline{J_f}(\mathbf{x}) \geq 0\} = \cap_{k \geq 1}\{\mathbf{x} : \overline{J_f}(\mathbf{x}) \geq 1/k\}$ and the fact that $p_k \to \overline{J_f}$ in $L_1(\mathbf{B})$ and hence in measure.)

17.5 Notes and sources

The material of this chapter is mostly from Lasserre (2013f). For elimination theory in algebraic geometry as well as (exact) algorithmic approaches, the interested reader is referred to for example Bochnak et al. (1998), Basu et al. (2003), the more recent Hong and Safey El Din (2012) and the many references therein.

Example 17.7 is from Henrion and Lasserre (2012) where the authors provide inner approximations of robust stability regions for a class of dynamical systems. Alternatively, for polynomial optimization problems with a PMI constraint $\mathbf{A(x)} \succeq 0$, one may still define an appropriate and ad hoc hierarchy of semidefinite relaxations (still in the spirit of the hierarchy (6.3)); see for example Scherer and Hol (2004), Hol and Scherer (2005), and Henrion and Lasserre (2006). But even if the latter approach is more economical than the hierarchy (6.3) (which uses a description of $\mathbf{S} := \{\mathbf{x} : \mathbf{A(x)} \succeq 0\}$ via m (high degree) polynomials), it still might not be ideal. In particular it is not clear how to detect (and then benefit from) some possible structured sparsity to reduce the computational cost.

Finally, sets defined with *two* quantifiers were also considered in Lasserre (2013f). That is, one now considers three types of variables $(\mathbf{x}, \mathbf{y}, \mathbf{u}) \in \mathbb{R}^n \times \mathbb{R}^m \times \mathbb{R}^s$, a box $\mathbf{B_x} \subset \mathbb{R}^n$, a box $\mathbf{B_y} \subset \mathbb{R}^m$, and a set $\mathbf{K} \subset \mathbf{B_x} \times \mathbf{B_y} \times \mathbf{U}$. It is assumed that for each $(\mathbf{x}, \mathbf{y}) \in \mathbf{B_{xy}} \,(:= \mathbf{B_x} \times \mathbf{B_y})$,

$$\mathbf{K_{xy}} := \{\mathbf{u} \in \mathbf{U} : (\mathbf{x}, \mathbf{y}, \mathbf{u}) \in \mathbf{K}\} \neq \emptyset.$$

Then again using the methodology developed in this chapter one can approximate sets of the form

$$\{\mathbf{x} \in \mathbf{B_x} : \exists\, \mathbf{y} \in \mathbf{B_y} \text{ such that } f(\mathbf{x}, \mathbf{y}, \mathbf{u}) \leq 0 \text{ for all } \mathbf{u} \in \mathbf{K_{xy}}\}$$

or sets of the form

$$\{\mathbf{x} \in \mathbf{B_x} : \text{ for all } \mathbf{y} \in \mathbf{B_y}, \exists\, \mathbf{u} \in \mathbf{K_{xy}} \text{ such that } f(\mathbf{x}, \mathbf{y}, \mathbf{u}) \geq 0\}$$

with simpler sublevel sets $\{\mathbf{x} : p_k(\mathbf{x}) \leq 0\}$, $k \in \mathbb{N}$, each associated with a single polynomial $p \in \mathbb{R}[\mathbf{x}]$.

18

Level sets and a generalization of the Löwner–John problem

18.1 Introduction

In this chapter we study some properties of the sublevel set $\mathbf{G} := \{\, \mathbf{x} \in \mathbb{R}^n :\ g(\mathbf{x}) \leq 1 \,\}$ associated with a polynomial $g \in \mathbb{R}[\mathbf{x}]$. Even though it is a topic of independent interest, we have seen in Chapter 17 that such sets are also very useful for approximating as closely as desired other semi-algebraic sets defined in a more complicated manner.

We will also consider the following nonconvex generalization of the Löwner–John ellipsoid problem.

> Let $\mathbf{K} \subset \mathbb{R}^n$ be a compact set not necessarily convex and d be an even integer. Find a nonnegative homogeneous polynomial g of degree d such that its sublevel set $\mathbf{G} := \{\, \mathbf{x} :\ g(\mathbf{x}) \leq 1 \,\}$ contains $\mathbf{K}$ and has minimum volume among all such sublevel sets with this inclusion property.

The set $\mathbf{G}$ is symmetric and so when $\mathbf{K}$ is a symmetric convex body and $d = 2$ then $\mathbf{G}$ is an ellipsoid, and one retrieves the Löwner–John ellipsoid problem in the symmetric case, a famous convex optimization problem. We will see that in view of the properties of $\mathbf{G}$, the above generalization of the Löwner–John problem where neither $\mathbf{G}$ nor $\mathbf{K}$ is required to be convex, is still a convex optimization problem with a unique optimal solution which is also characterized in terms of contact points in $\mathbf{G} \cap \mathbf{K}$.

18.2 Quasi-homogeneous polynomials and their level sets

In this section we describe some of the important properties of quasi-homogeneous polynomials and their associated sublevel sets.

Quasi-homogeneous polynomials   Recall that a polynomial $f \in \mathbb{R}[\mathbf{x}]$ is homogeneous of degree q if $f(\lambda \mathbf{x}) = \lambda^q f(\mathbf{x})$ for all $\mathbf{x} \in \mathbb{R}^n$ and all $\lambda \in \mathbb{R}$.

Quasi-homogeneity is a weaker notion of homogeneity, with respect to a rational multi-index $\mathbf{u} \in \mathbb{Q}^n$, as follows.

Definition 18.1   A polynomial $f \in \mathbb{R}[\mathbf{x}]$ is quasi-homogeneous if there exists $\mathbf{u} \in \mathbb{Q}^n$ such that $f(\lambda^{u_1} x_1 \cdots \lambda^{u_n} x_n)) = \lambda f(\mathbf{x})$ for all $\lambda > 0$ and all $\mathbf{x} \in \mathbb{R}$, in which case we say that f is $\mathbf{u}$-QH (or QH of type $\mathbf{u}$).

Example 18.2   Any homogeneous polynomial of degree $m > 0$ is $\mathbf{u}$-QH with $\mathbf{u} = (\frac{1}{m}, \cdots, \frac{1}{m})$.

The polynomials $\mathbf{x} \mapsto f(\mathbf{x}) := x_1^4 + x_1^2 x_2^6 + x_2^{12}$ and $\mathbf{x} \mapsto f(\mathbf{x}) := x_1^3 x_2^3 + x_1 x_2^9 + x_2^{12}$, as well as any of their monomials are $\mathbf{u}$-QHF with $\mathbf{u} = (\frac{1}{4}, \frac{1}{12})$.

The polynomial $\mathbf{x} \mapsto f(\mathbf{x}) := x_1 (x_2^3 + x_2^4 x_3)$ is $\mathbf{u}$-QH with $\mathbf{u} = (0, \frac{1}{3}, -\frac{1}{3})$ and with $\mathbf{u} = (\frac{k-1}{k}, \frac{1}{3k}, -\frac{1}{3k})$ for every $k \in \mathbb{N}$.

Notice that every quasi-homogeneous polynomial satisfies $g(0) = 0$. Given $f \in \mathbb{R}[\mathbf{x}]$ with vector of coefficients $\mathbf{f} = (f_\alpha)$, let $\operatorname{supp}(f) \subset \mathbb{N}^n$ denote the support of f, i.e., the set $\{\alpha \in \mathbb{N}^n : f_\alpha \neq 0\}$. In fact we have the following result.

Proposition 18.3   *A polynomial $f \in \mathbb{R}[\mathbf{x}]$ is quasi-homogeneous if and only if there exists some rational vector $\mathbf{u} \in \mathbb{Q}^n$ such that*

$$\langle \mathbf{u}, \alpha \rangle = 1, \qquad \forall \alpha \in \operatorname{supp}(f). \tag{18.1}$$

In particular, if $\# \operatorname{supp}(f) \leq n$ then f is quasi-homogeneous.

Proof   Suppose that (18.1) holds true. Then with $\lambda > 0$,

$$f(\lambda^{u_1} x_1, \ldots, \lambda^{u_n} x_n) = \sum_{\alpha \in \operatorname{supp}(f)} f_\alpha \lambda^{\langle \mathbf{u}, \alpha \rangle} \mathbf{x}^\alpha = \lambda f(\mathbf{x}),$$

which proves that f is $\mathbf{u}$-QH. Conversely, if f is $\mathbf{u}$-QH then

$$f(\lambda^{u_1} x_1, \ldots, \lambda^{u_n} x_n) = \sum_{\alpha \in \operatorname{supp}(f)} f_\alpha \lambda^{\langle \mathbf{u}, \alpha \rangle} \mathbf{x}^\alpha = \lambda f(\mathbf{x}).$$

As this should hold for every $\mathbf{x} \in \mathbb{R}^n$, $\lambda > 0$, one must have $\langle \mathbf{u}, \alpha \rangle = 1$ for all $\alpha \in \operatorname{supp}(f)$. Finally if $\# \operatorname{supp}(f) \leq n$ one may always find a rational vector $\mathbf{u} \in \mathbb{Q}^n$ such that (18.1) holds true. $\square$

So Proposition 18.3 states that quasi-homogeneous polynomials are *degenerate* in the sense that their Newton polytope[1] is not full-dimensional in $\mathbb{R}^n$ as it lies in the hyperplane $\mathbf{u}^T \alpha = 1$.

---

[1]   The Newton polytope of a polynomial $f \in \mathbb{R}[\mathbf{x}]$ is the convex hull of $\operatorname{supp}(f)$.

The celebrated Euler identity $\langle \mathbf{x}, \nabla f(\mathbf{x}) \rangle = d\, f(\mathbf{x})$, for homogeneous polynomials of degree d, generalizes to

$$\sum_{i=1}^{n} u_i\, x_i\, \frac{\partial f(\mathbf{x})}{\partial x_i} = f(\mathbf{x}), \tag{18.2}$$

for $\mathbf{u}$-QH polynomials.

18.2.1 Level sets of quasi-homogeneous polynomials

With $0 < \mathbf{u} \in \mathbb{Q}^n$ fixed, denote by $\mathbb{R}[\mathbf{x}]_{\mathbf{u}} \subset \mathbb{R}[\mathbf{x}]$ the vector space of $\mathbf{u}$-QH polynomials and by $\mathbf{P}[\mathbf{x}]_{\mathbf{u}} \subset \mathbb{R}[\mathbf{x}]_{\mathbf{u}}$ its (possibly empty) subset of $\mathbf{u}$-QH polynomials g whose sublevel set $\mathbf{G}_1 := \{\mathbf{x} : g(\mathbf{x}) \leq 1\}$ has finite Lebesgue volume (denoted $\mathrm{vol}(\mathbf{G}_1)$).

Notice that finite volume does not imply that the set $\mathbf{G}_1$ is bounded. For instance if $\mathbf{x} \mapsto g(\mathbf{x}) := x_1^4 x_2^2 + x_1^2 x_2^4$ then $\mathbf{G}_1$ has finite volume but is not bounded. On the other hand, if $g \in \mathbf{P}_{\mathbf{u}}[\mathbf{x}]$ then necessarily g is nonnegative. Indeed, suppose that $g(\hat{\mathbf{x}}) < 0$ for some $\hat{\mathbf{x}} \in \mathbb{R}^n$. Then $g < 0$ on the Euclidean ball $\mathbf{B}(\hat{\mathbf{x}}, \delta) := \{\mathbf{x} : \|\mathbf{x} - \hat{\mathbf{x}}\| < \delta\}$ for some $\delta > 0$. Therefore by quasi-homogeneity, $g < 0$ in the ball $\mathbf{B}(\hat{\mathbf{x}}_\lambda, \delta)$ for every $\lambda \geq 1$ (where $\hat{\mathbf{x}}_\lambda := (\lambda^{u_1}\hat{x}_1, \dots, \lambda^{u_n}\hat{x}_n)$). Hence $g < 0$ on the "tube" T generated by the balls $\mathbf{B}(\hat{\mathbf{x}}_\lambda, \delta)$, $\lambda \in [1, \infty)$, whose Lebesgue volume is not finite. As $T \subset \mathbf{G}_1$ this would yield a contradiction.

Proposition 18.4 *Let $0 < \mathbf{u} \in \mathbb{Q}^n$ with $\mathbf{P}[\mathbf{x}]_{\mathbf{u}} \neq \emptyset$. Then $\mathbf{P}[\mathbf{x}]_{\mathbf{u}}$ is a convex cone.*

Proof Let $g_1, g_2 \in \mathbf{P}[\mathbf{x}]_{\mathbf{u}}$ with associated sublevel sets $\mathbf{G}_1^i = \{\mathbf{x} : g_i(\mathbf{x}) \leq 1\}$, $i = 1, 2$. For $\lambda \in (0, 1)$, consider the $\mathbf{u}$-QH polynomial $\theta := \lambda g_1 + (1 - \lambda)\, g_2$ with associated sublevel set

$$\Theta_1 := \{\mathbf{x} : \theta(\mathbf{x}) \leq 1\} = \{\mathbf{x} : \lambda g_1(\mathbf{x}) + (1 - \lambda)\, g_2(\mathbf{x}) \leq 1\}.$$

Write $\Theta_1 = \Theta_1^1 \cup \Theta_1^2$ where $\Theta_1^1 = \Theta_1 \cap \{\mathbf{x} : g_1(\mathbf{x}) \geq g_2(\mathbf{x})\}$ and $\Theta_1^2 = \Theta_1 \cap \{\mathbf{x} : g_1(\mathbf{x}) < g_2(\mathbf{x})\}$. Observe that $\mathbf{x} \in \Theta_1^1$ implies $g_2(\mathbf{x}) \leq 1$ and so $\Theta_1^1 \subset \mathbf{G}_1^2$. Similarly $\mathbf{x} \in \Theta_1^2$ implies $g_1(\mathbf{x}) \leq 1$ and so $\Theta_1^2 \subset \mathbf{G}_1^1$. Therefore $\mathrm{vol}(\Theta_1) \leq \mathrm{vol}(\mathbf{G}_1^1) + \mathrm{vol}(\mathbf{G}_1^2) < \infty$. And so $\theta \in \mathbf{P}[\mathbf{x}]_{\mathbf{u}}$. $\qquad\square$

Notice that $0 \notin \mathbf{P}[\mathbf{x}]_{\mathbf{u}}$ and $\mathbf{P}[\mathbf{x}]_{\mathbf{u}}$ is not a closed cone. We next provide a crucial property of sublevel sets associated with $\mathbf{u}$-QH polynomials which is of independent interest. But we first need to recall the Laplace transform of a function $f : \mathbb{R}_+^n \to \mathbb{R}$. With $0 < \mathbf{u} \in \mathbb{Q}^n$ let $|\mathbf{u}| := \sum_i u_i$.

Laplace transform Given a measurable function $f : \mathbb{R} \to \mathbb{R}$ with $f(t) = 0$ for all $t < 0$, its one-sided (or unilateral) *Laplace transform* $\mathcal{L}[f] : \mathbb{C} \to \mathbb{C}$ is defined by

$$\lambda \mapsto \quad \mathcal{L}[f](\lambda) := \int_0^\infty \exp(-\lambda t) f(t) \, dt, \qquad \lambda \in D,$$

where its domain $D \subset \mathbb{C}$ is the set of $\lambda \in \mathbb{C}$ where the above integral is finite. Moreover $\mathcal{L}[f]$ is analytic on D and therefore if there exists an analytic function F such that $\mathcal{L}[f](\lambda) = F(\lambda)$ for all λ in a segment of the real line contained in D, then $\mathcal{L}[f](\lambda) = F(\lambda)$ for all $\lambda \in D$. This is a consequence of the Identity Theorem for analytic functions; see for example Freitag and Busam (2009, Theorem III.3.2, p. 125). For instance, let $f(t) = 0$ if $t < 0$ and $f(t) = t^a$ for $t \geq 0$ and $a > -1$. Then $\mathcal{L}[f](\lambda) = \frac{\Gamma(a+1)}{\lambda^{a+1}}$ and $D = \{\lambda : \Re(\lambda) > 0\}$.

Lemma 18.5   *With* $0 < \mathbf{u} \in \mathbb{Q}^n$, *let* $g \in \mathbf{P}[\mathbf{x}]_{\mathbf{u}}$ *and* $\mathbf{G}_y := \{\, \mathbf{x} : g(\mathbf{x}) \leq y \,\}$, $y \geq 0$. *Then for all* $y \geq 0$ *and all* $\alpha \in \mathbb{N}^n$:

$$\int_{\mathbf{G}_y} \mathbf{x}^\alpha \, d\mathbf{x} = \frac{y^{\sum_i u_i(1+\alpha_i)}}{\Gamma(1 + \sum_i u_i(1 + \alpha_i))} \int_{\mathbb{R}^n} \mathbf{x}^\alpha \exp(-g(\mathbf{x})) \, d\mathbf{x}, \quad (18.3)$$

if the left-hand side is finite, and in particular

$$\mathrm{vol}(\mathbf{G}_y) = \frac{y^{|\mathbf{u}|}}{\Gamma(1 + |\mathbf{u}|)} \int_{\mathbb{R}^n} \exp(-g(\mathbf{x})) \, d\mathbf{x}, \qquad y \geq 0. \quad (18.4)$$

Proof   As $g \in \mathbf{P}[\mathbf{x}]_{\mathbf{u}}$, the function $y \mapsto \mathrm{vol}(\mathbf{G}_y)$ is positively homogeneous of degree $|\mathbf{u}|$ and so $\mathrm{vol}(\mathbf{G}_1) < \infty$ implies $\mathrm{vol}(\mathbf{G}_y) < \infty$ for every $y \geq 0$. Let $v : \mathbb{R} \to \mathbb{R}$ be the function $y \mapsto v(y) := \int_{\mathbf{G}_y} \mathbf{x}^\alpha d\mathbf{x}$. Since g is nonnegative, the function v is positively homogeneous of degree $\sum_i u_i(1 + \alpha_i)$ (hence finite if $v(1)$ is finite) and vanishes on $(-\infty, 0]$. Its Laplace transform $\mathcal{L}[v] : \mathbb{C} \to \mathbb{C}$ is the function

$$\lambda \mapsto \mathcal{L}[v](\lambda) := \int_0^\infty \exp(-\lambda y) v(y) \, dy, \qquad \Re(\lambda) > 0,$$

whose domain is $D = \{\lambda \in \mathbb{C} : \Re(\lambda) > 0\}$. Next, for all $\lambda \in \mathbb{R}$ with $\lambda > 0$,

$$
\begin{aligned}
\mathcal{L}[v](\lambda) &= \int_0^\infty \exp(-\lambda y) \left(\int_{\{\mathbf{x}: g(\mathbf{x}) \le y\}} \mathbf{x}^\alpha d\mathbf{x} \right) dy \\
&= \int_{\mathbb{R}^n} \mathbf{x}^\alpha \left(\int_{g(\mathbf{x})}^\infty \exp(-\lambda y)\, dy \right) d\mathbf{x} \quad \text{[by the Fubini–Tonelli Theorem]} \\
&= \frac{1}{\lambda} \int_{\mathbb{R}^n} \mathbf{x}^\alpha \exp(-\lambda g(\mathbf{x}))\, d\mathbf{x} \\
&= \frac{1}{\lambda} \int_{\mathbb{R}^n} \mathbf{x}^\alpha \exp(-g(\lambda^{u_1} x_1, \ldots, \lambda^{u_n} x_n))\, d\mathbf{x} \quad \text{[by quasi-homogeneity]} \\
&= \frac{1}{\lambda^{1 + \sum_i u_i (1 + \alpha_i)}} \int_{\mathbb{R}^n} \mathbf{z}^\alpha \exp(-g(\mathbf{z}))\, d\mathbf{z} \quad \text{[by } \lambda^{u_i} x_i \to z_i \text{]} \\
&= \underbrace{\frac{\int_{\mathbb{R}^n} \mathbf{z}^\alpha \exp(-g(\mathbf{z}))\, d\mathbf{z}}{\Gamma(1 + \sum_i u_i (1 + \alpha_i))}}_{\text{constant } c} \frac{\Gamma(1 + \sum_i u_i (1 + \alpha_i))}{\lambda^{1 + \sum_i u_i (1 + \alpha_i)}}.
\end{aligned}
$$

Next, the function $\frac{c\, \Gamma(1 + \sum_i u_i (1 + \alpha_i))}{\lambda^{1 + \sum_i u_i (1 + \alpha_i)}}$ is analytic on D and coincides with $\mathcal{L}[v]$ on the real half-line $\{t : t > 0\}$ contained in D. Therefore by the Identity Theorem for analytic functions, $\mathcal{L}[v](\lambda) = \frac{c\, \Gamma(1 + \sum_i u_i (1 + \alpha_i))}{\lambda^{1 + \sum_i u_i (1 + \alpha_i)}}$ on D. Finally observe that whenever $a > -1$, $\frac{\Gamma(1 + a)}{\lambda^{1 + a}}$ is the Laplace transform of $t \mapsto h(t) = t^a$, which yields the desired result (18.3). $\qquad\square$

And we also conclude the following.

Corollary 18.6 *Let $0 < \mathbf{u} \in \mathbb{Q}^n$. Then $g \in \mathbf{P}[\mathbf{x}]_\mathbf{u}$, i.e., $\mathrm{vol}(\mathbf{G}_1) < \infty$, if and only if $\int_{\mathbb{R}^n} \exp(-g(\mathbf{x}))d\mathbf{x} < \infty$.*

Proof The implication $\Rightarrow$ follows from Lemma 18.5. For the reverse implication consider the function

$$
\lambda \mapsto h(\lambda) := \frac{1}{\lambda^{1 + |\mathbf{u}|}} \int_{\mathbb{R}^n} \exp(-g(\mathbf{x}))\, d\mathbf{x}, \qquad 0 < \lambda \in \mathbb{R}.
$$

Proceeding as in the proof of Lemma 18.5 but backward and with $\alpha = 0$, it follows that h is the Laplace transform of the function $y \mapsto \mathrm{vol}(\mathbf{G}_y)$ and the proof is complete. $\qquad\square$

Example 18.7 With $n = 2$ let $\mathbf{x} \mapsto g(\mathbf{x}) := (x_1^6 + x_2^4)/2.7 - 0.66\, x_1^3 x_2^2$ which is $\mathbf{u}$-QH with $\mathbf{u} := (1/6, 1/4)$. Then g is nonnegative on $\mathbb{R}^2$ and

$$
\mathrm{vol}(\mathbf{G}_1) = \frac{1}{\Gamma(17/12)} \int_{\mathbb{R}^2} \exp(-(x_1^6 + x_2^4)/2.7 - 0.66\, x_1^3 x_2^2)\, d\mathbf{x}.
$$

18.2.2 Sensitivity analysis and convexity.

Let $0 < \mathbf{u} \in \mathbb{Q}^n$ be fixed. By Proposition 18.3 $\mathbb{R}[\mathbf{x}]_{\mathbf{u}}$ is a finite-dimensional vector space whose dimension $\rho(n, \mathbf{u})$ is the cardinality of the set $\{\boldsymbol{\alpha} \in \mathbb{N}^n : \mathbf{u}^T \boldsymbol{\alpha} = 1\}$. We next investigate some properties of the function $f : \mathbf{P}[\mathbf{x}]_{\mathbf{u}} \to \mathbb{R}$ defined by:

$$g \mapsto f(g) := \mathrm{vol}\,(\mathbf{G}_1) = \int_{\{\mathbf{x}:g(\mathbf{x})\leq 1\}} d\mathbf{x}, \qquad g \in \mathbf{P}[\mathbf{x}]_{\mathbf{u}}, \qquad (18.5)$$

i.e., we now view vol $(\mathbf{G}_1)$ as a function of the vector $\mathbf{g} = (g_{\boldsymbol{\alpha}}) \in \mathbb{R}^{\rho(n,\mathbf{u})}$ of coefficients of g in the basis of $\mathbf{u}$-QH monomials.

Theorem 18.8 *The Lebesgue volume function* $f : \mathbf{P}[\mathbf{x}]_{\mathbf{u}} \to \mathbb{R}$ *defined in (18.5) is strictly convex. On* $\mathrm{int}(\mathbf{P}[\mathbf{x}]_{\mathbf{u}})$ *its gradient* ∇f *and Hessian* $\nabla^2 f$ *given by:*

$$\frac{\partial f(g)}{\partial g_{\boldsymbol{\alpha}}} = \frac{-1}{\Gamma(|\mathbf{u}| + 1)} \int_{\mathbb{R}^n} \mathbf{x}^{\boldsymbol{\alpha}} \exp(-g(\mathbf{x}))\,d\mathbf{x}, \qquad (18.6)$$

for all $\boldsymbol{\alpha} \in \mathbb{N}^n$, *with* $\boldsymbol{\alpha}^T \mathbf{u} = 1$.

$$\frac{\partial^2 f(g)}{\partial g_{\boldsymbol{\alpha}} \partial g_{\boldsymbol{\beta}}} = \frac{1}{\Gamma(|\mathbf{u}| + 1)} \int_{\mathbb{R}^n} \mathbf{x}^{\boldsymbol{\alpha}+\boldsymbol{\beta}} \exp(-g(\mathbf{x}))\,d\mathbf{x}, \qquad (18.7)$$

for all $\boldsymbol{\alpha}, \boldsymbol{\beta} \in \mathbb{N}^n$, *with* $\boldsymbol{\alpha}^T \mathbf{u} = \boldsymbol{\beta}^T \mathbf{u} = 1$. *Moreover, we also have*

$$\int_{\mathbb{R}^n} g(\mathbf{x}) \exp(-g(\mathbf{x}))\,d\mathbf{x} = |\mathbf{u}| \int_{\mathbb{R}^n} \exp(-g(\mathbf{x}))\,d\mathbf{x}. \qquad (18.8)$$

Proof We briefly provide a sketch of the proof. By Lemma 18.5, $f(g) = \Gamma(|\mathbf{u}|+1)^{-1} \int_{\mathbb{R}^n} \exp(-g(\mathbf{x}))\,d\mathbf{x}$. Let $p, q \in \mathbf{P}[\mathbf{x}]_{\mathbf{u}}$ and $\lambda \in [0, 1]$. By convexity of the function $u \mapsto \exp(-u)$,

$$f(\lambda p + (1-\lambda)\,q) \leq \Gamma(|\mathbf{u}|+1)^{-1} \int_{\mathbb{R}^n} [\lambda \exp(-p(\mathbf{x})) + (1-\lambda) \exp(-q(\mathbf{x}))]d\mathbf{x}$$

$$= \lambda f(p) + (1 - \lambda)f(q),$$

and so f is convex. Next, in view of the strict convexity of $t \mapsto \exp(-t)$, equality may occur only if $g(\mathbf{x}) = q(\mathbf{x})$ almost everywhere, which implies $p = q$ and which in turn implies strict convexity of f.

To obtain (18.6) – (18.7) when $g \in \mathrm{int}(\mathbf{P}[\mathbf{x}]_{\mathbf{u}})$, one takes partial derivatives under the integral sign, which in this context is allowed. To obtain (18.8) observe that $g \mapsto h(g) := \int \exp(-g)d\mathbf{x}$, $g \in \mathbf{P}[\mathbf{x}]_{\mathbf{u}}$, is a positively

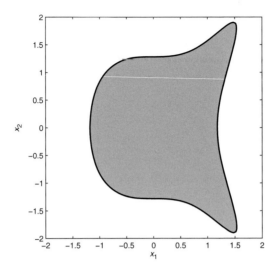

Figure 18.1 Example 18.7: $\mathbf{G}_1 := \{\mathbf{x} \in \mathbb{R}^2 : (x_1^6 + x_2^4)/2.7 - 0.66x_1^3 x_2^2 \leq 1\}$.

homogeneous function of degree $-|\mathbf{u}|$, continuously differentiable. And so combining (18.6) with Euler's identity $-|\mathbf{u}| \, h(g) = \langle \nabla h(g), g \rangle$, yields:

$$-|\mathbf{u}| \int_{\mathbb{R}^n} \exp(-g(\mathbf{x})) \, d\mathbf{x} = \langle \nabla h(g), g \rangle = - \int_{\mathbb{R}^n} g(\mathbf{x}) \, \exp(-g(\mathbf{x})) \, d\mathbf{x}.$$

□

> In other words, in $\mathbf{P}[\mathbf{x}]_{\mathbf{u}}$, the Lebesgue volume of the sublevel set $\{\mathbf{x} : g(\mathbf{x}) \leq 1\}$ is a strictly convex function of the coefficients of the quasi-homogeneous polynomial g.

Notice that convexity of f is not obvious at all from its definition (18.5), whereas it becomes almost transparent when using formula (18.4). The set $\mathbf{G}_1$ of Example 18.7 is displayed in Figure 18.1 to illustrate that $\mathbf{G}_1$ is not necessarily convex.

18.3 A generalization of the Löwner–John problem

Recall the nonconvex generalization $\mathbf{P}$ of the Löwner–John ellipsoid problem.

P: *Let $\mathbf{K} \subset \mathbb{R}^n$ be a (not necessarily convex) compact set with nonempty interior and d an even integer. Find an homogeneous polynomial g of degree d*

such that its sublevel set $\mathbf{G}_1 := \{\mathbf{x} : g(\mathbf{x}) \leq 1\}$ *contains* $\mathbf{K}$ *and has minimum volume among all such sublevel sets with this inclusion property.*

Homogeneous polynomials of degree d are $\mathbf{u}$-QH polynomials with $\mathbf{u} = \mathbf{e}/d$ where $\mathbf{e} \in \mathbb{R}^n$ is the vector of ones. So let $\mathbf{P}[\mathbf{x}]_d := \mathbf{P}[\mathbf{x}]_{\mathbf{u}}$ whenever $\mathbf{u} = \mathbf{e}/d$, and consider the optimization problem:

$$\mathcal{P}: \qquad \rho = \inf_{g \in \mathbf{P}[\mathbf{x}]_d} \left\{ \int_{\mathbb{R}^n} \exp(-g)\, d\mathbf{x} \; : \; 1 - g \in C_d(\mathbf{K}) \right\}, \qquad (18.9)$$

where $C_d(\mathbf{K}) \subset \mathbb{R}[\mathbf{x}]_d$ is the convex cone of polynomials of degree at most d that are nonnegative on $\mathbf{K}$.

Proposition 18.9 *Problem* **P** *has an optimal solution* $g^* \in \mathbf{P}[\mathbf{x}]_d$ *if and only if* g^* *is an optimal solution of problem* $\mathcal{P}$ *in (18.9). Moreover,* $\mathcal{P}$ *is a finite-dimensional convex optimization problem.*

Proof By Lemma 18.5, vol $(\mathbf{G}_1) = \frac{1}{\Gamma(1+n/d)} \int_{\mathbb{R}^n} \exp(-g)\, d\mathbf{x}$ whenever $\mathbf{G}_1$ has finite Lebesgue volume. Moreover, $\mathbf{G}_1$ contains $\mathbf{K}$ if and only if $1 - g \in C_d(\mathbf{K})$, and so **P** has an optimal solution $g^* \in \mathbf{P}[\mathbf{x}]_d$ if and only if g^* is an optimal solution of $\mathcal{P}$ (with value $\rho = \text{vol}(\mathbf{G}_1)\, \Gamma(1 + n/d)$). Now since $g \mapsto \int_{\mathbb{R}^n} \exp(-g) d\mathbf{x}$ is strictly convex (by Lemma 18.8) and both $C_d(\mathbf{K})$ and $\mathbf{P}[\mathbf{x}]_d$ are convex cones, problem $\mathcal{P}$ is a finite-dimensional convex optimization problem.   □

We now can state the main result of this chapter. Recall that $\mathcal{M}(\mathbf{K})_+$ is the convex cone of finite (positive) Borel measures on $\mathbf{K}$.

Theorem 18.10 *Let* $\mathbf{K} \subset \mathbb{R}^n$ *be compact with nonempty interior and consider the convex optimization problem* $\mathcal{P}$ *in (18.9).*

(a) $\mathcal{P}$ *has a unique optimal solution* $g^* \in \mathbf{P}[\mathbf{x}]_d$.
(b) *If* $g^* \in \text{int} (\mathbf{P}[\mathbf{x}]_d)$ *is the unique optimal solution of* $\mathcal{P}$ *then there exists a finite Borel measure* $\mu^* \in \mathcal{M}(\mathbf{K})_+$ *such that*

$$\int_{\mathbb{R}^n} \mathbf{x}^\alpha \exp(-g^*)\, d\mathbf{x} = \int_{\mathbf{K}} \mathbf{x}^\alpha\, d\mu^*, \qquad \forall \alpha \in \mathbb{N}, \; |\alpha| = d \qquad (18.10)$$

$$\int_{\mathbf{K}} (1 - g^*)\, d\mu^* = 0, \quad \mu^*(\mathbf{K}) = \frac{n}{d} \int_{\mathbb{R}^n} \exp(-g^*)\, d\mathbf{x}. \qquad (18.11)$$

In particular, μ^* *is supported on the set* $V := \{\mathbf{x} \in \mathbf{K} : g^*(\mathbf{x}) = 1\}$ *(equal to* $\mathbf{K} \cap \mathbf{G}_1^*$*) and, in fact,* μ^* *can be substituted with another measure* $\nu^* \in \mathcal{M}(\mathbf{K})_+$ *supported on at most* $\binom{n+d-1}{d}$ *contact points of* V.

(c) *Conversely, if $g^* \in$ int $(\mathbf{P}[\mathbf{x}]_d)$ with $1 - g^* \in C_d(\mathbf{K})$, and there exist points $(\mathbf{x}(i), \lambda_i) \in \mathbf{K} \times \mathbb{R}, \lambda_i > 0, i = 1, \dots, s$, such that $g^*(\mathbf{x}(i)) = 1$ for all $i = 1, \dots, s$, and*

$$\int_{\mathbb{R}^n} \mathbf{x}^\alpha \exp(-g^*) \, d\mathbf{x} = \sum_{i=1}^{s} \lambda_i \, \mathbf{x}(i)^\alpha, \qquad |\alpha| = d,$$

then g^* is the unique optimal solution of problem $\mathcal{P}$.

Proof (a) As $\mathcal{P}$ is a minimization problem, its feasible set $\{ g \in \mathbf{P}[\mathbf{x}]_d : 1 - g \in C_d(\mathbf{K}) \}$ can be replaced by the smaller set

$$F := \left\{ g \in \mathbf{P}[\mathbf{x}]_d : \int_{\mathbb{R}^n} \exp(-g(\mathbf{x})) \, d\mathbf{x} \le \int_{\mathbb{R}^n} \exp(-g_0(\mathbf{x})) \, d\mathbf{x}, \ 1 - g \in C_d(\mathbf{K}) \right\},$$

for some $g_0 \in \mathbf{P}[\mathbf{x}]_d$ such that $1 - g_0 \in C_d(\mathbf{K})$. The set F is a closed convex set since the convex function $g \mapsto \int_{\mathbb{R}^n} \exp(-g) d\mathbf{x}$ is continuous on the interior of its domain.

Next, let $\mathbf{z} = (z_\alpha), \alpha \in \mathbb{N}^n_d$, be a (fixed) element of int $(C_d(\mathbf{K})^*)$ (hence $z_0 > 0$). By Lemma 4.6 such an element exists. Then the constraint $1 - g \in C_d(\mathbf{K})$ implies $\langle \mathbf{z}, 1 - g \rangle \ge 0$, i.e., $\langle \mathbf{z}, g \rangle \le z_0$. On the other hand, being an element of $\mathbf{P}[\mathbf{x}]_d$, g is nonnegative and in particular $g \in C_d(\mathbf{K})$. But then by Corollary I.1.6 in Faraut and Korányi (1994, p. 4), the set $\{ g \in C_d(\mathbf{K}) : \langle \mathbf{z}, g \rangle \le z_0 \}$ (or equivalently, $\{ g \in C_d(\mathbf{K}) : \langle \mathbf{z}, g \rangle \le z_0 \}$) is compact (where again $\mathbf{g} \in \mathbb{R}^{s(d)}$ is the coefficient vector of $g \in C_d(\mathbf{K})$). Therefore, the set F is a compact convex set. Finally, since $g \mapsto \int_{\mathbb{R}^n} \exp(-g(\mathbf{x})) \, d\mathbf{x}$ is strictly convex, it is continuous on the interior of its domain and so it is continuous on F. Hence problem $\mathcal{P}$ has a unique optimal solution $g^* \in \mathbf{P}[\mathbf{x}]_d$.

(b) We may and will consider any homogeneous polynomial g as an element of $\mathbb{R}[\mathbf{x}]_d$ whose coefficient vector $\mathbf{g} = (g_\alpha)$ is such that $g^*_\alpha = 0$ whenever $|\alpha| < d$. And so Problem $\mathcal{P}$ is equivalent to the problem

$$\mathcal{P}' : \quad \begin{cases} \rho = \displaystyle\inf_{g \in \mathbb{R}[\mathbf{x}]_d} \int_{\mathbb{R}^n} \exp(-g(\mathbf{x})) \, d\mathbf{x} \\ \text{s.t.} \quad g_\alpha = 0, \quad \forall \alpha \in \mathbb{N}^n_d, |\alpha| < d \\ \quad\quad 1 - g \in C_d(\mathbf{K}), \end{cases} \tag{18.12}$$

where we have imposed that g is homogeneous of degree d, via the constraints $g \in \mathbb{R}[\mathbf{x}]_d$ and $g_\alpha := 0$ for all $\alpha \in \mathbb{N}^n_d$ with $|\alpha| < d$. Next, doing the change of variable $h = 1 - g$, $\mathcal{P}'$ reads:

$$
\mathcal{P}' : \quad
\begin{cases}
\rho = \displaystyle\inf_{h \in \mathbb{R}[\mathbf{x}]_d} \int_{\mathbb{R}^n} \exp(h(\mathbf{x}) - 1) \, d\mathbf{x} \\
\quad\text{s.t.}\quad h_{\boldsymbol{\alpha}} = 0, \quad \forall \boldsymbol{\alpha} \in \mathbb{N}_d^n, \ 0 < |\boldsymbol{\alpha}| < d \\
\qquad\qquad h_0 = 1 \\
\qquad\qquad h \in C_d(\mathbf{K}).
\end{cases}
\tag{18.13}
$$

As $\mathbf{K}$ is compact, there exists $\theta \in \mathbf{P}[\mathbf{x}]_d$ such that $1 - \theta \in \mathrm{int}\,(C_d(\mathbf{K}))$, i.e., Slater's condition holds for the convex optimization problem $\mathcal{P}'$. Indeed, choose $\mathbf{x} \mapsto \theta(\mathbf{x}) := M^{-1}\|\mathbf{x}\|^d$ for $M > 0$ sufficiently large so that $1 - \theta > 0$ on $\mathbf{K}$. Hence with $\|g\|_1$ denoting the ℓ_1-norm $|\mathbf{g}|_1$ of the coefficient vector of g (in $\mathbb{R}[\mathbf{x}]_d$), there exists $\epsilon > 0$ such that for every $h \in B(\theta, \epsilon)(:= \{ h \in \mathbb{R}[\mathbf{x}]_d : \|\theta - h\|_1 < \epsilon \})$, the polynomial $1 - h$ is (strictly) positive on $\mathbf{K}$.

Therefore, the unique optimal solution $(1 - g^*) =: h^* \in \mathbb{R}[\mathbf{x}]_d$ of $\mathcal{P}'$ in (18.13) satisfies the KKT optimality conditions which for problem (18.13) read:

$$
\int_{\mathbb{R}^n} \mathbf{x}^{\boldsymbol{\alpha}} \exp(h^*(\mathbf{x}) - 1) \, d\mathbf{x} = y_{\boldsymbol{\alpha}}^*, \qquad \forall |\boldsymbol{\alpha}| = d
\tag{18.14}
$$

$$
\int_{\mathbb{R}^n} \mathbf{x}^{\boldsymbol{\alpha}} \exp(h^*(\mathbf{x}) - 1) \, d\mathbf{x} + \gamma_{\boldsymbol{\alpha}} = y_{\boldsymbol{\alpha}}^*, \qquad \forall \, |\boldsymbol{\alpha}| < d
\tag{18.15}
$$

$$
\langle h^*, \mathbf{y}^* \rangle = 0, \quad h_0^* = 1; \ h_{\boldsymbol{\alpha}}^* = 0, \quad \forall 0 < |\boldsymbol{\alpha}| < d
\tag{18.16}
$$

for some $\mathbf{y}^* = (y_{\boldsymbol{\alpha}}^*)$, $\boldsymbol{\alpha} \in \mathbb{N}_d^n$, in the dual cone $C_d(\mathbf{K})^* \subset \mathbb{R}^{s(d)}$ of $C_d(\mathbf{K})$, and some vector $\boldsymbol{\gamma} = (\gamma_{\boldsymbol{\alpha}})$, $|\boldsymbol{\alpha}| < d$. By Lemma 4.7,

$$
C_d(\mathbf{K})^* = \left\{ \mathbf{y} \in \mathbb{R}^{s(d)} : \ \exists \mu \in \mathcal{M}(\mathbf{K})_+ \ \text{such that} \ y_{\boldsymbol{\alpha}} = \int_{\mathbf{K}} \mathbf{x}^{\boldsymbol{\alpha}} \, d\mu, \ \forall \boldsymbol{\alpha} \in \mathbb{N}_d^n \right\},
$$

and so (18.10) is just (18.14) restated in terms of μ^*.

Next, the condition $\langle h^*, \mathbf{y}^* \rangle = 0$ (or equivalently, $\langle 1 - g^*, \mathbf{y}^* \rangle = 0$), reads:

$$
\int_{\mathbf{K}} (1 - g^*) \, d\mu^* = 0,
$$

which combined with $1 - g^* \in C_d(\mathbf{K})$ and $\mu^* \in \mathcal{M}(\mathbf{K})_+$, implies that μ^* is supported on $\mathbf{K} \cap \{ \mathbf{x} : g^*(\mathbf{x}) = 1 \} = \mathbf{K} \cap \mathbf{G}_1^*$.

Next, let $s := \sum_{|\boldsymbol{\alpha}|=d} g_{\boldsymbol{\alpha}}^* y_{\boldsymbol{\alpha}}^* \ (=y_0^*)$. From $\langle 1 - g^*, \mu^* \rangle = 0$, the measure $s^{-1}\mu^* =: \psi$ is a probability measure supported on $\mathbf{K} \cap \mathbf{G}_1^*$, and satisfies $\int \mathbf{x}^{\boldsymbol{\alpha}} d\psi = s^{-1} y_{\boldsymbol{\alpha}}^*$ for all $|\boldsymbol{\alpha}| = d$.

Hence by Theorem 2.50 there exists an atomic probability measure $\nu^* \in \mathcal{M}(\mathbf{K})_+$ supported on $\mathbf{K} \cap \mathbf{G}_1^*$ such that

$$
\int_{\mathbf{K} \cap \mathbf{G}_1^*} \mathbf{x}^{\boldsymbol{\alpha}} d\nu^*(\mathbf{x}) = \int_{\mathbf{K} \cap \mathbf{G}_1^*} \mathbf{x}^{\boldsymbol{\alpha}} d\psi(\mathbf{x}) = s^{-1} y_{\boldsymbol{\alpha}}^*, \qquad \forall \, |\boldsymbol{\alpha}| = d.
\tag{18.17}
$$

By Theorem 2.50 the probability measure ν^* may be chosen to be supported on at most $1 + \binom{n+d-1}{d}$ points in $\mathbf{K} \cap \mathbf{G}_1^*$. But in fact, as $\langle 1 - g^*, \mathbf{y}^* \rangle = 0$, one of the moment conditions (18.17) is redundant and can be ignored; therefore only $\binom{n+d-1}{d}$ points in $\mathbf{K} \cap \mathbf{G}_1^*$ suffice. Hence in (18.10) the measure μ^* can be substituted with the atomic measure $s\,\nu^*$ supported on at most $\binom{n+d-1}{d}$ contact points in $\mathbf{K} \cap \mathbf{G}_1^*$.

To obtain $\mu^*(\mathbf{K}) = \frac{n}{d} \int_{\mathbb{R}^n} \exp(-g^*)$, multiply both sides of (18.14)–(18.15) by h_α^* for every $\alpha \neq 0$, sum up and use $\langle h^*, \mathbf{y}^* \rangle = 0$ to obtain

$$-y_0^* = \sum_{\alpha \neq 0} h_\alpha^* \, y_\alpha^* = \int_{\mathbb{R}^n} (h^*(\mathbf{x}) - 1) \exp(h^*(\mathbf{x}) - 1)\, d\mathbf{x}$$

$$= -\int_{\mathbb{R}^n} g^*(\mathbf{x}) \exp(-g^*(\mathbf{x}))\, d\mathbf{x} = -\frac{n}{d} \int \exp(-g^*(\mathbf{x}))\, d\mathbf{x},$$

where we have also used (18.8).

(c) Let $\mu^* := \sum_{i=1}^{s} \lambda_i \delta_{\mathbf{x}(i)}$ where $\delta_{\mathbf{x}(i)}$ is the Dirac measure at the point $\mathbf{x}(i) \in \mathbf{K}$, $i = 1, \ldots, s$. Next, let $y_\alpha^* := \int \mathbf{x}^\alpha d\mu^*$ for all $\alpha \in \mathbb{N}_d^n$, so that $\mathbf{y}^* \in C_d(\mathbf{K})^*$. In particular $\mathbf{y}^*$ and g^* satisfy

$$\langle 1 - g^*, \mathbf{y}^* \rangle = \int_{\mathbf{K}} (1 - g^*)\, d\mu^* = 0,$$

because $g^*(\mathbf{x}(i)) = 1$ for all $i = 1, \ldots, s$. Let $h^* := 1 - g^*$ and let

$$\gamma_\alpha := y_\alpha^* - \int_{\mathbb{R}^n} \mathbf{x}^\alpha \exp(h^*(\mathbf{x}) - 1)\, d\mathbf{x}, \quad \forall |\alpha| < d.$$

The triplet $(h^*, \gamma, \mathbf{y}^*)$ satisfies (18.14)–(18.16), i.e., the pair $(g^*, \mathbf{y}^*)$ satisfies the KKT optimality conditions associated with the convex problem $\mathcal{P}$. But since Slater's condition holds for $\mathcal{P}$, these conditions are also sufficient for g^* to be an optimal solution of $\mathcal{P}$, the desired result $\qquad\square$

Importantly, notice that neither $\mathbf{K}$ nor $\mathbf{G}_1^*$ is required to be convex. For instance Figure 18.2 displays two compact and nonconvex sublevel sets associated with homogeneous bivariate polynomials of degree 4 and 6.

Example

With $n = 2$ let $\mathbf{K} \subset \mathbb{R}^2$ be the box $[-1, 1]^2$ and let $d = 4, 6$, that is, one searches for the unique homogeneous polynomial $g \in \mathbb{R}[\mathbf{x}]_4$ or $g \in \mathbb{R}[\mathbf{x}]_6$ which contains $\mathbf{K}$ and has minimum volume among such sets.

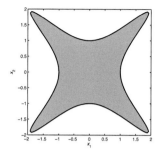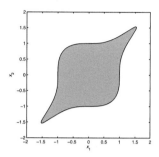

Figure 18.2 $\mathbf{G}_1$ with $x^4 + y^4 - 1.925\,x^2y^2$ and $x^6 + y^6 - 1.925\,x^3y^3$.

Theorem 18.11   *The sublevel set* $\mathbf{G}_1^4 = \{\,\mathbf{x} : g_4(\mathbf{x}) \leq 1\,\}$ *associated with the homogeneous polynomial*

$$\mathbf{x} \mapsto g_4(\mathbf{x}) = x_1^4 + x_2^4 - x_1^2 x_2^2, \qquad (18.18)$$

is the unique solution of problem $\mathbf{P}$ *with* $d = 4$. *That is,* $\mathbf{K} \subset \mathbf{G}_1^4$ *and* $\mathbf{G}_1^4$ *has minimum volume among all sets* $\mathbf{G}_1 \supset \mathbf{K}$ *defined with homogeneous polynomials of degree* 4.

Similarly, the sublevel set $\mathbf{G}_1^6 = \{\,\mathbf{x} : g_6(\mathbf{x}) \leq 1\,\}$ *associated with the homogeneous polynomial*

$$\mathbf{x} \mapsto g_6(\mathbf{x}) = x_1^6 + x_2^6 - (x_1^4 x_2^2 + x_1^2 x_2^4)/2 \qquad (18.19)$$

is the unique solution of problem $\mathbf{P}$ *with* $d = 6$.

Proof   Let g_4 be as in (18.18). We first prove that $\mathbf{K} \subset \mathbf{G}_1^4$, i.e., $1 - g_4(\mathbf{x}) \geq 0$ whenever $\mathbf{x} \in \mathbf{K}$. But observe that if $\mathbf{x} \in \mathbf{K}$ then

$$
\begin{aligned}
1 - g_4(\mathbf{x}) &= 1 - x_1^4 - x_2^4 + x_1^2 x_2^2 = 1 - x_1^4 + x_2^2(x_1^2 - x_2^2) \\
&\geq 1 - x_1^4 + x_2^2(x_1^2 - 1) \quad [\text{as } -x_2^2 \geq -1 \text{ and } x_2^2 \geq 0] \\
&\geq (1 - x_1^2)(1 + x_1^2 - x_2^2) \\
&\geq (1 - x_1^2)\,x_1^2 \geq 0 \quad [\text{as } -x_2^2 \geq -1 \text{ and } 1 - x_1^2 \geq 0].
\end{aligned}
$$

Hence $1 - g_4 \in C_d(\mathbf{K})$. Observe that $\mathbf{K} \cap \mathbf{G}_1^4$ consists of the eight contact points $(\pm 1, \pm 1)$ and $(0, \pm 1)$, $(\pm 1, 0)$. Next let ν^* be the measure defined by

$$\nu^* = a\,(\delta_{(-1,1)} + \delta_{(1,1)}) + b\,(\delta_{(1,0)} + \delta_{(0,1)}) \qquad (18.20)$$

where $\delta_{\mathbf{x}}$ denote the Dirac measure at $\mathbf{x}$ and $a, b \geq 0$ are chosen to satisfy

$$2a + b = \int_{\mathbb{R}^n} x_1^4 \exp(-g_4)\, d\mathbf{x}, \quad 2a = \int_{\mathbb{R}^n} x_1^2 x_2^2 \exp(-g_4)\, d\mathbf{x},$$

so that

$$\int \mathbf{x}^\alpha \, dv^* = \int_{\mathbb{R}^n} \mathbf{x}^\alpha \exp(-g_4(\mathbf{x}))\, d\mathbf{x}, \quad |\alpha| = 4.$$

Of course a unique solution $(a, b) \geq 0$ exists since

$$\int_{\mathbb{R}^n} x_1^4 \exp(-g_4)\, d\mathbf{x} \int_{\mathbb{R}^n} x_2^4 \exp(-g_4)\, d\mathbf{x} = \left(\int_{\mathbb{R}^n} x_1^4 \exp(-g_4)\, d\mathbf{x} \right)^2$$

$$\geq \left(\int_{\mathbb{R}^n} x_1^2 x_2^2 \exp(-g_4)\, d\mathbf{x} \right)^2.$$

Therefore the atomic measure v^* is indeed as in Theorem 18.10(c) and the proof is completed. Notice that as predicted by Theorem 18.10(b), v^* is supported on $4 \leq \binom{n+d-1}{d} = 5$ points. Similarly with g_6 as in (18.19) and $\mathbf{x} \in \mathbf{K}$,

$$1 - g_6(\mathbf{x}) = 1 - x_1^6 - x_2^6 + (x_1^4 x_2^2 + x_1^2 x_2^4)/2$$
$$= (1 - x_1^6)/2 + (1 - x_2^6)/2 - x_1^4(x_1^2 - x_2^2)/2 - x_2^4(x_2^2 - x_1^2)/2$$
$$\geq (1 - x_1^6)/2 + (1 - x_2^6)/2 - x_1^4(1 - x_2^2)/2 - x_2^4(1 - x_1^2)/2$$
$$[\text{as } -x_1^6 \geq -x_1^4 \text{ and } -x_2^6 \geq -x_2^4]$$
$$\geq (1 - x_1^2)(1 + x_1^2 + x_1^4 - x_2^4)/2 + (1 - x_2^2)(1 + x_2^2 + x_2^4 - x_1^4)/2$$
$$\geq (1 - x_1^2)(x_1^2 + x_1^4)/2 + (1 - x_2^2)(x_2^2 + x_2^4)/2 \geq 0$$
$$[\text{as } 1 - x_1^4 \geq 0 \text{ and } 1 - x_2^4 \geq 0].$$

So again the measure v^* defined in (18.20) where $a, b \geq 0$ are chosen to satisfy

$$2a + b = \int_{\mathbb{R}^n} x_1^6 \exp(-g_6)\, d\mathbf{x}, \quad 2a = \int_{\mathbb{R}^n} x_1^4 x_2^2 \exp(-g_6)\, d\mathbf{x},$$

is such that

$$\int \mathbf{x}^\alpha \, d\mu^* = \int_{\mathbb{R}^n} \mathbf{x}^\alpha \exp(-g_6(\mathbf{x}))\, d\mathbf{x}, \quad |\alpha| = 6.$$

Again a unique solution $(a, b) \geq 0$ exists because

$$\left(\int x_1^6 \exp(-g_6)\, d\mathbf{x} \right) \left(\int x_1^2 x_2^4 \exp(-g_6)\, d\mathbf{x} \right) \geq \left(\int x_1^4 x_2^2 \exp(-g_6)\, d\mathbf{x} \right)^2.$$

$\square$

With $d = 4$, the nonconvex sublevel set $\mathbf{G}_1^4 = \{ \mathbf{x} : g_4(\mathbf{x}) \leq 1 \}$ which is displayed in Figure 18.3 (left) is a much better approximation of $\mathbf{K} = [-1, 1]^2$ than the ellipsoid of minimum volume $\xi = \{ \mathbf{x} : \|\mathbf{x}\|^2 \leq 2 \}$ that contains $\mathbf{K}$.

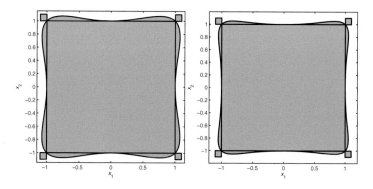

Figure 18.3 $\mathbf{K} = [-1, 1]^2$ and $\mathbf{G}_1^4 = \{\mathbf{x} : x_1^4 + x_2^4 - x_1^2 x_2^2 \leq 1\}$ (left), $\mathbf{G}_1^6 = \{\mathbf{x} : x_1^6 + x_2^6 - (x_1^4 x_2^2 + x_1^2 x_2^4)/2 \leq 1\}$ (right).

In particular, $\mathrm{vol}(\xi) = 2\pi \approx 6.28$ whereas $\mathrm{vol}(\mathbf{G}_1^4) \approx 4.298$. With $d = 6$, the nonconvex sublevel set $\mathbf{G}_1^6 = \{\mathbf{x} : g_6(\mathbf{x}) \leq 1\}$ which is displayed in Figure 18.3 (right) is again a better approximation of $\mathbf{K} = [-1, 1]^2$ than the ellipsoid of minimum volume $\xi = \{\mathbf{x} : \|\mathbf{x}\|^2 \leq 2\}$ that contains $\mathbf{K}$, and as $\mathrm{vol}(\mathbf{G}_1^6) \approx 4.1979$, it provides a better approximation than the sublevel set $\mathbf{G}_1^4$ with $d = 4$.

Finally if $\mathbf{K} = \mathbf{G}_1^6$ then $\mathbf{G}_1^4$ is an optimal solution of $\mathbf{P}$ with $d = 4$, that is $\mathbf{G}_1^4$ has minimum volume among all sets $\mathbf{G}_1 \supset \mathbf{G}_1^6$ defined by homogeneous polynomials $g \in \mathbb{P}[\mathbf{x}]_4$. Indeed first we have solved the polynomial optimization problem: $\rho = \inf_{\mathbf{x}}\{1 - g_4(\mathbf{x}) : 1 - g_6(\mathbf{x}) \geq 0\}$ via the hierarchy of semidefinite relaxations (6.3) of Chapter 6. At the fifth semidefinite relaxation we found $\rho = 0$ with the eight contact points $(\pm 1, \pm 1), (\pm 1, 0), (0, \pm 1) \in \mathbf{G}_1^4 \cap \mathbf{G}_1^6$ as global minimizers. This shows (up to 10^{-9} numerical precision) that $\mathbf{G}_1^6 \subset \mathbf{G}_1^4$. Then again the measure ν^* defined in (18.20) satisfies Theorem 18.10(b) and so g_4 is an optimal solution of problem $\mathbf{P}$ with $\mathbf{K} = \mathbf{G}_1^6$ and $d = 4$.

Finally, the ball $\mathbf{G}_1^2 = \{\mathbf{x} : (x_1^2 + x_2^2)/2 \leq 1\}$ is an optimal solution of $\mathbf{P}$ with $d = 2$ and we have $\mathbf{K} \subset \mathbf{G}_1^6 \subset \mathbf{G}_1^4 \subset \mathbf{G}_1^2$.

18.4 A numerical scheme

Even though $\mathcal{P}$ in (18.9) is a finite-dimensional convex optimization problem, it is hard to solve for the following two reasons.

- The convex cone $C_d(\mathbf{K})$ has no *exact* and *tractable* representation to handle efficiently the constraint $1 - g \in C_d(\mathbf{K})$ in an algorithm for solving problem (18.9).

- From Theorem 18.8, the gradient and Hessian of the (strictly) convex objective function $g \mapsto \int \exp(-g)$ requires evaluating integrals of the form

$$\int_{\mathbb{R}^n} \mathbf{x}^\alpha \exp(-g(\mathbf{x}))\, d\mathbf{x}, \quad \alpha \in \mathbb{N}^n_d,$$

or equivalently, by Lemma 18.5, integrals of the form

$$\int_{\mathbf{G}_1} \mathbf{x}^\alpha\, d\mathbf{x}, \quad \alpha \in \mathbb{N}^n_d,$$

a difficult and challenging problem. (And with $\alpha = 0$ one obtains the value of the objective function.)

However, below we outline a numerical scheme to approximate to any desired ϵ-accuracy (with $\epsilon > 0$):

- the optimal value ρ of (18.9),
- the unique optimal solution $g^* \in \mathbf{P}[\mathbf{x}]_d$ of $\mathcal{P}$ obtained in Theorem 18.10.

Brief sketch of a numerical scheme

We here assume that the compact (and not necessarily convex) set $\mathbf{K} \subset \mathbb{R}^n$ is a basic semi-algebraic set with nonempty interior, and defined by:

$$\mathbf{K} = \{ \mathbf{x} \in \mathbb{R}^n : h_j(\mathbf{x}) \geq 0, \ j = 1, \dots, m \}, \tag{18.21}$$

for some given polynomials $(h_j) \subset \mathbb{R}[\mathbf{x}]$, with h_0 being the constant polynomial equal to 1, and $v_j := \lceil \deg(h_j)/2 \rceil$, $j = 0, \dots, m$.

To handle the hard cone constraint $1 - g \in C_d(\mathbf{K})$ we can use the results of Chapter 4 where we have seen that $C_d(\mathbf{K})$ can be approximated from inside, and as closely as desired, by a sequence of inner approximations $Q_k(h) \subset C_d(\mathbf{K})$, $k \in \mathbb{N}$; see Theorem 4.1. Hence with k fixed, arbitrary, replace the condition $1 - g \in C_d(\mathbf{K})$ with the stronger condition $1 - g \in Q_k(h)$, i.e.,

$$1 - g = \sum_{j=0}^{m} \sigma_j\, h_j,$$

for some $\sigma_j \subset \Sigma[\mathbf{x}]_{k - v_j}$, $j = 0, \dots, m$, which in turn translates into LMI conditions. Then consider the following hierarchy of convex optimization problems $(\mathcal{P}_k)$, $k \in \mathbb{N}$, where for each fixed k:

$$\rho_k = \min_{g \in \mathbf{P}[\mathbf{x}]_d, \sigma_j} \int_{\mathbb{R}^n} \exp(-g) \, d\mathbf{x}$$

$$\text{s.t.} \qquad 1 - g = \sum_{j=0}^{m} \sigma_j \, h_j \qquad\qquad (18.22)$$

$$g_\alpha = 0, \quad \forall |\alpha| < d$$

$$g \in \mathbb{R}[\mathbf{x}]_d; \quad \sigma_j \in \Sigma_{k - v_j}, \quad j = 0, \ldots, m.$$

Of course the sequence (ρ_k), $k \in \mathbb{N}$, is monotone nonincreasing and $\rho_k \geq \rho$ for all k. Moreover, for each fixed $k \in \mathbb{N}$, $\mathcal{P}_k$ is a convex optimization problem which consists of minimizing a strictly convex function under LMI constraints.

By Corollary 18.6, $\int_{\mathbb{R}^n} \exp(-g) \, d\mathbf{x} < \infty$ if and only if $g \in \mathbf{P}[\mathbf{x}]_d$. That is, the objective function also acts as a barrier for the convex cone $\mathbf{P}[\mathbf{x}]_d$. Therefore, to solve $\mathcal{P}_k$ one may use first-order or second-order (local minimization) algorithms, starting from some initial guess $g_0 \in \text{int}\,(\mathbf{P}[\mathbf{x}]_d)$. For instance choose $\mathbf{x} \mapsto g_0(\mathbf{x}) := M^{-1} \sum_{i=1}^{n} x_i^d$ for some sufficiently large M.

Theorem 18.12 *Let $\mathbf{K}$ in (18.21) be compact with nonempty interior and let Assumption 2.14 hold. Then there exists k_0 such that for every $k \geq k_0$, problem $\mathcal{P}_k$ in (18.22) has a unique optimal solution $g_k^* \in \mathbf{P}[\mathbf{x}]_d$.*

Proof   Firstly, $\mathcal{P}_k$ has a feasible solution for sufficiently large k. Indeed consider the polynomial $\mathbf{x} \mapsto g_0(\mathbf{x}) = \sum_{i=1}^{n} x_i^d$ which belongs to $\mathbf{P}[\mathbf{x}]_d$. Then as $\mathbf{K}$ is compact, $M - g_0 > 0$ on $\mathbf{K}$ for some M and so by Theorem 2.15, $1 - g_0/M \in Q_k(h)$ for some k_0 (and hence for all $k \geq k_0$). Hence g_0/M is a feasible solution for $\mathcal{P}_k$ for all $k \geq k_0$. Of course, as $Q_k(h) \subset C_d(\mathbf{K})$, every feasible solution $g \in \mathbf{P}[\mathbf{x}]_d$ satisfies $0 \leq g \leq 1$ on $\mathbf{K}$. So proceeding as in the proof of Theorem 18.10 and using the fact that $Q_k(h)$ is closed, the set

$$\left\{ g \in \mathbf{P}[\mathbf{x}]_d \cap Q_k(h) : \int_{\mathbb{R}^n} \exp(-g) \, d\mathbf{x} \leq \int_{\mathbb{R}^n} \exp\left(-\frac{g_0}{M}\right) d\mathbf{x} \right\},$$

is compact. And as the objective function is strictly convex, the optimal solution $g_k^* \in \mathbf{P}[\mathbf{x}]_d \cap Q_k(h)$ is unique (but the representation of $1 - g_k^*$ in (18.22) is not unique in general).   □

We now consider the asymptotic behavior of the solution of (18.22) as $k \to \infty$.

Theorem 18.13 *Let $\mathbf{K}$ in (18.21) be compact with nonempty interior and let Assumption 2.14 hold. If ρ (respectively ρ_k) is the optimal value of $\mathcal{P}$ (respectively $\mathcal{P}_k$) then $\rho = \lim_{k \to \infty} \rho_k$. Moreover, for every $k \geq k_0$, let $g_k^* \in \mathbf{P}[\mathbf{x}]_d$*

be the unique optimal solution of $\mathcal{P}_k$. Then as $k \to \infty$, $g_k^* \to g^*$ where g^* is the unique optimal solution of $\mathcal{P}$.

Proof By Theorem 18.10, $\mathcal{P}$ has a unique optimal solution $g^* \in \mathbf{P}[\mathbf{x}]_d$. Let $\epsilon > 0$ be fixed, arbitrary. As $1 - g^* \in C_d(\mathbf{K})$, the polynomial $1 - g^*/(1+\epsilon)$ is strictly positive on $\mathbf{K}$, and so by Theorem 2.15, $1 - g^*/(1+\epsilon)$ belongs to $Q_k(h)$ for all $k \geq k_\epsilon$ for some integer k_ϵ. Hence the polynomial $g^*/(1 + \epsilon) \in \mathbf{P}[\mathbf{x}]_d$ is a feasible solution of $\mathcal{P}_k$ for all $k \geq k_\epsilon$. Moreover, by homogeneity,

$$\int_{\mathbb{R}^n} \exp\left(-\frac{g^*}{1+\epsilon}\right) d\mathbf{x} = (1+\epsilon)^{n/d} \int_{\mathbb{R}^n} \exp(-g^*) \, d\mathbf{x}$$
$$= (1+\epsilon)^{n/d} \rho.$$

This shows that $\rho_k \leq (1+\epsilon)^{n/d} \rho$ for all $k \geq k_\epsilon$. Combining this with $\rho_k \geq \rho$, and the fact that $\epsilon > 0$ was arbitrary, yields the convergence $\rho_k \to \rho$ as $k \to \infty$.

Next, let $\mathbf{y} \in \text{int}\,(C_d(\mathbf{K})^*)$ be as in the proof of Theorem 18.10. From $1 - g_k^* \in Q_k(h)$ we also obtain $\langle \mathbf{y}, 1 - g_k^* \rangle \geq 0$, i.e.,

$$y_0 \geq \langle \mathbf{y}, g_k^* \rangle, \quad \forall k \geq k_0.$$

Recall that the set $\{ g \in C_d(\mathbf{K}) : \langle \mathbf{y}, g \rangle \leq y_0 \}$ is compact. Therefore there exists a subsequence (k_ℓ), $\ell \in \mathbb{N}$, and $\tilde{g} \in C_d(\mathbf{K})$ such that $g_{k_\ell}^* \to \tilde{g}$ as $\ell \to \infty$. In particular, $1 - \tilde{g} \in C_d(\mathbf{K})$ and $\tilde{g}_\alpha = 0$ whenever $|\alpha| < d$ (i.e., $\tilde{g}$ is homogeneous of degree d). Moreover, one also has the pointwise convergence $\lim_{\ell \to \infty} g_{k_\ell}^*(\mathbf{x}) = \tilde{g}(\mathbf{x})$ for all $\mathbf{x} \in \mathbb{R}^n$. Hence by Fatou's lemma,

$$\rho = \lim_{\ell \to \infty} \rho_{k_\ell} = \lim_{\ell \to \infty} \int_{\mathbb{R}^n} \exp(-g_{k_\ell}^*(\mathbf{x})) \, d\mathbf{x}$$
$$\geq \int_{\mathbb{R}^n} \liminf_{\ell \to \infty} \exp(-g_{k_\ell}^*(\mathbf{x})) \, d\mathbf{x}$$
$$= \int_{\mathbb{R}^n} \exp(-\tilde{g}(\mathbf{x})) \, d\mathbf{x} \geq \rho,$$

which proves that $\tilde{g}$ is an optimal solution of $\mathcal{P}$, and by uniqueness of the optimal solution, $\tilde{g} = g^*$. As (g_{k_ℓ}), $\ell \in \mathbb{N}$, was an arbitrary converging subsequence, the whole sequence (g_k^*) converges to g^*.   □

Gradient and Hessian evaluation

Solving $\mathcal{P}_k$ with most standard minimization algorithms of convex optimization would require evaluating the gradient (and sometimes the Hessian as well) of the objective function which in this case requires evaluating finitely many moments of the finite Borel measure $d\mu = \exp(-g(\mathbf{x}))d\mathbf{x}$ on $\mathbb{R}^n$, a difficult task in general. However, by Lemma 18.5

$$z_\alpha := \int_{\mathbb{R}^n} \mathbf{x}^\alpha \exp(-g) \, d\mathbf{x} = \Gamma\left(1 + \frac{n + |\alpha|}{d}\right) \int_{\mathbf{G}_1} \mathbf{x}^\alpha \, d\mathbf{x}, \qquad (18.23)$$

that is, (18.23) relates in a very simple and explicit manner the sequence $\mathbf{z} = (z_\alpha)$, $\alpha \in \mathbb{N}^n$, of all moments of the Borel measure with density $\exp(-g)$ on $\mathbb{R}^n$, with the sequence $\lambda = (\lambda_\alpha)$, $\alpha \in \mathbb{N}^n$, of moments of the Lebesgue measure on the sublevel set $\mathbf{G}_1$.

It turns out that any finite subsequence (λ_α), $\alpha \in \mathbb{N}^n_\ell$, of λ can be approximated as closely as desired, by solving a hierarchy of semidefinite programs. See for instance Henrion et al. (2009a). Of course as the gradient and Hessian are only approximated, some care is needed to ensure convergence of a (first-order or second-order) minimization algorithm to solve $\mathcal{P}_k$. For instance, one might try to adapt ideas like those described in d'Aspremont (2008) where for certain optimization problems with noisy gradient information, first-order algorithms with convergence guarantees have been investigated in detail.

18.5 Exercises

Exercise 18.1 Show that the sublevel set $\mathbf{G}_1 = \{\mathbf{x} : g(\mathbf{x}) \leq 1\}$ associated with the homogeneous polynomial $\mathbf{x} \mapsto g(\mathbf{x}) := x_1^4 x_2^2 + x_1^2 x_2^4$ is not bounded but has finite Lebesgue volume.

Exercise 18.2 (a) Prove the following result (called *action independence* of integral discriminants in Morozov and Shakirov (2009a, 2009b)).

Lemma 18.14 *Let $0 < \gamma \in \mathbb{Q}$, $0 < \mathbf{u} \in \mathbb{Q}^n$ and let $\phi : \mathbb{R}_+ \to \mathbb{R}_+$ be a measurable function such that $\int_0^\infty t^{|\mathbf{u}|-1+1/\gamma} \phi(t) \, dt < \infty$. Let $g, h \in \mathbb{R}[\mathbf{x}]$ with g being $\mathbf{u}$-QH and positive on $\mathbb{R}^n \setminus \{0\}$ whereas h is $\gamma\mathbf{u}$-QH. Then if $\int_{\mathbb{R}^n} h(\mathbf{x}) \exp(-g(\mathbf{x}) \, d\mathbf{x} < \infty$,*

$$\int_{\mathbb{R}^n} \phi(g(\mathbf{x})) h(\mathbf{x}) \, d\mathbf{x} = \underbrace{\frac{\int_0^\infty t^{|\mathbf{u}|-1+1/\gamma} \phi(t) \, dt}{\Gamma(|\mathbf{u}| + 1/\gamma)}}_{cte(n,\gamma,|\mathbf{u}|,\phi)} \cdot \int_{\mathbb{R}^n} h \exp(-g) \, d\mathbf{x}, \quad (18.24)$$

and if h = 1,

$$\int_{\mathbb{R}^n} \phi(g(\mathbf{x}))\, d\mathbf{x} = \underbrace{\frac{\int_0^\infty t^{|\mathbf{u}|-1}\phi(t)\, dt}{\Gamma(|\mathbf{u}|)}}_{\text{cte}(n,|\mathbf{u}|,\phi)} \cdot \int_{\mathbb{R}^n} \exp(-g)\, d\mathbf{x}. \tag{18.25}$$

(*Hint:* In the integral do the change of variables: $x_1 = t^{u_1}$, $x_2 = t^{u_2} z_1, \ldots, x_n = t^{u_n} z_{n-1}$ on $[0, +\infty) \times \mathbb{R}^{n-1}$ and $x_1 = -t^{u_1}$, $x_2 = t^{u_2} z_1, \ldots, x_n = t^{u_n} z_{n-1}$ on $(-\infty, 0] \times \mathbb{R}^{n-1}$; then use the **u**-quasi-homogeneity of g and h.)

(b) Apply the result with $t \mapsto \phi(t) := I_{[0,y]}(t)$ and with $t \mapsto t\, I_{[0,y]}(t)$ (with $y > 0$).

Exercise 18.3 Consider problem **P** with $\mathbf{K} = [-1, 1]^2$ and $d = 8$. Show that an atomic measure as ν^* in (18.20) cannot be optimal for **P** as in Theorem 18.10(b).

18.6 Notes and sources

18.2 This section is mainly from Lasserre (2014a). Lemma 18.5 is a consequence of Lemma 18.14 proved in Morozov and Shakirov (2009a, 2009b) for homogeneous polynomials (whose proof does not use Laplace transform) and called *action independence* of integral discriminants; see Exercise 18.2. Such functional integrals are frequent in quantum physics and form a field of nonlinear algebra where the goal is to obtain an explicit expression in terms of algebraic invariants of the homogeneous polynomial in the integrand. Several nontrivial examples are provided in Morozov and Shakirov (2009a), Shakirov (2010) and some of their results are further analyzed in Fujii (2011). It is also worth noticing that non-Gaussian integrals are related to hypergeometric functions studied in several papers of Gelfand and his co-authors; for more details the interested reader is referred to Gelfand and Graev (1999) and references therein. Nonnegative homogeneous polynomials are also particularly interesting as they can be used to approximate norms; see for example Barvinok (2002, Theorem 3.4).

18.3-18.4 These sections are from Lasserre (2013b). Approximating data by relatively simple geometrical objects is a fundamental problem with many important applications (control, statistics, computer graphics, computer vision, to mention a few) and the ellipsoid of minimum volume is the most well-known of the associated computational techniques. In addition to its nice properties from the viewpoint of applications, the ellipsoid of minimum volume is also very interesting from a mathematical viewpoint. Indeed, if $\mathbf{K} \subset \mathbb{R}^n$ is a convex body, computing an ellipsoid of minimum volume that contains $\mathbf{K}$ is a classical and famous problem which has a unique optimal solution called the *Löwner–John* ellipsoid. In addition, John's theorem states that the optimal ellipsoid $\mathbf{\Omega}$ is characterized by s contacts points $u_i \in \mathbf{K} \cap \mathbf{\Omega}$, and positive scalars λ_i, $i = 1, \dots, s$, where s is bounded above by $n(n + 3)/2$ in the general case and $s \leq n(n + 1)/2$ when $\mathbf{K}$ is symmetric; see for example Ball (1992, 2001). In particular, and in contrast to other approximation techniques, computing the ellipsoid of minimum volume is a *convex* optimization problem for which efficient techniques are available. For a nice recent historical survey on the Löwner–John ellipsoid, the interested reader is referred to the recent paper by Henk (2012) and the many references therein.

Appendix A
Semidefinite programming

A.1 Semidefinite programming

Conic programming is a subarea of convex optimization that refers to linear optimization problems over general convex cones. Semidefinite programming (in short, SDP) is a particular case of conic programming when one considers the convex cone of positive semidefinite matrices, whereas linear programming considers the positive orthant of $\mathbb{R}^n$, a *polyhedral* convex cone.

Let $\mathcal{S}_p \subset \mathbb{R}^{p \times p}$ be the space of real $p \times p$ symmetric matrices. Whenever $\mathbf{A}, \mathbf{B} \in \mathcal{S}_p$, the notation $\mathbf{A} \succeq \mathbf{B}$ (respectively $\mathbf{A} \succ \mathbf{B}$) stands for $\mathbf{A} - \mathbf{B}$ is *positive semidefinite* (respectively positive definite). Also, the notation $\langle \mathbf{A}, \mathbf{B} \rangle$ stands for trace $(\mathbf{AB})$ (trace $(\mathbf{AB})$ = trace $(\mathbf{BA})$). In canonical form, a semidefinite program reads:

$$\mathbf{P}: \quad \inf_{\mathbf{x}} \left\{ \mathbf{c}^T \mathbf{x} : \mathbf{F}_0 + \sum_{i=1}^{n} \mathbf{F}_i \, x_i \succeq 0 \right\} \tag{A.1}$$

where $\mathbf{c} \in \mathbb{R}^n$, and $\{\mathbf{F}_i\}_{i=0}^{n} \subset \mathcal{S}_p$ for some $p \in \mathbb{N}$. Denote by $\inf \mathbf{P}$ the optimal value of $\mathbf{P}$ (possibly $-\infty$ if unbounded or $+\infty$ if $\mathbf{P}$ has no feasible solution $\mathbf{x} \in \mathbb{R}^n$). If the optimal value is attained at some $\mathbf{x}^* \in \mathbb{R}^n$ then write $\inf \mathbf{P} = \min \mathbf{P}$.

The semidefinite constraint $\mathbf{F}_0 + \sum_{i=1}^{n} \mathbf{F}_i \, x_i \succeq 0$ is also called a *linear matrix inequality* (LMI).

That $\mathbf{P}$ in (A.1) is a convex optimization problem can be seen as follows. Consider the mapping $\mathbf{F} : \mathbb{R}^n \to \mathcal{S}_p$ defined by:

$$\mathbf{x} \mapsto \mathbf{F}(\mathbf{x}) := \mathbf{F}_0 + \sum_{i=1}^{n} \mathbf{F}_i \, x_i, \qquad \mathbf{x} \in \mathbb{R}^n.$$

The constraint $\mathbf{F}(\mathbf{x}) \succeq 0$ is the same as $\lambda_{\min}(\mathbf{F}(\mathbf{x})) \geq 0$ where the function $\mathbf{x} \mapsto \lambda_{\min}(\mathbf{F}(\mathbf{x}))$ maps $\mathbf{x} \in \mathbb{R}^n$ to the *smallest* eigenvalue of the real symmetric matrix $\mathbf{F}(\mathbf{x}) \in \mathcal{S}_p$. But the smallest eigenvalue of a real symmetric matrix is a concave function of its entries, and the entries of $\mathbf{F}(\mathbf{x})$ are affine functions of $\mathbf{x}$. Hence the set $\{\mathbf{x} : \mathbf{F}(\mathbf{x}) \succeq 0\}$ is convex and therefore $\mathbf{P}$ consists of minimizing a linear functional on a convex set, i.e., a convex optimization problem.

Observe that if the matrices $(\mathbf{F}_i)$ are diagonal then $\mathbf{P}$ is just a finite-dimensional linear programming problem. Conversely, consider the linear programming problem

$$\inf_{\mathbf{x} \in \mathbb{R}^n} \{\mathbf{c}^T \mathbf{x} : \mathbf{a}_j^T \mathbf{x} \geq b_j, \ j = 1, \ldots, p; \quad \mathbf{x} \geq 0\},$$

and let $\mathbf{F}(\mathbf{x}) \in \mathcal{S}_p$ be the diagonal matrix defined by $\mathbf{F}(\mathbf{x})_{jj} = \mathbf{a}_j^T \mathbf{x} - b_j$ for every $j = 1, \ldots, p$. Then the above linear program is also the SDP $\inf_{\mathbf{x}} \{\mathbf{c}^T \mathbf{x} : \mathbf{F}(\mathbf{x}) \succeq 0\}$.

Semidefinite programming is a nontrivial extension of linear programming. Indeed, while the latter considers the positive orthant $\mathbb{R}_+^n$ (a polyhedral convex cone), the former considers the nonpolyhedral convex cone of positive semidefinite matrices.

A.1.1 Duality

The *dual* problem associated with $\mathbf{P}$ is the convex optimization problem:

$$\mathbf{P}^* : \quad \sup_{\mathbf{Z} \in \mathcal{S}_p} \{-\langle \mathbf{F}_0, \mathbf{Z} \rangle : \langle \mathbf{F}_i, \mathbf{Z} \rangle = c_i, \quad i = 1, \ldots, n; \ \mathbf{Z} \succeq 0\} \quad \text{(A.2)}$$

with optimal value denoted $\sup \mathbf{P}^*$ (possibly $+\infty$ if unbounded or $-\infty$ is there is no feasible solution $\mathbf{Z} \succeq 0$). If the optimal value is attained at some $\mathbf{Z}^* \succeq 0$ then write $\sup \mathbf{P}^* = \max \mathbf{P}^*$.

Weak duality states that $\inf \mathbf{P} \geq \sup \mathbf{P}^*$ and holds without any assumption. Indeed, let $\mathbf{x} \in \mathbb{R}^n$ and $0 \preceq \mathbf{Z} \in \mathcal{S}_p$ be any two feasible solutions of $\mathbf{P}$ and $\mathbf{P}^*$ respectively. Then

$$\mathbf{c}'\mathbf{x} = \sum_{i=1}^n \langle \mathbf{F}_i, \mathbf{Z} \rangle x_i = \left\langle \sum_{i=1}^n \mathbf{F}_i \, x_i \, , \, \mathbf{Z} \right\rangle \geq \langle -\mathbf{F}_0, \mathbf{Z} \rangle,$$

where we have used that if $\mathbf{Z} \succeq 0$ and $\mathbf{A} \succeq \mathbf{B}$ then $\langle \mathbf{A}, \mathbf{Z} \rangle \geq \langle \mathbf{B}, \mathbf{Z} \rangle$. And so $\inf \mathbf{P} \geq \sup \mathbf{P}^*$. The nonnegative quantity $\inf \mathbf{P} - \sup \mathbf{P}^*$ is called the *duality gap*.

The absence of a duality gap (i.e., $\inf \mathbf{P} = \sup \mathbf{P}^*$) does not hold in general. However, it happens under some strict-feasibility conditions.

Theorem A.1 (Strong duality) *Let $\mathbf{P}$ and $\mathbf{P}^*$ be as in (A.1) and (A.2) respectively.*

(a) *If there exists* $\mathbf{x} \in \mathbb{R}^n$ *such that* $\mathbf{F}(\mathbf{x}) \succ 0$ *then* $\inf \mathbf{P} = \sup \mathbf{P}^*$ *and* $\inf \mathbf{P} = \max \mathbf{P}^*$ *if the optimal value is finite.*

(b) *If there exists* $\mathbf{Z} \succ 0$ *feasible for* $\mathbf{P}^*$ *then* $\inf \mathbf{P} = \sup \mathbf{P}^*$ *and* $\min \mathbf{P} = \sup \mathbf{P}^*$ *if the optimal value is finite.*

(c) *If there exists* $\mathbf{Z} \succ 0$ *feasible for* $\mathbf{P}^*$ *and* $\mathbf{x} \in \mathbb{R}^n$ *such that* $\mathbf{F}(\mathbf{x}) \succ 0$, *then* $\min \mathbf{P} = \max \mathbf{P}^*$.

The strict-feasiblity condition in Theorem A.1(a) (respectively Theorem A.1(b)) is a specialization of Slater's condition in convex programming, when applied to the convex problem **P** (respectively the convex problem **P***).

A.1.2 Computational complexity

What makes semidefinite programming a powerful technique is its computational complexity when using some algorithms based on *interior point* methods. Indeed one may find an approximate solution within a prescribed accuracy $\epsilon > 0$, in a time that is polynomial in the input size of the problem. For instance, in using path following algorithms based on the Nesterov–Todd search directions, one only needs $O(\sqrt{n} \ln(1/\epsilon))$ iterations to find such an approximate solution; see for example Tuncel (2000).

A.2 Notes and sources

Most of this section is inspired by Vandenberghe and Boyd (1996). For recent results around semidefinite, conic and polynomial optimization, the interested reader is referred to Anjos and Lasserre (2012).

Appendix B
The GloptiPoly software

B.1 Presentation

In this chapter we describe and report on the GloptiPoly software that implements the methodology described in earlier chapters.

In fact GloptiPoly is a MATLAB[1] freeware for solving (or at least approximating) the Generalized Problem of Moments (GPM):

$$\min_{\mu_i} \left\{ \sum_{i=1}^{s} \int_{\mathbf{K}_i} f_i \, d\mu_i \; : \; \sum_{i=1}^{s} \int_{\mathbf{K}_i} g_{ik} \, d\mu_i \stackrel{=}{\underset{\leq}{}} b_k, \quad k = 1, \ldots, p \right\}, \quad \text{(B.1)}$$

where for each $i = 1, \ldots, s$, $\mathbf{K}_i \subset \mathbb{R}^{n_i}$ is a compact basic semi-algebraic set and all functions $f_i, g_{ik} : \mathbb{R}^{n_i} \to \mathbb{R}, k = 1, \ldots, p$, are polynomials.

For more details on the theory and applications of the GPM, the interested reader is referred to Lasserre (2009c). The polynomial optimization problem **P** in (1.1) is perhaps the simplest instance of the GPM (B.1). Indeed when **P** is written as the GPM (5.3), it involves only one unknown Borel measure μ and only one equality constraint, namely $\int_{\mathbf{K}} d\mu = 1$!

GloptiPoly3 is an extension of of the former version 2 of GloptiPoly described in Henrion and Lasserre (2003) dedicated to solving the global polynomial optimization problem (1.1). The software automatically generates and solves the hierarchy of semidefinite relaxations (6.3).

It is a user-friendly package that researchers and students can experiment with easily. It can also be used as tutorial material to illustrate the use of semidefinite relaxations for solving the GPM. So far, GloptiPoly aims at solving small and medium scale problems.

---

[1] MATLAB is a trademark of The MathWorks, Inc.

B.2 Installation

GloptiPoly3 is freeware subject to the General Public Licence (GPL) policy.
It is available for MATLAB 7.2 (Release 2006) and later versions. It can be
downloaded at

<code>www.laas.fr/~henrion/software/gloptipoly3</code>

The package, available as a compressed archive, consists of several m-files
and subdirectories, and it contains no binaries. Extracted files are placed in a
`gloptipoly3` directory that should be declared in the MATLAB working
path, for example using the MATLAB command

```
>> addpath gloptipoly3
```

To solve the semidefinite relaxation (6.3) GloptiPoly3 uses by default the
semidefinite programming solver SeDuMi of Sturm (1999); so this package
should be properly installed. Other semidefinite solvers can also be used provided they are installed and interfaced through YALMIP of Löfberg (2004).

B.3 Getting started

Please type the command

```
>> gloptipolydemo
```

to run interactively the basic example that follows.

Consider the classical problem of minimizing globally the two-dimensional
six-hump camelback function of Henrion and Lasserre (2003)

$$\min_{\mathbf{x}\in\mathbb{R}^2} g_0(\mathbf{x}) = 4x_1^2 + x_1x_2 - 4x_2^2 - 2.1x_1^4 + 4x_2^4 + \frac{1}{3}x_1^6.$$

The function has six local minima, two of them being global minima. Using
GloptiPoly3, this optimization problem can be modeled as an instance of the
moment problem as follows:

```
>> mpol x1 x2
>> g0 = 4*x1^2+x1*x2-4*x2^2-2.1*x1^4+4*x2^4+x1^6/3
Scalar polynomial
4x1^2+x1x2-4x2^2-2.1x1^4+4x2^4+0.33333x1^6
>> P = msdp(min(g0));
GloptiPoly 3.0
Define moment SDP problem
```

```
...
(GloptiPoly output suppressed)
...
Generate moment SDP problem
>> P
Moment SDP problem
  Measure label              = 1
  Relaxation order           = 3
  Decision variables         = 27
  Semidefinite inequalities  = 10x10
```

Once the moment problem is modeled, a semidefinite solver (here SeDuMi) can be used to solve it numerically.

```
>> [status,obj] = msol(P)
GloptiPoly 3.0
Solve moment SDP problem
*******************************************************
Calling SeDuMi
SeDuMi 1.1R3 by AdvOL, 2006 and Jos F. Sturm,
1998-2003.
...
(SeDuMi output suppressed)
...
2 globally optimal solutions extracted
>> status
status =
     1
>> obj
obj =
    -1.0316
>> x = double([x1 x2]);
x(:,:,1) =
    0.0898    -0.7127
x(:,:,2) =
   -0.0898     0.7127
```

The flag status = 1 means that the moment problem is solved successfully and that GloptiPoly3 can extract two globally optimal solutions reaching the objective function obj = -1.0316.

B.4 Description

GloptiPoly3 uses advanced MATLAB features for object-oriented programming and overloaded operators. The user should be familiar with the following basic objects.

B.4.1 Multivariate polynomials (mpol)

A multivariate polynomial is an affine combination of monomials, each monomial depending on a set of variables. Variables can be declared in the MATLAB working space as follows:

```
>> clear
>> mpol x
>> x
Scalar polynomial
x
>> mpol y 2
>> y
2-by-1 polynomial vector
(1,1):y(1)
(2,1):y(2)
>> mpol z 3 2
>> z
3-by-2 polynomial matrix
(1,1):z(1,1)
(2,1):z(2,1)
(3,1):z(3,1)
(1,2):z(1,2)
(2,2):z(2,2)
(3,2):z(3,2)
```

Variables, monomials and polynomials are defined as objects of class mpol.
   All standard MATLAB operators have been uploaded for mpol objects:

```
>> y*y'-z'*z+x^3
2-by-2 polynomial matrix
(1,1):y(1)^2-z(1,1)^2-z(2,1)^2-z(3,1)^2+x^3
(2,1):y(1)y(2)-z(1,1)z(1,2)-z(2,1)z(2,2)-z(3,1)
      z(3,2)+x^3
(1,2):y(1)y(2)-z(1,1)z(1,2)-z(2,1)z(2,2)-z(3,1)
      z(3,2)+x^3
(2,2):y(2)^2-z(1,2)^2-z(2,2)^2-z(3,2)^2+x^3
```

Use the instruction

```
>> mset clear
```

to delete all existing GloptiPoly variables from the MATLAB working space.

B.4.2 Measures (meas)

Variables can be associated with real-valued measures, and one variable is associated with only one measure. For GloptiPoly3, measures are identified with a label, a positive integer. When starting a GloptiPoly session, the default measure has label 1. By default, all created variables are associated with the current measure. Measures can be handled with the class meas as follows:

```
>> mset clear
>> mpol x
>> mpol y 2
>> meas
Measure 1 on 3 variables: x,y(1),y(2)
>> meas(y) % create new measure
Measure 2 on 2 variables: y(1),y(2)
>> m = meas
1-by-2 vector of measures
1:Measure 1 on 1 variable: x
2:Measure 2 on 2 variables: y(1),y(2)
>> m(1)
Measure number 1 on 1 variable: x
```

The above script creates a measure $\mu_1(dx)$ on $\mathbb{R}$ and a measure $\mu_2(d\mathbf{y})$ on $\mathbb{R}^2$. Use the instruction

```
>> mset clearmeas
```

to delete all existing GloptiPoly measures from the working space. Note that this does not delete existing GloptiPoly variables.

B.4.3 Moments (mom)

Linear combinations of moments of a given measure can be manipulated with the mom class as follows:

```
>> mom(1+2*x+3*x^2)
Scalar moment
I[1+2x+3x^2]d[1]
```

```
>> mom(y*y')
2-by-2 moment matrix
(1,1):I[y(1)^2]d[2]
(2,1):I[y(1)y(2)]d[2]
(1,2):I[y(1)y(2)]d[2]
(2,2):I[y(2)^2]d[2]
```

The notation I[p]d[k] stands for $\int p\,d\mu_k$ where p is a polynomial of the variables associated with the measure μ_k, and k is the measure label.

Note that it makes no sense to define moments over several measures, or nonlinear moment expressions:

```
>> mom(x*y(1))
??? Error using ==> mom.mom
Invalid partitioning of measures in moments
>> mom(x)*mom(y(1))
??? Error using ==> mom.times
Invalid moment product
```

Note also the distinction between a constant term and the mass of a measure:

```
>> 1+mom(x)
Scalar moment
1+I[x]d[1]
>> mom(1+x)
Scalar moment
I[1+x]d[1]
>> mass(x)
Scalar moment
I[1]d[1]
```

Finally, let us mention three equivalent notations to refer to the mass of a measure:

```
>> mass(meas(y))
Scalar moment
I[1]d[2]
>> mass(y)
Scalar moment
I[1]d[2]
>> mass(2)
Scalar moment
I[1]d[2]
```

The first command refers explicitly to the measure, the second command is a handy shortcut to refer to a measure via its variables, and the third command refers to GloptiPoly's labeling of measures.

B.4.4 Support constraints (supcon)

By default, a measure on n variables is defined on the whole $\mathbb{R}^n$. We can restrict the support of a mesure to a given semi-algebraic set as follows:

```
>> 2*x^2+x^3 == 2+x
Scalar measure support equality
2x^2+x^3 == 2+x
>> disk = (y'*y <= 1)
Scalar measure support inequality
y(1)^2+y(2)^2 <= 1
```

Support constraints are modeled by objects of class supcon. The first command which also reads $x^3 + 2x^2 - x - 2 = (x - 1)(x + 1)(x + 2) = 0$, means that the measure μ_1 must be discrete, a linear combination of three Dirac measures at 1, -1 and -2. The second command restricts the measure μ_2 to be supported on the unit disk.

Note that it makes no sense to define a support constraint on several measures:

```
>> x+y(1) <= 1
??? Error using ==> supcon.supcon
Invalid reference to several measures
```

B.4.5 Moment constraints (momcon)

We can constrain linearly the moments of several measures:

```
>> mom(x^2+2) == 1+mom(y(1)^3*y(2))
Scalar moment equality constraint
I[2+x^2]d[1] == 1+I[y(1)^3y(2)]d[2]
>> mass(x)+mass(y) <= 2
Scalar moment inequality constraint
I[1]d[1]+I[1]d[2] <= 2
```

Moment constraints are modeled by objects of class momcon.

For GloptiPoly an objective function to be minimized or maximized is considered as a particular moment constraint:

```
>> min(mom(x^2+2))
Scalar moment objective function
min I[2+x^2]d[1]
>> max(x^2+2)
Scalar moment objective function
max I[2+x^2]d[1]
```

The latter syntax is a handy shortcut which directly converts an `mpol` object into a `momcon` object.

B.4.6 Floating point numbers (`double`)

Variables in a measure can be assigned numerical values:

```
>> m1 = assign(x,2)
Measure 1 on 1 variable: x
supported on 1 point
```

which is equivalent to enforcing a discrete support for the measure. Here μ_1 is set to the Dirac measures at the point 2.

The `double` operator converts a measure or its variables into a floating point number:

```
>> double(x)
ans =
     2
>> double(m1)
ans =
     2
```

Polynomials can be evaluated in a similar fashion:

```
>>double(1-2*x+3*x^2)
ans =
     9
```

Discrete measure supports consisting of several points can be specified in an array:

```
>> m2 = assign(y,[-1 2 0;1/3 1/4 -2])
Measure 2 on 2 variables: y(1),y(2)
supported on 3 points
>> double(m2)
ans(:,:,1) =
```

```
   -1.0000
    0.3333
ans(:,:,2) =
    2.0000
    0.2500
ans(:,:,3) =
        0
       -2
```

B.5 Solving polynomial optimization (msdp)

Once a moment problem is defined, it can be solved numerically with the
instruction msol. In the sequel we give several examples of GPMs handled
with GloptiPoly3.

B.5.1 Unconstrained minimization

In Section B.3 we already encountered an example of an unconstrained polynomial optimization solved with GloptiPoly3. Let us revisit this example:

```
>> mset clear
>> mpol x1 x2
>> g0 = 4*x1^2+x1*x2-4*x2^2-2.1*x1^4+4*x2^4+x1^6/3
Scalar polynomial
4x1^2+x1x2-4x2^2-2.1x1^4+4x2^4+0.33333x1^6
>> P = msdp(min(g0));
...
>> msol(P)
...
2 globally optimal solutions extracted
Global optimality certified numerically
```

This indicates that the global minimum is attained with a discrete measure
supported on two points. The measure can be constructed from the knowledge
of its first moments of degree up to 6:

```
>> meas
Measure 1 on 2 variables: x1,x2
  with moments of degree up to 6, supported on
  2 points
>> double(meas)
```

```
ans(:,:,1) =
     0.0898
    -0.7127
ans(:,:,2) =
    -0.0898
     0.7127
>> double(g0)
ans(:,:,1) =
    -1.0316
ans(:,:,2) =
    -1.0316
```

When converting to floating point numbers with the operator double, it is essential to make the distinction between mpol and mom objects:

```
>> v = mmon([x1 x2],2)'
1-by-6 polynomial vector
(1,1):1
(1,2):x1
(1,3):x2
(1,4):x1^2
(1,5):x1x2
(1,6):x2^2
>> double(v)
ans(:,:,1) =
     1.0000    0.0898   -0.7127    0.0081   -0.0640    0.5079
ans(:,:,2) =
     1.0000   -0.0898    0.7127    0.0081   -0.0640    0.5079
>> double(mom(v))
ans =
     1.0000    0.0000   -0.0000    0.0081   -0.0640    0.5079
```

The first instruction mmon generates a vector of monomials v of class mpol, so the command double(v) calls the convertor @mpol/double which evaluates a polynomial expression on the discrete support of a measure (here two points). The last command double(mom(v)) calls the convertor @mom/double which returns the value of the moments obtained after solving the moment problem.

Note that when inputting moment problems on a unique measure whose mass is not constrained, GloptiPoly assumes by default that the measure has mass one, i.e., that we are seeking a probability measure. Therefore, if g0 is the polynomial defined previously, the two instructions

```
>> P = msdp(min(g0));
```

and

```
>> P = msdp(min(g0), mass(meas(g0))==1);
```

are equivalent.

B.5.2 Constrained minimization

Consider the nonconvex quadratic problem

$$
\begin{aligned}
\min \quad & -2x_1 + x_2 - x_3 \\
\text{s.t.} \quad & 24 - 20x_1 + 9x_2 - 13x_3 + 4x_1^2 - 4x_1x_2 + 4x_1x_3 + 2x_2^2 - 2x_2x_3 + 2x_3^2 \geq 0 \\
& x_1 + x_2 + x_3 \leq 4, \quad 3x_2 + x_3 \leq 6 \\
& 0 \leq x_1 \leq 2, \quad 0 \leq x_2, \quad 0 \leq x_3 \leq 3
\end{aligned}
$$

Each constraint in this problem is interpreted by GloptiPoly3 as a support constraint on the measure associated with **x**.

```
>> mpol x 3
>> x(1)+x(2)+x(3) <= 4
Scalar measure support inequality
x(1)+x(2)+x(3) <= 4
```

The whole problem can be entered as follows:

```
>> mpol x 3
>> g0 = -2*x(1)+x(2)-x(3);
>> K = [24-20*x(1)+9*x(2)-13*x(3)+4*x(1)^2-4*x(1)
  *x(2) ...
  +4*x(1)*x(3)+2*x(2)^2-2*x(2)*x(3)+2*x(3)^2 >= 0,...
  x(1)+x(2)+x(3) <= 4, 3*x(2)+x(3) <= 6, ...
  0 <= x(1), x(1) <= 2, 0 <= x(2), 0 <= x(3),
  x(3) <= 3];
>> P = msdp(min(g0), K)
...
Moment SDP problem
  Measure label           = 1
  Relaxation order        = 1
  Decision variables      = 9
  Linear inequalities     = 8
  Semidefinite inequalities = 4x4
```

The moment problem can then be solved:

```
>> [status,obj] = msol(P)
GloptiPoly 3.0
Solve moment SDP problem
...
Global optimality cannot be ensured
status =
      0
obj =
    -6.0000
```

Since `status=0` the moment SDP problem can be solved but it is impossible to detect global optimality. The value `obj=-6.0000` is then a lower bound on the global minimum of the quadratic problem.

The measure associated with the problem variables can be retrieved as follows:

```
>> mu = meas
Measure 1 on 3 variables: x(1),x(2),x(3)
   with moments of degree up to 2
```

Its vector of moments can be built as follows:

```
>> mv = mvec(mu)
10-by-1 moment vector
(1,1):I[1]d[1]
(2,1):I[x(1)]d[1]
(3,1):I[x(2)]d[1]
(4,1):I[x(3)]d[1]
(5,1):I[x(1)^2]d[1]
(6,1):I[x(1)x(2)]d[1]
(7,1):I[x(1)x(3)]d[1]
(8,1):I[x(2)^2]d[1]
(9,1):I[x(2)x(3)]d[1]
(10,1):I[x(3)^2]d[1]
```

These moments are the decision variables of the SDP problem solved with the above `msol` command. Their numerical values can be retrieved as follows:

```
>> double(mv)
ans =
    1.0000
    2.0000
   -0.0000
```

```
    2.0000
    7.6106
    1.4671
    2.3363
    4.8335
    0.5008
    8.7247
```

The numerical moment matrix can be obtained using the following commands:

```
>> double(mmat(mu))
ans =
    1.0000    2.0000   -0.0000    2.0000
    2.0000    7.6106    1.4671    2.3363
   -0.0000    1.4671    4.8335    0.5008
    2.0000    2.3363    0.5008    8.7247
```

As explained in Chapter 6, one can build up a hierarchy of relaxations, whose associated monotone sequence of optimal values converges to the global optimum, under mild technical assumptions. By default the command msdp builds the relaxation of lowest order, equal to half the degree of the highest degree monomial in the polynomial data. An additional input argument can be specified to build higher order relaxations:

```
>> P = msdp(min(g0), K, 2)
...
Moment SDP problem
  Measure label             = 1
  Relaxation order          = 2
  Decision variables        = 34
  Semidefinite inequalities = 10x10+8x(4x4)
>> [status,obj] = msol(P)
...
Global optimality cannot be ensured
status =
     0
obj =
    -5.6922
>> P = msdp(min(g0), K, 3)
...
Moment SDP problem
  Measure label             = 1
```

```
   Relaxation order            = 3
   Decision variables          = 83
   Semidefinite inequalities = 20x20+8x(10x10)
>> [status,obj] = msol(P)
...
Global optimality cannot be ensured
status =
     0
obj =
    -4.0684
```

Observe that the semidefinite programming problems involve an increasing number of variables and constraints. They generate a monotone nondecreasing sequence of lower bounds on the global optimum, which is eventually reached numerically at the fourth relaxation:

```
>> P = msdp(min(g0), K, 4)
...
Moment SDP problem
   Measure label               = 1
   Relaxation order            = 4
   Decision variables          = 164
   Semidefinite inequalities = 35x35+8x(20x20)
>> [status,obj] = msol(P)
...
2 globally optimal solutions extracted
Global optimality certified numerically
status =
     1
obj =
    -4.0000
>> double(x)
ans(:,:,1) =
    2.0000
    0.0000
    0.0000
ans(:,:,2) =
    0.5000
    0.0000
    3.0000
>> double(g0)
```

```
ans(:,:,1) =
   -4.0000
ans(:,:,2) =
   -4.0000
```

B.6 Notes and sources

The material of this chapter is taken from Henrion et al. (2009b). To our knowledge, GloptiPoly3 is the first software package to solve (or approximate) the generalized problem of moments (B.1).

The software SOSTOOLS of Prajna et al. (2002) is dedicated to solving problems involving sums of squares by building up a hierarchy of semidefinite programs in the spirit of (6.4) but allowing products of the polynomials (g_j) as in Schmüdgen's Positivstellensatz in Theorem 2.13.

SparsePOP described in Waki et al. (2008) is software that implements the sparse semidefinite relaxations (8.6)–(8.7). In particular it can also build up a sparsity pattern $I = \{1, \ldots, n\} = \cup_{j=1}^{p} I_j$ from only the initial data f, g_j, and with *no* a priori knowledge of any sparsity pattern.

Finally, YALMIP developed by J. Löfberg, is a MATLAB toolbox for rapid prototyping of optimization problems, which also implements the moment and SOS approach, see for example `htp://control.ee.ethz.ch/~joloef/yalmip.php`

References

Acquistapace, F., Andradas, C., and Broglia, F. 2000. The strict Positivstellensatz for global analytical functions and the moment problem for semianalytic sets. *Math. Ann.*, **316**, 609–616.

Acquistapace, F., Andradas, C., and Broglia, F. 2002. The Positivstellensatz for definable functions on O-minimal structures. *Illinois J. Math.*, **46**, 685–693.

Adams, W. W. and Loustaunau, P. 1994. *An Introduction to Gröbner Bases*. Providence, RI: American Mathematical Society.

Ahmadi, A. A. and Parrilo, P. A. 2010. On the equivalence of algebraic conditions for convexity and quasi convexity of polynomials. In: *49th IEEE Conference on Decision and Control*, pp. 3343–3348. New York: IEEE.

Ahmadi, A. A., Olshevsky, A., Parrilo, P. A., and Tsitsiklis, J. N. 2013. NP-hardness of deciding convexity of quartic polynomials and related problems. *Math. Prog.*, **137**, 453–476.

Ahmed, S. and Guan, Yongpei. 2005. The inverse optimal value problem. *Math. Prog. Ser. A*, **102**, 91–110.

Ahuja, R. K̃. and Orlin, J. B. 2001. Inverse optimization. *Oper. Res.*, **49**, 771–783.

Akhiezer, N. I. 1965. *The Classical Moment Problem*. New York: Hafner.

Ali, M. M., Khompatraporn, C., and Zabinsky, Z. B. 2005. A numerical evaluation of several stochastic algorithms on selected continuous global optimization test problems. *J. Global Optim.*, **31**, 635–672.

Anastassiou, G. A. 1993. *Moments in Probability and Approximation Theory*. New York: Longman Scientific and Technical.

Andronov, V. G., Belousov, E. G., and Shironin, V. M. 1982. On solvability of the problem of polynomial programming. *Izv. Akad. Nauk SSSR, Teckh. Kibern.*, **4**, 194–197.

Androulakis, I. P., Maranas, C. D., and Floudas, C. A. 1995. alphaBB: a global optimization method for general constrained nonconvex problems. *J. Global Optim.*, **7**, 337–363.

Anjos, M. 2001. *New Convex Relaxations for the Maximum Cut and VLSI Layout Problems*. Ph.D. thesis, University of Waterloo, Ontario, Canada. orion.math. uwaterloo.ca/~hwolkowi.

Anjos, M. and Lasserre, J. B. (editors) 2012. *Handbook on Semidefinite, Conic and Polynomial Optimization*. International Series in Operations Research and Management Science, vol. 166. New York: Springer.

Ash, R. B. 1972. *Real Analysis and Probability*. San Diego, CA: Academic Press.

Balas, E., Ceria, S., and Cornuéjols, G. 1993. A lift-and-project cutting plane algorithm for mixed 0/1 programs. *Math. Prog.*, **58**, 295–324.

Baldoni, V., Berline, N., De Loera, J. A., Köppe, M., and Vergne, M. 2011. How to integrate a polynomial over a simplex. *Math. Comp.*, **80**, 297–325.

Ball, K. 1992. Ellipsoids of maximal volume in convex bodies. *Geom. Dedicata*, **41**, 241–250.

Ball, K. 2001. Convex geometry and functional analysis. In: Johnson, W. B. and Lindenstrauss, J. (editors), *Handbook of the Geometry of Banach Spaces I*, pp. 161–194. Amsterdam: North Holland.

Barvinok, A. 2002. *A Course in Convexity*. Providence, RI: American Mathematical Society.

Basu, S., Pollack, R., and Roy, M.-F. 2003. *Algorithms in Real Algebraic Geometry*. Algorithms and Computations in Mathematics, vol. 10. Berlin: Springer.

Bayer, C. and Teichmann, J. 2006. The proof of Tchakaloff's theorem. *Proc. Amer. Math. Soc.*, **134**, 3035–3040.

Becker, E. and Schwartz, N. 1983. Zum Darstellungssatz von Kadison–Dubois. *Arch. Math. (Basel)*, **40**, 421–428.

Belousov, E. G. 1977. *Introduction to Convex Analysis and Integer Programming*. Moscow: Moscow University Publications.

Belousov, E. G. and Klatte, D. 2002. A Frank–Wolfe type theorem for convex polynomial programs. *Comp. Optim. Appl.*, **22**, 37–48.

Benabbas, S. and Magen, A. 2010. Extending SDP integrality gaps to Sherali–Adams with applications to quadratic programming and MaxCutGain. In: Eisenbrand, F. and Shepherd, F. B. (editors), *Integer Programming and Combinatorial Optimization*, pp. 299–312. Lecture Notes in Computer Science, vol. 6080. Berlin: Springer.

Benabbas, S., Georgiou, K., Magen, A., and Tulsiani, M. 2012. SDP gaps from pairwise independence. *Theory Comp.*, **8**, 269–289.

Ben-Tal, A. and Nemirovski, A. 2001. *Lectures on Modern Convex Optimization*. Philadelphia, PA: SIAM.

Ben-Tal, A., El Ghaoui, L., and Nemirovski, A. 2000. Robustness. In: Wolkowicz, H., Saigal, R., and Vandenberghe, L. (editors), *Handbook of Semidefinite Programming: Theory, Algorithms, and Applications*. Boston, MA: Kluwer Academic.

Ben-Tal, A., Boyd, S., and Nemirovski, A. 2006. Extending scope of robust optimization: comprehensive robust counterparts of uncertain problems. *Math. Prog. Ser. B*, **107**, 63–89.

Berg, C. 1987. The multidimensional moment problem and semi-groups. In: Landau, H. J. (editor), *Moments in Mathematics*, pp. 110–124. Proceedings of Symposia in Applied Mathematics, vol. 37. Providence, RI: American Mathematical Society.

Bernstein, S. 1921. Sur la représentation des polynômes positifs. *Math. Z.*, **9**, 74–109.

Bertsimas, D. 1995. *Nonlinear Programming*, second edition. Boston, MA: Athena Scientific.

Bertsimas, D. and Sim, M. 2004. The price of robustness. *Oper. Res.*, **52**, 35–53.

Bertsimas, D. and Sim, M. 2006. Tractable approximations to robust conic optimization problems. *Math. Prog. Ser. B*, **107**, 5–36.

Blekherman, G. 2006. There are significantly more nonnegative polynomials than sums of squares. *Isr. J. Math.*, **153**, 355–380.

Blekherman, G. 2012. Nonnegative polynomials and sums of squares. *J. Amer. Math. Soc.*, **25**, 617–635.

Blekherman, G. 2014. Positive Gorenstein ideals. *Proc. Amer. Math. Soc.* To appear.

Blekherman, G. and Lasserre, J. B. 2012. The truncated **K**-moment problem for closure of open sets. *J. Func. Anal.*, **263**, 3604–3616.

Bochnak, J., Coste, M., and Roy, M.-F. 1998. *Real Algebraic Geometry*. New York: Springer.

Bomze, I. M. 2012. Copositive optimization – recent developments and applications. *Eur. J. Oper. Res*, **216**, 509–520.

Bomze, I. M. and de Klerk, E. 2002. Solving standard quadratic optimization problems via linear, semidefinite and copositive programming. *J. Global Optim.*, **24**, 163–185.

Bomze, I. M., Dürr, M., de Klerk, E., Roos, C., Quist, A. J., and Terlaky, T. 2000. On copositive programming and standard quadratic optimization problems. *J. Global Optim.*, **18**, 301–320.

Bomze, I. M., Schachinger, W., and Uchida, G. 2012. Think co(mpletely) positive! – matrix properties, examples and a clustered bibliography on copositive optimization. *J. Global Optim.*, **52**, 423–445.

Bonnans, J. F. and Shapiro, A. 2000. *Perturbation Analysis of Optimization Problems*. New York: Springer.

Buelens, P. F. and Hellinckx, L. J. 1974. Optimal control of linear multivariable systems with quadratic performance index, and the inverse optimal control problem. *Int. J. Control*, **20**, 113–127.

Bugarin, F., Henrion, D., and Lasserre, J. B. 2011. *Minimizing the Sum of Many Rational Functions*. Technical Report. LAAS-CNRS, Toulouse, France. arXiv: 1102.4954.

Burer, S. 2009. On the copositive representation of binary and continuous nonconvex quadratic programs. *Math. Prog. Ser. A*, **120**, 479–495.

Burer, S. 2012. Copositive programming. In Anjos, M. and Lasserre, J. B. (editors), *Handbook on Semidefinite, Conic and Polynomial Optimization*, pp. 201–218. International Series in Operations Research and Management Science, vol. 166. New York: Springer.

Burgdorf, S., Scheiderer, C., and Schweighofer, M. 2012. Pure states, nonnegative polynomials and sums of squares. *Comment. Math. Helv.*, **87**, 113–140.

Burton, D. and Toint, P. 1992. On an instance of the inverse shortest path problem. *Math. Prog.*, **63**, 45–61.

Cafieri, S., Lee, J., and Liberti, L. 2010. On convex relaxations of quadrilinear terms. *J. Global Optim.*, **47**, 661–685.

Carleman, T. 1926. *Les Fonctions Quasi-analytiques*. Paris: Gauthier-Villars.

Cassier, G. 1984. Problème des moments sur un compact de $\mathbb{R}^n$ et représentation de polynômes à plusieurs variables. *J. Func. Anal.*, **58**, 254–266.

Castaing, C., de Fitte, P. R., and Valadier, M. 2004. *Young Measures on Topological Spaces. With Applications in Control Theory and Probability Theory*. Mathematics and its Applications, vol. 571. Dordrecht: Kluwer Academic.

Chlamtac, E. 2007. Approximation algorithms using hierarchies of semidefinite programming relaxations. In: *48th Annual IEEE Symposium on Foundations of Computer Science (FOCS'07)*, pp. 691–701. New York: IEEE.

Chlamtac, E. and Singh, G. 2008. Improved approximation guarantees through higher levels of SDP hierarchies. In: *Approximation, Randomization and Combinatorial Optimization Problems*, pp. 49–62. Lecture Notes in Computer Science, vol. 5171. Berlin: Springer.

Chlamtac, E. and Tulsiani, M. 2012. Convex relaxations and integrality gaps. In: Anjos, M. and Lasserre, J. B. (editors), *Handbook on Semidefinite, Conic and Polynomial Optimization*, pp. 139–170. International Series in Operations Research and Management Science, vol. 166. New York: Springer.

Curto, R. and Fialkow, L. 1991. Recursiveness, positivity, and truncated moment problems. *Houston Math. J.*, **17**, 603–635.

Curto, R. and Fialkow, L. 1996. *Solution of the Truncated Complex Moment Problem for Flat Data*. Memoirs of the American Mathematical Society, vol. 119. Providence, RI: American Mathematical Society.

Curto, R. and Fialkow, L. 1998. *Flat Extensions of Positive Moment Matrices: Recursively Generated Relations*. Memoirs of the American Mathematical Society, vol. 136. Providence, RI: American Mathematical Society.

Curto, R. and Fialkow, L. 2000. The truncated complex **K**-moment problem. *Trans. Amer. Math. Soc.*, **352**, 2825–2855.

Curto, R. E. and Fialkow, L. A. 2008. An analogue of the Riesz–Haviland theorem for the truncated moment problem. *J. Funct. Anal.*, **255**, 2709–2731.

d'Aspremont, A. 2008. Smooth optimization with approximate gradient. *SIAM J. Optim.*, **19**, 1171–1183.

de Klerk, E. and Laurent, M. 2011. On the Lasserre hierarchy of semidefinite programming relaxations of convex polynomial optimization problems. *SIAM J. Optim.*, **21**, 824–832.

de Klerk, E. and Pasechnik, D. V. 2002. Approximation of the stability number of a graph via copositive programming. *SIAM J. Optim.*, **12**, 875–892.

de Klerk, E. and Pasechnik, D. V. 2007. A linear programming formulation of the standard quadratic optimization problem. *J. Global Optim.*, 75–84.

de Klerk, E., Laurent, M., and Parrilo, P. A. 2006. A PTAS for the minimization of polynomials of fixed degree over the simplex. *Theor. Comp. Sci.*, **361**, 210–225.

de Klerk, E., Pasechnik, D. V., and Schrijver, A. 2007. Reduction of symmetric semidefinite programs using the regular $\star$-representation. *Math. Prog.*, **109**, 613–624.

Diaconis, P. and Freedman, D. 2006. The Markov moment problem and de Finetti's Theorem Part I. *Math. Z.*, **247**, 183–199.

Dunkl, C. F. and Xu, Y. 2001. *Orthogonal Polynomials of Several Variables*. Encyclopedia of Mathematics and its Applications, vol. 81. Cambridge: Cambridge University Press.

Dürr, M. 2010. Copositive programming – a survey. In: Diehl, M., Glineur, F., Jarlebring, E., and Michiels, W. (editors), *Recent Advances in Optimization and its Applications in Engineering*, pp. 3–20. New York: Springer.

Faraut, J. and Korányi, A. 1994. *Analysis on Symmetric Cones*. Oxford: Clarendon Press.

Fekete, M. 1935. Proof of three propositions of Paley. *Bull. Amer. Math. Soc.*, **41**, 138–144.

Feller, W. 1966. *An Introduction to Probability Theory and Its Applications,* second edition. New York: John Wiley & Sons.

Fidalgo, C. and Kovacec, A. 2011. Positive semidefinite diagonal minus tail forms are sums of squares. *Math. Z.*, **269**, 629–645.

Floudas, C. A. 2000. *Deterministic Global Optimization Theory, Methods and Applications*. Dordrecht: Kluwer Academic.

Floudas, C. A. and Pardalos, P. (editors) 2001. *Encyclopedia of Optimization*. Dordrecht: Kluwer Academic.

Floudas, C. A., Pardalos, P. M., Adjiman, C. S., Esposito, W. R., Gümüs, Z. H., Harding, S. T., Klepeis, J. L., Meyer, C. A., and Schweiger, C. A. 1999. *Handbook of Test Problems in Local and Global Optimization*. Boston, MA: Kluwer. titan.princeton.edu/TestProblems.

Freeman, R. A. and Kokotovic, P. V. 1996. Inverse optimality in robust stabilization. *SIAM J. Control Optim.*, **34**, 1365–1391.

Freitag, E. and Busam, R. 2009. *Complex Analysis*, second edition. Berlin: Springer.

Fujii, K. 2011. Beyond the Gaussian. *SIGMA*, **7**, arXiv: 0912,2135.

Gaddum, J. W. 1958. Linear inequalities and quadratic forms. *Pacific J. Math.*, **8**, 411–414.

Gaterman, K. and Parrilo, P. A. 2004. Symmetry group, semidefinite programs and sums of squares. *J. Pure Appl. Alg.*, **192**, 95–128.

Gelfand, I. M. and Graev, M. I. 1999. GG-Functions and their relations to general hypergeometric functions. *Lett. Math. Phys.*, **50**, 1–28.

Ghasemi, M. and Marshall, M. 2010. Lower bounds for polynomials using geometric programming. *Arch. Math. (Basel)*, **95**, 343–353.

Ghasemi, M. and Marshall, M. 2012. Lower bounds for a polynomial in terms of its coefficients. *SIAM J. Optim.*, **22**, 460–473.

Goemans, M. X. and Williamson, D. P. 1995. Improved approximation algorithms for maximum cut and satisfiability problems using semidefinite programming. *J. ACM*, **42**, 1115–1145.

Golub, G. and Loan, C. F. Van. 1996. *Matrix Computations*, third edition. New York: John Hopkins University Press.

Gounaris, C. E. and Floudas, C. A. 2008. Tight convex underestimators for C^2-continuous problems: II. Multivariate functions. *J. Global Optim.*, **42**, 69–89.

Gouveia, J. and Netzer, T. 2011. Positive polynomials and projections of spectrahedra. *SIAM J. Optim.*, **21**, 960–976.

Gouveia, J. and Thomas, R. 2012. Convex hulls of algebraic sets. In: Anjos, M. and Lasserre, J. B. (editors), *Handbook on Semidefinite, Conic and Polynomial Optimization*, pp. 113–138. International Series in Operations Research and Management Science, vol. 166. New York: Springer.

Gouveia, J., Parrilo, P. A., and Thomas, R. 2010. Theta bodies for polynomial ideals. *SIAM J. Optim.*, **20**, 2097–2118.

Gouveia, J., Laurent, M., Parrilo, P. A., and Thomas, R. 2012. A new hierarchy of semidefinite programming relaxations for cycles in binary matroids and cuts in graphs. *Math. Prog. Ser. A.* To appear.

Güler, O. 2010. *Foundations of Optimization.* New York: Springer.

Guzman, Y. A., Hasan, M. M. F., and Floudas, C. A. 2014. Computational comparison of convex underestimators for use in a branch-and-bound global optimization framework. In: Rassias, T. M., Floudas, C. A., and Butenko, S. (editors), *Optimization in Science and Engineering: In Honor of the 60th Birthday of Panos Pardalos.* New York: Springer.

Handelman, D. 1988. Representing polynomials by positive linear functions on compact convex polyhedra. *Pacific J. Math.*, **132**, 35–62.

Hanzon, B. and Jibetean, D. 2003. Global minimization of a multivariate polynomial using matrix methods. *J. Global Optim.*, **27**, 1–23.

Hausdorff, F. 1915. Summationsmethoden und Momentfolgen I. *Soobshch. Kharkov matem. ob-va Ser. 2*, **14**, 227–228.

Haviland, E. K. 1935. On the momentum problem for distributions in more than one dimension, I. *Amer. J. Math.*, **57**, 562–568.

Haviland, E. K. 1936. On the momentum problem for distributions in more than one dimension, II. *Amer. J. Math.*, **58**, 164–168.

Helton, J. W. and Nie, J. 2009. Sufficient and necessary conditions for semidefinite representability of convex hulls and sets. *SIAM J. Optim.*, **20**, 759–791.

Helton, J. W. and Nie, J. 2010. Semidefinite representation of convex sets. *Math. Prog. Ser. A*, **122**, 21–64.

Helton, J. W. and Nie, J. 2012. Semidefinite representation of convex sets and convex hulls. In: Anjos, M. and Lasserre, J. B. (editors), *Handbook on Semidefinite, Conic and Polynomial Optimization*, pp. 77–112. International Series in Operations Research and Management Science, vol. 166. New York: Springer.

Helton, J. W. and Putinar, M. 2007. Positive polynomials in scalar and matrix variables, the spectral theorem and optimization. In: Bakonyi, M., Gheondea, A., and Putinar, M. (editors), *Operator Theory, Structured Matrices and Dilations*, pp. 229–306. Bucharest: Theta.

Helton, J. W., Lasserre, J. B., and Putinar, M. 2008. Measures with zeros in the inverse of their moment matrix. *Ann. Prob.*, **36**, 1453–1471.

Henk, M. 2012. Löwner–John ellipsoids. *Doc. Math.*, 95–106. extra volume: Optimization stories, 95–106.

Henrion, D. and Lasserre, J. B. 2003. GloptiPoly: global optimization over polynomials with Matlab and SeDuMi. *ACM Trans. Math. Software*, **29**, 165–194.

Henrion, D. and Lasserre, J. B. 2005. Detecting global optimality and extracting solution in GloptiPoly. In: Henrion, D. and Garulli, A. (editors), *Positive Polynomials in Control*, pp. 293–310. Lecture Notes in Control and Information Science, vol. 312. Berlin: Springer.

Henrion, D. and Lasserre, J. B. 2006. Convergent relaxations of polynomial matrix inequalities and static output feedback. *IEEE Trans. Autom. Control*, **51**, 192–202.

Henrion, D. and Lasserre, J. B. 2012. Inner approximations for polynomial matrix inequalities and robust stability regions. *IEEE Trans. Autom. Control*, **57**, 1456–1467.

Henrion, D., Lasserre, J. B., and Savorgnan, C. 2009a. Approximate volume and integration for basic semi-algebraic sets. *SIAM Review*, **51**, 722–743.

Henrion, D., Lasserre, J. B., and Löfberg, J. 2009b. GloptiPoly 3: moments, optimization and semidefinite programming. *Optim. Methods Software*, **24**, 761–779.

Hernández-Lerma, O. and Lasserre, J. B. 1996. *Discrete-Time Markov Control Processes: Basic Optimality Criteria*. New York: Springer.

Heuberger, C. 2004. Inverse combinatorial optimization: a survey on problems, methods and results. *J. Comb. Optim.*, **8**, 329–361.

Hol, C. W. J. and Scherer, C. W. 2005. A sum of squares approach to fixed-order H_∞ synthesis. In: Henrion, D. and Garulli, A. (editors), *Positive Polynomials in Control*, pp. 45–71. Lecture Notes in Control and Information Science, vol. 312. Berlin: Springer.

Hong, H. and Safey El Din, M. 2012. Variant quantifier elimination. *J. Symb. Comput.*, **47**, 883–901.

Huang, S. and Liu, Z. 1999. On the inverse problem of linear programming and its application to minimum weight perfect k-matching. *Eur. J. Oper. Res.*, **112**, 421–426.

Jacobi, T. and Prestel, A. 2001. Distinguished representations of strictly positive polynomials. *J. Reine. Angew. Math.*, **532**, 223–235.

Jibetean, D. and de Klerk, E. 2006. Global optimization of rational functions: a semidefinite programming approach. *Math. Prog.*, **106**, 93–109.

Jibetean, D. and Laurent, M. 2005. Semidefinite approximations for global unconstrained polynomial optimization. *SIAM J. Optim.*, **16**, 490–514.

John, F. 1948. Extremum problems with inequalities as subsidiary conditions. In: Friedrichs, K. O., Neugebauer, O. E., and Stoker, J. J. (editors), *Studies and Essays Presented to R. Courant on his 60th Birthday*, pp. 187–204. New York: Interscience.

Jordan, M. I. 2004. Graphical models. *Stat. Sci.*, **19**, 140–155. Special Issue on Bayesian Statistics.

Karlin, A. R., Mathieu, C., and Nguyen, C. Thach. 2011. Integrality gaps of linear and semi-definite programming relaxations for knapsack. In: Günlük, O. and Woeginger, G. J. (editors), *Integer Programming and Combinatorial Optimization*, pp. 301–314. Lecture Notes in Computer Science, vol. 6655. Berlin: Springer.

Karush, W. 1939. *Minima of Functions of Several Variables with Inequality as Side Constraints*. Ph.D. thesis, Department of Mathematics, University of Chicago, Chicago, IL.

Kemperman, J. H. B. 1968. The general moment problem, a geometric approach. *Ann. Math. Stat.*, **39**, 93–122.

Kemperman, J. H. B. 1987. Geometry of the moment problem. In: Landau, H. J. (editors), *Moments in Mathematics*, pp. 110–124. Proceedings of Symposia in Applied Mathematics, vol. 37. Providence, RI: American Mathematical Society.

Kim, S., Kojima, M., and Waki, H. 2009. Exploiting sparsity in SDP relaxation for sensor network localization. *SIAM J. Optim.*, **20**, 192–215.

Kojima, M. and Maramatsu, M. 2007. An extension of sums of squares relaxations to polynomial optimization problems over symmetric cones. *Math. Prog.*, **110**, 315–336.

Kojima, M. and Maramatsu, M. 2009. A note on sparse SOS and SDP relaxations for polynomial optimization problems over symmetric cones. *Comp. Optim. Appl.*, **42**, 31–41.

Kojima, M., Kim, S., and Maramatsu, M. 2005. Sparsity in sums of squares of squares of polynomials. *Math. Prog.*, **103**, 45–62.

Krivine, J. L. 1964a. Anneaux préordonnés. *J. Anal. Math.*, **12**, 307–326.

Krivine, J. L. 1964b. Quelques propriétés des préordres dans les anneaux commutatifs unitaires. *C. R. Acad. Sci. Paris*, **258**, 3417–3418.

Kuhlmann, S. and Putinar, M. 2007. Positive polynomials on fibre products. *C. R. Acad. Sci. Paris, Ser. 1*, **1344**, 681–684.

Kuhlmann, S. and Putinar, M. 2009. Positive polynomials on projective limits of real algebraic varieties. *Bull. Sci. Math.*, **133**, 92–111.

Kuhlmann, S., Marshall, M., and Schwartz, N. 2005. Positivity, sums of squares and the multi-dimensional moment problem II. *Adv. Geom.*, **5**, 583–606.

Landau, H. (editor) 1987. *Moments in Mathematics*. Proceedings of Symposia in Applied Mathematics, vol. 37. Providence, RI: American Mathematical Society.

Lasserre, J. B. 2000. Optimisation globale et théorie des moments. *C. R. Acad. Sci. Paris, Ser. 1*, **331**, 929–934.

Lasserre, J. B. 2001. Global optimization with polynomials and the problem of moments. *SIAM J. Optim.*, **11**, 796–817.

Lasserre, J. B. 2002a. An explicit equivalent positive semidefinite program for nonlinear 0-1 programs. *SIAM J. Optim.*, **12**, 756–769.

Lasserre, J. B. 2002b. Polynomials nonnegative on a grid and discrete optimization. *Trans. Amer. Math. Soc.*, **354**, 631–649.

Lasserre, J. B. 2002c. Semidefinite programming vs. LP relaxations for polynomial programming. *Math. Oper. Res.*, **27**, 347–360.

Lasserre, J. B. 2004. Polynomial programming: LP-relaxations also converge. *SIAM J. Optim.*, **15**, 383–393.

Lasserre, J. B. 2005. SOS approximations of polynomials nonnegative on a real algebraic set. *SIAM J. Optim.*, **16**, 610–628.

Lasserre, J. B. 2006a. Convergent SDP-relaxations in polynomial optimization with sparsity. *SIAM J. Optim.*, **17**, 822–843.

Lasserre, J. B. 2006b. Robust global optimization with polynomials. *Math. Prog. Ser. B*, **107**, 275–293.

Lasserre, J. B. 2006c. A sum of squares approximation of nonnegative polynomials. *SIAM J. Optim*, **16**, 751–765.

Lasserre, J. B. 2007. Sufficient conditions for a real polynomial to a sum of squares. *Arch. Math. (Basel)*, **89**, 390–398.

Lasserre, J. B. 2008. Representation of nonnegative convex polynomials. *Arch. Math. (Basel)*, **91**, 126–130.

Lasserre, J. B. 2009a. Convex sets with semidefinite representation. *Math. Prog. Ser. A*, **120**, 457–477.

Lasserre, J. B. 2009b. Convexity in semi-algebraic geometry and polynomial optimization. *SIAM J. Optim.*, **19**, 1995–2014.

Lasserre, J. B. 2009c. *Moments, Positive Polynomials and Their Applications*. London: Imperial College Press.

Lasserre, J. B. 2010a. Certificates of convexity for basic semi-algebraic sets. *Appl. Math. Lett.*, **23**, 912–916.

Lasserre, J. B. 2010b. A "joint+marginal" approach to parametric polynomial optimization. *SIAM J. Optim.*, **20**, 1995–2022.

Lasserre, J. B. 2011. A new look at non negativity on closed sets and polynomial optimization. *SIAM J. Optim.*, **21**, 864–885.

Lasserre, J. B. 2013a. Borel measures with a density on a compact semi-algebraic set. *Arch. Math. (Basel)*, **101**, 361–371.

Lasserre, J. B. 2013b. A generalization of Löwner–John's ellipsoid problem. *Math. Prog. Ser. A*. To appear.

Lasserre, J. B. 2013c. Inverse polynomial optimization. *Math. Oper. Res.*, **38**, 418–436.

Lasserre, J. B. 2013d. The K-moment problem with continuous linear functionals. *Trans. Amer. Math. Soc.*, **365**, 2489–2504.

Lasserre, J. B. 2013e. A Lagrangian relaxation view of linear and semidefinite hierarchies. *SIAM J. Optim.*, **23**, 1742–1756.

Lasserre, J. B. 2013f. Tractable approximations of sets defined with quantifiers. *Math. Prog. Ser. B*. To appear.

Lasserre, J. B. 2014a. Level sets and non Gaussian integrals of positively homogeneous functions. *Int. Game Theory Rev*. To appear.

Lasserre, J. B. 2014b. New approximations for the cone of copositive matrices and its dual. *Math. Prog.*, **144**, 265–276.

Lasserre, J. B. and Netzer, T. 2007. SOS approximations of nonnegative polynomials via simple high degree perturbations. *Math. Z.*, **256**, 99–112.

Lasserre, J. B. and Putinar, M. 2010. Positivity and optimization for semi-algebraic functions. *SIAM J. Optim.*, **20**, 3364–3383.

Lasserre, J. B. and Putinar, M. 2012. Positivity and optimization: beyond polynomials. In: Anjos, M., and Lasserre, J. B. (editors), *Handbook on Semidefinite, Conic and Polynomial Optimization*, pp. 407–436. International Series in Operations Research and Management Science, vol. 166. New York: Springer.

Lasserre, J. B. and Thanh, T. P. 2011. Convex underestimators of polynomials. In: *Proceedings of the 50th IEEE CDC Conference on Decision and Control, December 2011*, pp. 7194–7199. New York: IEEE.

Lasserre, J. B. and Thanh, T. P. 2012. A "joint+marginal" heuristic for 0/1 programs. *J. Global Optim.* **54**, 729–744.

Lasserre, J. B. and Thanh, T. P. 2013. Convex underestimators of polynomials. *J. Global Optim.*, **56**, 1–25.

Laurent, M. 2003. A comparison of the Sherali–Adams, Lovász–Schrijver and Lasserre relaxations for 0-1 programming. *Math. Oper. Res.*, **28**, 470–496.

Laurent, M. 2005. Revisiting two theorems of Curto and Fialkow on moment matrices. *Proc. Amer. Math. Soc.*, **133**, 2965–2976.

Laurent, M. 2007a. Semidefinite representations for finite varieties. *Math. Prog. Ser. A*, **109**, 1–26.

Laurent, M. 2007b. Strengthened semidefinite programming bounds for codes. *Math. Prog. Ser. B*, **109**, 239–261.

Laurent, M. 2008. Sums of squares, moment matrices and optimization over polynomials. In: Putinar, M. and Sullivant, S. (editors), *Emerging Applications of Algebraic Geometry*, pp. 157–270. IMA Volumes in Mathematics and its Applications, vol. 149. New York: Springer.

Löfberg, J. 2004. YALMIP: a toolbox for modeling and optimization in MATLAB. In: *Proceedings of the IEEE Symposium on Computer-Aided Control System Design (CACSD), Taipei, Taiwan*. New York: IEEE.

Lombardi, H., Perrucci, D., and Roy, M.-F. 2014. *An Elementary Recursive Bound for Effective Positivstellensatz and Hilbert 17th Problem*. Technical Report IRMAR, Rennes, France. arXiv:1404.2338v1.

Lovász, L. 2003. Semidefinite programs and combinatorial optimization. In: Reed, B. A. and Sales, C. L. (editors), *Recent Advances in Algorithms and Combinatorics*, pp. 137–194. CMS Books Math./Ouvrages Math. SMC, 11. New York: Springer.

Lovász, L. and Schrijver, A. 1991. Cones of matrices and set-functions and 0-1 optimization. *SIAM J. Optim.*, **1**, 166–190.

Marshall, M. 2006. Representation of non-negative polynomials with finitely many zeros. *Ann. Fac. Sci. Toulouse*, **15**, 599–609.

Marshall, M. 2008. *Positive Polynomials and Sums of Squares*. AMS Mathematical Surveys and Monographs, vol. 146. Providence, RI: American Mathematical Society.

Marshall, M. 2009. Representation of non-negative polynomials, degree bounds and applications to optimization. *Canad. J. Math.*, **61**, 205–221.

Marshall, M. 2010. Polynomials non-negative on a strip. *Proc. Amer. Math. Soc.*, **138**, 1559–1567.

Marshall, M. and Netzer, T. 2012. Positivstellensätze for real function algebras. *Math. Z.*, **270**, 889–901.

Morozov, A. and Shakirov, S. 2009a. Introduction to integral discriminants. *J. High Energy Phys.*, **12**(12).

Morozov, A. and Shakirov, S. 2009b. *New and Old Results in Resultant Theory*. Technical Report ITEP, Moscow. *Theor. Math. Phys.* To appear.

Moylan, P. J. and Anderson, B. D. O. 1973. Nonlinear regulator theory and an inverse optimal control problem. *IEEE Trans. Autom Control*, **18**, 460–465.

Mulholland, H. P. and Rogers, C. A. 1958. Representation theorems for distribution functions. *Proc. London Math. Soc.*, **8**, 177–223.

Nesterov, Y. 2000. Squared functional systems and optimization problems. In: Frenk, H., Roos, K., Terlaky, T., and Zhang, S. (editors), *High Performance Optimization*, pp. 405–440. New York: Springer.

Netzer, T., Plaumann, D., and Schweighofer, M. 2010. Exposed faces of semidefinitely representable sets. *SIAM J. Optim.*, **20**, 1944–1955.

Nie, J. 2014. Optimality conditions and finite convergence of Lasserre's hierarchy. *Math. Prog. Ser. A*, **146**, 97–121.

Nie, J. and Schweighofer, M. 2007. On the complexity of Putinars' Positivstellensatz. *J. Complexity*, **23**, 135–150.

Nie, J., Demmel, J., and Sturmfels, B. 2006. Minimizing polynomials via sum of squares over the gradient ideal. *Math. Prog. Ser. A*, **106**, 587–606.

Nussbaum, A. E. 1966. Quasi-analytic vectors. *Ark. Mat.*, **5**, 179–191.

Park, J. G. and Lee, K. Y. 1975. An inverse optimal control problem and its application to the choice of performance index for economic stabilization policy. *IEEE Trans. Syst. Man Cybern.*, **5**, 64–76.

Parrilo, P. A. 2000. *Structured Semidefinite Programs and Semialgebraic Geometry Methods in Robustness and Optimization*. Ph.D. thesis, California Institute of Technology, Pasadena, CA.

Parrilo, P. A. 2002. *An Explicit Construction of Distinguished Representations of Polynomials Nonnegative over Finite Sets*. Technical Report Auto-02, IfA, ETH, Zurich, Switzerland.

Parrilo, P. A. 2003. Semidefinite programming relaxations for semialgebraic problems. *Math. Prog.*, **96**, 293–320.

Pedregal, P. 1999. Optimization, relaxation and Young measures. *Bull. Amer. Math. Soc.*, **36**, 27–58.

Pena, J., Vera, J., and Zuluaga, L. 2007. Computing the stability number of a graph via linear and semidefinite programming. *SIAM J. Optim.*, 87–105.

Pólya, G. 1974. *Collected Papers*, vol. II. Cambridge, MA: MIT Press.

Pólya, G. and Szegö, G. 1976. *Problems and Theorems in Analysis II*. Berlin: Springer.

Powers, V. and Reznick, B. 2000. Polynomials that are positive on an interval. *Trans. Amer. Math. Soc.*, **352**, 4677–4692.

Prajna, S., Papachristodoulou, A., and Parrilo, P. A. 2002. Introducing SOSTOOLS: a general purpose sum of squares programming solver. In: *Proceedings of the 41st IEEE Conference on Decision and Control*. New York: IEEE.

Prestel, A. and Delzell, C. N. 2001. *Positive Polynomials*. Berlin: Springer.

Putinar, M. 1993. Positive polynomials on compact semi-algebraic sets. *Indiana Univ. Math. J.*, **42**, 969–984.

Putinar, M. 2000. A note on Tchakaloff's theorem. *Proc. Amer. Math. Soc.*, **125**, 2409–2414.

Reznick, B. 1995. Uniform denominators in Hilbert's seventeenth problem. *Math. Z.*, **220**, 75–98.

Reznick, B. 2000. Some concrete aspects of Hilbert's 17th problem. In: Delzell, C. N. and Madden, J. J. (editors), *Real Algebraic Geometry and Ordered Structures*. Contemporary Mathematics, vol. 253. Providence, RI: American Mathematical Society.

Richter, H. 1957. Parameterfreie Abschätzung und Realisierung von Erwartungswerten. *Bl. Dtsch. Ges. Versicherungsmath.*, **3**, 147–161.

Riener, C., Theobald, T., Jansson, L., and Lasserre, J. B. 2013. Exploiting symmetries in SDP-relaxations for polynomial optimization. *Math. Oper. Res.*, **38**, 122–141.

Rockafellar, R. T. 1970. *Convex Analysis*. Princeton, NJ: Princeton University Press.

Rogosinsky, W. W. 1958. Moments of non-negative mass. *Proc. R. Soc. London Ser. A*, **245**, 1–27.

Royden, H. L. 1988. *Real Analysis*, third edition. New York: Macmillan.

Rugh, W. J. 1971. On an inverse optimal control problem. *IEEE Trans. Autom. Control*, **16**, 87–88.

Schaefer, A. 2004. Inverse integer programming. *Optim. Lett.*, **3**, 483–489.

Scheiderer, C. 2008. Positivity and sums of squares: a guide to some recent results. In: Putinar, M. and Sullivant, S. (editors), *Emerging Applications of Algebraic Geometry*, pp. 271–324. IMA Volumes in Mathematics and its Applications, vol 149. New York. Springer.

Scherer, C. W. and Hol, C. W. J. 2004. Sum of squares relaxation for polynomial semidefinite programming. *Proceedings of the 16th International Symposium on Mathematical Theory of Networks and Systems, Leuven*, pp. 1–10.

Schichl, H. and Neumaier, A. 2006. Transposition theorems and qualification-free optimality conditions. *SIAM J. Optim.*, **17**, 1035–1055.

Schmid, J. 1998. *On the Degree Complexity of Hilbert's 17th Problem and the Real Nullstellensatz*. Ph.D. thesis, University of Dortmund. Habilitationsschrift zur Erlangug der Lehrbefignis für das Fach Mathematik an der Universität Dortmund.

Schmüdgen, K. 1991. The K-moment problem for compact semi-algebraic sets. *Math. Ann.*, **289**, 203–206.

Schneider, R. 1994. *Convex Bodies: The Brunn–Minkowski Theory*. Cambridge: Cambridge University Press.

Schoenebeck, G. 2008. Linear level Lasserre lower bounds for certain k-CSPs. In: *49th Annual IEEE Symposium on Foundations of Computer Science (FOCS'08)*, pp. 593–602. New York: IEEE.

Schrijver, A. 2005. New codes upper bounds from the Terwilliger algebra and semidefinite pogramming. *IEEE Trans. Inf. Theory*, **51**, 2859–2866.

Schweighofer, M. 2005. On the complexity of Schmüdgen's Positivstellensatz. *J. Complexity*, **20**, 529–543.

Schweighofer, M. 2006. Global optimization of polynomials using gradient tentacles and sums of squares. *SIAM J. Optim.*, **17**, 920–942.

Shakirov, S. R. 2010. Nonperturbative approach to finite-dimensional non-Gaussian integrals. *Theor. Math. Phys.*, **163**, 804–812.

Sherali, H. D. and Adams, W. P. 1990. A hierarchy of relaxations between the continuous and convex hull representations for zero-one programming problems. *SIAM J. Discr. Math.*, **3**, 411–430.

Sherali, H. D. and Adams, W. P. 1999. *A Reformulation-Linearization Technique for Solving Discrete and Continuous Nonconvex Problems*. Dordrecht: Kluwer.

Sherali, H. D., Adams, W. P., and Tuncbilek, C. H. 1992. A global optimization algorithm for polynomial programming problems using a reformulation-linearization technique. *J. Global Optim.*, **2**, 101–112.

Shor, N. Z. 1987. Quadratic optimization problems. *Tekh. Kibern.*, **1**, 128–139.

Shor, N. Z. 1998. *Nondifferentiable Optimization and Polynomial Problems*. Dordrecht: Kluwer.

Simon, B. 1998. The classical moment problem as a self-adjoint finite difference operator. *Adv. Math.*, **137**, 82–203.

Stengle, G. 1974. A Nullstellensatz and a Positivstellensatz in semialgebraic geometry. *Math. Ann.*, **207**, 87–97.

Sturm, J. F. 1999. Using SeDuMi 1.02, a MATLAB toolbox for optimizing over symmetric cones. *Optim. Methods Software*, **11-12**, 625–653.

Tawarmalani, M. and Sahinidis, N. V. 2002. Convex extensions and envelopes of lower semi-continuous functions. *Math. Prog.*, **93**, 247–263.

Tawarmalani, M. and Sahinidis, N. V. 2005. A polyhedral branch-and-cut approach to global optimization. *Math. Prog.*, **103**, 225–249.

Tchakaloff, V. 1957. Formules de cubature mécanique à coefficients non négatifs. *Bull. Sci. Math.*, **81**, 123–134.

Tuncel, L. 2000. Potential reduction and primal-dual methods. In: Wolkowicz, H., Saigal, R., and Vandenberghe, L. (editors), *Handbook of Semidefinite Programming: Theory, Algorithms, and Applications*, pp. 235–265. Boston, MA: Kluwer Academic.

Vallentin, F. 2007. Symmetry in semidefinite programs. *Linear Alg. Appl.*, **430**, 360–369.

Vandenberghe, L. and Boyd, S. 1996. Semidefinite programming. *SIAM Rev.*, **38**, 49–95.

Vasilescu, F.-H. 2003. Spectral measures and moment problems. *Spectral Theory Appl.*, 173–215.

Vui, H. H. and Pham, T. S. 2008. Global optimization of polynomials using the truncated tangency variety and sums of squares. *SIAM J. Optim.*, **19**, 941–951.

Waki, S., Kim, S., Kojima, M., and Maramatsu, M. 2006. Sums of squares and semidefinite programming relaxations for polynomial optimization problems with structured sparsity. *SIAM J. Optim.*, **17**, 218–242.

Waki, H., Kim, S., Kojima, M., Muramatsu, M., and Sugimoto, H. 2008. SparsePOP : a sparse semidefinite programming relaxation of polynomial optimization problems. *ACM Trans. Math. Software*, **35**, article 15.

Zhang, J. and Liu, Z. 1996. Calculating some inverse linear programming problems. *J. Comp. Appl. Math.*, **72**, 261–273.

Zhang, J. and Liu, Z. 1999. A further study on inverse linear programming problems. *J. Comp. Appl. Math.*, **106**, 345–359.

Zhang, J., Ma, Z., and Yang, C. 1995. A column generation method for inverse shortest path problems. *Math. Methods Oper. Res.*, **41**, 347–358.

Index

Printed in the United States
By Bookmasters